P. Malfertheiner H. Ditschuneit (Eds.)

Helicobacter pylori, Gastritis and Peptic Ulcer

With 141 Figures and 94 Tables

Springer-Verlag
Berlin Heidelberg New York
London Paris Tokyo Hong Kong

Malfertheiner, P., M. D., Priv.-Doz.
Ditschuneit, H., M. D., Prof.
Department of Internal Medicine II
Robert-Koch-Straße 8
D-7900 Ulm, W-Germany

ISBN-13: 978-3-642-75317-6 e-ISBN-13: 978-3-642-75315-2
DOI: 10.1007/ 978-3-642-75315-2

Typesetting: Fotosatz-Service Köhler, 8700 Würzburg
2119/3140/543210 – Printed on acid-free paper

Preface

Helicobacter pylori has recently been recognized as a new genus according to specific taxonomic criteria; the "popular" name *Campylobacter pylori* has been corrected by scientific progress. Following the discovery of the spiral microorganism in gastric mucosa by Marshall and Warren in 1982, it took only a few years for *H. pylori* to become established as a factor in the pathogenesis of gastritis and peptic ulcer disease. Interest in different aspects of *H. pylori* has grown continuously and has attracted scientists from various medical and biological disciplines such as gastroenterology, microbiology, pathology, immunology, and pharmacology. Indeed *H. pylori* provides an excellent model for interdisciplinary interaction and cooperation. To promote this concept of interdisciplinary research and exchange of knowledge, a European Campylobacter (Helicobacter) Pylori Study Group was founded in 1987 in Copenhagen. The second meeting of this expanding group was held from October 12–14, 1989 in Ulm, FRG. The fact that more than 500 participants attended the conference and that 187 original contributions from all five continents were presented clearly confirmed that *H. pylori* has "scientifically infected" the whole world. Our understanding of the microbiological and pathogenetic aspects of *H. pylori* is continuously being challenged as new results follow swifthy from different research areas. This book includes an update and progress report on the various aspects of *H. pylori* presented and discussed in special workshops held during the meeting in Ulm. The topics covered in the book, written by leading scientists in this field, include microbiological features of *H. pylori*, its pathogenic mechanisms, interactions with the immune system, the response of the gastroduodenal mucosa to infection, morphological patterns of gastritis, the role of *H. pylori* in peptic ulcer disease, and attempts at curative treatment. Active researchers in this field and clinicians operating in the area of gastroduodenal diseases should find this book a source of practical and stimulating information.

Finally, we would like to thank all those colleagues who contributed to this work for their efforts in helping to provide current knowledge in this rapidly evolving area in time.

Ulm, 15th March 1990 PETER MALFERTHEINER · HANS DITSCHUNEIT

Contents

Local and Systemic Immune Response

Morphological Aspects of Gastritis

Helicobacter pylori in Duodenal Ulcer Pathogenesis

Nonulcer Dyspepsia and Motility

Therapeutic Approaches

List of Contributors

You will find the addresses at the beginning of the respective contribution.

Aleljung, R. 96
Ambrosini, L. 245
Arends, J. W. 188
Ascencio, F. 104
Axon, A. T. R. 154, 345, 434

Baczako, K. 206, 292
Barberi, A. 86, 159
Barbosa, A. J. A. 438
Bayerdörffer, E. 327
Beardshall, K. 317
Beattie, S. 395
Bell, G. D. 172
Blaser, M. J. 38, 110
Blendinger, C. 327
Bode, G. 63, 292
Börsch, G. 441
Borschein, W. 327
Bosi, F. 245
Büchler, M. 335
Bugnoli, M. 86

Cadranel, S. 402
Cafferky, M. 395
Calam, J. 317
Capella, C. 245
Castro, L. P. 438
Chausson, Y. 438
Clayton, G. L. 74, 167
Coelho, L. G. V. 438
Collins, S. M. 370
Cornaggia, M. 245
Costas, S. M. 23
Courcux, P. 19
Curry, A. 29
Cusia, M. G. 86, 159
Cussac, V. 19

Cussi, V. 19
Czinn, S. J. 181

DeKoster, E. 381
Deltenre, M. 381, 458
Demagnistris, T. 86
Ditommaso, A. 86
Ditschuneit, H. 63, 292
Dixon, M. F. 154

Eidt, S. 228
Engstrand, L. 122

Fabritius, P. 206
Fanti, O. 159
Fauchere, J. L. 110
Figura, N. 86, 159
Fiocca, R. 245
Fischbach, W. 213
Flendrig, J. A. 188
Foulkes, W. 317
Fraser, J. 23

Gant, P. 172
Geis, G. 41
Ghosh, P. 317
Guglielmetti, P. 159
Glupczynski, J. 49, 381
Goodwin, C. S. 3
Graham, D. Y. 463
Guarna, M. 86
Guglielmetti, P. 159
Guruge, J. L. 96, 104

Haddad, G. 317
Haot, J. 237
Harrison, G. 172
Hawtin, P. R. 81, 162

Helicobacter pylori –
Taxonomy and Biology

Taxonomy of *Helicobacter pylori* and Related Bacteria

C. S. GOODWIN

In April 1982, in the Microbiology Department of Royal Perth Hospital, Perth, Western Australia, endoscopic biopsy specimens of human gastric mucosa were cultured and yielded a microaerophilic, curved organism which was named *Campylobacter pyloridis* [9]. The specific epithet was grammatically incorrect, and the name was corrected to *Campylobacter pylori* [8]. However, the ultrastructure of *C. pylori* and its fatty acid composition were found to be remarkably different from Campylobacters [3]. Four studies of rRNA sequencing have shown clearly that *C. pylori* does not belong in the genus *Campylobacter* [6, 11–13]. These studies indicated that *C. pylori* is nearer to *Wolinella succinogenes*, but two of the studies noted sufficient differences between *W. succinogenes* and *C. pylori* to justify separate genera [6, 12]. Owen [10] has identified 14 phenotypic differences between *C. pylori* and *W. succinogenes* which indicate that these species should be in separate genera.

Very recently, detailed analysis of five major groups of taxonomic features of these bacteria (ultrastructural features, cellular fatty acid profiles, respiratory quinones, growth characteristics, and enzyme capabilities) have revealed many differences between *C. pylori* and *W. succinogenes*, so that a new genus name has been suggested, *Helicobacter pylori* [4]. In this paper I will delineate these studies.

Ribonucleic acid sequencing of *"Flexispira rappini"* has indicated a close relationship between this organism and *H. pylori*. However, *"F. rappini"* is not a spiral organism, although it is urease positive. We have also described chemotaxonomic differences between *H. pylori* and *"F. rappini"* which confirm that the latter should not be in the same genus as *H. pylori* [4].

From the stomachs of ferrets a spiral organism similar to *H. pylori* has been isolated which is also urease positive, and was originally named *Campylobacter pylori* subsp. *mustelae*. However, in Perth, our DNA-DNA hybridization studies yielded a low degree of homology, indicating a more distant relationship than at the subspecies level. Other important taxonomic features distinguish *H. pylori* and the ferret organism, and the latter was elevated to species status as *Campylobacter mustelae* [2]. This organism is now the second member of the *Helicobacter* genus as *Helicobacter mustelae*. The new genus name *Helicobacter* reflects the two morphological appearances of the organism, helical in vivo, but often rodlike in vitro (*bakter*, a staff). The name *Helicobacter* is partially similar to *Campylobacter*, and is euphonious.

Microbiology Department, Faculty of Medical and Health Sciences, UAE University, PO Box 15551, AL AIN, United Arab Emirates

Helicobacter pylori, Gastritis and Peptic Ulcer
Edited by P. Malfertheiner and H. Ditschuneit

From Perth we published the cellular fatty acid profiles of spiral, urease-positive, microaerophilic organisms from the stomachs of monkeys and a pig [5]. The profiles of all these organisms except one were not markedly dissimilar from *H. pylori* and suggest a relationship at a subspecies level. The nonidentity with *H. pylori* from humans has been shown by us in Perth by the fact that growth on brain-heart infusion agar with 0.3% urea is achieved by animal organisms but not by the human *H. pylori*. However, not all strains of isolates from *Macaca mulatta*, the rhesus monkey, grow on 0.3% urea. Furthermore, the soluble and cell-associated haemagglutins of these animal organisms differ from human *H. pylori*. This would seem to justify the designation of *H. pylori* subsp. *simiae*.

The type B organism obtained from *M. nemestrina* has an unusual cellular fatty acid composition [4] and a G+C content of 24%. A proposal will be made that this organism should be designated as "*Helicobacter nemistrinae*".

Materials and Methods

These were fully described in our earlier paper [4]. We performed DNA hybridization using two methods; and ultrastructural studies included the use of tannic acid in conjunction with glutaraldehyde-osmium fixation, which provides stabilization and contrast enhancement of polysaccharide-rich bacterial surface material.

Growth Conditions and Enzyme Capabilities

Growth was tested on brain-heart infusion agar containing 7% horse blood, 1% Isovitalex and 0.25% yeast extract supplemented with various components shown in Table 3. In addition, growth was tested on the peptone-starch-dextrose (PSD) medium of Dunkelberg et al. [1]. Enzyme capabilities were tested in a microaerophilic atmosphere.

Results

DNA base Composition

The G + C content of a strain of *H. mustelae* from the USA (ATCC 43772) was 36.5 ± 0.8 mol% and the G + C content of an isolate from the UK, NCTC 1203T was $40.6 + 0.57$ mol%.

DNA Hybridization

The levels of DNA relatedness of *H. pylori* and *H. mustelae* strains are shown in Table 1.

Table 1. Percentage of DNA relatedness of *H. pylori* and *H. mustelae*

Test strain	Reference DNA[a]				
	H. mustelae			*H. pylori*	
	ATCC 43772^{T} Method P	NCTC 12032 Method P	NCTC 12032 Method Q	NCTC 11637^{T} Method P	NCTC 11637^{T} Method Q
H. mustelae ATCC 43772^{T}	100			7	30
H. mustelae NCTC 12032	75		100	1	44
H. mustelae FP1	96			8	
H. pylori NCTC 11637^{T}	2	<5	49	100	100

[a] Hybridization analyses were conducted at optimal renaturation temperatures. T_{OR}, calculated according to Gillis et al. [16], were 68 °C for NCTC 12032 and 66 °C for NCTC 11637^{T}

Ultrastructural Features

The major ultrastructural features of *H. pylori*, *H. mustelae*, *C. jejuni* and *W. succinogenes* are shown in Table 2.

Table 2. Ultrastructural features of *H. pylori*, *H. mustelae*, *C. jejuni* and *W. succinogenes*

	C. jejuni NCTC 11351	*H. pylori* NCTC 11637^{T}	*H. mustelae* NCTC 12032	*W. succinogenes* NCTC 11488
Cell-wall membrane	Rugose	Smooth	Smooth	Variable
Flagella	Single bipolar	Multiple unipolar	Multiple bipolar and lateral	Single unipolar
Flagellar sheath	Absent	Present	Present	Absent
Flagellar bulb	Absent	Present	Present	Absent
Glycocalyx (in vitro in shaken broth)	Present	Present	Present	Scanty

Cellular Fatty Acid Analysis

H. pylori strains had a high percentage of 14:0 (36% – 41%), only a trace of 16:1, a small percentage of 3–OH18:0 and a moderate amount of 19:0 cyc (16% – 17%). *H. mustelae* strains had a lower concentration of 14:0 (13% – 17%) with a relatively high concentration of 16:0 (28% – 36%) and no 3–OH18:0. *W. succinogenes* had only 3% of 14:0, 33% of 16:1 and did not have 3–OH18:0 or 19:0 cyc.

Table 3. Biochemical and cultural features of *C. jejuni*, *H. pylori*, *H. mustelae*, *Wolinella succinogenes* and *"Flexispira rappini"*

	C. jejuni NCTC 11351	*H. pylori* NCTC 11637	*H. mustelae* NCTC 12032	*W. succinogenes* NCTC 11488	*"F. rappini"* 1893
Catalase	+	+	+	−	+
Urease	−	+	+	−	+
Hippurate hydrolysis	+	−	−	−	−
Nitrate reduction (microaerophilic)	+	−	+	+	−
H_2S production in triple sugar iron agar	−	−	−	W[a]	−
Gammaglutamyltranspeptidase	−	+	+	−	+
Alkaline phosphatase	W	+	+	−	−
Motility in brain heart infusion broth	+	+	+	+	+
Motility from agar plate	+	−	−	−	−
Growth microaerophilically at: 25 °C	−(2d)	−(4d)	−(4d)	−(2d)	−(2d)
30 °C	+(1d)	+(2d)	+(2d)	−(2d)	−(2d)
37 °C	+(1d)	+(2d)	+(2d)	+(1d)	+(1d)
42 °C	+(1d)	−(4d)	+(2d)	W(2d)	+(1d)
Growth in CO_2 incubator	+(1d)	+(2d)	−(4d)	−(2d)	−(2d)
Growth anaerobically at 37 °C	+	−	W	+	+
0.5% Glycine	+	+	+	−	+
1% Glycine	+	−	+	−	W
1% Bile	+	−	−	+	+
Growth on peptone-starch-dextrose agar	+	+	−	+	+
Growth on charcoal/casein/deoxycholate 0.1%	+	−	−	+	−

W, weak

Growth Conditions and Enzyme Capabilities

The major biochemical characteristics of *H. pylori*, *H. mustelae*, *C. jejuni*, *W. succinogenes* and *"F. rappini"* are shown Table 3.

Discussion

Ultrastructurally, *H. mustelae* is shorter than *H. pylori* and has both bipolar and lateral flagella (Table 2). *H. pylori* has only unipolar flagella. The differences in growth characteristics and enzyme capabilities between *H. pylori* and *H. mustelae* (Table 3) include nitrate reduction by *H. mustelae* and not by *H. pylori*. *H. pylori* grows on starch-containing media and in air enriched with CO_2, buth not anaerobically and variably at 42 °C. *H. mustelae* does not grow on starch-containing agar, nor in air enriched with CO_2, but grows well at 42 °C and weakly in an anaerobic atmosphere.

The phylogenic position of *H. pylori* as determined by 16S ribonucleic acid sequencing clearly indicates that this species should be excluded from the genus *Campylobacter*. In Perth we have delineated five major groups of taxonomic features which differentiate *H. pylori* and *H. mustelae* from *W. succinogenes* and *C. jejuni*. We conclude that these phenotypic differences outweigh at the genus level the genomic relatedness of *H. pylori* and *W. succinogenes*, suggested by 16S ribosomal ribonucleic acid sequencing, and thus justify the transfer of *H. pylori* and *H. mustelae* to the new genus we suggest.

"F. rappini" is a straight, not a spiral organism with a G + C content of 33 mol%, and it grows at 43 °C. *"F. rappini"* does not possess alkaline phosphatase nor does it grow at 30 °C, whereas *H. pylori* and *H. mustelae* possess this enzyme and grow at 30 °C.

Spiral Organisms Other than *H. pylori*

Tightly spiralled "corkscrew" organisms have been found in the stomachs of 0.3% of humans, sometimes with gastritis, and in the stomachs of cats, dogs and many monkeys. These organisms have not yet been cultured but have been transferred to the stomachs of mice, where they multiply. Although they have not yet been cultured, a provisional name of *"Gastrospirillum hominis"* has been suggested. Another spiral organism has been cultured from the stomach of the cat and fully described by Lee [7]; this latter organism may also be incorporated in the *Helicobacter* genus. This cat spiral organism has a marked feature of periplasmic fibres. Another spiral organism which does not have periplasmic fibres nor a glycocalyx shown by tannic acid has been seen by us in the stomachs of baboons. It has not yet been cultured.

Conclusion

We have therefore described a new genus, *Helicobacter* (Goodwin et al. gen. nov.) and we have described *Helicobacter pylori* comb. nov. as the type species with *Helicobacter mustelae* comb. nov. as another species in the genus [4].

References

1. Dunkelberg WE, Skaggs R, Kellogg DS (1970) Method for isolation and identification of Corynebacterium vaginale (Haemophilus vaginalis). Appl Microbiol 19:47–52
2. Fox JG, Chilvers T, Goodwin CS, Taylor NS, Edmonds P, Sly LI, Brenner DJ (1989) Campylobacter mustelae, a new species resulting from the elevation of Campylobacter pylori subsp. mustelae to species status. Int J Syst Bacteriol 39:301–303
3. Goodwin CS, McCulloch RK, Armstrong JA, Wee SH (1985) Unusual cellular fatty acids and distinctive ultrastructure in a new spiral bacterium (Campylobacter pyloridis) from the human gastric mucosa. J Med Microbiol 19:257–267
4. Goodwin CS, Armstrong JA, Chilvers T, Peters M, Collins DM, Sly LI, McConnell W, Harper WES (1989) Transfer of Campylobacter pylori and Campylobacter mustelae to Helicobacter gen. nov. as Helicobacter pylori comb. nov. and Helicobacter mustelae comb. nov., respectively. Int J Syst Bacteriol 39:397–405
5. Goodwin CS, McConnell W, McCulloch RK, McCullough C, Hill R, Bronsdon MA, Kasper G (1989) Cellular fatty acid composition of Campylobacter pylori from primates and ferrets compared with those of other campylobacters. J Clin Microbiol 27:938–943
6. Lau PP, DeBrunner-Vossbrinck B, Dunn B, Miotto K, MacDonnell MT, Rollins DM, Pillidge CJ, Hespell RB, Colwell RR, Mitchell SL, Fox GE (1987) Phylogenetic diversity and position of the genus Campylobacter. Syst Appl Microbiol 9:231–238
7. Lee A (1989) Human gastric spirilla other than C. pylori. In: Blaser MJ (ed) Campylobacter pylori in gastritis and peptic ulcer disease. Igaku-Shoin, New York, pp 225–240
8. Marshall BJ, Goodwin CS (1987) Revised nomenclature of Campylobacter pyloridis. Int J Syst Bacteriol 37:68
9. Marshall BJ, Royce H, Annear DI, Goodwin CS, Pearman JW, Warren JR, Armstrong JA (1984) Original isolation of Campylobacter pyloridis from human gastric mucosa. Microbios Lett 25:83–88
10. Owen RJ (1989) Taxonomy of Campylobacter pylori. In: Rathbone BJ, Heatley RV (eds) Campylobacter pylori and gastroduodenal disease. Blackwell, Oxford, pp 12–23
11. Paster BJ, Dewhirst FE (1988) Phylogeny of campylobacters, wolinellas, Bacteroides gracilis, and Bacteroides ureolyticus by 16S ribosomal ribonucleic acid sequencing. Int J Syst Bacteriol 38:56–62
12. Romaniuk PJ, Zoltowaska B, Trust TJ, Lane DJ, Olsen GJ, Pace NR, Stahl DA (1987) Campylobacter pylori, the spiral bacterium associated with human gastritis, is not a true Campylobacter spp. J Bacteriol 169:2137–2141
13. Thompson LM, Smibert RM, Johnson JL, Kreig N (1988) Phylogenetic study of the genus Campylobacter. Int J Syst Bacteriol 38:190–200

Helicobacter pylori: Microbiological Aspects

A Plea for a More Critical Approach to Laboratory Investigation of this Gastroduodenal Pathogen*

A. LEE

The Change of Name to *Helicobacter pylori*

The spiral shaped bacterium first isolated in Perth, Western Australia by Barry Marshall in 1983 [13, 25] was called *Campylobacter pyloridis* because on light microscopy it appeared similar to other campylobacters, e.g. *C. jejuni*, and similar media and gaseous environments were required to grow it. Just as workers in the field were accepting this name it was changed to *Campylobacter pylori* so as to be more grammatically correct [14]. Now, yet again we are asked to use a different name, *Helicobacter pylori* [8] as is explained elsewhere in this publication. The reasons for these changes are well founded. From the first laboratory investigation of this gastric bacterium, it was clear that the organism did not sit easily in the genus *Campylobacter*. Indeed, attempts to draw analogies with *Campylobacter jejuni* resulted in some unrealistic investigations such as screening flocks of chickens as a likely reservoir. Gastroenterologists should rest assured that it is very unlikely we, the microbiologists, will continue to inflict new nomenclature on them in the future. *H. pylori* now sits rightly as a distinct genus and provided this name comes into common usage it will remain. However, the more we learn about gastric bacteria the greater the chances that other species within the genus will be discovered.

Stages in the Discovery of a New Pathogen

It is interesting to briefly review the stages of investigation into *H. pylori* and predict what may happen in the future. A similar pattern of scientific investigation has occurred with many other unexpected discoveries in biomedical science.

Stage 1: Disbelief

Following the publication of the original isolation of this bacterium and the suggestion it could be involved in duodenal pathology, the gastroenterological world remained highly sceptical. Barry Marshall would attest to the frustration of

School of Microbiology, University of New South Wales, Sydney, Australia

* The research work of our team is supported by the National Health and Medical Research Council of Australia

Helicobacter pylori, Gastritis and Peptic Ulcer
Edited by P. Malfertheiner and H. Ditschuneit

this stage of the discovery as he tramped the world trying to spread his message with almost evangelical zeal.

Stage 2: Growing Conviction

As a limited number of gastroenterologists asked their microbiologists to look for this new spiral bacterium, the conviction that there might be something to the story started to grow and further publications began to appear.

Stage 3: Excitement but Little Funding

A group of enthusiastic believers now started to confirm Marshall's results and hypothesize on the basis of pathogenesis of the organism. Scepticism of the value of this work meant that there was very little funding from the granting agencies. Thus, although the logarithmic publication phase now began, most of this work was done on a limited budget and so the temptation was to publish early in order to be able to more easily obtain support.

In vivo, the result of this phase of investigation was the publication of many small uncontrolled trials. Results of the majority of these trials were consistent with the original hypothesis of Marshall and it was becoming apparent that eradication of the organism was difficult.

In vitro, the era of "phenomenology" was born as will be described below.

Stage 4: Growing Funding and Better Science

The stage of excitement is an essential phase of scientific discovery and the publication of many trials, albeit often as abstracts from meetings, is the catalyst to an era in which more funds become available, trials become more controlled and the standard of scientific investigation is raised.

Stage 5: General Acceptance and Medical Progress

When the results of high quality double-blind trials are published together with well-controlled laboratory investigations then the sceptical community of biomedical scientists and gastroenterologists becomes convinced and medical progress is achieved.

I believe the time has come in *Helicobacter* research to move from Stage 3, the stage of hasty science, to Stage 4 the stage of good science. As this is a talk with a microbiological rather than clinical focus, I will concentrate on what is referred to above as "the era of phenomenology" in order to explain my meaning.

The Era of Phenomenology

As we study a new pathogen in the laboratory we ascribe importance to phenomena we have observed at the bench often without enough investigation of control organisms or attention to the human disease. This is best illustrated with two examples. My aim is not to be solely critical of past studies but hopefully to highlight questions that need to be addressed so we can identify the true importance of much of this work.

Phenomenon 1: Haemagglutination and Adhesion of *H. pylori* to Tissue Culture Cells

Many workers have demonstrated that *H. pylori* will adhere to red cells from many animal species and a wide range of tissue culture cells [5, 18, 19]. Indeed there are very few mammalian cells that are not attractive to this organism. An active research area has been spawned, that is, the identification of these cell adhesins.

The question is what relevance does this work have to the human disease?

Certainly *H. pylori* in vivo does adhere strongly to the epithelial surface and so-called adhesion pedestals are seen on transmission electron microscopy [7, 16], also this adhesion is very tissue specific as is seen in specimens of duodenal tissue infected with *H. pylori* (Fig. 1). The organism only adheres to the gastric

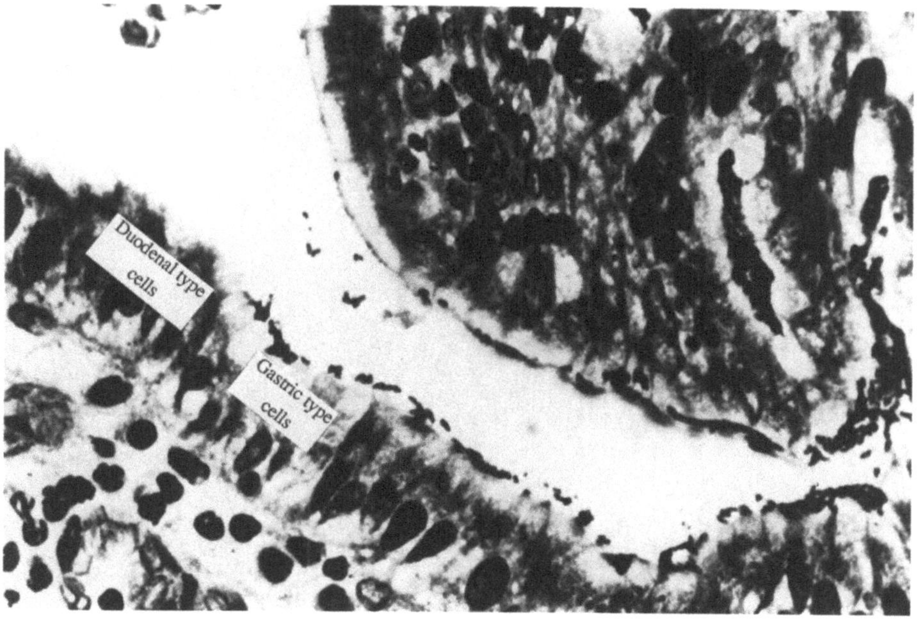

Fig. 1. Warthin Starry stain of a duodenal biopsy specimen from a patient with a duodenal ulcer showing an area of superficial gastric metaplasia and intestinal tissue with *H. pylori* (*arrow*) only associated with the surface of a single layer of gastric-type epithelium

metaplastic cells and stops as soon as the cell surface is composed of duodenal cells [1, 27]. The converse is found in stomachs where intestinal metaplasia is seen. The organism avidly attaches to the gastric cells but not to the metaplastic cells.

Thus, there must be adhesins involved but before one concludes that either the haemagglutins or receptors to tissue culture cells are responsible for gastric specificity, closer attention should be given to the in vivo appearance of *H. pylori*. Association of this bacterium to gastric surfaces is very different to the picture seen with the enterotoxigenic *Escherichia coli* on which much of the classical adhesin work has been based. The main features of the *H. pylori* gastric surface association are:

1. *H. pylori* association is very cell specific.
2. Not all bacteria adhere to the tissue surface. Many are found free in the mucus layer.
3. The majority of tissue-associated organisms do not attach over the whole surface but congregate at and between intercellular junctions and appear to attach to the edges of gastric cells only.

This latter observation of association with the junctions is often neglected when considering gastric colonisation and it may have important implications with regard to pathogenicity. Figure 2 illustrates these three different locations of *H. pylori* in gastric biopsy specimens but to gain a real appreciation of the tissue association that needs to be explained by any in vitro adhesins, readers are invited to compare them with the excellent illustrations in the following publications: Steer [24], Wyatt and Gray [26]; Rauws and Tytgat [20]; Morgan and Leunk [16].

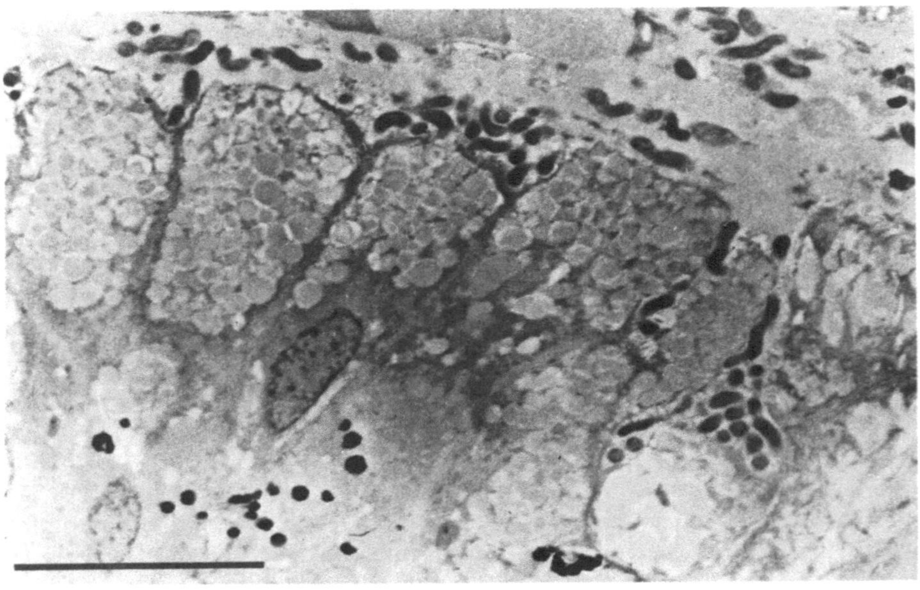

Fig. 2. Back scattered scanning electron micrograph of a gastric biopsy specimen from a patient with *H. pylori* associated gastritis showing organisms in the mucus, congregating at an intercellular junction, and adhering to the surface of epithelial cell. *Bar*, 10 μm. [From 10]

Phenomenon 2: The Effect on Mucus of Incubation In Vitro with *H. pylori*

A series of experiments that are increasingly being quoted as evidence for mucus degradation being a major determinant of pathogenicity of *H. pylori* are those of Sarosiek et al. [21] and Slomiany et al. [22, 23]. These authors took pure cultures of the bacterium and incubated them in vitro with pig gastric mucus and purified mucus glycoprotein polymer. After 48 h of incubation mucus viscosity was decreased by 36%. Significant changes were not detected until after 24 h incubation. These results were claimed as evidence that similar degenerative changes may be a contributing factor in pathogenesis. Before this conclusion becomes enshrined as *H. pylori* dogma one needs to critically evaluate the data and answer the following questions:

1. What effects occur with other nongastric control organisms, for example *E. coli*?
2. What is the turnover rate of gastric mucus? For a significant in vitro effect 24 h is required. Does the mucus remain on the gastric surface for that length of time in vivo?
3. If *H. pylori* has this effect on mucus in vivo and the gastric epithelium is thus more susceptible to acid attack, why do not all persons with antral gastritis get gastric ulcers?

Given the natural location of the organism and the nature of the ulcerogenic process it is likely that *H. pylori* may have some local effect on gastric mucus but the evidence to date is not strong enough to explain what that effect is and certainly it is not good enough to warrant the extremely high citation rate this work is receiving.

Towards the Era of General Acceptance

Having been critical of a number of the laboratory investigations on *H. pylori* in the past it is only reasonable that I be asked to suggest how can we improve our experimental design and increase the usefulness of our data so that valid conclusions can be drawn? Two suggestions of areas in which more care can be taken are illustrated with examples below.

Determining If Phenomena Are Relevant to the Human Disease

In the discussion following a presentation by Leunk et al. [12] at a workshop in Colorado in 1987 on in vitro cytotoxin production by *H. pylori* there was considerable criticism of the work and its relevance was challenged. However, since that time these workers have put much effort into demonstrating that relevance. The original observation was that cell-free extracts of *H. pylori* cultures caused a characteristic vacuolating cytopathic effect on tissue cultures [16]. Now similar vacuolating cytopathic effects have been demonstrated in human biopsy material in cells closely associated with *H. pylori*. Toxic strains have been shown to cause a similar cytopathic effect in gnotobiotic pigs. Antibody against the toxin has been demonstrated in patients with *H. pylori* infection. Thus although the importance of cytotoxin in the pathogenesis of gastroduodenal disease is not finally established, these in vivo correlations make the investigations much more promising.

Checking Phenomena with Control Organisms, i.e., Investigating the Uniqueness of the Phenomenon to *H. pylori*

Recently an exciting new finding was published in *The Lancet* by Cave and Vargas [2]. These investigators showed that a protein product of *H. pylori* switched off acid secretion in an in vitro assay using isolated rabbit parietal cells. Incorporation of C^{14} aminopyrine was taken as a marker of acid secretion. This antisecretory activity of *H. pylori* could be the explanation for the hypochlorhydria observed in early infection [17] and the occasional presence of the organism in parietal cells [3]. However, the authors only included one control organism in their study, that is *C. jejuni*.

The inclusion of control organisms to demonstrate the uniqueness of *H. pylori* phenomenon has been suggested but what organisms are appropriate? A suggested set are listed in Table 1. First, at least two related organisms such as *C. jejuni* and *C. coli* should be tested. But is this enough? Is the observation simply a general phenomenon that is caused by almost any metabolically active bacterium such as *E. coli*? In the case of the Cave experiments could it be that almost any bacterium in the in vitro cell system would produce metabolic byproducts that damage the parietal cells? Finally, a group of gastric organisms that inhabit similar ecological niches to *H. pylori* have recently been described. It would appear important to check these bacteria if possible for any phenomenon observed with *H. pylori* to identify those factors both in vitro and in vivo that are shared by these bacteria as this may give pointers to significant aspects of *H. pylori* pathogenesis. These related organisms are:

Helicobacter mustelae. This is the bacterium, first grown by Fox et al. [6] that naturally colonises ferret gastric tissue causing an associated gastritis with the

Table 1. Some control organisms that are appropriate to study when trying to assess the relevance of *H. pylori* phenomena

Related but not gastric
Campylobacter jejuni
Campylobacter coli
Related gastric organisms
Helicobacter mustelae
"Helicobacter felis"
Unrelated gastrointestinal bacteria
Proteus vulgaris
Escherichia coli

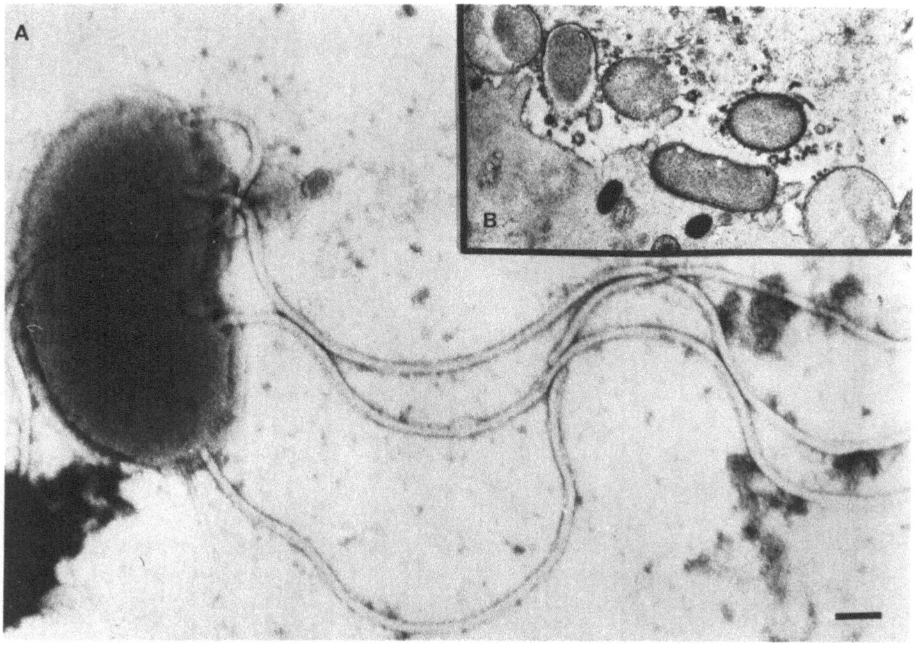

Fig. 3 A, B. Transmission electron micrograph of **A** a negatively stained preparation of *Helicobacter mustelae* isolated from a ferret stomach (*bar*, 0.2 μm) and **B** thin section of ferret gastric mucosa showing adherent *H. mustelae*

occasional ulcerating lesion. The organism appears to form adhesion pedestals as does *H. pylori* (Fig. 3).

"Helicobacter felis". This highly spiralled organism which has been shown by ribosomal RNA sequencing to be the closest relative to *H. pylori* was grown from the cat stomach [11] and can induce gastritis in germ-free mice and dogs. (Paster et al.; Lee et al., unpublished observations; Fig. 4).

"Gastrospirillum hominis". Bacteria with the morphology of this bacterium have been shown to infect humans causing acute or chronic gastritis [15]; the source of infection is probably cats or dogs [9]. *"Gastrospirillum"* has recently been cultured in vivo by administration to mice [4]. In vitro culture has not been successful so comparative studies are difficult but in some cases are valid.

The value of using other organisms when investigating in vitro phenomena found with *H. pylori* was shown in a collaborative investigation in Boston where the Cave and Vargas experiments were repeated using the organisms shown in Table 1. Only one bacterium other than *H. pylori* was found to almost completely switch off parietal cell activity. This was *"H. felis"*. Significantly this is the only other organism that has been found within the canaliculus of parietal cells (except *"Gastrospirillum hominis"* which could not be tested because it could not be grown

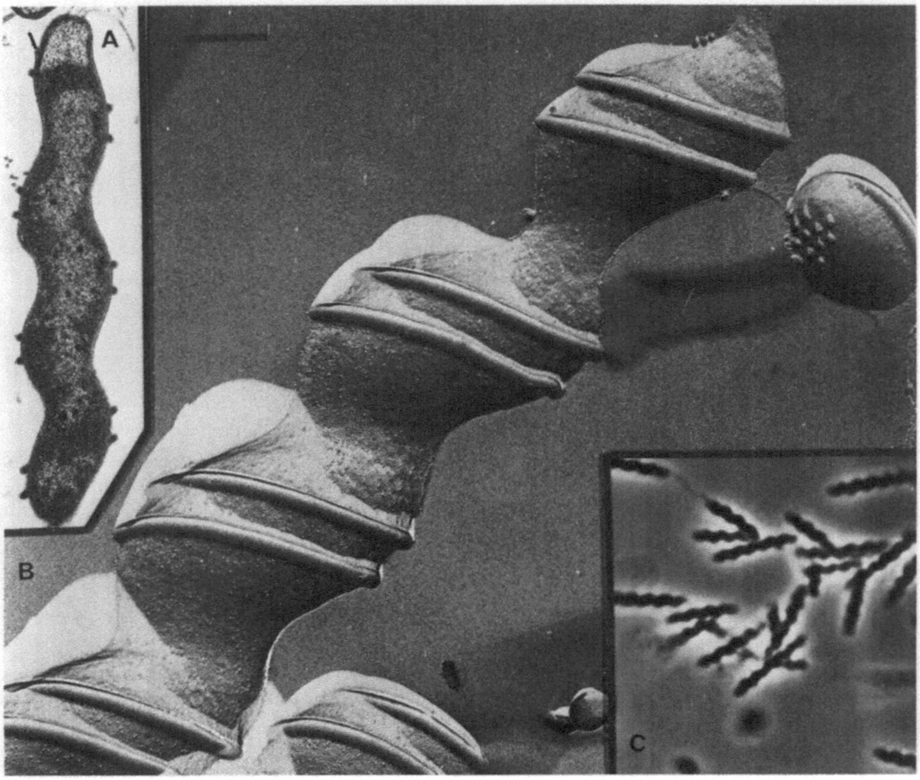

Fig. 4A–C. Photomicrographs of a pure culture of *"Helicobacter felis"* isolated from the stomach of a cat. **A** Cross section of a single organism showing charactristic periplasmic fibrils. **B** Freeze fracture preparation; *bar*, 0.2 μm. **C** Culture viewed under phase contrast microscopy showing the spiral morphology

in vitro). These results would appear to support the conclusion that antisecretory activity is a real phenomenon and that this property has been acquired by the bacteria in order that they are better able to survive in the stomach.

Conclusion

The early phase of *H. pylori* research is over. We are convinced of the importance of this organism as a significant gastroduodenal pathogen. We now have to convince the world. The *H. pylori* "nonbelievers" in the world of gastroenterology will only be convinced by good science, by well-conducted double-blind randomized trials and by well-controlled high quality laboratory investigation. We must be more critical of our work as it is presented at conferences and ensure that the quality of papers submitted to peer review journals improves. It is only excellent science that will allow the field of *H. pylori* research to be classed as medical progress with its potential impact on the morbidity of peptic ulcer disease and possibly gastric cancer.

Acknowledgements. I would like to thank Jani O'Rourke for her help in preparing this manuscript and producing her usual excellent micrographs. I also thank Nerissa Lee for help with preparation of the manuscript.

References

1. Carrick J, Lee A, Hazell S, Ralston M, Daskalopoulos G (1989) Campylobacter pylori, duodenal ulcer and gastric metaplasia: possible role of heterotopic tissue in ulcerogenesis. Gut 30:790–797
2. Cave DR, Varges M (1989) Effect of a Campylobacter pylori protein on acid secretion by parietal cells. Lancet 2:187–189
3. Chen XG, Correa P, Offerhaus J, Rodriguez E, Janney F, Hoffman E, Fox J, Hunter F, Diavolitsis S (1986) Ultrastructure of the gastric mucosa harbouring Campylobacter-like organisms. Am J Clin Pathol 86:575–582
4. Dick E, Lee A, Watson G, O'Rourke J (1989) Use of the mouse for the isolation and investigation of stomach-associated, spiral/helical shaped bacteria from man and other animals. J Med Microbiol 29:55–62
5. Evans DG, Evans DJ Jr, Moulds JJ, Graham DY (1988) N-Acetylneuraminyllactose-binding fibrillar hemagglutinin of Campylobacter pylori: a putative colonization factor antigen. Infect Immun 56:2896–2906
6. Fox JG, Edrise BM, Cabot EB, Beaucage C, Murphy JC, Prostak KS (1986) Campylobacter-like organisms isolated from gastric mucosa of ferrets. Am J Vet Res 47:236–239
7. Goodwin CS, Armstrong JA, Marshall BJ (1986) Campylobacter pyloridis, gastritis and peptic ulceration. J Clin Pathol 39:353–365
8. Goodwin CS, Armstrong JA, Chilvers T, Peters M, Collins DM, Sly L, McConnell W, Harper WES (1989) Transfer Campylobacter pylori and Campylobacter mustelae to Helicobacter gen. nov. as Helicobacter pylori comb. nov. and Helicobacter mustelae comb. nov., respectively. Int J Syst Bacteriol 39:397–405
9. Lee A (1989) Human gastric spirilla other than C. pylori. In: Blaser MJ (ed) Campylobacter pylori in gastritis and peptic ulcer disease. Igaku-Shoin, New York, pp 225–240
10. Lee A, Hazell SL (1988) Campylobacter in health and disease. Microbiol Ecol Health Dis 1:1–16
11. Lee A, Hazell SL, O'Rourke J, Kouprach S (1988) Isolation of a spiral-shaped bacterium from the cat stomach. Infect Immun 56:2843–2850
12. Leunk RD, Johnson PJ, David BC, Kraft WG, Morgan DR (1988) Cytotoxic activity in broth-culture filtrates of Campylobacter pylori. J Med Microbiol 26:93–99
13. Marshall BJ (1983) Unidentified curved bacilli on gastric epithelium in active chronic gastritis. Lancet 1:1273–1275
14. Marshall BJ, Goodwin CS (1987) Revised nomenclature of Campylobacter pyloridis. Int J Syst Bacteriol 37:68
15. McNulty CAM, Dent J, Curry A, Uff JS, Ford GA, Gear MWL, Wilkinson SP (1989) New spiral bacterium in gastric mucosa. J Clin Pathol 42:585–591
16. Morgan DR, Leunk RD (1989) Pathogenesis of infection by C. pylori. In: Blaser MJ (ed) Campylobacter pylori in gastritis and peptic ulcer disease. Igaku-Shoin, New York, pp 115–133
17. Morris A, Nicholson G (1987) Ingestion of Campylobacter pyloridis causes gastritis and raised fasting gastric pH. Am J Gastroenterol 82:192–197
18. Nakazawa T, Ishibashi M, Konishi H, Takemoto T, Shigeeda M, Kochiyama T (1989) Hemagglutination activity of Campylobacter pylori. Infect Immun 57:989–991

19. Neman-Simha V, Megraud F (1988) In vitro model for Campylobacter pylori adherence properties. Infect Immun 56:3329–3333
20. Rauws EAJ, Tytgat GNT (1989) Campylobacter pylori. Academic Medical Centre, Amsterdam
21. Sarosiek J, Gabryelewicz A, Slomiany BL (1987) Changes in the macromolecular organisation in physical properties of gastric mucus with peptic ulcer. Gastroenterology 92:1615
22. Slomiany BL, Bilski J, Murty VLN, Sarosiek J, Dworkin B, van Horn K, Slomiany A (1987) Campylobacter pyloridis degrades mucin and undermines mucosal integrity. Gastroenterology 92:1645
23. Slomiany BL, Bilski J, Sarosiek J, Murty VLN, Dworkin B, van Horn K, Zielenski, Slomiany A (1987) Campylobacter pyloridis degrades mucin and undermines gastric mucosal integrity. Biochem Biophys Res Commun 144:307–314
24. Steer HW (1989) Ultrastructure of Campylobacter pylori in vivo. In: Rathbone BJ, Heatley RV (eds) Campylobacter pylori and gastroduodenal disease. Blackwell, Oxford, pp 146–154
23. Slomiany BL, Bilski J, Sarosiek J, Murty VLN, Dworkin B, van Horn K, Zielenski, Slomiany A (1987) Campylobacter pyloridis degrades mucin and undermines gastric mucosal integrity. Biochem Biophys Res Commun 144:307–314
24. Steer HW (1989) Ultrastructure of Campylobacter pylori in vivo. In: Rathbone BJ, Heatley RV (eds) Campylobacter pylori and gastroduodenal disease. Blackwell, Oxford, pp 146–154
25. Warren JR (1983) Unidentified curved bacilli on gastric epithelium in active chronic gastritis. Lancet 1:1273
26. Wyatt JI, Gray SF (1989) Detection of Campylobacter pylori by histology. In: Rathbone BJ, Heatley RV (eds) Campylobacter pylori and gastroduodenal disease. Blackwell, Oxford, pp 63–68
27. Wyatt JI, Rathbone BJ, Dixon MF, Heatley RV (1987) Campylobacter pyloridis and acid induced gastric metaplasia in the pathogenesis of duodenitis. J Clin Pathol 40:841–848

Development of Genetic and Molecular Approaches for the Diagnosis and Study of the Pathogenesis of *Helicobacter pylori*

A. Labigne, V. Cussac, and P. Courcoux

Introduction

In addition to the clinical, bacteriological, immunological, and epidemiological approaches that have been developed and extensively used for the study of *H. pylori* since its discovery in 1983, genetic and molecular approaches should be actively developed in the coming years. Such approaches should be useful as well as informative in three distinct domains.

First they should provide us with powerful tools, specific enough to allow the detection of *H. pylori* in human specimens such as biopsy material as well as gastric juice, saliva and, potentially, stools or sera of patients. With this view, molecular tools may consist of specific DNA sequences used as probes in hybridization techniques, or as a pair of synthetic oligonucleotides (i.e. primers specifying *H. pylori* target DNA) in the "polymerase chain reaction" (PCR) for the amplification of rare molecules. Alternatively, they may consist of monoclonal or polyclonal antibodies raised against *H. pylori* specific epitopes. These epitopes may be prepared as synthetic peptides according to the amino acid sequence deduced from the DNA sequence analysis of genes encoding the major antigens of *H. pylori*, or they can be expressed via DNA recombinant techniques as exposed epitopes in bacteria easy to grow such as *E. coli*. In both cases the capability to easily prepare large amounts of the immunogenic epitopes will provide us with *H. pylori* specific antigenic material usable for serological tests.

Second, these approaches should allow a better understanding of the pathogenesis of *H. pylori* infections by identifying via cloning experiments the genes responsible for the expression of virulence determinants attributed to *H. pylori* such as genes allowing the expression of urease, cytotoxin and adhesins. The contribution of each putative determinant to the pathogenic process will be then addressed by constructing stable *H. pylori* isogenic mutants, each modified in a single determinant. In addition, the cloned stretches of DNA responsible for the expression of the various determinants can then be used as probes to determine their distribution among clinical isolates with the view to find out whether or not various pathovars of *H. pylori* can be encountered among clinical isolates which might be responsible for various symptoms or diseases.

Unité des Entérobactéries, U199 INSERM, 28 Rue du Docteur Roux, 75724 PARIS Cedex 15, France

Helicobacter pylori, Gastritis and Peptic Ulcer
Edited by P. Malfertheiner and H. Ditschuneit

Finally, as a long-term goal, the approaches should facilitate the engineering of living oral vaccines either using attenuated *H. pylori* strains or colonizing strains expressing *H. pylori* major immunogenic and protective epitopes.

As an attempt to reach these general goals, we have studied the genes responsible for the expression of the urease activity in *H. pylori*.

Cloning of the Genes Required for the Expression of Urease

As we previously reported [1], this cloning was made possible by using a shuttle cosmid vector that we constructed for developing *Campylobacter* genetic material [2, 3]; it allows replication and maintenance of cloned DNA sequences either in an *E. coli* or a C. jejuni background and to move these sequences from one to the other bacteria. With this cloning approach, we were able to clone a large portion of *H. pylori* chromosomal DNA (44 kilobases (kb) in length), which, when introduced into *E. coli* did not express urease activity, but when transfered by conjugation to *C. jejuni* led to the expression of urease activity.

In order to localize more precisely the DNA region responsible for the urease activity, the 44 kb DNA fragment was randomly restricted in DNA fragments with an average size of 10 kb; each resulting fragment was cloned into the pILL550 shuttle vector [2]. As a result, a 8.1 kb fragment was shown to contain the genes required for the full expression of the urease (Fig. 1). This 8.1 kb insert was then mutagenized using two different methods. First a series of deletions starting at either end of the 8.1 kb insert and extending up to 3.6 kb in the insert were performed using the *Bal* 31 enzyme. Second, the 8.1 kb fragment was randomly disrupted by inserting transposable elements. In both cases, the 8.1 kb mutated fragment was cloned into the pILL550 shuttle vector and transfered to *C. jejuni* to look for urease activity.

Only 4.2 kb of the 8.1 kb insert fragment were shown to be required for the expression of the urease (Fig. 1).

Nucleotide Sequence of the Urease Region

The nucleotide sequence of the 4.2 kb DNA fragment involved in the urease expression has been determined using the dideoxynucleotide chain termination method of Sanger following cloning of appropriate restriction fragments into M13mp18 and M13mp19 phage DNA.

From the sequence analysis (not shown here) we found three large open reading frames: two correspond to the genes designated *ureA* and *ureB: ureA* consists of 717 base pairs encoding a 239 amino acid peptide, with a calculated molecular weight of 26.5 kDa; *ureB* consists of 1710 base pairs encoding a 570 amino acid peptide with a calculated molecular weight of 61.6 kDa. The *ureA* and *ureB* genes are not only in the same frame but they are juxtaposed in such a way that the stop codon of the *ureA* gene is separated from the methionine initiator codon of the *ureB* gene by a single codon. These two polypeptides correspond to the two subunits of the *H. pylori* urease enzyme.

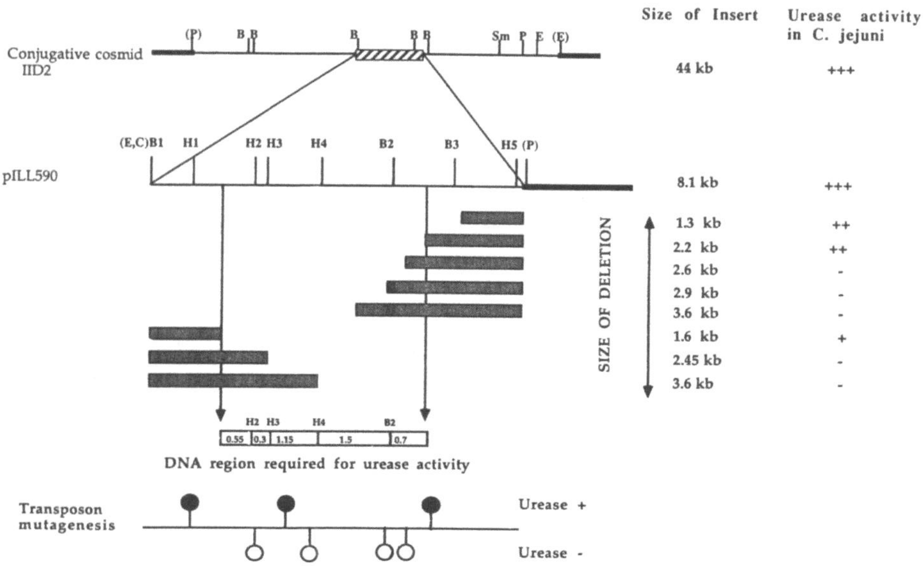

Fig. 1. Restriction maps of the recombinant cosmid and plasmid derivative allowing the expression of *Helicobacter pylori* urease genes in *Campylobacter jejuni*. Sizes are expressed in kilobases (kb). The lengths of the *black boxes* correspond to the lengths of the deletions generated at either end of the pILL590 insert. The *dark circles* indicate transposable element insertions that abolish the urease activity; the *open circles* indicate insertion that did not alter the urease expression. *P, PstI; B, Bam*HI; *Sm, Sma*I; *E, Eco*RI; *H, Hin*dIII

The third open reading frame, upstream of the *ureA* and *ureB* gene, corresponds to a gene that we designated *ureC*. This gene is absolutely required for the urease activity but no polypeptide corresponding to the predicted peptide deduced from the DNA sequence could be visualized on acrylamide gel following expression in *E. coli*, whereas the *ureA* as well as the *ureB* genes were fully expressed. The role of this putative polypeptide is still unknown. *UreC* and *ureA* genes are separated by a noncoding region 420 base pairs in length.

The most interesting result comes from the comparison of the deduced amino acid sequence of the two subunits of the *H. pylori* urease enzyme with that of the well-known jack bean urease whose amino acid sequence was determined biochemically [4]. The jack bean enzyme consists of a unique polypeptidic chain of 840 amino acid residues. By comparing the amino acid sequence of the small subunit linked to the amino acid sequence of the large subunit of *H. pylori* urease with that of the jack bean urease, we found an incredible degree of conservation (50% conservation with the 26.5 kDa polypeptide and 60% conservation with the 61.5 kDa) and a perfect alignment of the two sequences whether isolated from bacteria or from plant.

This high degree of conservation allowed us to localize the active site of the *H. pylori* urease; this site corresponds to the presence of eight histidine residues as well as a cysteine residue which, respectively, in the jack bean urease are thought to play a major role in the binding of the nickel ions and in the enzymatic activity [4]. The fact that these residues are conserved and perfectly aligned between the two sequences suggests a similar role in the *H. pylori* urease.

These results allow us to draw important conclusions:

1. The urease enzyme of *H. pylori* consists solely of two polypeptidic subunits, namely the 26.5 and 61.6 kDa peptides.
2. Such a high degree of conservation between the urease of a plant and that of a bacteria leads to the presumption that the degree of conservation of the amino acid sequence among bacterial ureases such as those of *Proteus, Klebsiella, and Staphylococcus* would be even higher. We should therefore be prepared to find common epitopes between the different bacterial urease enzymes and be careful when using the whole urease fraction as antigen for serological tests. Comparison of the complete sequences of these bacterial ureases, once they will be available, with that of *H. pylori* urease should allow us to carefully choose short amino acid sequences specific of *H. pylori* as potential immunogenic epitopes usable either for diagnostic purposes or as potentially protective epitopes.
3. The identification of the active site of the *H. pylori* urease and the knowledge of its complete amino acid sequence should enable us to find compounds capable of blocking the urease activity of *H. pylori*.

References

1. Labigne-Roussel A, Courcoux P (1989) Cloning and expression of the urease genes of Campylobacter pylori. In: Mégraud F, Lamouliatte H (eds) Gastroduodenal pathology and Campylobacter pylori. Elsevier, Amsterdam, pp 119–122
2. Labigne-Roussel A, Harel J, Tompkins L (1987) Gene transfer from Escherichica coli to Campylobacter species: development of shuttle vectors for genetic analysis of Campylobacter jejuni. J Bacteriol 169:5320–5323
3. Labigne-Roussel A, Courcoux P, Tompkins L (1988) Gene disruption and replacement as a feasible approach for mutagenesis of Campylobacter jejuni. J Bacteriol 170:1704–1708
4. Takishima K, Suga T, Mamiya G (1988) The structure of jack bean urease. The complete amino acid sequence, limited proteolysis and reactive cysteine residues. Eur J Biochem 175:151–165

Assessment of DNA and Protein Molecular Fingerprinting Methods for Strain Identification of *Helicobacter pylori*

J. Fraser[1], R. J. Owen[1], D. D. Morgan[1], M. Costas[1], and D. R. Morgan[2]

Introduction

The increasing evidence linking *Helicobacter pylori* (*H. pylori*) with gastro-duodenal disease has led to the need for sensitive methods for identifying *H. pylori* isolates. The study of *H. pylori* ecology and its role in infection is hindered by the lack of reliable, easily applicable methods of strain identification. Differences between strains are, however, revealed by molecular fingerprinting techniques and the purpose of the present study was to assess the utility of genomic DNA restriction endonuclease digests, restriction fragment length polymorphism (RFLP) using Southern blot hybridization, and one-dimensional polyacrylamide gel electrophoresis (1-D PAGE) protein patterns for high resolution identification.

Materials and Methods

Bacterial Strains

The 27 clinical isolates (patient sets A–H) used in the study were from endoscopic biopsy specimens of gastric mucosa of patients with gastritis. These were provided by Dr. D. M. Jones (Public Health Laboratory, Withington Hospital, Manchester, UK). Most of the strain sets comprised two isolates collected on the same day (date 1) from two adjacent antral sites in the stomach and two further isolates from the same sites on subsequent biopsy 1 month later (date 2) after treatment with nitrofurantoin. The reference strain used was NCTC 11638 from Perth (Australia), which was obtained from the National Collection of Type Cultures (NCTC), but originated from human gastric mucosa.

Organisms were grown routinely on Oxoid brain-heart infusion agar supplemented with 5% v/v horse blood and 1% Isovitalex for 2 days at 37 °C in microaerobic conditions (5%, O_2; 5%, CO_2; 2%, H_2; 88%, N_2) in a Variable Atmosphere Incubator (Don Whitley Scientific Ltd, Shipley, Yorks, UK). Strains were preserved at −196 °C on glass beads in Nutrient Broth No. 2 (Oxoid: CM 67) containing 10% v/v glycerol.

[1] National Collection of Type Cultures, Central Public Health Laboratory, London NW9 5HT, UK
[2] The Procter & Gamble Company Miami Valley Laboratories, Cincinnati, Ohio USA

Helicobacter pylori, Gastritis and Peptic Ulcer
Edited by P. Malfertheiner and H. Ditschuneit

DNA Preparation

Chromosomal DNA was isolated and purified using the guanidium thiocyanate reagent method [1].

DNA Digestion and Electrophoresis

DNA from *H. pylori* NCTC 11638 was incubated with 20 different restriction endonucleases using the buffers and the conditions recommended by the manufacturer (Northumbria Biologicals Limited, UK). All DNA samples (5 µg) were then digested for 4 h at 37 °C with *Hae*III (2–3 U/µg of DNA). The digested DNA was electrophoresed at 25 V for 16 h in a horizontal 0.7% (w/v) agarose gel in a buffer containing 89 m*M* Tris hydrochloride, 89 m*M* boric acid, and 2 m*M* disodium ethylenediaminetetra-acetic acid (EDTA) (pH 8.3). After electrophoresis, the gels were stained with ethidium bromide and photographed with Polaroid 667 film. Tri-X pan professional film was used to obtain negatives for densitometry.

Scanning of Gels

Patterns were scanned and analyzed with a laser densitometer (LKB) interfaced to a Compaq Deskpro 386 microcomputer. Data were processed (segmented linear correction and background trend removal), corrected partial profiles were compared by stepwise alignment to obtain best fit, and the Pearson product moment correlation coefficient was calculated. Strains were then clustered and a dendrogram plotted [2].

Preparation of Biotinylated Probe cDNA

The biotinylated cDNA probe was prepared from *Escherichia coli* 16 *S* and 23 *S* rRNA (BDH Ltd., UK) using Moloney mouse leukaemia virus reverse transcriptase (Gibco-BRL). Biotinylation was achieved by the incorporation of biotin-16-dUTP [3].

Southern Blotting and Hybridization

After electrophoresis and photography the gels were transferred to nylon membranes (Hybond-N, Amersham International) using either the standard capillary transfer method of Southern [4] or by a vacuum method. The membranes were then hybridized by standard procedures for 18 h at 42 °C using the biotinylated cDNA probe. The membranes were washed and the hybridized probe detected colorimetrically using a nonradioactive detection kit – BluGENE (Gibco-BRL) [5].

Protein Electrophoresis

Extracts of total protein were prepared by suspending cell material in extraction buffer. Samples were electrophoresed in 10% polyacrylamide–0.1% sodium dodecyl sulphate (SDS) gels for 4 h at a constant current (30 mA) under carefully controlled conditions. The gels were stained using 0.1% Page Blue 83 (BDH Ltd.) and then dried. Patterns were scanned as described above. Full details of the methods are published elsewhere [2].

Plasmid DNA Extraction and Electrophoresis

Organisms were cultured as described above. Single colonies were picked and plasmid DNA was prepared by the method of Kado and Liu [6]. Samples were electrophoresed in a horizontal 0.7% agarose gel for 2 h at 100 V. After staining with ethidium bromide, the gels were viewed in a UV transilluminator and photographed. Standard plasmids of *Escherichia coli* K12 hosts were used to determine molecular weights.

Results

DNA from *H. pylori* NCTC 11638 was cut to varying degrees by nine restriction endonucleases (*Bam*HI, *Bst*EII, *Eco*RI, *Hae*III, *Hin*dIII, *Msp*I, *Pvu*II, *Sac*I, *Stu*I) but only *Hae*III (recognition sequence GG.CC) and *Hin*dIII (recognition sequence A.AGCTT) gave clear and distinctive patterns of bands. *Hae*III was selected to study all strains.

With a set of 27 isolates from eight patients in the UK, it was found that the DNAs of all were cut with *Hae*III and that isolates from different patients had quite distinct patterns. These observations were confirmed by Southern blot hybridization and PAGE protein analyses. All three methods revealed a high degree of similarity between multiple isolates from the same patient, although DNA digest and protein analyses also provided evidence of minor variations between isolates in some patient sets. There was no association found between the presence of plasmid DNA and genome digest pattern.

These results are illustrated in Figs. 1 and 2.

Discussion

Valid molecular fingerprinting methods should fulfil certain criteria, namely typability, reproducibility and the ability to discriminate [7] in this case, among strains of *H. pylori*. DNA fingerprinting based on the use of *Hae*III digest patterns is a very sensitive and highly discriminatory method of typing *H. pylori* isolates. Typability of strains is good with *Hae*III since most, but not all strains are cut with this enzyme. Flexibility in this method of typing allows for the use of other

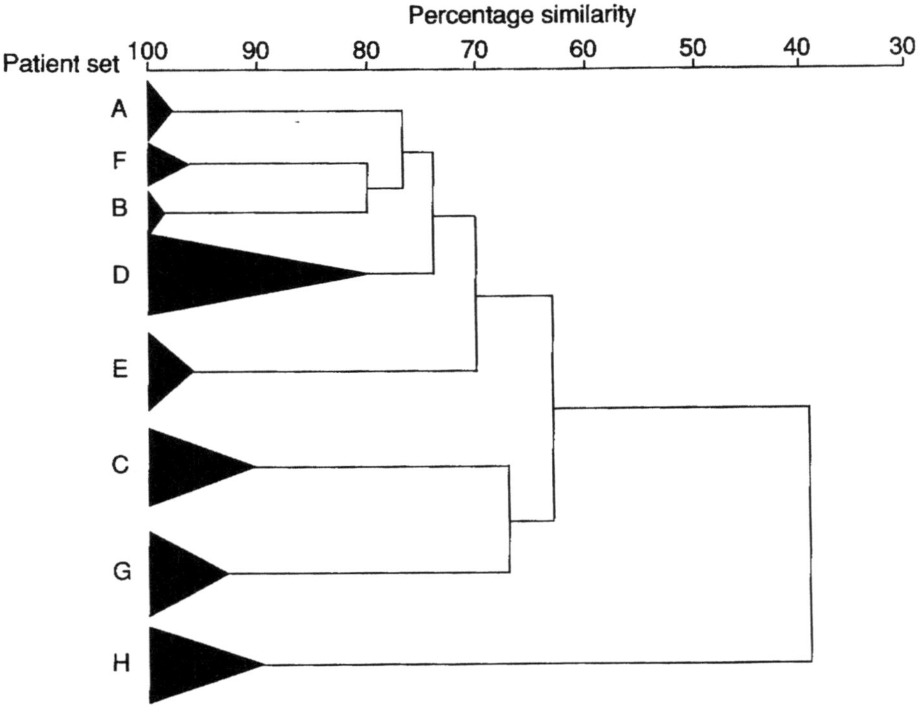

Fig. 1. Simplified dendrogram of the cluster analysis based on *Hae*III digest patterns of *Helicobacter pylori* strains from patients A–H

enzymes when *Hae*III is not suitable, but a number of enzymes should be employed to fully ensure strain identity.

Readily detectable differences among DNA digests enable electrophoretic gels to be visually compared for instant comparison of strains, particularly of multiple isolates from different patients, where there are clear differences between patients. Laser densitometry may be used for numerical analysis of digest patterns, yielding a matrix of similarities with relationships shown in a dendrogram. This type of analysis requires high quality band patterns and is more useful for typing of similar strains, for example, those from the same patient. DNA restriction digest patterns obtained with a particular restriction endonuclease are highly reproducible and between gel reproducibility is also very good. This method is relatively quick and easy, with the availability of rapid DNA extraction methods providing pattern results after 24 h.

Southern blot hybridization with a biotin-labelled cDNA probe synthesized from a mixture of 16 S and 23 S rRNA from *E. coli* yields simpler patterns for interpretation and shows clear differences among isolates from different patients. *E. coli* yields simpler patterns for interpretation and shows clear differences among isolates from different patients. RFLP patterns support DNA digest

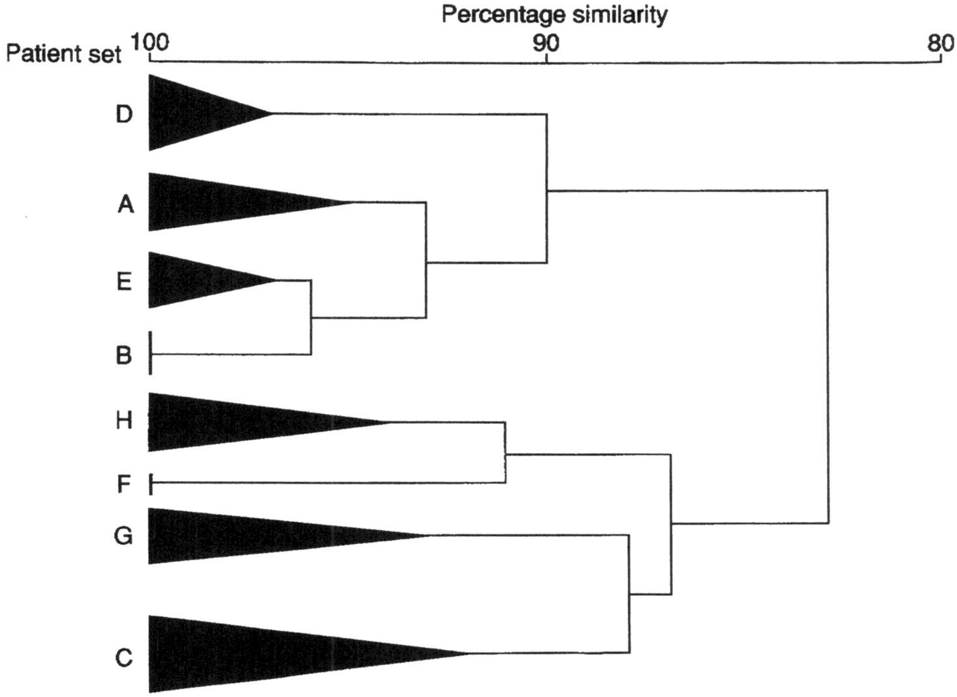

Fig. 2. Simplified dendrogram of the cluster analysis based on total protein content of *Helicobacter pylori* strains from patients A–H

results and protein profiles and are of particular use for typing into groups, for example, isolates from the same patient. Reproducibility is good, but with the combination of the rRNA gene probe and *Hae*III digestion, sensitivity and discrimination is less than with the other methods.

The RNA ribosomal genes are present in the smaller restriction fragments which migrate further down the gel, whereas the minor differences in digest patterns are to be seen in the higher molecular weight region. A more specific probe is required to highlight differences throughout the genome. Southern blot hybridization is technically more complicated and takes longer than DNA restriction endonuclease digestion alone.

Analysis of 1-D SDS–PAGE protein patterns reveals that most variation among strains is attributable to band differences in the 47–56 kDa region, and so-called hypervariable region. Determination of protein profiles provides the basis of a reproducible method for typing of *H. pylori* isolates, so long as rigorous standardization is adhered to. The hypervariable region highlights differences among strains from different patients, or indeed similarities.

The typing potential of 1-D SDS–PAGE protein patterns is limited by a proportion of strains that do not give analyzable patterns. Those patterns that are obtained require computerized analysis. This technique has the advantage that contamination is immediately obvious. DNA restriction endonuclease digests and SDS–PAGE protein profiles are comparable in discriminatory ability, although digest patterns are easier to compare visually, whilst protein patterns require computer analysis. RFLP patterns compare favourably with the other two techniques, but do not emphasize minor variations. Although technically more complicated to produce, such patterns are simpler to interpret. All three methods require further evaluation on larger numbers of strains.

References

1. Pitcher DG, Saunders NA, Owen RJ (1989) Rapid extraction of bacterial genomic DNA with guanidium thiocyanate. Lett Appl Microbiol 8:151–156
2. Owen RJ, Costas M, Morgan DD, On SLW, Hill LR, Pearson AD, Morgan DR (1989) Strain variation in Helicobacter pylori detected by numerical analysis of one-dimensional electrophoretic protein patterns. Antonie van Leeuwenhoek 55:253–267
3. Pitcher DG, Owen RJ, Dyal P, Beck A (1987) Synthesis of a biotinylated DNA probe to detect ribosomal RNA cistrons in Providencia stuartii. FEMS Microbiol Lett 48:283–287
4. Maniatis T, Fritsch EF, Sambrook J (1982) Molecular cloning. A laboratory manual. Cold Spring Harbor Laboratory, Cold Spring Harbor, USA, pp 382–386
5. Owen RJ, Costas M, Dawson C (1989) Application of different chromosomal DNA restriction digest fingerprints to specific and subspecific identification of *Campylobacter* isolates. J Clin Microbiol 27:2338–2343
6. Kado CI, Liu ST (1981) Rapid procedure for detection and isolation of large and small plasmids. J Bacteriol 145:1365–1373
7. Hunter PR, Gaston MA (1988) Numerical index of the discriminatory ability of typing systems: an application of Simpson's index of diversity. J Clin Microbiol 26:2465–2466

The Genesis of Coccal Forms of *Helicobacter pylori*

D. M. Jones and A. Curry

Introduction

In prolonged culture the morphology of *Helicobacter pylori* [2] alters from "the normal" helical form to a spherical cell via U- and V-shaped forms. This morphological transformation may be the result of less than optimal cultural conditions, a reaction to waste products, or it may be a natural phenomenon. This paper describes the formation of these atypical forms with an explanation of the mechanism of their genesis.

Materials and Methods

Culture Technique

Stored strains of *Helicobacter pylori* were cultivated on blood agar. At various time intervals from 2 to 8 days, organisms were removed from the culture plates and prepared either for negative staining transmission electron microscopy, thin-sectioning transmission electron microscopy or scanning electron microscopy.

Negative Contrast Transmission Electron Microscopy

Pioloform-coated (Agar Scientific) 400 mesh copper grids were "dipped" into colonies, washed in distilled water and stained with phosphotungstic acid, pH 6.5. Such grids were examined in either an AEI EM 801 or Philips EM 420 electron microscope.

Thin Section Transmission Electron Microscopy

For thin sectioning, plates were flooded with 3% cacodylate-buffered glutaraldehyde (pH 7.2) and colonies were transferred using a cotton bud to a centrifuge tube containing fixative. Following overnight primary fixation, the cells were washed in 0.1 *M* cacodylate buffer, postfixed in 1% osmium tetroxide,

Public Health Laboratory, Withington Hospital, Manchester M20 8LR, U.K.

Helicobacter pylori, Gastritis and Peptic Ulcer
Edited by P. Malfertheiner and H. Ditschuneit

dehydrated in a graded series of alcohols and embedded in Agar 100 resin (Agar Scientific). Ultrathin sections were cut on a Reichert OMU 4 Ultracut ultramicrotome, mounted on 400-mesh copper grids and stained with uranyl acetate and lead citrate. Grids were examined as above.

Scanning electron microscopy (SEM)

For scanning electron microscopy, culture plates were flooded in glutaraldehyde fixative and small pieces of agar removed with overlying bacterial growth. These were allowed to air dry before being attached by a conductive adhesive to SEM specimen stubs. The specimens were sputter coated in gold and examined in a Cambridge S 360 scanning electron microscope.

Results

Illustrated in Fig. 1, for comparison with the atypical forms, is the more usual morphology of *Helicobacter pylori*. Atypical forms are seen in cultures of various age and passage and 4- or 5-day cultures will usually contain many such forms although this is also dependent on the culture medium. Appropriate cultures show a predominance of U-shaped and V-shaped forms (Fig. 2). However, close examination of negatively stained or thin-sectioned preparations show earlier stages in the genesis of these forms and later cultures contain mainly spherical

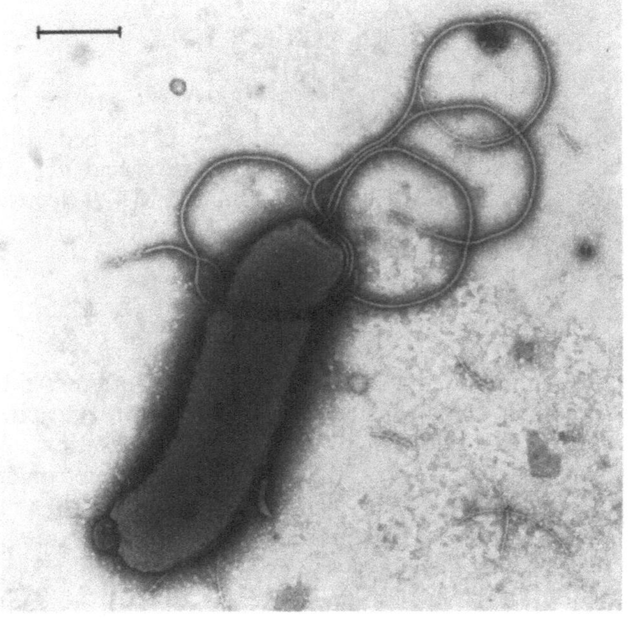

Fig. 1. Negatively stained preparation of *Helicobacter pylori* showing gently sinusoidal form and sheathed flagellar filaments from one pole. *Scale bar*, 600 nm

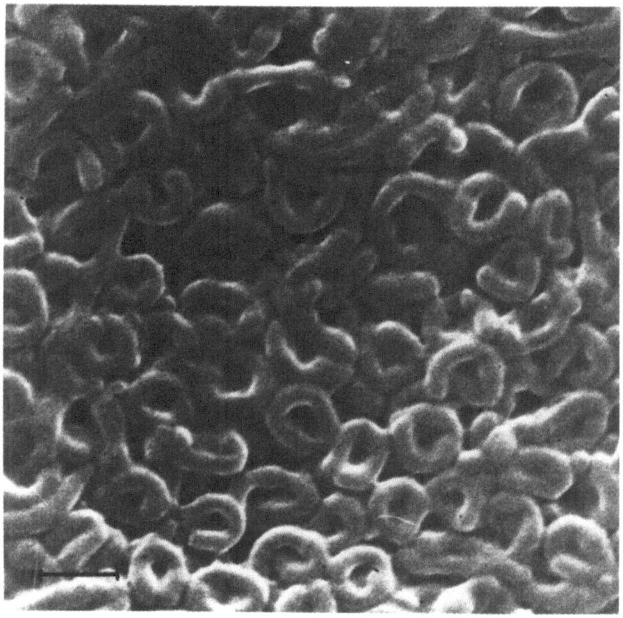

Fig. 2. Scanning electron micrograph of a 4-day culture of *H. pylori*. Note preponderance of U-shaped forms. Some of these may be spherical forms since the air drying probably causes collapse of the outer envelope. *Scale bar*, 2 μm

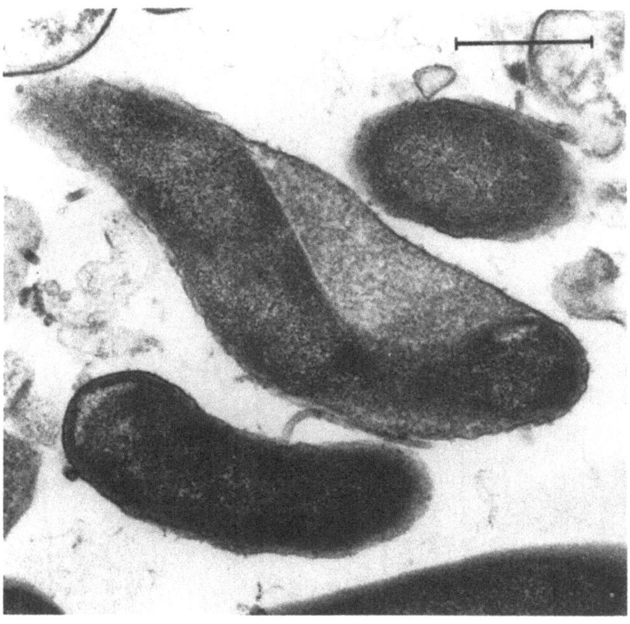

Fig. 3. Thin section of an organism showing more than normal bending. Note detachment of the outer envelope on the inside of the bend. A normal organism with the terminal specializations associated with flagellation is also seen in this micrograph. *Scale bar*, 600 nm

forms. The earliest atypical forms are recognisable because of bending unassociated with the normal helical form. These organisms show detachment of the outer envelope on the inner side of the bend (Fig. 3). Both non-dividing (Fig. 3) and dividing organisms (Fig. 4) can show this feature. Later stages show organisms bent into U- or V-shaped configurations within a loosened outer envelope (Fig. 5). One arm is nearly always flagellate with the normal complement of sheathed flagellar filaments (Fig. 6), but the other arm may or may not be flagellate. If flagella are present on the other pole, then they are usually fewer in number with a division septum to be seen at the point of the V. This septum does not incorporate the outer membrane and separation of the cells will not therefore occur.

Spherical forms are the ultimate stage of this morphological transformation, and feature a completely lossened membrane envelope within which a U-shaped or complexly folded cell may be apparent (Fig. 7). As before, spherical forms may have one or two groups of flagellar filaments (Fig. 7).

By thin sectioning, the atypical forms described above initially show detachment of the outer membrane which seems to progressively loosen in the more advanced forms. Inside this membrane "bag" the bacterial cell, enclosed within the plasma membrane, retains a cylindrical form. U-shaped forms show the loose membrane between the arms of the U (Fig. 8). The space between the arms of the protoplasmic cylinder and the externally loosened membrane-like wall component sometimes, but not always, fills with a homogeneous cytoplasmic material (Fig. 9). When present, this cytoplasm contains both bacterial ribosomes and filamentous material reminiscent of the bacterial nucleoid. It is, however, more lucent than the cytoplasm within the protoplasmic cylinder. Disruption of the

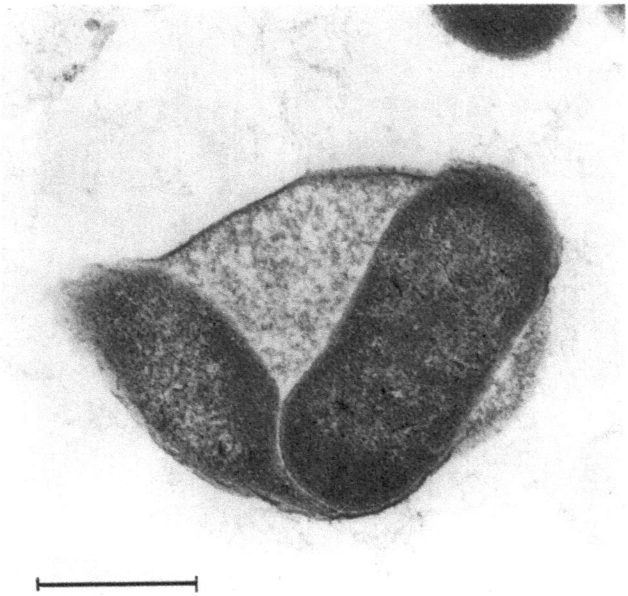

Fig. 4. Thin section of a V-shaped organism showing division septum and detached outer membrane. *Scale bar*, 600 nm

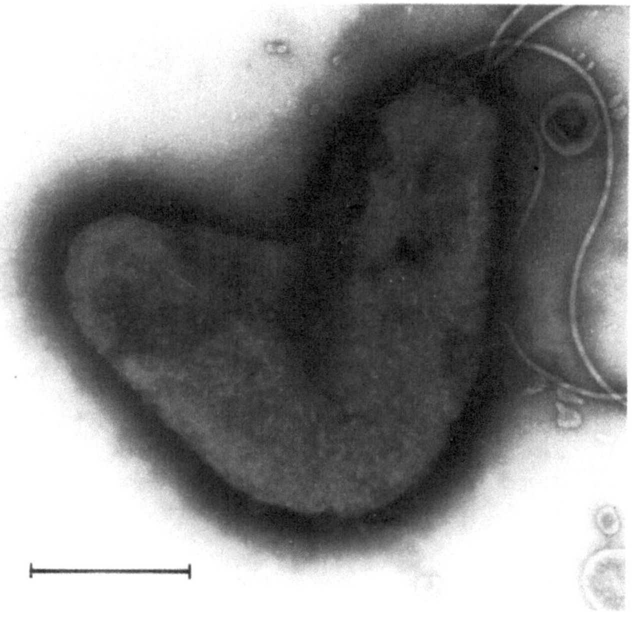

Fig. 5. Negatively stained preparation of a V-shaped organism showing detached membrane between the arms and a single bunch of flagella. *Scale bar*, 600 nm

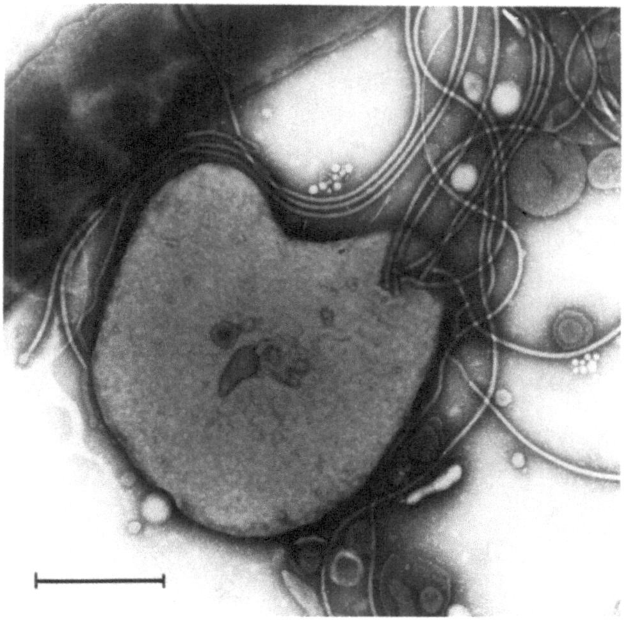

Fig. 6. Negatively stained preparation of a U-shaped organism with detached membrane between the arms and a single bunch of flagella. *Scale bar*, 600 nm

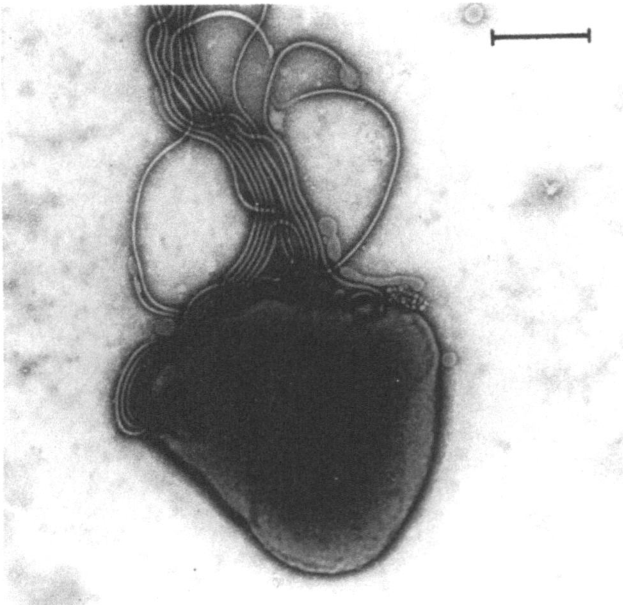

Fig. 7. Negatively stained preparation of a spherical form showing complex folding of the internal protoplasmic cylinder (just visible) and two bunches of flagellar filaments. *Scale bar*, 600 nm

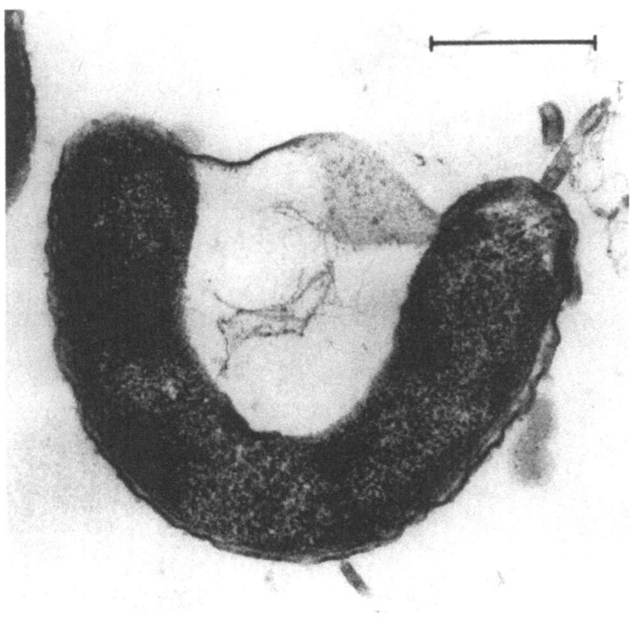

Fig. 8. Thin section of a U-shaped form showing outer membrane stretched between the arms. Note sheathed flagella filament on one pole and electron lucency of the space between the arms. *Scale bar*, 600 nm

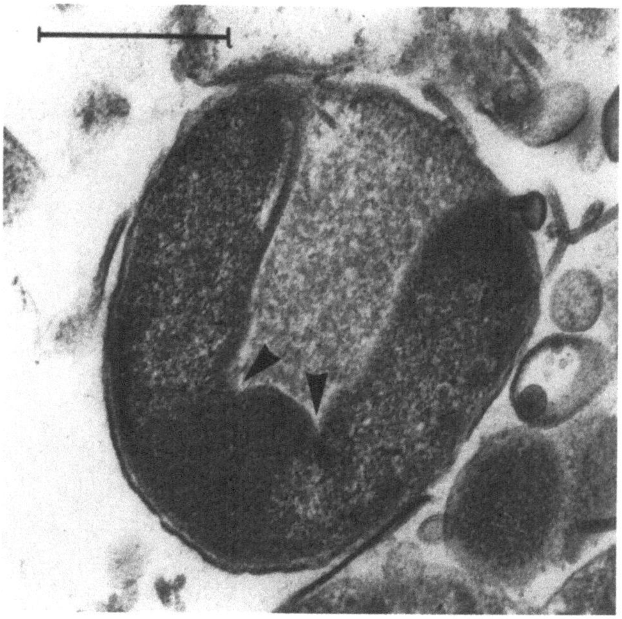

Fig. 9. Thin section of a U-shaped form showing a homogeneous granular material filling the space between the arms. Note the bulge and the buckling of the plasma membrane (*arrowheads*). *Scale bar*, 600 nm

plasma membrane has not been observed although it does buckle and bulge in the concavity of the bend (Fig. 9). V-shaped forms show similar features to the U-shaped forms but a division septum can sometimes be found at the narrowest point of the V.

In spherical forms, the internal protoplasmic cylinder almost produces a ring with the inner space being filled with "leaked" cytoplasm. Other spherical forms may show a complex folding of the protoplasmic cylinder. This can be demonstrated in negatively stained preparations when the stain penetrates the outer membrane (Fig. 7), and in thin sections where profiles cut transversely and longitudinally can be identified within a single membrane covering (Fig. 10).

None of the forms described above seems degenerate. Degenerating cells show obvious signs of disintegration, particularly the fragmentation of the sheathed flagellar filaments. In addition degenerate cells absorb the negative stain and therefore stain positively rather than negatively.

Discussion

The normal ultrastructure of *H. pylori* has been described previously [3, 4]. The U, V and spherical forms described here have, thus far, only been seen in culture. Ultrastructurally Gram-negative bacteria, such as *H. pylori*, are enclosed within two membranes, the inner of which is the plasma or cell membrane and the outer, a membrane-like lipoprotein structure. the space between these two membranes

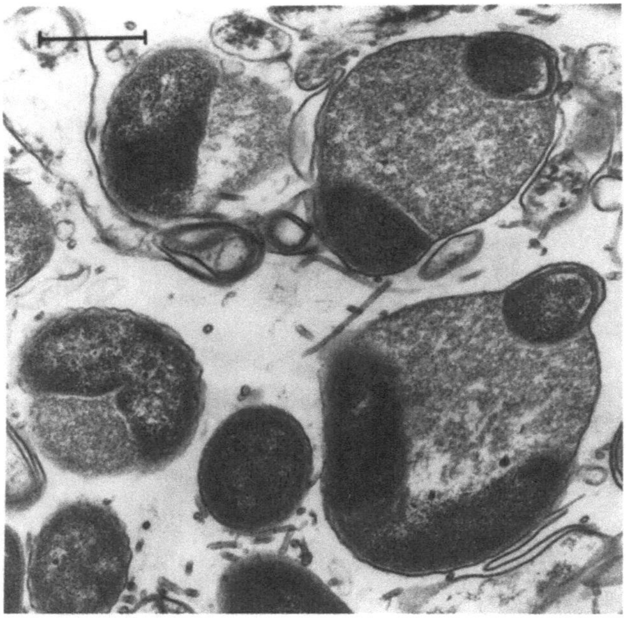

Fig. 10. Thin section of a number of atypical forms. The most complex is a spherical form with the protoplasmic cylinder cut in both transverse and longitudinal section, indicating extensive folding (*bottom right*). A U-shaped form is also cut transversely through both arms and shows expansion of the outer envelope and the granular filling (*top right*). Also visible is an organism showing protoplasmic buckling at the point of maximum bending (*bottom left*). *Scale bar*, 600 nm

contains a thin peptidoglycan layer which helps to maintain cell shape [6]. In addition some Gram-negative bacteria have large (on the molecular scale), regularly-spaced protein structures incorporated within or on the outer membrane [5, 8, 9]. Such structures have been identified in *H. pylori* [4] and, indeed, are easily detached. The peptidoglycan layer and the wall proteins probably both have a role in maintaining cell shape. Any weakening or deficiencies in these components or the links that hold the membrane layers together would, perhaps, cause loosening of the surface membrane. In *Vibrio sp* [1] changes in cell morphology and phase of culture were correlated with degradation but not complete removal of the peptidoglycan layer in the wall. The spherical forms produced in *Vibrio sp* were not thought to be viable. In true *Campylobacter* species (not to be confused with *H. pylori*) degenerative coccal forms have been found in the centre of colonies whereas the young, actively growing spiral forms were at the periphery of the colony [7]. Intermediate ring-shaped cells were also observed in this study of *Campylobacter* species. Spherical forms of *Campylobacter jejuni* have been observed in natural waters (pearson, personal communication) and may be an adaption to adverse conditions. Assumption of a spherical form would minimise the surface in contact with external conditions and may therefore be a survival adaptation in natural environments. It follows that it may be the coccal form of *H. pylori* that is involved in transmission through the environment. The atypical forms of *H. pylori* described here are not merely spherical forms of normal helical vegetative cells, but are formed by protoplasmic cylinder growth within a restricted envelope. All the various forms described here can have their mechanism of genesis explained if protoplasmic growth and division can continue within a loosened envelope that cannot expand in synchrony with this internal protoplasmic cylinder. Apart from the bending and twisting, the protoplasmic cylinder retains its shape, thus leading one to assume that perhaps the peptidoglycan layer remains attached to the plasma membrane. If this is the case then the loosening of the outer membrane may be the result of deficiencies in the links that hold the inner and outer membranes together. extreme bending would cause the "wall" of the protoplasmic cylinder to buckle and bulge and it may be this strain that leads to some leakage of cytoplasm although the plasma membrane appears to re-anneal. Spherical forms ultimately result from the outer envelope losing the last vestiges of rigidity and the protoplasmic cylinder bending and twisting in what may be a complex fashion within a restrictive envelope. These forms can be identified by both negative staining and thin-sectioning transmission electron microscopy, but because of drying artefact and subsequent collapse of the outer membrane, mainly U and V forms are seen by scanning electron microscopy.

Although the coccal forms we have described have the ultrastructural appearance of viability, this remains to be shown by cultural methods.

Acknowledgements. We would like to acknowledge the excellent technical help given to us by Patricia Rowland, Hilary Cotterill, Joan Eldridge and Sarah Hunt.

References

1. Baker DA, Park RWA (1975) Changes in morphology and cell wall structure that occur during growth of Vibrio sp NCTC 4716 in batch culture. J Gen Mcrobiol 86:12–28
2. Goodwin CS, Armstrong JA, Chilvers T, et al. (1989) Transfer of Campylobacter pylori and Campylobacter mustelae to Helicobacter gen. nov. as Helicobacter pylori comb. nov. and Helicobacter mustelae comb. nov., respectively. Int J Syst Bacteriol 39:397–405
3. Jones DM, Curry A (1989) The ultrastructure of Campylobacter pylori. In: Rathbone BJ, Heatley RV (eds) Campylobacter pylori and gastroduodenal disease
4. Jones DM, Curry A, Fox AJ (1985) An ultrastructural study of the gastric campylobacter-like organism "Campylobacter pyloridis". J Gen Microbiol 131:2335–2341
5. Kist ML, Murray RGE (1984) Components of the regular surface array of Aquaspirillum serpens MW5 and their assembly in vitro. J Bacteriol 157:599–606
6. Meadow PM (1978) The chemistry and composition of microorganisms. In: Norris JR, Richmond MH (eds) Essays in microbiology. Wiley, Chichester
7. Ng L-K, Sherburne R, Taylor DE, Stiles ME (1985) Morphological forms and viability of Campylobacter species studied by electron microscopy. J Bacteriol 164:338–343
8. Osborn MJ, Wu HCP (1980) Proteins of the outer membrane of gram-negative bacteria. Annu Rev Microbiol 34:369–422
9. Steven AC, Heggeler TEN, Muller R, Kistler J, Rosenbusch JP (1977) Ultrastructure of a periodic protein layer in the outer membrane of Escherichia coli. J Cell Biol 72:292–301

Epidemiology of *Helicobacter pylori* Infections

M. J. Blaser

Helicobacter pylori (formerly known as *Campylobacter pylori*) has been strongly associated with type B gastritis and with peptic ulcer disease. Although this organism was first discovered only 7 years ago, an increasing body of evidence now indicates that it contributes to the pathogenesis of these common and important disorders of the upper gastrointestinal tract. In this report, I will briefly outline how the development of an appropriate diagnostic test has facilitated our understanding of the epidemiology of this infection, and then turn to the specific epidemiologic characteristics that have been defined.

Development of Accurate Serologic Methods for Diagnosis of *H. pylori* Infections

It is now clear that *H. pylori* infections persist for years, if not decades. For other chronic bacterial infections, such as syphilis and brucellosis, for example, it has been possible to use the host response to the offending organism as a means for diagnosing the presence of the infection. Early on, it was found that persons infected with *H. pylori* develop serum antibodies to the organism [1]. Since then, several different diagnostic techniques have been utilized to detect these serum antibodies. The techniques include bacterial agglutination, hemagglutination, complement fixation, indirect immunofluorescence, and enzyme-linked immunosorbent assays (ELISA) [reviewed in 2]. Even in the ELISA format, several different assays have been described, which vary considerably in their diagnostic accuracy. Knowledge of the antigenic characteristics of *H. pylori* has enabled my laboratory to develop a novel ELISA, which has improved diagnostic power over the previously published assays. Use of this assay permits diagnosis of *H. pylori* infection with both sensitivity and specificity exceeding 95% [3–5]. In more recent work, we have found that the sensitivity of the assay exceeds that for histologic examination and culture. This is not surprising since gastritis is often a patchy phenomenon, and biopsy samples only a small part of the gastric antrum, whereas serology in essence samples the entire stomach. The availability of a noninvasive method for diagnosis of *H. pylori* infections permits epidemiologic surveys as well as evaluation of individual patients.

Division of Infectious Diseases, Department of Medicine, Vanderbilt University School of Medicine, Nashville, TN

Helicobacter pylori, Gastritis and Peptic Ulcer
Edited by P. Malfertheiner and H. Ditschuneit

Epidemiologic Characteristics of *H. pylori* Infection

The epidemiologic characteristics of *H. pylori* infection have recently been reviewed [6]. We now know that *H. pylori* infection is present in all parts of the world. Since infection is persistent, it is not surprising that its prevalence rises with age. Among Caucasian populations in the United States and other industrialized countries, *H. pylori* infection is infrequent during childhood, but rises in prevalence during adulthood, reaching about 50% in those persons who are 60 years or older. The prevalence data and a prospective study of epidemiologists in the United States suggests an incidence of infection of 0.5%–2% annually. Seroprevalence appears to reach a plateau after the sixth decade of life. Prevalence rates appear to be higher in Blacks and Hispanics than in non-Hispanic Caucasians but whether this is related to differences in socioeconomic status or to other differences is not yet resolved.

In developing countries, *H. pylori* infection begins much earlier in life. In rural Thailand there was little infection under the age of 10 years but in a Bangkok orphange, 75% of children were positive by the age of 2 years. In several other populations in developing countries, up to 50% of children are positive by the age of 10 years [7]. Prevalence of infection reaches much higher levels in adults as well, with from 70% to 90% of the population positive by the age of 30. In all studies, there is little difference in infection rates between men and women.

Thus, although age has emerged as the most important risk factor for infection, there are marked geographic and ethnic differences in the distribution of this infection.

Transmission of *H. pylori* Infection

There is no known reservoir for *H. pylori* in the environment or among the animals with which humans most often come into contact. These observations, in addition to the widespread nature of human infection with *H. pylori* suggest that humans are in fact the ultimate reservoir for the organism. Recent studies of family members of *H. pylori*-infected children indicate that parents and siblings are infected at significantly higher rates than are controls [5]. These data, for which the strongest association is with maternal infection, indicate that there is intrafamilial clustering of *H. pylori*.

How might the organism be transmitted? Population-based studies in general show that those groups with the lowest socioeconomic standards have the highest *H. pylori* infection rates. Children in orphanages where hygienic standards are low, and institutionalized mentally retarded children have very high infection rates suggesting the possibility of fecal-oral transmission. The association of *H. pylori* positivity with evidence of previous hepatitis A infection further points toward fecal-oral transmission.

We have examined the hypothesis that there is sexual transmission of *H. pylori*. Among those attending a sexually transmitted disease clinic in Colorado, there was no association of the presence of *H. pylori* infection with lifetime number of sexual partners, either for homosexual or heterosexual men. In another study, involving heterosexual couples undergoing evaluation at an infertility clinic, there was no association of the *H. pylori* status of the partners that could not be explained by chance alone. In total, these data suggest that sexual transmission of *H. pylori* infection is uncommon at best.

Conclusions

The development of noninvasive diagnostic methodologies based on serology has opened the door to detailed examination of the epidemiology of *H. pylori* infection. Use of appropriate serologic tests will facilitate diagnosis and management of *H. pylori* infection as well. It is now clear that *H. pylori* is one of the bacteria with which humans most frequently become infected during their lifetimes. The major remaining epidemiologic problems are to understand transmission of the organism, and to determine which host or bacterial factors contribute to the considerable variation in the natural history of infection. Work is proceeding rapidly in these areas and the important questions should be resolved within the coming years.

References

1. Jones DM, Lessells AM, Eldridge J (1984) Campylobacter-like organisms on the gastric mucosa: culture, histological, and serological studies. J Clin Pathol 37:1002–1006
2. Perez-Perez GI, Dunn BE (1989) Serologic methods to diagnose C. pylori infections. In: Blaser MJ (ed) Campylobacter pylori in gastritis and peptic ulcer disease. Igaku-Shoin, New York
3. Perez-Perez GI, Dworkin B, Chodos J, Blaser MJ (1988) Campylobacter pylori antibodies in humans. Ann Intern Med 109:11–17
4. Dooley CP, Cohen H, Fitzgibbons PL, Bauer M, Marshall B, Appleman MD, Perez-Perez GI, Blaser MJ (1989) Prevalence of Helicobacter pylori infection and histologic gastritis in asymptomatic persons. N Engl J Med
5. Drumm B, Perez-Perez GI, Blaser MJ, Sherman P (1990) Intrafamilial clustering of Campylobacter pylori infection. N Engl J Med 322:359–363
6. Blaser MJ (1990) Epidemiology and pathophysiology of Campylobacter pylori infections. Rev Infect Dis 12:S99–S106
7. Megraud F, Brassens-Rabbe MP, Denis F, Belbouri A, Hoa DQ (1989) Seroepidemiology of Campylobacter pylori infection in various populations. J Clin Microbiol 27:1870–1873

Physiology of *Helicobacter pylori*

W. Opferkuch, G. Geis, H. Leying, and S. Suerbaum

Our knowledge about the basic physiology of *Helicobacter pylori* (*H. pylori*) is still very fragmentary and based on a number of studies investigating physiological properties such as enzyme profiles or growth conditions mainly for diagnostic or taxonomic purposes. However, more detailed information has started to accumulate about the composition and biology of the *H. pylori* cell envelope and its putative virulence factors.

Growth conditions

There are no systematic studies on the optimal culture conditions for *H. pylori*, but the following conclusions can be extracted from the available data [5, 10, 21, 27, 35, 37]. Chocolate agar is widely considered the most suitable solid medium. Similar results, however, can also be achieved with other media supplemented with horse serum, sheep blood, hemin, starch, yeast, or cholesterol [5]. These data suggest that the function of these supplements is rather the binding of toxic substances than the supplementation of nutrients. However, their exact functions are still not known.

Recently Hazell et al. [21] conducted experiments which were designed to determine the role of heme and the importance of other factors for the growth of *H. pylori*. They were able to show that heme is not required as a source of porphyrin and suggested that bovine serum albumin and catalase provided protection from toxic fatty acids and inhibited the formation of other toxic products.

Antibiotic supplements are widely used to isolate *H. pylori* from gastric biopsies. Dent and McNulty [10] proposed a modification of the Skirrow formula replacing polymyxin B by cefsulodin. They estimated that an additional 5% of *H. pylori* strains which are sensitive to polymyxin B can be isolated on such a medium. Nalidixic acid should also be avoided in supplements used for primary isolation of *H. pylori*, since it was shown to inhibit the growth of up to 14% of strains [10, 27, 28].

Cultivation in liquid media is now feasible in many laboratories and several formulas for liquid media have been suggested [37]. We used the medium

Ruhr-Universität Bochum, Medizinische Mikrobiologie und Immunologie, Universitätsstr. 150D, 4630 Bochum 1, FRG

Helicobacter pylori, Gastritis and Peptic Ulcer
Edited by P. Malfertheiner and H. Ditschuneit

described by Majewski and Goodwin [33], which is brain-heart infusion broth (Oxoid, Wesel, F.R.G.) supplemented with horse serum, yeast, and hemin solution.

Coudron and Kirby [8] have addressed the question of which transport media are necessary for the optimum recovery of *H. pylori* from biopsies. They conclude that biopsies can be stored at 4 °C in sodium chloride solutions for up to 72 h. We observed that under these conditons viable cell counts dropped by 1–1.5 log steps. We used Fildes broth and obtained 96% positive cultures in patients with duodenal ulcer [4].

Cell envelope composition

Interesting data have been gathered about the composition of the *H. pylori* cell envelope. Evans et al. [15] demonstrated that *H. pylori* cells are enveloped by a fibrillar hemagglutinin which binds to *N*-acetyl-neuraminyl-lactose. *H. pylori* is able to attach to certain cell lines [14, 39] and to adhere to erythrocytes of different species [13, 23, 38]. Huang was able to divide *H. pylori* strains into two phenotypes according to their hemagglutination of either human group A and O erythrocytes. The second phenotype could be further subtyped by hemagglutination patterns using ox, sheep, goat, and guinea pig erythrocytes. He raised the question whether these hemagglutination patterns might reflect different degrees of strain virulence. These data also indicate that more than the adhesin described by Evans et al. [15] must be responsible for *H. pylori* hemagglutination properties. Recently, Lingwood et al. [32] isolated a glycerolipid from gastric mucosal tissue that mediates specific attachment of *H. pylori* and might be the receptor for a yet unidentified further adhesin.

Up to now, most data about *H. pylori* proteins are derived from whole cell protein preparations. The separation of outer and inner membranes has not yet been conclusively demonstrated. Perez-Perez and Blaser [43] and Dunn et al. [11] found the protein patterns of different *H. pylori* strains to be remarkably similar; however, there were marked differences from the patterns found in other Campylobacter species. Newell [40] also published results on *H. pylori* protein patterns and found a low variability from strain to strain. However, her data differ from those of Perez-Perez in certain details.

Using numerical analysis of SDS-PAGE patterns, Owen et al. [41] found ten different electropherotypes, which mainly differed in the region of 47–56 Kda proteins but overall had high similarities. From data obtained by Apel et al. [1], it can be concluded that a 120-Kda protein is not expressed by all strains of *H. pylori*. These studies and our own observations show that proteins are similar in all *H. pylori* strains, but that there is a variety of strain specific minor changes in the protein profile.

As for *H. pylori* lipopolysaccharide (LPS), only few data are available, but there seems to be a significant variability, which might contribute to the serologic heterogeneity of the species that some groups have described [43]. Using immunoblot analysis Burnie et al. [6] divided *H. pylori* strains into nine serological groups. Whether these serotypes reflect differences in the LPS structure or not needs further investigation. Serologic heterogeneity was also found by Danielsson et al. [9] using a coagglutination test with rabbit antibodies. In contrast, the same group found extensive antigenic similarities when analyzing rabbit immune responses against *H. pylori* strains with an enzyme immunoassay technique [22].

Analysis of the total fatty acid composition of *H. pylori* has been published by several groups, and the results of the recently published papers correspond well to each other (Table 1). The main fatty acids are myristic acid and C_{19} cyclopropane fatty acid, unusual is the appearance of 3-hydroxypalmitic and 3-hydroxystearic acid. We have studied the fatty acid substitution of isolated phospholipids and lipopolysaccharides and found the major fatty acids of *H. pylori* phospholipids to be myristic acid and C_{19} cyclopropane fatty acid, while the major fatty acids of *H. pylori* lipopolysaccharides are stearic acid, 3-3-hydroxypalmitic acid and 3-hydroxystearic acid (Fig. 1). This distribution of fatty acids in *H. pylori* is highly unusual. Myristic acid is the typical major fatty acid of enterobacterial lipopolysaccharides and not a typical constituent of bacterial phospholipids. Its presence in *H. pylori* phospholipids as their major fatty acid is highly remarkable. This must lead to very peculiar physicochemical properties of *H. pylori* membranes [18a].

Another very distinctive feature of *H. pylori* is the structure of its flagella. *H. pylori* is lophotrichously flagellated, and each flagellum is enveloped by a flagellar sheath which at its end dilates into a terminal club-shaped structure [26]. We have purified *H. pylori* flagellin; its molecular weight is 51 Kda [18]. So far there is no information about the structure and function of the *H. pylori* flagellar sheath, this is currently under investigation.

The reviewed data and also other studies such as those on exoenzyme profiles [34, 35] indicate some degree of diversity within the species *H. pylori*. Reina and Alomar [44] were able to divide *H. pylori* into four biotypes using the Api ZYM-system. In addition to this, several groups [30, 33] have found marked differences in the DNA restriction endonuclease patterns among various strains of *H. pylori*. Whether these differences do really reflect genomic variability or just different enzyme susceptibilities due to different degrees of methylation has yet to be clarified. We and other investigators [42] have screened larger groups of strains for plasmid carriage; we found plasmids in 17 out of our 75 strains tested. The size varied between 2 and 40 kilobasepairs (unpublished data). The functions of these plasmids so far are unknown.

Virulence factors

The question of the heterogeneity of the species *H. pylori* is not at all academic. For a bacterium that is such a common parasite in man, criteria that would permit separation of pathogenic from non-pathogenic strains would be of the highest therapeutic and economical importance. In order to achieve this goal for *H. pylori*, the identification of factors making *H. pylori* virulent is a prerequisite, and the knowledge about possible *H. pylori* virulence factors will be reviewed very briefly in this light. At the moment, there are six major putative virulence factors in

Table 1. Cellular fatty acid composition of *Helicobacter pylori* (% of total fatty acids)

Reference		Fatty acids								
		14:0	16:0	16:1	18:0	18:1	18:2	19:0 cyc	3-OH-16:0	3-OH-18:0
Itoh et al. [25]	NCTC 11637	42	2	–	11	9	2	31	–	–
	11638	39	2	–	13	2	2	34	–	–
	11639	36	14	–	10	4	5	25	–	–
	strains 1–20	31	10	–	15	8	8	24	–	–
Lambert et al. [29]	strains 1–3	31	4	–	12	11	–	23	3	8
Goodwin et al. [20]	NCTC 11637	41	5	tr	5	16	4	17	5	2
	11638	41	7	–	7	9	2	16	5	3
	Q188	37	5	tr	8	17	3	17	5	2
	72R	36	6	tr	11	17	2	17	7	3
Geis et al. [18a]	NCTC 11637	45	2	–	7	5	tr	24	4	7
	NCTC 11639	31	5	–	12	5	3	20	10	7
	713 8981	43	3	–	7	6	tr	22	6	7
	713 9015	44	4	–	9	2	tr	21	4	6

tr (traces): 0.5–1.9%

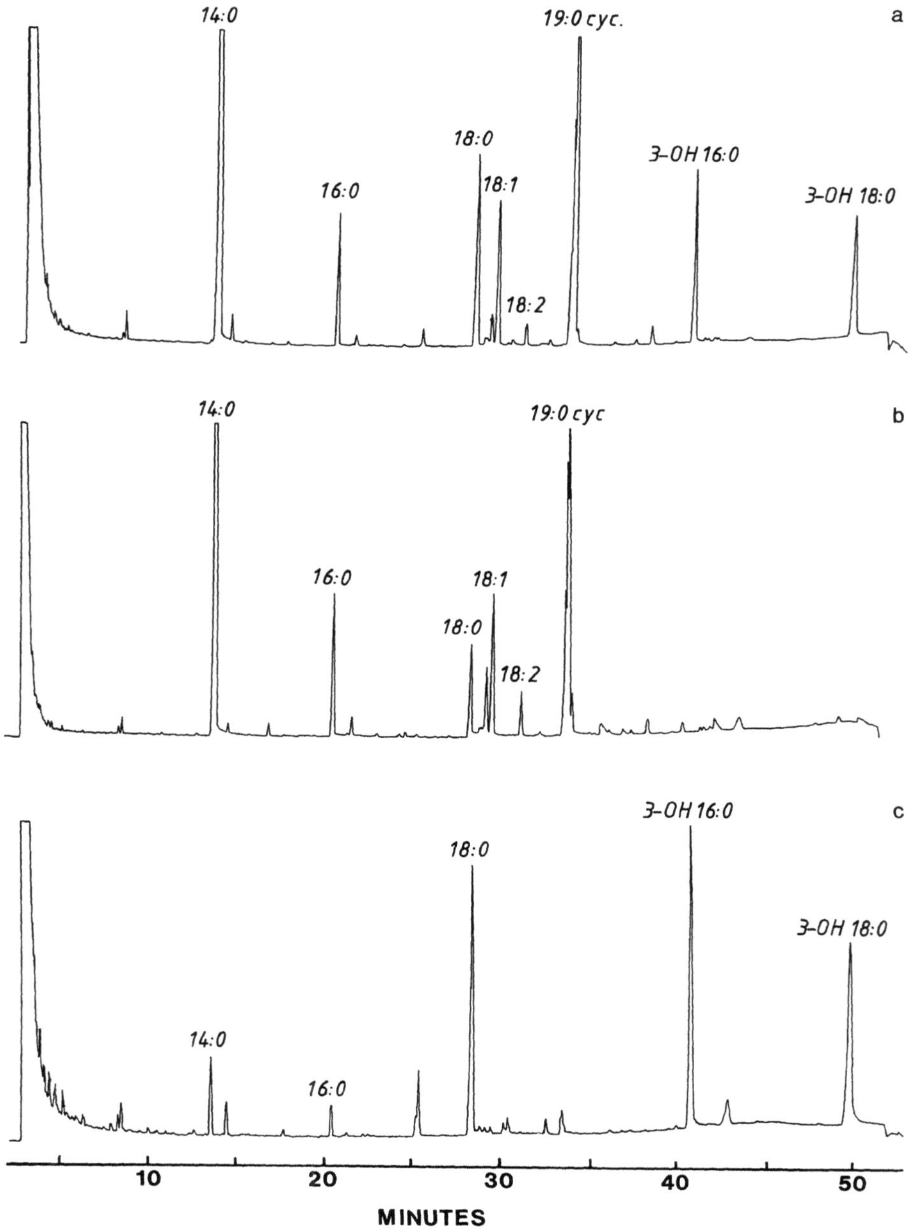

Fig. 1a–c. Gas chromatography of *H. pylori* fatty acid methyl esters (FAME) of **a** whole cells, **b** phospholipids, and **c** lipopolysaccharides (strain 7138981). FAME of whole cells, isolated phospholipids and lipopolysaccharides were prepared by method A as described by Gmeiner and Martin [19]. Samples were analyzed on a gas chromatograph (model 3700, Varian, Palo Alto, CA) equipped with a flame ionization detector and a fused silica capillary column (FS-FFAP-CB 0.25, Macherey-Nagel, Düren, FRG). The column was temperature programmed from 140 °C to 220 °C. FAME were identified by comparing retention time to methylester standards

H. pylori, two of which are structural parts of the bacterium (flagella, adhesins), the four other factors being extracellular products, namely urease, cytotoxins, mucin-degrading proteases, and an acid secretion-inhibiting protein.

The *H. pylori* urease is a 500–600 Kda protein with a temperature optimum of 43 °C–45 °C, a pH optimum at 8.2, and high substrate affinity [16, 36, 47, 48]. It has been claimed that the urease is localized in the outer membrane and periplasmic space [3, 11]. Several groups have reported that culture filtrates or lysates of *H. pylori* have cytotoxic effects on different cell lines [17, 24, 31]. Sarosiek et al. [45] and Slomiany et al. [46] demonstrated that *H. pylori* secretes a mucin-degrading protease, but this activity was not found by Baxter et al. [2]. Cave and Vargas [7] showed that *H. pylori* inhibits acid secretion by gastric parietal cells, probably by a protein factor.

There is only one published study about the relevance of these factors as virulence factors in an animal model. Eaton et al. [12] compared the virulence of *H. pylori* strains, which differed in motility and cytotoxin production, in their gnotobiotic piglet model and found a good correlation between motility and virulence, and a questionable correlation between cytotoxin production and virulence. The prevalence of these possible virulence factors has been studied systematically by several groups only for cytotoxin production. Leunk et al. [31] screened 200 strains for cytotoxin production and found 110 strains to produce cytotoxin. Urease production seems to be an almost invariable property of clinical *H. pylori* isolates, although the amount seems to vary. The prevalence of the other possible virulence factors is not known.

From the reviewed data, it is likely that strains of *H. pylori* may differ in virulence, and that it may be feasible to establish criteria for guiding the decision to institute therapy, not only according to clinical data but also according to the results of strain analysis, especially concerning their expression of virulence factors.

References

1. Apel I, Jacobs E, Kist M, Bredt W (1988) Antibody response of patients against a 120 kDa surface protein of Campylobacter pylori. Zbl Bakt Hyg A 268:271–276
2. Baxter A, Campbell CJ, Cox DM, Grinham CJ, Pendlebury JE (1989) Proteolytic activities of human Campylobacter pylori and ferret gastric Campylobacter-like organisms. Biochem Biophys Res Commun 163:1–7
3. Bode G, Malfertheiner P, Nilius M, Lehnhardt G, Ditschuneit H (1989) Ultrastructural localization of urease in outer membrane and periplasm of Campylobacter pylori. J Clin Pathol 42:778–779
4. Börsch G, Mai U, Reitemeyer E, Leverkus F, Opferkuch W (1989) Cympylobacter pylori: Klinische Korrelationen und prospektive vergleichende Untersuchungen zur Wertigkeit mehrerer Testverfahren. Immun Infekt 17:83–90
5. Buck GE, Smith JS (1987) Medium supplementation for growth of Campylobacter pyloridis. J Clin Microbiol 25:597–599
6. Burnie JP, Lee W, Dent JC, McNulty CAM (1988) Immunoblot fingerprinting of Campylobacter pylori. J Med Microbiol 27:153–159
7. Cave DR, Vargas M (1989) Effect of a Campylobacter pylori protein on acid secretion by parietal cells. Lancet ii:187–189

8. Coudron PE, Kirby DF (1989) Comparison of rapid urease tests, staining techniques, and growth on different solid media for detection of Campylobacter pylori. J Clin Microbiol 27:1527–1530
9. Danielsson D, Blomberg B, Järnerot G, Kosunen TU (1988) Heterogeneity of Campylobacter pylori as demonstrated by co-agglutination testing with rabbit antibodies. Scand J Gastroenterol 23:58–63
10. Dent JC, McNulty CAM (1988) Evaluation of a new selective medium for Campylobacter pylori. Eur J Clin Microbiol Infect Dis 7:555–558
11. Dunn BE, Perez-Perez GI, Blaser MJ (1989) Two-dimensional gel electrophoresis and immunoblotting of Campylobacter pylori proteins. Infect Immun 57:1825–1833
12. Eaton KA, Morgan DR, Krakowka S (1989) Campylobacter pylori virulence factors in gnotobiotic piglets. Infect Immun 57:1119–1125
13. Emody L, Carlsson A, Ljungh A, Wadström T (1988) Mannose-resistant haemagglutination by Campylobacter pylori. Scand J Infect Dis 20:353–354
14. Evans DG, Evans DJ jr, Graham DY (1989) Receptor-mediated adherence of Campylobacter pylori to mouse Y-1 adrenal cell monolayers. Infect Immun 57:2272–2278
15. Evans DG, Evans DJ, Moulds JJ, Graham DY (1988) N-acetylneuraminyllactose-binding fibrillar hemagglutinin of Campylobacter pylori: a putative colonization factor antigen. Infect Immun 56:2896–2906
16. Ferrero RL, Hazell SL, Lee A (1988) The urease enzymes of Campylobacter pylori and a related bacterium. J Med Microbiol 27:33–40
17. Figura N, Guglielmetti P, Rossolini A, Barberi A, Cusi G, Musmanno RA, Russi M, Quaranta S (1989) Cytotoxin production by Campylobacter pylori stains isolated from patients with peptic ulcers and from patients with chronic gastritis only. J Clin Microbiol 27:225–226
18. Geis G, Leying H, Suerbaum S, Mai U, Opferkuch W (1989) Ultrastructure and chemical analysis of Campylobacter pylori flagella. J Clin Microbiol 27:436–441
18a. Geis G, Leying H, Suerbaum S, Opferkuch W (1990) Unusual fatty acid substitution in lipids and lipopoly. Saccharides of Helicobacter pylori. J Clin Microbiol 28:(in press)
19. Gmeiner J, Martin HH (1976) Phospholipid and lipopolysaccharide in Proteus mirabilis and its stable protoplast L-form. Eur J Biochem 67:487–494
20. Goodwin CS, McConnell W, McCulloch RK, McCullough C, Hill R, Bronsdon MA, Kasper G (1989) Cellular fatty acid composition of Campylobacter pylori from primates and ferrets compared with those of other Campylobacters. J Clin Microbiol 27:938–943
21. Hazell SL, Markesich DC, Evans DJ, Evans DG, Graham DY (1989) Influence of media supplements on growth and survival of Campylobacter pylori. Eur J Clin Microbiol Infect Dis 8:597–602
22. Höök J, Blomberg B, Danielsson D, Kosunen TU (1989) Detection of antibody responses in rabbits hyperimmunized with Cympylobacter pylori. Enzyme immunoassay indicates extensive antigenic similarities. APMIS 97:56–60
23. Huang J, Smyth CJ, Kennedy NP, Arbuthnott JP, Keeling PWN (1988) Haemagglutinating activity of Campylobacter pylori. FEMS Microbiol Lett 56:109–112
24. Hupertz V, Czinn S (1988) Demonstration of a cytotoxin from Campylobacter pylori. Eur J Clin Microbiol Infect Dis 7:576–578
25. Itoh T, Yanagawa Y, Shingaki M, Takahashi M, Kai A, Ohashi M, Hamana G (1987) Isolation of Campylobacter pyloridis from human gastric mucosa and characterization of the isolates. Microbiol Immunol 31:603–614
26. Jones DM, Curry A, Fox AJ (1985) An ultrastructural study of the gastric Campylobacter-like organism "Campylobacter pyloridis". J Gen Microbiol 131:2335–2341
27. Krajden S, Bohnen J, Anderson J, Kempston J, Fuksa M, Matlow A, Marcon N, Haber G, Kortan P, Karmall M, Corey P, Petrea C, Babida C, Hayman S (1987) Comparison of selective and nonselective media for recovery of Campylobacter pylori from antral biopsies. J Clin Microbiol 25:1117–1118
28. Lambert T, Mégraud F, Gerbaud G, Courvalin P (1986) Susceptibility of Campylobacter pyloridis to 20 antimicrobial agents. Antimicrob Agents Chemother 30:510–511

29. Lambert MA, Patton CM, Barrett TJ, Moss CW (1987) Differentiation of Campylobacter and Campylobacter-like organisms by cellular fatty acid composition. J Clin Microbiol 25:706–713
30. Langenberg W, Rauws EAJ, Widmomokusumo A, Tytgat GNJ, Zanen HC (1986) Identification of Campylobacter pyloridis isolates by restriction endonuclease DNA analysis. J Clin Microbiol 24:414–417
31. Leunk RD, Johnson PT, David BC, Kraft WG, Morgan DR (1988) Cytotoxic activity in broth-culture filtrates of Campylobacter pylori. J Med Microbiol 26:93–99
32. Lingwood CA, Pellizzari A, Law H, Sherman P (1989) Gastric glycerolipid as a receptor for Campylobacter pylori. Lancet ii:238–241
33. Majewski SIH, Goodwin CS (1988) Restriction endonuclease analysis of the genome of Campylobacter pylori with a rapid extraction method: evidence for considerable genomic variation. J Infect Dis 157:465–471
34. McNulty CAM, Dent JC (1987) Rapid identification of Campylobacter pylori (C. pyloridis) by preformed enzymes. J Clin Microbiol 25:1683–1686
35. Mégraud F, Bonnet F, Garnier M, Lamouliatte H (1985) Characterization of "Campylobacter pyloridis" by culture, enzymatic profile, and protein content. J Clin Microbiol 22:1007–1010
36. Mobley HLT, Cortesia MJ, Rosenthal LE, Jones BD (1988) Characterization of urease from Campylobacter pylori. J Clin Microbiol 26:831–836
37. Morgan DR, Freedman R, Depew CE, Kraft WG (1987) Growth of Campylobacter pylori in liquid media. J Clin Microbiol 25:2123–2125
38. Nakazawa T, Ishibashi M, Konishi H, Takemoto T, Shigeeda M, Kochiyama T (1989) Hemagglutination activity of Campylobacter pylori. Infect Immun 57:989–991
39. Neman-Simha V, Mégraud F (1988) In vitro model for Cympylobacter pylori adherence properties. Infect Immun 56:3329–3333
40. Newell DG (1987) Identification of the outer membrane proteins of Campylobacter pyloridis and antigenic cross-reactivity between C. pyloridis and C. jejuni. J Gen Microbiol 133:163–170
41. Owen RJ, Costas M, Morgan DD, On SLW, Hill LR, Pearson AD, Morgan DR (1989) Strain variation in Campylobacter pylori detected by numerical analysis of one-dimensional electrophoretic protein patterns. Antonie Van Leeuwenhoek 55:253–268
42. Penfold SS, Lastovica AJ, Elisha BG (1988) Demonstration of plasmids in Campylobacter pylori. J Infect Dis 157:850–851
43. Perez-Perez GI, Blaser MJ (1987) Conservation and diversity of Campylobacter pyloridis major antigens. Infect Immun 55:1256–1263
44. Reina J, Alomar P (1989) Biotypes of Campylobacter pylori isolated in gastroduodenal biopsies. Eur J Clin Microbiol Infect Dis 8:175–176
45. Sarosiek J, Slomiany A, Slomiany BL (1988) Evidence for weakening of gastric mucus integrity by Campylobacter pylori. Scand J Gastroenterol 23:585–590
46. Slomiany BL, Bilski J, Sarosiek J, Murty VLN, Dworkin B, Van Horn K, Zielenski J, Slomiany A (1987) Campylobacter pyloridis degrades mucin and undermines gastric mucosal integrity. Biochem Biophys Res Commun 144:307–314
47. Taylor MB, Goodwin CS, Karim QN (1988) Two urease activities with different pH optima in Campylobacter pylori and similar organisms. FEMS Lett 55:259–262
48. Tompkins DS, Millar MR, West AP (1988) Isoelectric focusing of ureases from Campylobacter pylori and related organisms. J Clin Microbiol 26:2678–2679

In Vitro Susceptibility of *Helicobacter pylori* to Antibiotics and Bismuth Salts and the Importance of Acquired Resistance to Antibiotics in Treatment Failures of *H. pylori* Infection

Y. GLUPCZYNSKI

There is now overwhelming evidence that *Helicobacter pylori* (formerly *Campylobacter pylori*) is the major etiological agent of chronic active (type B) gastritis and that it may further predispose to peptic ulceration. Challenge studies in gnotobiotic piglets [1, 2] and in two human volunteers [3, 4] have reproduced the typical features of gastritis. Clearance of the organism with amoxicillin, furazolidone, or bismuth salts leads to improvement or even total resolution of gastritis [5–8]. Moreover, well-designed controlled studies have shown that patients with healed peptic ulcer who remain persistently infected with *H. pylori* relapse much more frequently than patients cleared of the organism [9, 10]. In vitro, *H. pylori* is sensitive to most antibiotics (penicillin, ampicillin, cephalosporins, macrolides, quinolones, aminoglycosides, tetracyclines, and nitraimidazoles) except vancomycin, trimethoprim, and sulfonamides [11–15]. Among the various antiulcer drugs, bismuth salts are the only compounds that possess an antibacterial activity, though only a moderate one (minimal whibitory, concentration MIC_{90}, 8–32 mg/l), against *H. pylori* [12, 13]. There does not seems to be any difference in the levels of antimicrobial activity of the numerous existing forms of bismuth salts (subsalicylate, colloidal bismuth subcitrate, subcarbonate, nitrate, subnitrate, or oxychloride) (Glupczynski and Thibaumont 1989, unpublished data).

In the clinic, however, experience with various antimicrobials has been particularly disappointing, especially when these were administered as a single agent. Amoxicillin – the most extensively studied antibiotic – has been shown to be effective in clearing *H. pylori* in 70%–90% of patients, but the relapse rate shortly after treatment is very high and long-term eradication has not been achieved in more than 20% of individuals [8, 16, 17]. In one study, administration of furazolidone resulted in encouraging results, with an initial clearance rate of 93% and a 6-week eradication rate of about 40% [7]. Monotherapy with other agents has been almost totally ineffective with clearance rates ranging between 0% and 30% [8, 16–18].

Different factors have been suggested as being responsible for the lack of clinical efficacy: insufficient drug concentration in the gastric mucosa and crypts due to poor penetration, local inactivation of the drug, decreased antibacterial activity of some antibiotics at low pH, inappropriate formulation or modality of administration of the drug, inadequate duration of treatment, or poor compliance

Department of Microbiology, Brugmann University Hospital, 4, Place Van Gehuchten, B-1020, Brussels, Belgium

Helicobacter pylori, Gastritis and Peptic Ulcer
Edited by P. Malfertheiner and H. Ditschuneit

[19–22]. Although acquired resistance of *H. pylori* to antibiotics has been previously reported [23, 24], its exact role in the treatment failures has not been thoroughly discussed. In this paper we will review the various antibiotics which proved ineffective in the eradication of *H. pylori*, possibly because of the development of acquired resistance during treatment.

Nitroimidazoles

Several early studies performed before *H. pylori* was recognized demonstrated some efficacy of metronidazole in the treatment of peptic ulcer [25, 26]. However, these studies essentially focused on the initial ulcer healing rate and did not mention the frequency with which relapses occurred after initial healing. Moreover, these investigators did not differentiate between duodenal and gastric ulcer, so that their results cannot be further evaluated concerning a pathogenic role of *H. pylori* in either gastric or duodenal ulcer.

In vitro, *H. pylori* is moderately sensitive to metronidazole and to tinidazole (MIC range 0.5–32 mg/l), and, according to different studies [11, 12, 15, 24, 27], between 15% and 50% of strains tested have been found to be primarily resistant to metronidazole or tinidazole (MIC > 8 mg/l). Both compounds have proved unsuccessful in clearing *H. pylori* from the gastric mucosa when used as a single agent (clearance rate, 3%–20%) [10, 18, 28, 29]. In a double-blind placebo-controlled prospective study in patients with duodenal ulceration, Goodwin et al. [24] found that isolates from 19/27 (70%) patients after combined treatment with cimetidine and tinidazole had become resistant to tinidazole (see Table 1). By contrast, when colloidal bismuth subcitrate (CBS) was administered with tinidazole, acquired resistance associated with failure to eradicate *H. pylori* was observed in only 2/22 (9%) patients. Spontaneous resistance to nitroimidazoles was induced in vitro in three of 22 sensitive cultures after overnight subculture in liquid medium and plating on agar containing 8 mg/l of metronidazole [24]. Combination of a nitroimidazole with a bismuth salt not only prevents the emergence of resistance, but it also enables long-term eradication (6–12 months after completion of treatment) of *H. pylori* in about 75% of the patients [10]. Although direct comparative studies have not been performed, various reports from the literature suggest that the nitroimidazole/bismuth regimen achieves a higher and more sustained eradication rate than a combination of bismuth and amoxicillin [8, 10, 16].

The fact that metronidazole – but not ampicillin – is secreted in the stomach and reaches high levels in the gastric juices after parenteral administration [30] is an argument for using this compound in clinical trials for *H. pylori* and might possibly account for the better results achieved with the nitroimidazole/bismuth combinations. Because bismuth salts are not licensed in all countries and because these compounds are contraindicated in children, alternative regimens aimed at the prolonged eradication of *H. pylori* have been sought.

The efficacy of tinidazole plus amoxicillin was recently assessed in a group of 32 children with nonspecific abdominal pain and *H. pylori*-associated gastritis. Out of these 32 children, 30 (94%) were cleared of *H. pylori* after a 6-week treatment and 9/12 (75%) of these still remained free of *H. pylori* after 6 months [31]. Recently, we treated 41 dyspeptic adult patients with *H. pylori*-associated gastritis with the same amoxicillin/tinidazole regimen, but administered the drugs for only 7 days (A. Burette and Y. Glupczynski, unpublished data). Overall eradication of *H. pylori* was found in 25/41 (61%) patients, 6 weeks after the end of treatment. However, significantly different results were observed according to the sensitivity of *H. pylori* to tinidazole before therapy.

All ten patients originally infected with a *H. pylori* strain naturally resistant to tinidazole failed to respond, whereas 25/31 (81%) patients with an initial isolate sensitive to tinidazole responded to treatment ($\chi^2 = 17.4$, $p < 3 \times 10^{-5}$). In this subgroup of patients, 5/6 patients not cleared of *H. pylori* had an isolate resistant to tinidazole at the control endoscopy. Emergence of resistance to tinidazole occurred in only 5/31 (18%) patients under the double amoxicillin/tinidazole combination therapy. It seems thus clear from all the aforementioned studies that a combination of tinidazole or metronidazole with another active anti-*H. pylori* agent (either a bismuth salt or amoxicillin) may be both clinically effective and reduce the risk of development of resistance. So far, a triple association including bismuth, metronidazole, and amoxicillin has not resulted in better eradication of *H. pylori* than the double association without amoxicillin (about 85% eradication 4 weeks after treatment with both regimens) nor does it further reduce the risk of the emergence of resistance to metronidazole [32].

Above all, when considering combination therapy with a nitroimidazole compound, the sensitivity of the *H. pylori* strain before treatment may efficiently predict the outcome of the therapy. As far as the mechanisms of resistance are concerned, it is hypothesized that a small subpopulation of resistant mutants may be selected by metronidazole or may preexist, and that combination with a second active agent will lower the inoculum size of the infecting organisms, hence preventing the emergence of resistant mutants. Resistance to metronidazole has not been extensively investigated and the genetic modifications by which bacteria acquire resistance to this drug remain as yet poorly understood. The resistance is usually ascribed to decreased uptake through the bacterial cell wall and/or to decreased nitroreduction by bacterial enzymes inside the bacteria [33].

Fluoroquinolones

Despite good in vitro activity, several clinical studies with fluoroquinolones (ciprofloxacin, ofloxacin, and norfloxacin) have so far proved unsuccessful in eradicating *H. pylori* from the gastric mucosa [18, 23, 34, 35]. In most instances (between 70% and 100% according to various studies), when the susceptibility of the infecting strains was determined, the rapid development of resistance to quinolones was noted [18, 23, 34, 35] (Table 1). A 16- to 64-fold increase in the MIC (from 0.25–1 mg/l before treatment up to 16–32 mg/l after treatment) of ofloxacin or ciprofloxacin were reported in two studies after the failures observed with these agents [18, 23]. This resistance extended to other 4-quinolone derivatives but not to amoxicillin or to other antibiotics (tetracycline, metronidazole) [23]. Besides this, it has been stated that the lack of clinical efficacy of quinolones might be due to decreased activity at low pH [20, 21]. However, despite a 4- to 16-fold decrease in antibacterial activity against *H. pylori* at pH 5 (MIC, 2 mg/l) as compared to neutral pH (MIC, 0.12 mg/l), ciprofloxacin achieves gastric mucosal concentration largely above the MIC even 6 h after a single 500-mg oral dose [21].

This observation and the fact that resistance also frequently emerged in patients receiving concomitant treatment with H2 antagonists indicate that decreased activity at a more acidic pH cannot completely explain the lack of clinical efficacy of the quinolones.

In vitro, it has been possible to select the development of resistant mutants (at a frequency of about 10^{-8}) after several passages of organisms in the presence of low concentration of various quinolones [20]. The resistant mutants thus selected were 32- to 64-fold less susceptible to quinolones (MIC, 32–64 mg/l) than their sensitive counterparts (MIC_{90}, 0.25–0.5 mg/l). As for various *Campylobacter* species, two different mechanisms may account for this resistance. It may either involve a modification in the DNA gyrase – the target enzyme of quinolones – or it may be due to a modification in the bacterial outer membrane proteins, rendering the drug unable to penetrate the bacteria [36]. It is thus probable that despite the low frequency of spontaneous mutation, this event may be of clinical significance in patients heavily infected with *H.* pylori ($\geqq 10^9$ colonies at some sites of infection). Quinolones would thus eliminate most organisms from the gastric mucosa and select for the few resistant mutants, which would subsequently replace the susceptible population of bacteria. Additionally, this selection of resistant organisms may be further facilitated by low drug concentrations in some areas of the colonized stomach and by reduced antibacterial activity in the presence of a low pH.

However, unlike with nitroimidazoles, there is at present insufficient evidence to support the hypothesis that the combination of a quinolone with a second active anti-*H. pylori* agent will effectively increase the clinical efficacy.

In one study, Bayerdörffer et al. [37] reported a 78% clearance rate and an eradication rate of 33% (4 weeks after the end of treatment) in 14 patients with duodenal ulcer who had received bismuth subsalicylate and ofloxacin for 1 month. After treatment, 3/14 (21%) isolates from these patients were resistant to ofloxacin (MIC > 4 mg/l). However, this study compared the efficacy of bismuth and offloxacin versus ranitidine on ulcer healing and *H. pylori* eradication, so that no patients received either bismuth or ofloxacin alone.

In another small noncomparative pilot study of 11 patients with *H. pylori*-associated gastritis, a combination of amoxicillin (4 × 500 mg/day) and ofloxacin (2 × 200 mg/day) administered for 7 days proved ineffective in eradicating *H. pylori* from the stomach in ten out of 11 patients despite the fact that no resistance to ofloxacin emerged during treatments. The one patient who was temporarily cleared of *H. pylori* relapsed after 2 months of follow-up (Y. Glupczynski and A. Burette, unpublished data).

Besides disappointing clinical results, another point strongly arguing against the use of quinolones in treating *H. pylori*-associated gastritis is the fact that these antibiotics bind to cations and form inactive chelate complexes in the intestinal lumen, and hence have a reduced absorption in the presence of antacids containing Al 3+ and Mg 2+ [38]. Such a situation is likely to occur frequently in dyspeptic patients who often take these medications.

Is summary, the rapid development or resistance combined with the decreased activity of quinolones at low pH make these agents unsuitable for the treatment of *H. pylori* infection, even in association with bismuth salts or other antibiotics, because better, alternative agents are available.

Macrolides

In vitro, macrolides demonstrate excellent levels of activity against *H. pylori* (as active as amoxicillin, the MICs of erythromycin being in the range of 0.06–0.5 mg/l [14, 20, 39]. In a placebo-controlled trial, however, erythromycin ethylsuccinate has been found ineffective in clearing *H. pylori* from the gastric mucosa of patients with gastritis (only one of 15 or 7% patients cleared) [6]. Similar studies with other macrolides or related agents such as osamycin, spiramycin, or clindamycin have not been more successful [17, 28, 40].

Since the activity of erythromycin is markedly affected by the reduction in pH (about 100 times less active at pH 5 than at pH 7.5) [22], it can be expected to be even less active in vivo at the pH of the gastric mucosa and crypts. Moreover, comparative measurements of gastric mucosal concentrations after a single 500-mg oral dose of erythromycin ethylsuccinate have shown lower levels than after a similar dose of other antibiotics (amoxicillin, pivampicillin, and ciprofloxacin) [21]. Although these local concentrations of erythromycin were above the MIC_{90} value for *H. pylori* at pH 7.5 (0.12 mg/l) in 17/18 assessed patients, they were found insufficient to achieve inhibitory concentrations at a pH level less than 6, which thus explained the lack of clinical efficacy.

In a recent small pilot study in dyspeptic patients with *H. pylori* gastritis, we evaluated the efficacy of azythromycin. This new macrolide antibiotic has a similar level of activity as erythromycin against *H. pylori* [38] but better acid stability, and it achieves 10- to 40-fold higher tissue levels than erythromycin following oral administration [41]. Nevertheless, azythromycin did not prove more effective than other macrolides; only 2/12 (16%) patients were cleared of *H. pylori* after a 1-week course of treatment, and the two patients temporarily cleared relapsed 4 weeks after the end of treatment. Studies of the susceptibility of *H. pylori* after treatment showed that in 9/12 patients, the *H. pylori* strain isolated at the follow-up endoscopies was resistant to azythromycin (MIC > 64 mg/l) and that in all cases it showed cross-resistance to erythromycin and also to clindamycin, but not to amoxicillin, tetracyclines, or metronidazole.

The development of resistance was documented in four patients in whom both the pre- and post-treatment matching isolates were available (pretreatment strain, MIC 0.25 mg/l; post-treatment, MIC > 64 mg/l). This resistance was stable and persisted for up to 3 months in three patients followed up after treatment.

Although more acid stable than erythromycin, the in vitro activity of azythromycin is also significantly decreased when the pH level drops from 8 (MIC, 0.25 mg/l) to 5.5 (MIC, 1 mg/l) [20]. It is, however, improbable that this factor accounted on its own for the failures since the clinical efficacy of azythromycin was not enhanced by the coadministration of H2 antagonists in two patients.

A combination of CBS and erythromycin base (500 mg every 6 h) for 14 days has been administered in an open-mode study to 17 patients with duodenal ulcer and *H. pylori*-associated antral gastritis [24]. This regimen seemed more efficacious than single therapy with erythromycin alone and resulted in ulcer healing in 15/17 (88%) subjects and in the elimination of *H. pylori* in 11 (65%) of these patients. However, no long-term follow-up was reported and the sensitivity to erythromycin was apparently not determined. Indeed, acquired resistance to erythromycin and related agents has not been reported before in *H. pylori* studies, most probably because sensitivity was not determined in the earlier clinical trials in which these antibiotics were used.

Resistance to erythromycin and to other macrolides – usually associated with cross-resistance to clindamycin – is however well known in related organism such as *Campylobacter jejuni* or *Campylobacter coli*. Although the mechanism of resistance in these bacteria has not been fully elucidated, it seems unrelated to the presence of a plasmid, but rather to be constitutive and chromosomally mediated [36]. As with the quinolones, spontaneous mutants of *H. pylori* resistant to macrolides may be selected in vitro at low frequency (10^{-8}) after successive exposure of *H. pylori* to growing concentrations of erythromycin [20].

Other Factors Implicated in the Failure of *H. pylori* Infection Treatment

Among other antimicrobials that have been used in the treatment of *H. pylori* infection, no acquired resistance of *H. pylori* to ampicillin, amoxicillin, furazolidone, or tetracyclines has yet been described (Table 1). Moreover, no resistance or increase in the MIC to bismuth salts was observed in a group of 15 patients who relapsed after having received CBS [42]. The fact that recrudescence of infection rapidly takes place after single therapy with amoxicillin or CBS without the development of resistance to these compounds clearly indicates that other factors may be important in explaining the failures to treat *H. pylori* infection observed with these agents.

Besides gastric mucosa concentrations and the stability of the drugs over a wide range of pH, the physicochemical properties of a molecule (lipid solubility, degree of ionisation) are probably of crucial importance to the antibacterial activity at the site of infection. The respective importance of the local and systemic effects of antibiotics are still poorly understood; the optimal drug formulation (syrup versus powder, capsules, or tablets) is as yet unknown, as no direct comparative studies with the same agent in different forms have been conducted. However, such studies are needed since different formulations could result in variable local or systemic concentrations. It is indeed possible that patchy distribution of a locally active antimicrobial over the gastric mucosa may cause wide variation in the concentration in the esophagus, stomach, or duodenum, and hence subsequently lead to relapse or to the development of resistance. Similarly, the best modality of administration (time of administration, number of daily doses) as well as the

Table 1. Activity of various antibiotics against *H. pylori* before and after treatment

Agent	(MIC_{90}) (mg/l)		% *H. pylori* resistant		Reference
	Pre-treatment	Post-treatment	Pre-treatment	Post-treatment	
Quinolones					
ofloxacin	0.5	32	0	100	23
ofloxacin	1	16	0	80	18
ciprofloxacin	0.5	18	0	70	18
ciprofloxacin	?	?	0	100	35
norfloxacin	?	?	5	88	34
Nitroimidazoles					
Metronidazole	1	64	0	50	18
tinidazole	4	>16	17	70	24
Beta-lactams					
amoxicillin	0.03	0.06	0	0	5
Ampicillin	0.12	0.06	0	0	18
Penicillin	0.12	0.12	0	0	18
Others					
Azythromycin	0.5	>64	0	66	A. Burette, 1989 (unpublished data)
minocycline	0.5	0.5	0	0	17
Paromomycin	4	4	0	0	17
Furazolidone	12.5	12.5	0	0	7
Bismuth subcitrate	16–32	16–32	0	0	42

optimal duration of treatment will have to be determined and adapted individually for each suitable therapeutic agent. All these parameters should also be investigated in the future in experimental animal models in order to gain a more comprehensive approach to treatment in humans. Finally, poor compliance of the patients to treatment should not be overlooked and may account for a significant proportion of failures, especially with prolonged antibiotic treatment [19].

Conclusion

This paper summarizes the clinical trials in which failures to eradicate *H. pylori* from the stomach possibly occurred because of the development of resistance during treatment. It confirms that the in vitro susceptibility of this bacterium to antibiotics is not at all predictive of the in vivo eradication rate. There is however, good evidence from in vitro data that resistance to antibiotics may have occurred during treatment because of the selection of naturally resistant mutants. Such an event which occurs at low frequency, may be avoided by combining the antibiotic with a second active anti-*H. pylori* agent (either a bismuth salt or another antibiotic).

Indeed, almost all reports dealing with acquired resistance in *H. pylori* have been in patients who had received a single antibiotic or combination therapy with H2 antagonists. This last observation casts some doubts on the hypothesis that the clinical efficacy of some antimicrobials (quinolones, macrolides, ampicillin) would be enhanced at higher pH by combining them with an H2 antagonist. On the contrary, treatment with anti-H2 blockers might increase the degree of stomach colonization by *H. pylori* and subsequently facilitate the selection of resistant mutants. Furthermore, by increasing the colonization of the stomach with a mouth or intestinal flora, there is a theoretical risk that antibiotic resistance encoding plasmids might be transferred to *H. pylori* from other bacteria, although this has not yet been described.

On the other hand, combination therapy with a bismuth salt and an antibiotic, or two different antibiotics alone, will both improve the eradication rate of *H. pylori* and reduce the risk of resistance development. Until the "optimal" drug is found, monotherapy should not be recommended for the treatment of patients with *H. pylori* infection. At present, it seems that a double combination including a nitroimidazole drug and a bismuth salt or amoxicillin is suitable for the long-term eradication of *H. pylori* provided the initial infecting strain is susceptible to nitroimidazoles. Alternative regimens need to be found however for those patients infected by a naturally nitroimidazole-resistant bacterium, a situation which occurs in about 20% of the cases.

Whatever the therapeutic combination proposed, it is strongly advisable for all therapeutic protocols of *H. pylori* infection to include a culture of *H. pylori* both before and after treatment in order to monitor its susceptibility to antimicrobials. As the importance of *H. pylori* in the pathogenesis of various gastroduodenal disorders begins to be recognized, antibacterial chemotherapy will be more frequently administered in the future for the management of patients with gastroduodenal ulcers. We wish to emphasize the importance of the rational use of antibiotics in this clinical setting in order to avoid their misuse frequently leading to antimicrobial resistance and hence having a negative ecological impact on the gastric and intestinal microflora.

References

1. Krakowka S, Morgan DR, Kraft WG, Leunk RD (1987) Establishment of gastric Campylobacter pylori infection in the neonatal gnotobiotic piglet. Infect Immun 55:2789–2796
2. Lambert JR, Borromeo M, Pinkard KJ, Turner H, Chapman CB, Smith ML (1987) Colonization of gnotobiotics piglets with Campylobacter pyloridis, an animal model? J Infekt Dis 155:1344
3. Marshall BJ, Armstrong JA, McGechte DB, Glancy RJ (1985) Attempts to fulfil Koch's postulates for pyloric campylobacter. Med J Aust 142:436–439
4. Morris A, Nicholson G (1987) Ingestion of Campylobacter pyloridis causes gastritis and raised fasting gastric pH. Am J Gastroenterol 82:192–199
5. Glupczynski Y, Burette A, Labbé M, De Prez C, Dereuck M, Deltenre M (1988) Campylobacter pylori associated-gastritis: a double-blind placebo controlled trial with amoxycillin. Am J Gastroenterol 83:365–372
6. McNulty CAM, Gearty JC, Crump B, Davis M, Donovan IA, Melikian V, Lister DM, Wise R (1986) Campylobacter pyloridis and associated gastritis: investigator blind, placebo

controlled trial of bismuth salicylate and erythromycin ethylsuccinate. Br Med J 293:645–649

7. Morgan D, Kraft W, Bender M, Pearson A, The Gastrointestinal Physiology Working Group of Cayetano Heredia and the John Hopkins Universities (1988) Nitrofurans in the treatment of gastritis associated with *Campylobacter pylori*. Gastroenterol 95:1178–1184
8. Rauws FAJ, Langenberg W, Houthoff HJ, Zanen HC, Tytgat GNJ (1988) Campylobacter pyloridis-associated chronic active antral gastritis-A prospective study of its prevalence and the effects of antiulcer treatment. Gastroenterology 94:33–40.
9. Coghlan JG, Gilligan D, Humphries H, McKenna D, Dooley C, Sweeney E, Keane C, O'Morain C (1987) Campylobacter pylori and recurrence of duodenal ulcers – a 12 month follow up study. Lancet 11:1109–1111
10. Marshall BJ, Goodwin CS, Warren JR, Murray R, Blincow ED, Blackbourn SJ, Phillips M, Waters TE, Sanderson CR (1988) Prospective double-blind trial of duodenal relapse after eradication of Campylobacter pylori. Lancet 11:1437–1441
11. Kasper G, Dickgiesser N (1984) Antibiotic sensitivity of Campylobacter pylori. Eur J Clin Microbiol 3:444
12. McNulty CAM, Dent JC, Wise R (1985) Susceptibility of clinical isolates of Campylobacter pyloridis to 11 antimicrobial agents. Antimicrob Agents Chemother 28:837–838
13. Goodwin CS, Blake P, Blincow E (1986) The minimum inhibitory and bactericidal concentrations of antibiotics and anti-ulcer agents against Campylobacter pyloridis. J Antimicrob Chemother 17:309–314
14. Lambert T, Mégraud F, Gerbaud G, Courvalin P (1986) Susceptibility of Campylobacter pyloridis to 20 antimicrobial agents. Antimicrob Agents Chemother 30:510–511
15. Glupczynski Y, Delmée M, Brück C, Labbé M, Avesani V, Burette A (1987) Susceptibility of clinical isolates of Campylobacter pylori to 24 antimicrobial and anti-ulcer agents. Eur J Epidemiol 4:154–157
16. Bayerdörffer E, Ottenjann R (1988) The role of antibiotics in Campylobacter pylori associated peptic ulcer disease. Scand J Gastroenterol 23 [Suppl 142]:93–100
17. Glupczynski Y, Burette A, Nyst JF, De Prez C, De Koster E, Deltenre M (1988) Campylobacter pylori-associated gastritis: attempts to eradicate the bacteria by various antibiotics and anti-ulcer regimens. Acta Gastroenterol Belg 51:329–337
18. Hirschl AM, Hentschel E, Schütze K, Nemec H, Pötzi R, Gangl A, Weiss W, Pletschette M, Stanek G, Rotter ML (1988) The efficacy of antimicrobial treatment in Campylobacter pylori-associated gastritis and duodenal ulcer. Scand J Gastroenterol 23 [Suppl 142]:76–81
19. Glupczynski Y, Labbé M, Van der Linden MP, Nysl JF, Burette A (1989) Lack of antibiotic compliance in patients treated for Campylobacter pylori associated gastritis. Am J Gastroenterol 84:1126
20. Hardy DJ, Hanson CW, Hensey DM, Beyer JM, Fernandes PB (1988) Susceptibility of Campylobacter pylori to macrolides and fluoroquinolones. J Antimicrob Chemother 22:631–636
21. McNulty CAM, Dent JC, Ford GA, Wilkinson SP (1988) Inhibitory antimicrobial concentrations against Campylobacter pylori in gastric mucosa. J Antimicrob Chemother 22:729–738
22. McNulty CAM (1989) Bacteriological and pharmacological basis for the treatment of Campylobacter pylori infection. Gastroenterol Clin Biol 13:96B–100B
23. Glupczynski Y, Labbé M, Burette A, Delmée M, Avesani V, Brück C (1987) Treatment failure of ofloxacin in Campylobacter pyloridis infection. Lancet 1:1096
24. Goodwin CS, Marschall BJ, Blincow ED, Wilson DH, Blackbourn S, Phillips M (1988) Prevention of nitroimidazola resistance in Campylobacter pylori by coadministration of colloidal bismuth subcitrate: clinical and in vitro studies. J Clin Pathol 41:207–210
25. Quintero Diaz M, Sotto Escobar A (1986) Metronidazole versus cimetidine in treatment of gastroduodenal ulcer. Lancet i:907
26. Shirokova KI, Filomonov RM, Polkova LV (1981) Metronidazole in the treatment of peptic ulcer. Klin Med (Mosk) 59:48–50
27. Novo Y, Gavinet AM, Megraud F, Bebear C (1989) Susceptibility of Campylobacter pylori to nitroimidazole compounds in vitro. In: Megraud F, Lamouliatte F (eds) Gastroduodenal pathology and Campylobacter pylori. Elsevier, Amsterdam, pp 165–167

28. Rauws EAJ, Tytgat GNJ (1989) Therapeutic trials of eradication of Campylobacter pylori. In: Rauws EAJ, Tytgat GNJ (eds) Campylobacter pylori. Ouden, Amsterdam, pp 89–103
29. Patey O, Chaplain C, Dublanchet A, Malkin JE, Roucayrol AM, Charasz N (1989) Failure of treatment with metronidazole for Campylobacter pylori infection. In: Megraud F, Lamouliatte F (eds) Gastroduodenal pathology and Campylobacter pylori. Elsevier, Amsterdam, pp 629–632
30. Hollingsworth JA, Goldie J, Silletti CF, Li Y, Richardson H, Hunt RH (1987) Gastric secretion of antibiotics used for Campylobacter pyloridis. Gut 28 (10): A1409 (Abstract)
31. Oderda G, Vaira D, Holton J, Ainley C, Altare F, Ansaldi N (1989) Amoxycillin plus tinidazole for Campylobacter pylori-gastritis in children: assessment by serum IgG antibody, pepsinogen I, and gastrin levels. Lancet i: 690–692
32. O'Riordan T, Tobin A, Beattie S, Sweeney E, Keane C, O'Morain C (1989) Adjuvant metronidazole improves eradication of Campylobacter pylori in duodenal ulcer treated with colloidal bismuth subcitrate. Gut 30(5): A733 (Abstract)
33. Rosenblatt JE, Edson RS (1987) Symposium on antimicrobial agents. Metronidazole. Mayo Clin Proc 62: 1013–1017
34. Mertens JCC, Dekker W, Ligtvoet EEJ, Blok P (1989) Treatment failure of norfloxacin against Campylobacter pylori and chronic gastritis in patients with non ulcerative dyspepsia. Antimicrob Agents Chemother 33: 256–257
35. Stone JW, Wise R, Donovan IA, Gearty J (1988) Failure of ciprofloxacin to eradicate Campylobacter pylori from the stomach. J Antimicrob Chemother 22: 92–93
36. Taylor DE, Courvalin P (1988) Mechanisms of antibiotic resistance in Campylobacter species. Antimicrob Agents Chemother 32: 1107–1112
37. Bayerdörffer E, Simon T, Bästlein C, Ottenjann R, Kasper G (1987) Bismuth/ofloxacin combination for duodenal ulcer. Lancet ii: 1467–1468
38. Preheim LC, Cuevas TA, Roccaforte JS, Mellencamp MA, Bittner MJ (1986) Ciprofloxacin and antacids. Lancet ii: 46
39. Czinn S, Carr H, Aronoff S (1986) Susceptibility of Campylobacter pylori to three macrolide antibiotics (erythromycin, roxithromycin [RU 28965], and CP 62,993) and rifampin. Antimicrob Agents Chemother 30: 328–329
40. Lamouliatte H, Mégraud F, DeMascarel A, Quinton A (1988) Etude controlée de la josamycine dans les gastrites associées au Campylobacter pylori. Gastroenterol Clin Biol 12: A156 (Abstract)
41. Girard AE, Girard D, English AR, Gootz TD, Cimochowski CR, Faiella JA, Haskell SL, Retsema JA (1987) Pharmacokinetic and in vivo studies with azythromycin (CP-62,993), a new macrolide with extented half-life and excellent tissue distribution. Antimicrob Agents Chemother 31: 1948–1954
42. Goodwin CS, Bell B, McCullough C, Turner M (1989) Sensitivity of Campylobacter pylori to colloidal bismuth subcitrate. J Clin Pathol 42: 216–219

Taxonomy and Biology of *Helicobacter pylori* – a Comment

F. MÉGRAUD

A great deal of information has already been published concerning the taxonomy and biology of *Helicobacter* (*Cambylobacter*) *pylori*. From a taxonomic point of view, the final decision with regard to the classification of this species had not been made. Knowledge of the biology of the microorganism is still unclear regarding the coccoidal forms, and molecular biology is at an early stage. New data concerning these three topics have been presented here.

Taxonomy

Genotypic data of rRNA hybridization and sequencing have shown that *H. pylori* is very different from the true Campylobacters and is similar to *Wolinella succinogenes*. The point then became to decide if both were to be included in the same genus *Wolinella* or if a new genus had to be created. Goodwin presented arguments supporting the latter proposal, showing that the ultrastructure, fatty acid composition, and biochemical tests are different for *H. pylori* and *W. succinogenes*.

The new genus created is *Helicobacter* from the Greek "helix" which means spiral. Two species have been assigned to this new genus: *H. pylori* and *H. mustelae*, the ferret organism. Since these data have been published in the *International Journal of Systematic Bacteriology* (1989, 39, 397–405), the change in genus is definitive.

Coccoidal Forms

These forms are without doubt a key point in the physiology of the bacterium. The existence of such nonculturable forms helps in understanding the lack of positive results while looking for *H. pylori* in the environment, and the difficulty in the eradication of these bacteria.

Using transmission electron microscopy, Jones was able to observed the morphological transformations from spiral to coccoidal forms, an event which he thinks is a response to less favorable conditions. In these transformations, the problem seems to be at the cell wall level. While the nucleus is dividing, the

Laboratoire de Bactériologie, Hôspital des Enfants, 168, cours de l'Argonne,
33077 Bordeaux Cedex, France

Helicobacter pylori, Gastritis and Peptic Ulcer
Edited by P. Malfertheiner and H. Ditschuneit

peptidoglycan fails to do so and a spherical form develops. This anomaly may be linked to nutritional requirements which are not met. The presence of flagella at two sites supports this mechanism because two-site flagella are only seen on dividing organisms.

Quantitative analyses performed on both spiral and coccoidal forms were reported by Opferkuch. Coccoidal forms contain more lipopolysaccharides, more protein, and less phospholipids.

Molecular Biology

Labigne and colleagues presented important new data on the gene for urease. The gene was cloned by using a shuttle vector. The urease activity was originally found in a 44 kb fragment and, subsequently, in an 8.1 kb subfragment. A 4.2 kb sequence is required for the expression of the urease: two genes encode the two subunits of the molecule, another gene is absolutely required for urease activity. When they looked at the amino acid sequence derived from DNA analysis in comparison with that of the jack bean urease, they found a high degree of conservation and a perfect alignment of the amino acids. Not only is the N terminal sequence conserved but also the whole sequence of the molecule. Such a high degree of homology between plant and bacterial ureases suggests that the conservation of polypeptides will be higher with bacterial ureases. This could be a source of cross reactions in serology if urease is used as the antigen.

Need for Culture

Given the present state of our knowledge, culture of the organism when studying patient gastric biopsy material is an absolute necessity. Opferkuch highlighted the potential differences in the virulence factors between strains. Glupczynski addressed the question of antimicrobial resistance in *H. pylori*. Acquisition of resistance concerns at least three groups of antimicrobial agents: nitroimidazole compounds, quinolones, and macrolides. This is the first report of resistance to the latter group. Resistance is a major problem for nitroimidazoles because they are widely used in the treatment of *H. pylori* infection. If a strain is resistant, treatment failure can be predicted by an in vitro susceptibility test. It is also important to culture the organism in order to be able to apply typing methods of epidemiological value. Fraser shared with us her experience using three techniques: DNA restriction endonuclease analysis, Southern blot hybridization, and sodium dodecylsulfate polyacrylamide gel electrophoresis. The second technique gave very good results but was technically more complicated.

These important findings have been presented in the foregoing pages. We are now looking forward, in particular, to seeing a molecular biology approach to the problem of coccoidal forms.

Pathogenic Mechanisms of *Helicobacter pylori*

Virulence Factors of *Helicobacter pylori* – Ultrastructural Features

G. Bode, P. Malfertheiner, G. Lehnhardt, and H. Ditschuneit

The detection of *Helicobacter pylori* in gastric biopsy specimens of patients with gastritis, and gastric or duodenal ulcers has provided the basis for a new concept in the etiology of gastroduodenal diseases [1–4].

Little is known concerning the factors that allow the colonization of the gastric mucosa and the induction of gastritis. *H. pylori* entering the human digestive system encounter a battery of nonspecific and specific chemical and cellular defense mechanisms. The role of these mechanisms is to inactivate and eliminate foreign materials. By overruling the various defenses *H. pylori* is able to establish itself within the mucosa. As this occurs, the bacteria start to exert their pathogenic potential.

The phenomenon of mucosal infection by *H. pylori* is complex and multifactorial. Several biochemical properties may act individually or together to produce infection and disease. After identification of microbial factors in various strains of *H. pylori* exhibiting virulence-associated properties, it is necessary to evaluate their actual contribution to bacterial pathogenicity.

This paper deals in detail with the virulence factors of *H. pylori* and their influence on ultrastructural alterations of surface mucus cells. Additionally an attempt is made to identify strains of high and low virulence on the basis of the morphological features. The investigated virulence factors, their proposed function, and their relation to ultrastructural alterations are summarized in Table 1.

Table 1. Virulence factors of *H. pylori* and their relation to pathological alterations

Virulence factor	Reference	Pathological alterations
Motility (shape, flagella)	[5, 7, 8]	Favors colonization
Urease	[12, 13]	Neutralizes gastric acid
Protease	[20, 23]	Breakdown of mucus network
Lipolytic activity	[29, 30]	Generation of lysolecithin
Cytotoxins	[34, 35]	Intraplasmatic vacuolization
Adhesins	[37, 39]	Binding to the receptor

University of Ulm, Dept. of Internal Medicine II, Robert-Koch-Straße 8, D-7900 Ulm, FRG

Helicobacter pylori, Gastritis and Peptic Ulcer
Edited by P. Malfertheiner and H. Ditschuneit

Shape – Flagella – Motility

H. pylori appears to be specifically adapted to the viscous milieu of the gastric mucus. The spiral shape and the active flagella enable the bacteria to move easily in the viscous environment [5]. Motility is an established virulence factor for bacterial pathogens [6, 7] and it was shown by Eaton et al. [8] that the most motile strain of *H. pylori* is the most virulent with 100% infection rate in gnotobiotic piglets. In recent studies it was shown that flagella do not only provide motility, but they bear adhesins and are thus more directly involved in the actual adhesion event [9]. In addition, flagella of *H. pylori* have bulbs on their ends (Fig. 1) similar to those seen in Vibrio species. These bulbs may essentially favor the adhesion.

In several isolated strains of *H. pylori* we were able to demonstrate a so-called polar organelle (Fig. 2). This organelle presumably is part of the flagellar

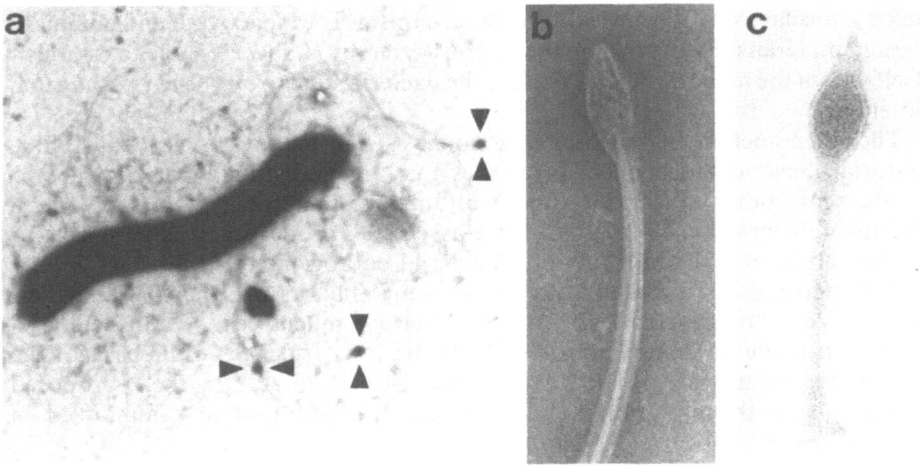

Fig. 1a–c. The flagellar characteristics of *H. pylori*. **a** 4–6 flagella originate from one pole. *Arrows* point at the typical "terminal bulb." Negative staining; ×20000. **b** Negative staining; ×160000. **c** Ultrathin section; ×160000

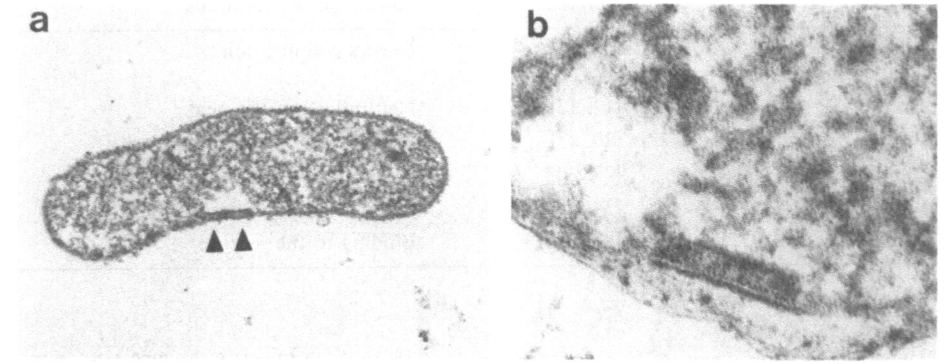

Fig. 2a, b. The "polar" organelle of *H. pylori*. **a** This organelle is occasionally found along the membrane; ×35000. **b** Ultrastructural characteristics; ×170000

apparatus and contains ATPase and cytochrome oxidase [10, 11]. As this organelle has been observed only in flagellated bacteria and often in close association with the flagellar insertion area, but also along the bacterial membrane, it is speculated that this organelle works as a "proton-translocating pumping system."

Shape, flagellar organelles, and motility enhance the virulence of *H. pylori* by facilitating colonization of the antral mucosa.

The Urease System

The ability of *H. pylori* to survive the bactericidal acidity in the stomach is primarily attributed to a high bacterial urease activity. The hydrolysis of urea molecules present physiologically in the gastric juice leads to production of ammonia which acts as an acceptor for H^+ ions, and finally increases the local pH [5, 12]. Compared with other urease-producing bacterial species *H. pylori* demonstrates the highest activity [13]. If urease activity represents an important factor for *H. pylori*, the enzyme must be rapidly and constantly available. To validate this hypothesis we studied the localization and characteristics of urease using electron microscopic techniques [14, 15]. In these experiments 12 *H. pylori* strains of patients with chronic active gastritis were incubated with sodium tetraphenylboron, an ammonia-precipitating agent in the presence of urea. This precipitate was replaced by silver ions and thin sections were analyzed in the electron microscope.

The analyzed strains were divided into three groups according to the localization of urease activity. The urease-positive strains (11/12 = 91.5%) showed urease activity in the periplasmatic region and at the outer membrane (5/12 = 41.5%), and in six strains additionally in the cytoplasm (6/12 = 50.0%). One strain had lost its urease activity during various culturing passages (Fig. 3). There are no data available at the moment concerning the relation between the site of urease localization and the degree of virulence of the tested strains. However, in strains with urease activity in both locations, there is a more pronounced atrophy of gastric mucosa.

In numerous studies, it has been shown that ammonia can produce functional and morphological alterations in gastric epithelium in vitro, including (a) reduction of transmucosal potential difference [12, 16, 17], (b) enhanced back-diffusion of H^+ to the epithelial cells [18], (c) increase of bacterial adherence [19], and (d) inactivation of complement.

Proteolytic Activity

Many microorganisms produce extracellular proteases. Protease activity is implicated as a critical factor for the colonization of *H. pylori* in the mucus layer of the stomach. This virulence factor first described for *H. pylori* by Slomiany [20] is capable of rapid degradation of gastric mucin. This process impairs the viscoelastic and hydrogen ion retardation capacity of gastric mucus [21–23].

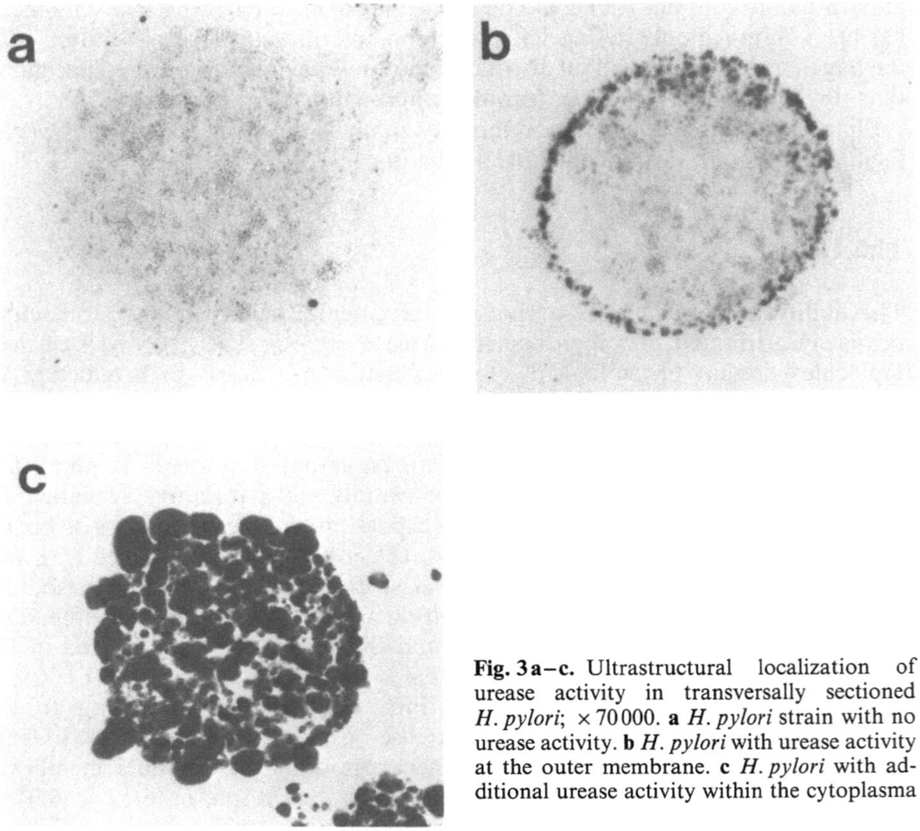

Fig. 3a–c. Ultrastructural localization of urease activity in transversally sectioned *H. pylori*; ×70000. **a** *H. pylori* strain with no urease activity. **b** *H. pylori* with urease activity at the outer membrane. **c** *H. pylori* with additional urease activity within the cytoplasma

Recently we were able to demonstrate by electron microscopy that the mucus layer is built up by a three-dimensional network [5]. This netlike structure (Fig. 4) stabilizes the unstirred layer and builds up a pH gradient together with bicarbonate [24]. Figure 4 shows an ultrathin section through the mucus layer of a normal control person without *H. pylori* infection. Globular particles are radially combined with thin filamentous structures (arrows). This is also shown by scanning electron microscopy.

In patients with *H. pylori* infection the threadlike structure is destroyed and the globular particles are clumped together, demonstrating a breakdown of the stable and protective mucus layer in this area (Fig. 4). After eradication of *H. pylori*, the rapid regeneration of this netlike structure is observed.

Similar results and improvement of the physiologic properties of mucus were obtained by blocking the proteolytic activity of *H. pylori* by colloidal bismuth subcitrate (CBS) [25].

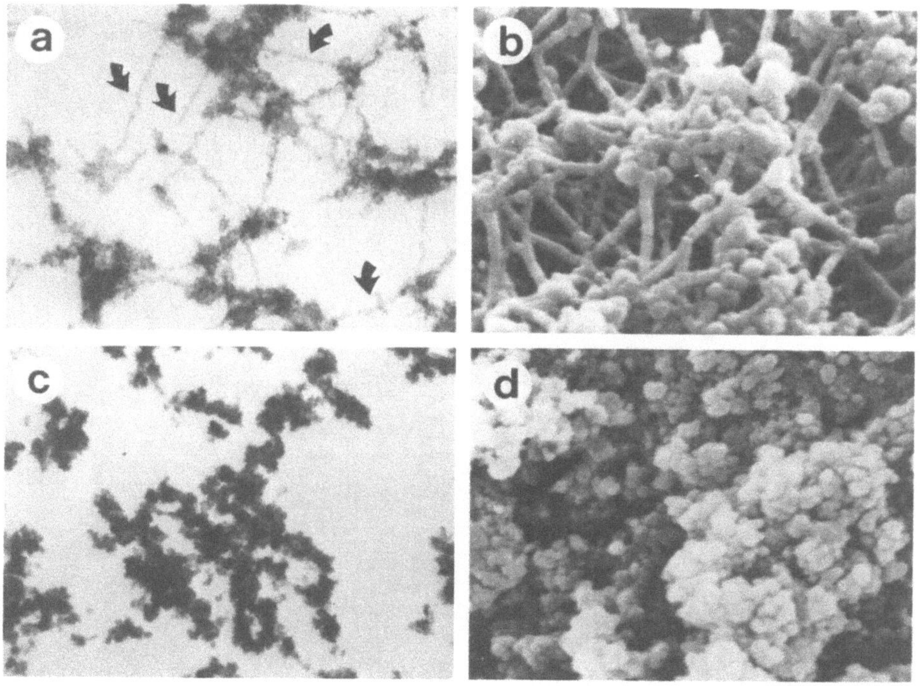

Fig. 4a–d. Ultrastructural set-up of the mucous layer. **a** Netlike structure; *arrows* point at thin filaments; × 100000. **b** Scanning electron microscopy; × 120000. **c** Clumped globular particles of the disrupted mucous layer; × 100000. **d** Same preparation by scanning electron microscopy; × 120000

Lipolytic Activity

Many pathogenic microorganisms are known to produce lipases and phospholipases which are regarded as virulence factors [26]. It was recently shown that epithelial mucous cells of the stomach are able to synthesize and store surface-active phospholipids, which may be an essential component of the stomach's protective hydrophobic lining [27, 28]. Slomiany et al. [29] and Rädsch et al. [30] have shown that *H. pylori* is capable of inducing mucosal lipid degradation and that the main activity is a phospholipase A.

Ultrastructural studies are in progress to investigate if the hydrophobic lining of the surface mucous cells is diminished or distorted by *H. pylori* produced phospholipase [31].

Cytotoxins

There is recent data concerning the cytotoxin production by *H. pylori* strains [32, 33]. Some morphological and ultrastructural alterations in cultured cells have

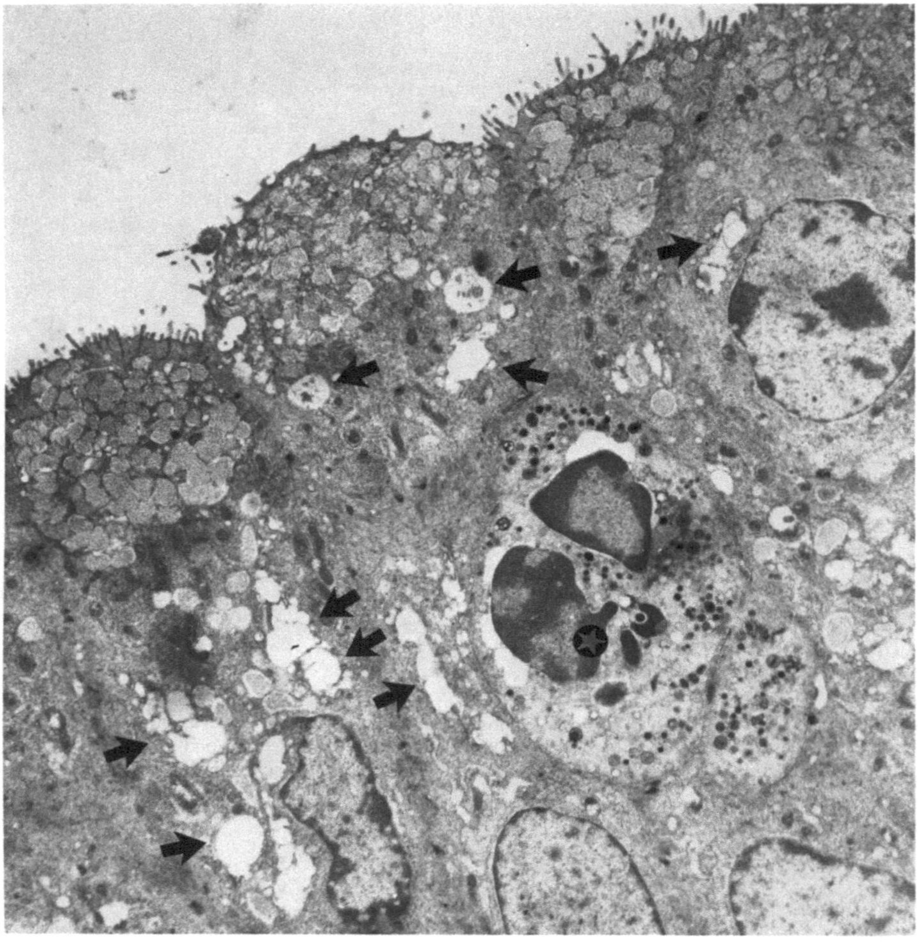

Fig. 5. Intracytoplasmatic vacuolization (*arrows*) of surface mucous cells; ×11 000

been observed in the presence of cytotoxins in vitro. The actual role of cytotoxins in *H. pylori* infection has not yet been clarified. A cytotoxic but not cytolethal factor was shown able to induce intracellular vacuolization in HeLa cells [34].

We found intracytoplasmatic vacuolization of surface mucous cells in antral biopsy specimens obtained from patients with active chronic gastritis and patients with active duodenal ulceration (Fig. 5). This is likely to be an effect of cytotoxin because Leunk et al. [35] did not report such alterations to be a consequence of urease activity.

The ultrastructural data shown in Table 2 may be related to some "toxic effects" elaborated by *H. pylori*. The thickness of the glycocalyx of the surface mucous cells is significantly reduced ($P < 0.001$) compared with a control group (*H. pylori* negative). This is also shown in Fig. 6. The length of the microvilli in the

Table 2. Ultrastructural data of surface mucous cells of patients with active chronic gastritis (ACG) compared with a control group (C); mean ±SD

	Glycocalyx		Microvilli
	Thickness (μm)	Length (μm)	Diameter (μm)
C ($n=8$)	0.24±0.04	0.66±0.08	0.21±0.03
ACG ($n=18$)	0.12±0.01 **	0.59±0.08 *	0.21±0.02

** $P<0.001$ (C vs ACG); * $P<0.01$ (C vs ACG)

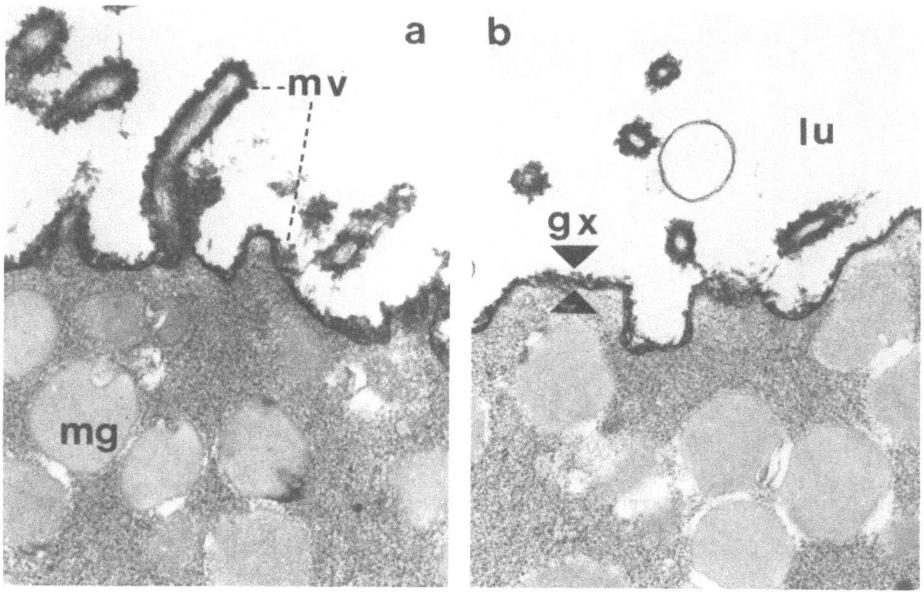

Fig. 6a, b. Ultrastructural appearance of microvilli (*mv*) and glycocalyx (*gx*). **a** Normal surface mucous cell; ×80000 (*mg*, mucous granule). **b** Surface mucous cell of a patient with active chronic gastritis; ×80000 (*lu*, lumen)

patient group with active chronic gastritis (*H. pylori* positive) is also significantly reduced ($P<0.01$).

In some strains of *H. pylori* grown in liquid media we were able to demonstrate an intracellular crystalline structure with distinct subunits (Fig. 7). Several morphological forms of these crystalline structures were observable depending on the plane of section. The nature of these structures remains unclear, however, they show some similarity with crystalline structures in *Bacteroides nodosus* which may have some relation to the virulence properties of that organism [36].

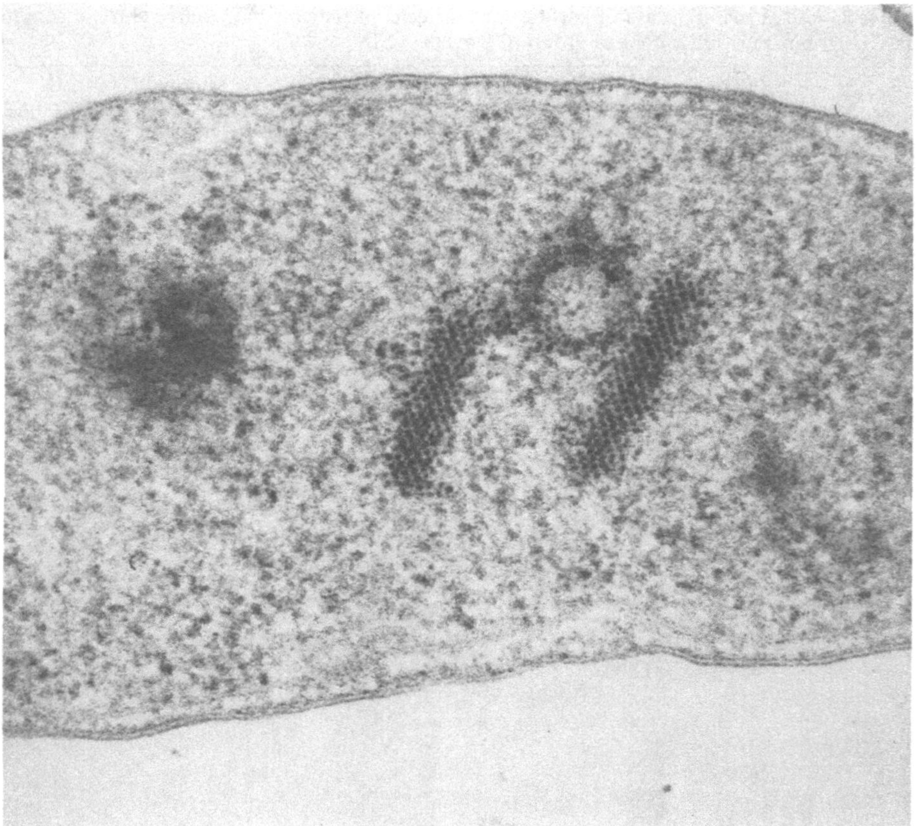

Fig. 7. Intracytoplasmatic crystal-like inclusions; × 200 000

Adhesion

Adhesion is an important factor in the initiation of many bacterial infections. The adhesion of *H. pylori* to the apical cell membranes of surface mucous cells represents the final step in the process of association with the mucosa.

The surface of *H. pylori* may contain both nonspecific factors (differences in surface charges, hydrophobic interactions) and specific factors (adhesins, agglutinins) which influence its adherence to the membrane of surface mucous cells [37–39].

Besides the intimate contact of the bacterial cell wall with the membrane of the surface mucous cell, there is often a "fibrillar" connection observed between *H. pylori* and its "target" cell (Fig. 8). At the ultrastructural level two possibilities of contact appear to exist.

The precise molecular structure and function of these specialized adherence factors have to be established.

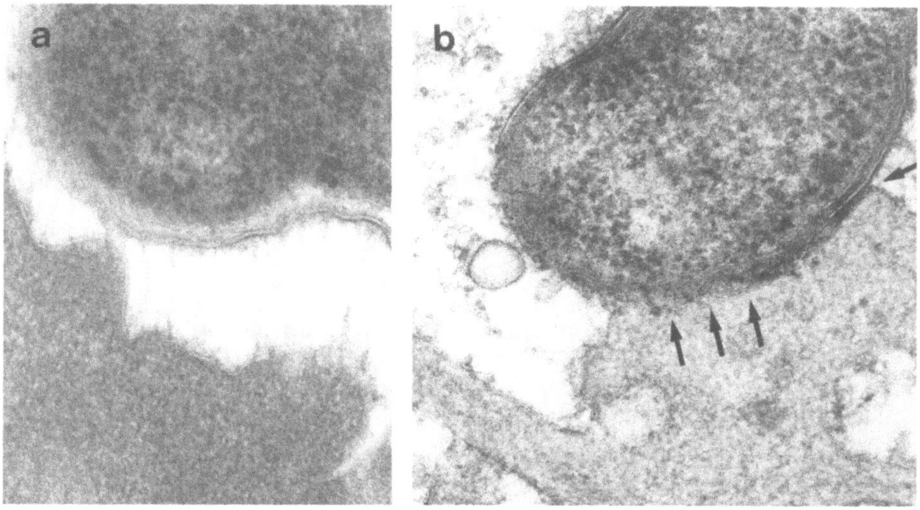

Fig. 8a, b. Contact characteristics of *H. pylori* with the cell membrane of a surface mucous cell. **a** Thin fibrillar connection between both membranes; × 120000. **b** Direct contact of *H. pylori* and the surface mucous cell (*arrows*); × 100000

Outlook

Not all strains of *H. pylori* have an equal potential for causing infection, disease, and probably symptoms.

We are just beginning to unterstand the molecular basis of *H. pylori* pathogenicity. At the present, there are few examples for which a biochemical mechanism is thought to be known. These factors are mainly all surface structures that cause *H. pylori* to exhibit one or more of several pathogenicity-associated properties, namely urease, protease, lipolytic activity, cytotoxins, and adhesins.

The virulence in *H. pylori* strains is obviously multifactorial and there is a great deal yet to be learned about the relative importance of the various factors discussed.

References

1. Marshall BJ, McGechie DB, Rogers PA, Glancy RJ (1985) Pyloric Campylobacter infection and gastroduodenal disease. Med J Aust 142:439–444
2. Rathbone BJ, Wyatt JI, Heatley RV (1986) Campylobacter pyloridis: a new factor in peptic ulcer disease. Gut 27:635–641
3. Blaser MJ (1987) Gastric Campylobacter-like organisms, gastritis and peptic ulcer disease. Gastroenterology 93:371–383
4. Graham DY, Klein PD, Evans DJ, Evans DG, Alpert LC, Opekun AR, Boutton TW (1987) Campylobacter pylori detected noninvasively by the 13 C-urea breath test. Lancet I:1174–1177

5. Bode G, Malfertheiner P, Ditschuneit H (1988) Pathogenetic implications of ultrastructural findings in Campylobacter pylori related gastroduodenal disease. Scand J Gastroenterol [Suppl 142] 23:25–39
6. Guentzel MN, Berry LJ (1975) Motility as a virulence factor for Vibrio cholerae. Infect Immun 11:890–897
7. Drake D, Montie TC (1988) Flagella, motility and invasive virulence of Pseudomonas aeruginosa. J Gen Microbiol 134:43–52
8. Eaton KA, Morgan DR, Krakowka S (1989) Campylobacter pylori virulence factors in gnotobiotic piglets. Infect Immun 57:1119–1125
9. Attridge SR, Rowley D (1983) The role of the flagellum in the adherence of Vibrio cholerae. J Infect Dis 147:864–872
10. Tauschel HD (1985) ATPase and cytochrome oxidase activities at the polar organelle in swarm cells of Sphaerotieus natans: an ultrastructural study. Arch Microbiol 141:303–308
11. Ritchie AE, Keeler RF, Bryner JH (1966) Anatomical features of Vibrio fetus: electron microscopic survey. J Gen Microbiol 43:427–438
12. Thomson L, Tasman-Yones C, Morris A, Wiggins P, Lee S, Forlong C (1989) Ammonia produced by Campylobacter pylori neutralizes H^+ moving through gastric mucus. Scand J Gastroenterol 24:761–768
13. Mobley HLT, Cortesia MJ, Rosenthal LE, Jones BD (1988) Characterization of urease from Campylobacter pylori. J Clin Microbiol 26:831–836
14. McLean RJC, Cheng KJ, Gould WD, Costerton YW (1985) Cytochemical localization of urease in a rumen staphylococcus sp. by electron microscopy. Appl Environ Microbiol 49:253–255
15. Bode G, Malfertheiner P, Nilius M, Lehnhardt G, Ditschuneit H (1989) Ultrastructural localization of urease in outer membrane and periplasm of Campylobacter pylori. J Clin Pathol 42:778–779
16. Murakami M, Yoo JK, Inada M, Miyake T (1988) Effect of ammonia on the gastric mucosa in rats: pathophysiological importance of urease in gastric ulcer disease. Jpn J Pharmacol 47:330–332
17. Thomson L, Tasman-Jones C, Morris A (1989) Na^+/H^+ Ion-exchange property of postmortem human gastric mucus. The effects of Campylobacter pylori infection and sucralfate. Scand J Gastroenterol 24:781–786
18. Hazell SL, Borody TJ, Lee A (1987) Campylobacter pyloridis gastritis. I. Detection of urease as a marker of bacterial colonization and gastritis. Am J Gastroenterol 82:292–296
19. Parsons CL, Stauffer C, Mulholland SG, Griffith DP (1984) Effect of ammonium on bacterial adherence to bladder transitional epithelium. J Urol 132:365–366
20. Slomiany BL, Bilski J, Sarosiek J, Murty VLN, Dworkin B, Van Horn K, Zielenski J, Slomiany A (1987) Campylobacter pyloridis degrades mucin and undermines gastric mucosal integrity. Biochem Biophys Res Commun 144:307–314
21. Turner NC, Martin GP, Marriott C (1985) The influence of native porcine gastric mucus gel on hydrogen ion diffusion: the effect of potentially ulcerogenic agents. J Pharm Pharmacol 37:776–780
22. Quigley EMM, Turnberg LA (1987) pH of the microclimate lining human gastric and duodenal mucosa in vivo. Studies in control subjects and in duodenal ulcer patients. Gastroenterology 92:1876–1884
23. Sarosiek J, Slomiany A, Slomiany BL (1988) Evidence for weakening of gastric mucus integrity by Campylobacter pylori. Scand J Gastroenterol 23:585–590
24. Smith GW, Tasman-Jones C, Wiggins PM, Lee SP (1985) Pig gastric mucus: a one-way barrier for H^+. Gastroenterology 89:1313–1318
25. Sarosiek J, Bilski J, Murty VLN, Slomiany A, Slomiany BL (1989) Colloidal bismuth subcitrate (De-Nol) inhibits degradation of gastric mucus by Campylobacter pylori protease. Am J Gastroenterol 84:506–510
26. Barret-Bee K, Hayes Y, Wilson RG, Ryler JF (1985) A comparison of phospholipase activity, cellular adherence and pathogenicity of yeasts. J Gen Microbiol 131:1217–1221
27. Lichtenberger LM, Graziani LA, Dian EJ, Buttler BD, Hills BA (1983) Role of surface-active phospholipids in gastric cytoprotection. Science 219:1327–1329

28. Kao YJ, Lichtenberger LM (1987) Localization of phospholipid-rich zones in rat gastric mucosa: possible origin of a protective hydrophobic luminal lining. J Histochem Cytochem 35:1285–1298
29. Slomiany BL, Nishikawa H, Piotrowski J, Okazaki K, Slomiany A (1989) Lipolytic activity of Campylobacter pylori: effect of sofalcone. Digestion 43:33–40
30. Raedsch R, Stiehl A, Pohl S, Plachky J (1989) Quantification of phospholipase A_2-activity of Campylobacter pylori. Gastroenterology 96:A405
31. Slomiany BL, Kasinathan C, Slomiany A (1989) Lipolytic activity of Campylobacter pylori: effect of colloidal bismuth subcitrate (De-Nol). Am J Gastroenterol 84:1273–1277
32. Fumarola D, Miragliotta (1988) Possible pathogenic mechanisms of Campylobacter pylori. Eur J Clin Microbiol Infect Dis 7:579–580
33. Figura N, Guglielmetti P, Rossolini A, Barberi A, Cusi G, Musmanno RA, Russi M, Quaranta S (1989) Cytotoxin production by Campylobacter pylori strains isolated from patients with peptic ulcers and from patients with chronic gastritis only. J Clin Microbiol 27:225–226
34. Hupertz V, Czinn S (1988) Demonstration of a cytotoxin from Campylobacter pylori. Eur J Clin Microbiol Infect Dis 7:576–578
35. Leunk RD, Johnson PT, David BC, Kraft WG, Morgan DR (1988) Cytotoxic activity in broth-culture filtrates of Campylobacter pylori. J Med Microbiol 26:93–99
36. Hine PM (1988) Intracellular crystal formation in Bacteroides nodosus. J Ultrastruct Mol Struct Res 100:263–268
37. Evans DG, Evans DJ, Moulds JJ, Graham DY (1988) N-Acetylneuraminyllactose-binding fibrillar hemagglutinin of Campylobacter pylori: a putative colonization factor antigen. Infect Immun 56:2896–2906
38. Neman-Simha V, Megraud F (1988) In vitro model for Campylobacter pylori adherence properties. Infect Immun 56:3329–3333
39. Lingwood CA, Law H, Pellizzari A, Sherman P, Drumm B (1989) Gastric glycerolipid as a receptor for Campylobacter pylori. Lancet I:238–241

Molecular Cloning and Sequencing of *Helicobacter pylori* Urease Genes and Detection of *H. pylori* Using the Polymerase Chain Reaction Technique

C. L. CLAYTON, M. J. PALLEN, H. KLEANTHOUS, B. W. WREN and S. TABAQCHALI

The genes encoding the two structural subunits of the *Helicobacter* (*Campylobacter*) *pylori* urease enzyme were cloned and sequenced. The derived amino acid sequences of the two *H. pylori* urease genes showed marked homology to both jack bean and soybean ureases. Almost 57% of residues were identical in the bacterial and plant ureases. This degree of similarity between a eucaryotic protein and a bacterial protein is almost unprecedented. The close similarity of the bacterial protein to the plant enzymes is an evolutionary anomaly that is perhaps best explained by horizontal gene transfer. The polymerase chain reaction technique using oligonucleotide primers to the *H. pylori* urease DNA sequence was found to sensitively detect *H. pylori* cells and may allow determination of the source and route of infection of the organism.

H. pylori constitutively produces large amounts of the enzyme urease, which is an immunodominant antigen in the mammalian immune response to the organism and is strongly implicated as the major virulence factor of *H. pylori* λEMBL3 recombinant clone on screening a λEMBL3 gene library with *H. pylori* whole cell rabbit antisera [5]. A 2.7 kilobase (kb) DNA fragment subcloned from the positive λEMBL3 clone into pUC18 was shown to encode two polypeptides (66 and 31 kDa). These polypeptides reacted on immunoblots with polyclonal antibody raised against purified *H. pylori* urease [5], and a monoclonal antibody to the *H. pylori* urease, capable of inhibiting the enzyme's activity was also found to react with the 31 kDa polypeptide [4] suggesting that the cloned antigens may be part of the urease enzyme. The nucleotide sequence of the 2.7 kb DNA fragment was determined and also enabled the construction of oligonucleotide primers for the sensitive detection of *H. pylori* by polymerase chain reaction amplification.

* This work was supported by grants from the Wellcome Trust and the British Medical Research Council

Department of Medical Microbiology, St. Bartholomew's Hospital Medical College, West Smithfield, London EC1A 7BE, UK

Helicobacter pylori, Gastritis and Peptic Ulcer
Edited by P. Malfertheiner and H. Ditschuneit

Methods

DNA Sequencing

The restriction enzyme map of the 2.7 kb *Taq*1 DNA fragment was determined. The entire *Taq*1 insert and two *Hind*III fragments (1.4 kb, 1.3 kb) were cloned into M13 vector DNA. Single stranded template was produced and both strands were sequenced using the Sanger dideoxy chain termination technique [14] with ^{35}S dATP and T7 polymerase. Oligonucleotide primers were also synthesised for determining the nucleotide sequence of the 2.7 kb insert.

Polymerase Chain Reaction

Oligonucleotide primers to sequences within the 2.7 kb *Taq*1 fragment were used to perform the polymerase chain reaction amplification technique to give amplified DNA fragments of specific size. This process allows amplification of DNA segments by heat denaturation of the double stranded DNA, annealing of two primers at low temperature and extending at an intermediate temperature. The nucleotide extension is achieved by the activity of heat stable *Taq* polymerase enzyme and repetition of the cycle can amplify DNA segments by at least 10^5-fold. The PCR technique was performed on both purified DNA and crude DNA released from boiled bacterial cells. PCR amplification was carried out according to manufacturer's recommendations (Cetus) and the temperature cycles controlled by an intelligent heating block (Hybaid). Twenty-six cycles of amplification were performed in total. The reactions were performed in 100-µl volumes and 20-µl samples were run on agarose gels against standard DNA molecular weight markers.

Results

The nucleotide sequence of the cloned 2.7 kb DNA fragment was determined. Two large open reading frames were present in the sequence which encodes proteins of calculated molecular weight 26657 (alpha-subunit) and 60473 (beta-subunit; Fig. 1). The two open reading frames were found to be separated by three nucleotides. Comparison of the predicted amino acid sequences of the two proteins with a sequence of a urease purified from the jack bean *Canavalia ensiformis* [15] shows that the single subunit jack bean urease (predicted molecular weight 90770) is directly homologous to the two subunits of the *H. pylori* urease (Fig. 1). Some 57% of predicted amino acid residues in the *H. pylori* sequences are identical to those in the corresponding sites in the jack bean urease sequence (Fig. 1). The homology occurs along the entire length of both subunits, except for the final few carboxy-terminal amino acids in each case. Comparison of the *H. pylori* protein sequence with a sequence of 130 amino acids predicted from a cloned DNA fragment for the soybean (*Glycine max*) urease gene [10] shows a

```
      *
   1' MKLTPKELDKLMLHYAGELAKKRKEKGIKLNYVEAVRLISAHIMEEARRGKKTAAELMQE
      :::.:.:..:: :: :: ::.:: ..:..:::.::: ::...::: :: :.::.:.::
   1" MKLSPREVEKLGLHNAGYLAQKRLARGVRLNYTEAVALIASQIMEYARDGEKTVAQLMCL

  61' GATLLKPDDVMDGVASMIHEVGIEAMFPDGTKLVTVHTPIEA-NG---------------
      :. :: . .:. .:. ....: .:: :::::::::::.::.  ::
  61" GQHLLGARQVLPAVPHLLNAVQVEATFPDGTKLVTVHDPISRENGELQEALFGSLLPVPS

 105' ----------KLVPGELFLKNEDITINEGKKAVSVKVPPVGDRPVQIGSHFHFFEVNRCL
                . .:::.. ..: .:.: :.::: .::.. ::::.:.:::.::.:::..:
 121" LDKFAETKEDNAIPGEILCEDECLTLNIGAKAVILKVTSKGDRPIQVGSHYHFIEVNPYL

 155' DFDREKTFGKRLDIASGTAVRFEPGEEKSVELIDIGGNRRIFGFNALVDRQADNESKKIA
      .::: :..:.::.::.:::::::::. :::.:..:.::. : : ::..: ......  .:
 181" TFDRAKAYGMRLNIAAGTAVRFEPGDCKSVTLVSIEGNKVIRGGNAIADGPVNETNLEAA
                                     *
 215' LHRAKERGF---HGAKSDDNYVKTIKE   MKKISRKEYASMYGPTTGDKVRLGDTDLIA
      .: ...:::   .. .. ....:.  .   .  : :::::..::::::::.:::::.:.:
 241" MHAVRSAGFGHEEEKDAPEGFTKEDPNCSFNTFIHRKEYANKYGPTTGDKIRLGDTNLLA

  31' EVEHDYTIYGEELKFGGGKTLREGMSQS-NNPSKEELDLIITNALIVDYTGIYKADIGIK
      :.:.::..::.:  :::::..:.::.:: ..:.  .:: .::::.:.::::: :::::::
 301" EIEKDYALYGDECVFGGGKVIRDGMGQSCGHPPAISLDTVITNAVIIDYTGIIKADIGIK

  90' DGKIAGIGKGGNKDTQDGVKNNLSVGPATEALAGEGLIVTAGGIDTHIHFISPQQIPTAF
      :: ::.:::.:: :. .:: .:. .:..::..::::::::::.:: :.:.:.:: . .:.
 361" DGLIASIGKAGNPDIMNGVFSNMIIGANTEVIAGEGLIVTAGAIDCHVHYICPQLVYEAI

 150' ASGVTTMIGGGTGPADGTNATTITPGRANLKFMLRAAEEYSMNFGFLAKGNASNDASLAD
      .::.::..:::::::.::.::: ::.. ....::..... ..:::: .::..: ..: : .
 421" SSGITTLVGGGTGPAAGTAATTCTPSPTQMALMLQSTDDLPLNFGFTGKGSSSKPDELHE

 210' QIEAGAIGLKIHEDWGTTPSAINHALDVADKYDVQVAIHTDTLNEAGCVEDTMAAIAGRT
       :.:::.:::.:::::.::.::.. :..:...:.:..:::::::::: ::...::. :::
 481" IIKAGAMGLKLHEDWGSTPAAIDNCLTIAEHHDIQINIHTDTLNEAGFVEHSIAAFKGRT

 270' MHTYHTEGAGGGHAPDIIKVAGEHNILPASTNPTIPFTVNTEAEHMDMLMVCHHLDKSIK
      .::::.:::::::::::::: :  :.::.::::: :.: :: .::.::::::::::..:
 541" IHTYHSEGAGGGHAPDIIKVCGIKNVLPSSTNPTRPLTSNTIDEHLDMLMVCHHLDREIP
                                                I
 330' EDVQFADSRIAPQTIAAEDTLHDMGIFSITSSDSQAMGRVGEVITRTWQTADKNKKEFGR
      ::..::.:.:  .::::::.:.:.:.  ::.::::::::::::::.::::::::.: . :.
 601" EDLAFAHSAIRKKTIAAEDVLNDIGAISIISSDSQAMGRVGEVISRTWQTADKMKAQTGP
                                                I
 390' LKEEKGDNDNFRIKRYLSKYTINPAIAHGISEYVGSVEVGKVADLVLWSPAFFGVKPNMI
      :: ...:::::::.::..:::::::::.:.:.:::::::::.::::.:.:.:::.::.:.
 661" LKCDSSDNDNFRIRRYIAKYTINPAIANGFSQYVGSVEVGKLADLVMWKPSFFGTKPEMV
                                                        I
 450' IKGGFIALSQMGDANASIPTPQPVYYREMFAHHGKAKYDANITFVSQAAYDKGIKEELGL
      ::::..: ...::.:::::::.::  : :..  :::   . :.:::.:: :. ..   ::
 721" IKGGMVAWADIGDPNASIPTPEPVKMRPMYGTLGKAGGALSIAFVSKAALDQRVNVLYGL
                                                        I
 510' ERQVLPVKNCRNITKKDMQFNDTTAHIEVNSETYHVFVDGKEVTLNQPIK
      ...: .:.: . .::.::..::....:.:. :.: : .::: . ....
 781" NKRVEAVSNVAKLTKLDMKLNDALPEITVDPESYTVKADGKLLCVSEATTVPLSANYFLF
```

Fig. 1

similar degree of similarity as between the *H. pylori* sequence and the corresponding sequence from jack bean enzyme (Fig. 1).

The large subunits of the bacterial ureases of *Klebsiella aerogenes* and *Proteus mirabilis* have been partially characterised [11] and comparison of the amino terminal sequence of the beta-subunit of the *H. pylori* enzyme with the amino terminii of the two other enzymes shows that the *H. pylori* enzyme shares 12/19 amino acids (61 %) in common with *K. aerogenes* urease and 15/21 amino acids (71 %) in common with the *P. mirabilis* enzyme (Fig. 2).

Klebsiella aerogenes			S	N	I	S	R	Q	A	Y	A	D	M	F	G	P	T	V	G	D	K
Proteus mirabilis	M	K	T	N	I	S	R	Q	A	Y	A	D	M	F	G	P	T	T	G	D	K
Helicobacter pylori	M	K	K	-	I	S	R	K	E	Y	A	S	M	Y	G	P	T	T	G	D	K
Jackbean (*C. ensiformis*)		N	T	F	I	H	R	K	E	Y	A	N	K	Y	G	P	T	T	G	D	K

Fig. 2. Amino terminii of three bacterial ureases and an internal sequence from the jack bean urease (residues 271–290). Sequence identities are *enclosed*

Studies using the polymerase chain reaction technique showed that at least 1 ng of purified *H. pylori* DNA was required to give an amplified DNA fragment. DNA isolated from *Campylobacter jejuni*, *Campylobacter coli* and *Helicobacter mustelae* was not amplified by the oligonucleotide primer pairs. *H. pylori* DNA was also amplified from $10^6 - 10^2$ *H. pylori* cells and an amplified product of the correct size was not obtained with *C. jejuni* or *C. coli* cells (Fig. 3 A, B).

Discussion

DNA sequencing of the 2.7 kb *Taq*1 DNA fragment and the deduced amino acid sequence has revealed that this DNA insert encodes 61 and 27 kDa polypeptides rather than the 66 and 31 kDa polypeptides that had been previously reported [5]. The significant amino acid sequence homology of these *H. pylori* polypeptides with the single subunit of jack bean urease and the close similarity of the N-terminal 61 kDa polypeptide sequence with the N-terminal sequences of two other bacterial major urease subunits confirms that the cloned *H. pylori* polypeptides are subunits of urease. The recombinant clone did not show any urease activity [5] because other genes in addition to those encoding the structural polypeptides are probably required for synthesis of the active *H. pylori* urease. This is the case for other bacterial ureases [11]. The subunit structure of the *H. pylori* urease has been

Fig. 1. Deduced amino acid sequence of the two *H. pylori* urease subunits (*top line*) compared with jack bean urease (*bottom line*), showing 56.8 % identity in an 803 amino acid overlap. Identical residues are marked by *two dots* (:), conserved substitutions by a *single dot* (.). A 25 amino acid gap in the alpha-subunit was necessary to give the best alignment. The two N-terminal methionines are *marked**. The conserved histidines and reactive cysteine residues are *underlined*. The region homologous to the soy bean fragment is *underlined* and delineated by *vertical lines* II

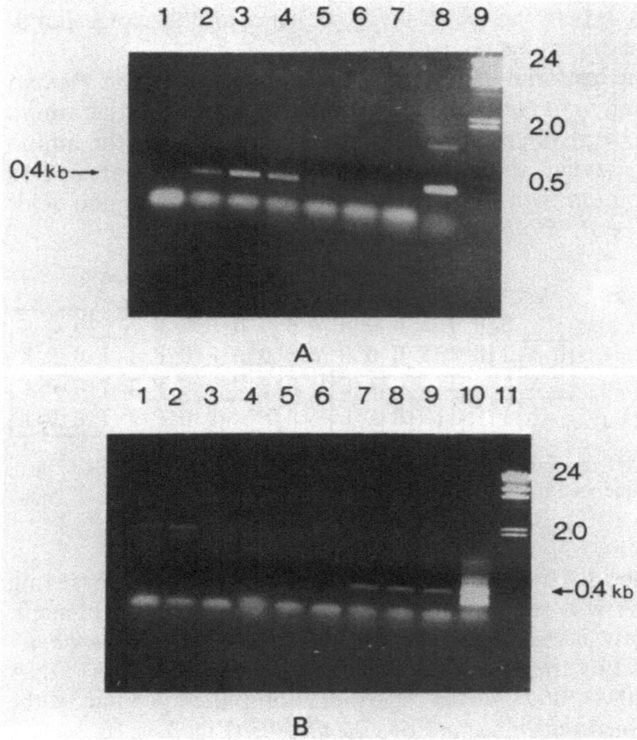

Fig. 3A, B. Agarose gel electrophoresis of a 0.4 kb DNA fragment amplified from boiled *H. pylori* cells using oligonucleotides 040 and 039. **A** DNA released and amplified from *H. pylori* CP145 using 10^6 (*lane 1*), 10^5 (*2*), 10^4 (*3*), 10^3 (*4*), 10^2 (*5*), 10^1 (*6*) and 1 (*7*) bacterial cells; 0.4 kb fragment amplified from control pTCP3 (1 ng) *lane 8*. **B** DNA released and amplified from: *C. jejuni* 50452 10^7 (*lane 1*), 10^6 (*2*), 10^5 (*3*); *C. coli* 50453 10^7 (*lane 4*), 10^6 (*5*), 10^5 (*6*); *H. pylori* CP268 10^6 (*lane 7*), 10^5 (*8*), 10^4 (*9*); control pTCP3 (*lane 10*)

a subject of dispute [6]. Our findings support a two component enzyme, setting the *H. pylori* urease apart from most other bacterial ureases, which consist of three subunits [11]. Takishima et al. noted that 13 of the 25 histidines in the jack bean urease are crowded into a stretch of 128 residues and suggested that this histidine-rich region contains the nickel-binding site of the enzyme [15]. The corresponding region in the *H. pylori* sequence lies in the beta-subunit. Eight of the 13 histidine residues are conserved in the *H. pylori* sequence (Fig. 1). Also conserved in the *H. pylori* sequence is the cysteine residue (Cys-592 in the jack bean enzyme) known to be essential for enzymatic activity [15] (Fig. 1).

Despite the close similarities between the N-terminal sequences of the major urease subunits of *K. aerogenes*, *P. mirabilis* and *H. pylori* the 2.7 kb fragment encoding the *H. pylori* urease did not hybridise to colony blots of other urease-producing bacteria under conditions of high stringency (data not shown), confirming similar findings reported by others [11]. However, Blanchard and Barile [2] have recently shown that a DNA fragment carrying the urease genes of

Ureaplasma urealyticum hybridises to the urease genes of *Providencia stuartii*, *E. coli* and *H. pylori* under conditions of low stringency, suggesting close sequence similarity between urease genes from phylogenetically distant procaryotes.

Immunological cross-reactivity between jack bean urease and bacterial ureases has been described [13], and it has been reported that immunisation with jack bean urease will enhance growth in farm animals, presumably by suppressing gut bacterial urease activity [13]. However, we have found no cross-reactivity between *H. pylori* urease antibody and jack bean urease on immunoblots (data not shown).

The relationship between the *H. pylori* urease and those of the two legumes *C. ensiformis* and *G. max*, is one of the closest yet described between a procaryotic and a eucaryotic protein. Some of the proteins most highly conserved between eucaryotes and procaryotes include mitochondrial and chloroplast proteins (the beta-subunits of the ATP synthetases from barley chloroplasts and *E. coli* share 68% identical residues [16], while EF-Tu from yeast mitochondria is identical in 64% of sites to the equivalent protein in *E. coli* [12]). However, jack bean urease is known to be a cytosolic protein [7]. Heat-shock proteins are also highly conserved between bacteria and eucaryotes, but the degree of sequence similarity between the eucaryotic proteins and their bacterial counterparts is less than that seen with the ureases [1, 9]. We speculate that the most likely explanation for the close similarity between the plant and bacterial ureases is the horizontal transfer of urease genes from a plant to a bacterium. Interestingly, the apparent eucaryotic origin of another bacterial enzyme involved in nitrogen assimilation has been described – glutamate synthetase II, which enables ammonia produced by urease to be incorporated into glutamate. A gene encoding this enzyme is thought to have spread from a plant (directly or via another bacterium) into the ancestor of *Bradyrhizobium japonicum*, the soy-bean symbiont [3]. Plant and bacterial glutamate synthetase II amino acid sequences are identical in up to 47% of sites [3].

The PCR technique using oligonucleotide primer pairs internal to the urease encoding polypeptide genes enabled the specific detection of *H. pylori* with a level of sensitivity of 100 bacterial cells. The technique is currently being optimised and double rounds of 26 cycles particularly using nested primers should greatly increase the sensitivity of the method. This technique may enable the detection of nonculturable *H. pylori*, that could exist in saliva and faecal samples of patients, thus allowing determination of the source and route of infection of the organism.

In conclusion, molecular biology techniques have enabled the urease genes of *H. pylori* to be characterised which will allow defined urease negative mutant *H. pylori* strains to be constructed and the role of *H. pylori* urease to be evaluated in pathogenesis tests in animal models. The PCR method should also enable the epidemiology of this organism to be determined in more detail.

Acknowledgements. Some of the above information is included in a U.K. Patent application No. 8928625.6. We are grateful to J. Keyte (Biochemistry Department, University of Leicester) for the synthesis of oligonucleotides. Thanks to Ann Carey and Sandy Gale for typing of the manuscript.

References

1. Bardwell JCA, Craig EA (1987) Eucaryotic Mr 83000 heat shock protein has a homologue in Escherichia coli. Proc Natl Acad Sci USA 84:5177–5181
2. Blanchard A, Barile MF (1989) Cloning of Ureaplasma Urealyticum DNA sequences showing genetic homology with urease genes from gram negative bacteria. Res Microbiol 140:281–290
3. Carlson TA, Chelm BK (1986) Apparent eukaryotic origin of glutamine synthetase II from the bacterium Bradyrhizobium japonicum. Nature 322:568–570
4. Clayton CL, Topping AJ, Kleanthous H, Wren BW, Mullany P, Tabaqchali S (1989) Molecular cloning and characterisation of Campylobacter pylori species-specific antigens in Escherichia coli K 12. In: Megraud F, Lamoulliatte P (eds) Gastroduodenal pathology and Campylobacter pylori. Elsevier, Amsterdam pp 115–118
5. Clayton CL, Wren BW, Mullany P, Topping A, Tabaqchali S (1989) Molecular cloning and expression of Campylobacter pylori species-specific antigens in Escherichia coli K 12. Infect Immun 57:623–629
6. Dunn BE, Perez-Perez GI, Blaser MJ (1989) Two-dimensional gel electrophoresis and immunoblotting of Campylobacter pylori proteins. Infect Immun 57:1825–1833
7. Faye L, Greenwood JS, Chrispeels MJ (1986) Urease in jack bean (Canavalia ensiformis (L) DC) seeds is a cytoplasmic protein. Planta 168:579–585
8. Hazell SL, Lee A (1986) Campylobacter pyloridis, urease, hydrogen ion back diffusion and gastric ulcers. Lancet 2:15–17
9. Hemmingsen SM, Woolford C, van der Vies SM, Tilly K, Dennis DT, Georgopolous CP, Hendrix RW, Ellis RJ (1988) Homologous plant and bacterial proteins chaperone oligomeric protein assembly. Nature 333:330–334
10. Krueger RW, Holland MA, Chisholm D, Polacco JC (1987) Recovery of soy bean urease genomic clone by sequential library screening with two synthetic oligodeoxynucleotides. Gene 54:41–50
11. Mobley HLT, Hausinger RP (1989) Microbial urease: significance, regulation and molecular characterisation. Microbiol Rev 53:85–108
12. Nagata S, Tsunetsugu-Yokota Y, Naito A, Kaziro Y (1983) Molecular cloning and sequence determination of the nuclear gene coding for mitochondrial elongation factor Tu of Saccharomyces cerevisiae. Proc Natl Acad Sci USA 80:6192–6196
13. Pimenta JL, Cook ME (1988) Improved growth in the progeny of hens immunised with jack bean uease. Poult Sci 67:434–439
14. Sanger FS, Nicklen S, Coulson AR (1977) DNA sequencing with chain-terminating inhibitors. Proc Natl Acad Sci USA 74:5463–5467
15. Takishima K, Suga T, Mamiya G (1988) The structure of jack bean urease. Eur J Biochem 175:151–165
16. Zurawski G, Clegg MT (1984) The barley chloroplast DNA atpBE, trnMZ and ErnVl loci. Nucleic Acids Res 12:2549–2559

The Structure of *Helicobacter pylori* Urease

P. R. HAWTIN, A. R. STACEY, and D. G. NEWELL

The urease is a prominent differential characteristic of *Helicobacter pylori* which is employed in diagnostic tests such as the C^{13}/C^{14} breath test, serodiagnosis and the rapid CLO tests. The role of the enzyme is, as yet, unclear. The number of different ureolytic bacteria being found on the gastric mucosa of various hosts is increasing, raising the question of whether the ureases of these organisms are related.

We have looked at the structure of *H. pylori* urease by first partially purifying the enzyme by size exclusion on fast liquid protein chromatography (FPLC, Pharmacia; Fig. 1) and then producing monoclonal antibodies directed against the urease. The material from the FPLC fraction 14 (M_r 512000) which contains the peak urease activity was examined by polyacrylamide gel electrophoresis (PAGE). Under denaturing conditions by treatment with sodium dodecyl

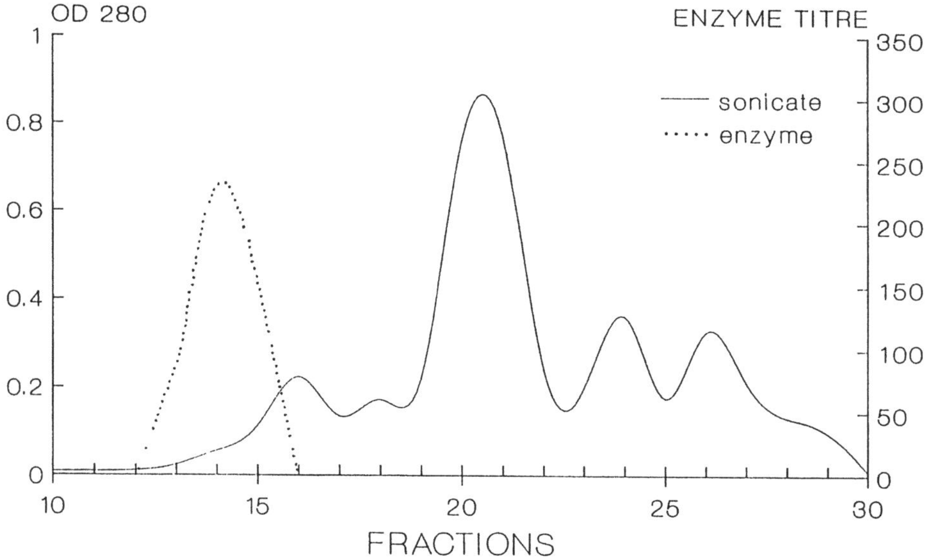

Fig. 1. FPLC trace of *H. pylori* sonicate supernatant

PHLS, Level B, South Academic Block, Southampton General Hospital, Tremona Road, Southampton, SO9 4XY
Division of Pathology, PHLS CAMR, Porton Down, Salisbury, SP4 0JG

Helicobacter pylori, Gastritis and Peptic Ulcer
Edited by P. Malfertheiner and H. Ditschuneit

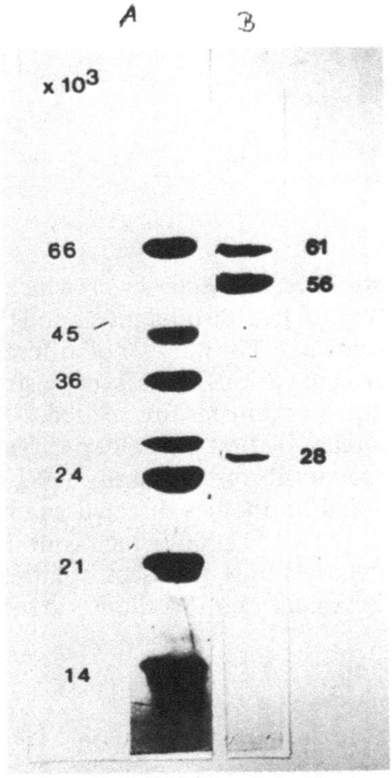

Fig. 2. SDS-PAGE gradient (10%–25%) of fraction 14. *Lane A*, molecular weight markers; *lane B*, fraction 14

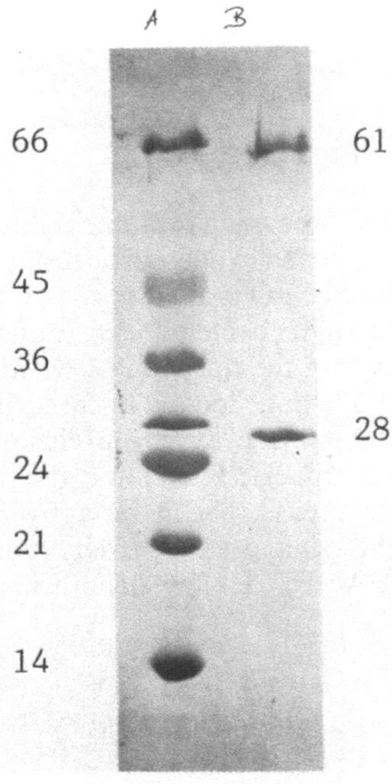

Fig. 3. SDS-PAGE (10%–25%) of urease from 5% linear PAGE. *Lane A*, molecular weight markers, $\times 10^3$; *lane B*, urease preparation

sulphate and β-mercaptoethanol, 10%–25% gradient SDS-PAGE of fraction 14 resolved three major polypeptides at 61 kDa, 56 kDa and 28 kDa (Fig. 2).

The urease was also separated and analysed by loading a 5% linear PAG with a whole cell sonicate supernatant without β-mercaptoethanol treatment and locating the enzyme band by substrate overlay. This band was then cut out and homogenized in buffer containing β-mercaptoethanol and run on 10%–25% gradient denaturing PAG (Fig. 3). The 56 kDa polypeptide was absent in this preparation. It seems, therefore, that this polypeptide is unnecessary for enzyme activity. It is still possible that either the 56 kDa polypeptide is closely associated with the active enzyme or that it is simply copurifying with the native enzyme on the FPLC column. The FPLC fraction 14 also contained 11 nm structures termed "doughnuts" (Fig. 4). Preliminary work has been unable to separate these structures from the urease but, when sedimented by ultracentrifugation, they have the same protein profile as the soluble fraction by SDS-PAGE. Similar structures have been described in outer membrane preparations of *H. pylori* by Newell [1].

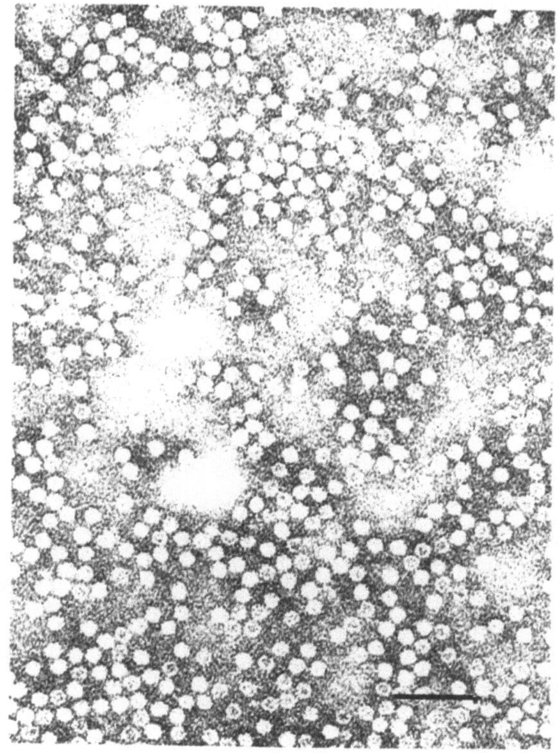

Fig. 4. Electron micrograph of 11 nm structures from FPLC fraction 14; *bar*, 50 nm

The FPLC fraction 14 urease preparation was used to hyperimmunise F1 BALB/c × CBA mice and the immune splenocytes fused with NS1 myeloma cells. Hybridomas were screened for specific antibody production by enzyme linked immunosorbent assay (ELISA) using fraction 14 for antibody capture. The monoclonal antibodies produced were directed predominantly against the smallest polypeptide (28 kDa). Two antibodies directed against the 56 kDa polypeptide were obtained (Fig. 5). The antibodies were tested for inhibition of enzyme activity. Only one, CP11, inhibited urease activity by about 70%. CP11 recognised the 28 kDa polypeptide as shown by western blotting (lane 1, Fig. 5). Moreover, this antibody captured the active enzyme. These results suggest that the epitope for this antibody is close to the enzyme active site so that antibody binding prevents substrate binding, and/or recognition, by competing with the substrate for the active site. Alternatively, the antibody binding may alter the conformation of the native urease sufficiently to change the effective binding of the substrate.

The anti-urease monoclonal antibody CP11 has been used to identify the enzyme on the bacterial cell by indirect immunogold labelling (Fig. 6). The antigen was found to cover the whole of the surface of the organism of intact bacteria and was also contained in amorphous material seen associated with *H. pylori*. In cryosectioned unfixed bacteria, the antigen was located almost solely on the outer membrane.

The antigenic relationship between the ureases of various gastric spiral bacteria and other spiral ureolytic and nonureolytic bacteria was investigated by immunoperoxidase staining techniques using the anti-urease monoclonal antibody CP11 (Table 1). From these results it can be concluded that the ureases from the gastric spiral bacteria are related antigenically [2]. However, the ureases from other ureolytic bacteria are antigenically distinct. Moreover, the antibody did not

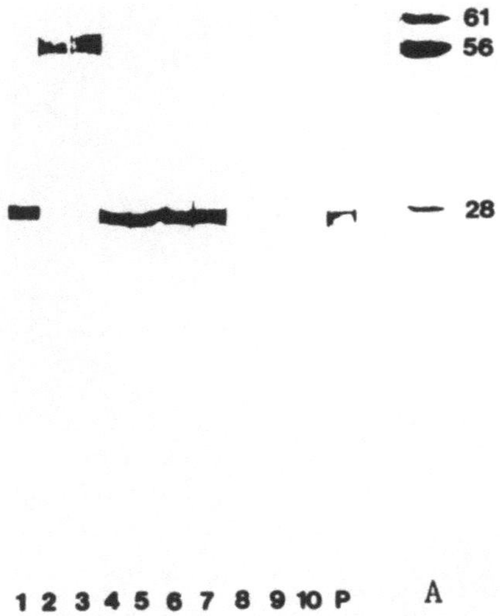

Fig. 5. Western blotting of monoclonal antibodies against FPLC fraction 14. *Lanes 1–10*, monoclonal antibodies; *lane P*, polyclonal serum; *lane A*, protein stain of fraction 14

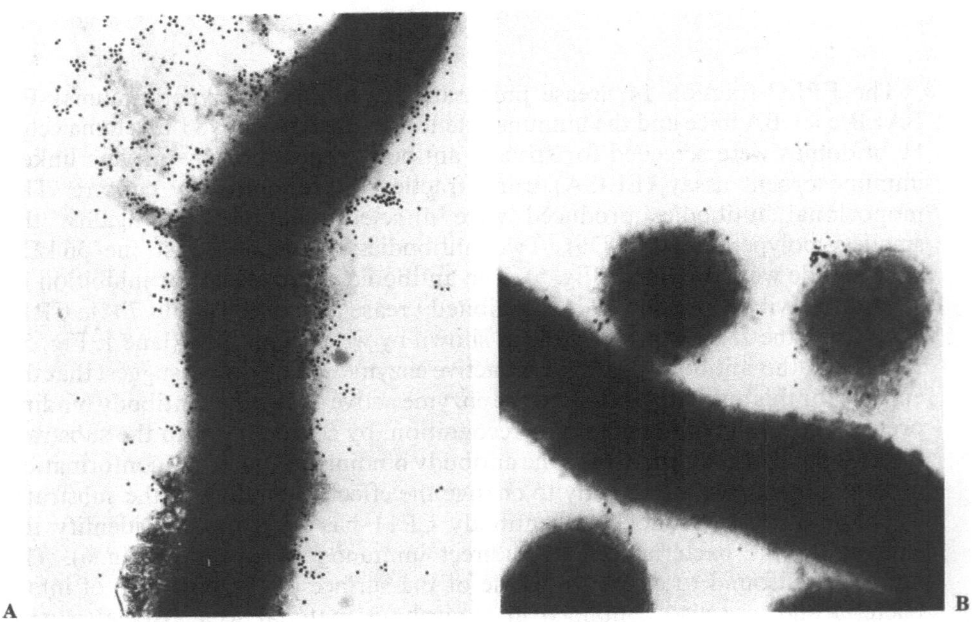

Fig. 6 A, B. Electron micrographs of *H. pylori* immunogold labelled with CP11. **A** Intact *H. pylori*; **B**, cryosectioned *H. pylori*

Table 1. Reactivity of CP11 against various spiral-shaped and helical-shaped bacteria

Bacteria	*Source*	*Reactivity with CP11*
H. pylori	Human stomach	+++
H. pylori	Rhesus stomach	+++
CS1	Cat stomach	+++
Gastric spirals	Rhesus stomach	+++
Gastric spirals	Baboon stomach	+++
B1	Human stomach [a]	++
C. jejuni	Human faeces	–
C. coli	Human faeces	–
UPTC	River water	–
C. nitrofigilis	Brackish mud	–
H. mustelae	Ferret stomach	–
ST1	Rat ileum	–

[a] Maintained in the mouse stomach
–, Negative; ++, positive; +++, strongly positive

recognise urease negative, but related spiral bacteria. Interestingly, the urease of *H. mustelae* did not react with CP11 even though the organism is very similar to *H. pylori* biochemically, and specifically colonises the ferret gastric mucosa. Further work is required to establish relationships between these two ureases.

In conclusion, the urease of *H. pylori* has a molecular weight of 512000 as estimated by size exclusion chromatography and can be resolved into two major polypeptides of 61 kDa and 28 kDa by denaturing PAGE. The enzyme is located on the outer membrane and is surface exposed. *H. pylori* urease shows antigenic conservation with the ureases of other gastric spiral organisms, but not with those of nongastric spiral bacteria and thus may be a prerequisite for colonisation of gastric mucosa, i.e. a virulence factor.

A functional role of the enzyme has yet to be established but this may or may not be incidental to the disease associated with infection by *H. pylori*. Gastritis is the infiltration of inflammatory cells into the lamina propria, but which antigens stimulate this host response is not clear. Our work has shown that the urease of *H. pylori* is highly antigenic and immunodominant. It is therefore likely that the urease contributes significantly to the bacterial antigenic load on the gastric mucosa inducing a chronic, but apparently ineffective local immune response by the host. Such a response requires proliferation of immunocompetent cells to enable regulation. The immunopathological events observed in *H. pylori*-associated gastritis may, therefore, be a consequence of these immunoregulatory responses.

References

1. Newell DG (1987) Identification of the outer membrane proteins of Campylobacter pyloridis and antigenic cross-reactivity between C. pyloridis and C. jejuni. J Gen Microbiol 133:163–170
2. Newell DG, Lee A, Hawtin PR, Hudson MJ, Stacey AR and Fox J (1989) Antigenic conservation of the ureases of spiral – and helical – shaped bacteria colonising the stomachs of man and animals. 65:183–186

Pathogenic Mechanisms of *Helicobacter pylori*: Production of Cytotoxin

N. FIGURA, M. BUGNOLI, M. G. CUSI, A. M. PUCCI, P. LUSINI, S. QUARANTA, A. BARBERI, A. ROSSOLINI, A. L. DI TOMMASO, T. DE MAGISTRIS, R. RAPPUOLI, L. MARRI, R. A. MUSMANNO, M. RUSSI, M. GUARNA, and M. LOSI

Introduction

Bacteria associated with mucosal infections of the digestive system generally produce toxins, especially when they cause inflammatory lesions. Illnesses due to thermotolerant campylobacters, to enterohemorrhagic *Escherichia coli*, and to *Clostridium difficile* are only some examples. It would be surprising if *Helicobacter pylori* (HP) did not produce any toxic substances. The difficulty consists in attributing a pathogenic meaning to the toxin, since the range is quite wide of clinical and histological presentation of gastroduodenal inflammatory diseases linked to the presence in the stomach of *H. pylori* organisms [1]. Johnson and Lior [2] firstly reported the production of heat-labile cytotoxin by 80.6% of 36 HP strains they tested. However, most of our knowledge of the cytotoxigenicity of HP is from Leunk et al. [3] whose work has inspired us in part. They found that about 55% of 201 HP strains isolated in four different parts of the world produced a substance which caused intracellular vacuolization in cells of several lines in vitro, not only in lines generally employed in toxigenicity tests, like chinese hamster ovary (CHO) cells, Vero cells, and Y-1 cells, but also in human tumoral cells like HeLa, KATO III, and HEp-2, as well as in human embryonic intestinal cells which were the most responsive. They also inferred that the toxin was proteinaceous in nature being heat labile (destroyed at 70 °C for 30 min), protease sensitive, and ammonium-sulfate precipitable. Its molecular weight ought to be higher than 100 kDa since cytotoxic activity could be found only in the retentate of a concentrated broth culture filtrate (CBCF) passed through 100 kDa molecular weight limit ultrafiltration membrane.

The pathogenic significance of cytotoxin is still mostly obscure. In a recent study, we observed that HP strains isolated from patients with ulcers produced the toxin significantly more frequently than strains isolated from patients with chronic gastritis only. This suggested that the toxic substance could be involved in the development of peptic ulcer [4].

The aims of the present study were to investigate some aspects of the genetic basis of HP cytotoxin production (i.e., whether cytotoxic HP strains contain plasmids and temperate mitomycin C inducible phages) and some of the biological

Istituto di Malattie Infettive Università di Siena, Via P.A. Mattioli 10, 53100 Siena, Italy; Sclavo, Centro Ricerche, Siena; Istituto di Microbiologica, Università di Siena; Istituto di Istologia ed Embriologia Generale, Università di Siena; Istituto di Chimica Biologica, Università di Siena; Servizio di Gastroenterologia, Div. Medica 2, Ospedale di Grosseto, Italy

Helicobacter pylori, Gastritis and Peptic Ulcer
Edited by P. Malfertheiner and H. Ditschuneit

activities of cytotoxic filtrates (i.e., if they stimulate cAMP formation in vitro, influence the replication of cells in vitro, and impair certain normal cellular functions, i.e., the phagocytosis and the intracellular oxidative metabolism of polymorphonuclear cells in vitro). We tested whether the toxin was also produced in vivo, was immunogenic, and if so what the distribution of specific antibodies was in infected and non-infected patients; we then determined whether patients could be simultaneously infected by cytotoxic and noncytotoxic HP strains. Finally, selected cases were studied with the electron microscope to see whether the in vitro production of toxins corresponded to particular aspects of mucosa.

Material and Methods

Extrachromosomal DNA was searched for in ten cytotoxic HP strains isolated during a previous study [4] using a rapid alkaline extraction procedure [5]. Samples were subjected to electrophoresis in Tris-borate buffer at 80 V for 6 h, and at 20 V for 18 h in a 0.75% agarose horizontal slab gel. Gels were then stained with ethidium bromide (1 μg/ml for 30 min), illuminated by a short-wave ultraviolet transilluminator, and photographed.

Three cytotoxic HP strains were screened for phages by induction with mitomycin C. Washings of agar in the growth inhibition haloes determined by discs charged with 1 μg mitomycin C were examined by electron microscopy. One strain was also cultured in brucella broth with 10% fetal calf serum (FCS) for 48 h. Then it was treated with 1 μg/ml mitomycin C for 18 h. Cells were removed by centrifugation and the supernatant was examined by electron microscopy.

The adenylate cyclase activity assay was carried out on 11 cytotoxic and two noncytotoxic HP broth-culture filtrates (BCF), obtained as further described. The enzyme was assayed in incubation mixtures of 0.2 ml containing 50 m*M* Tris-HCl, 20 m*M* $MgSO_4$, 1 m*M* dithiothreitol, 2 m*M* ATP, and an appropriate amount of BCF. The mixture, prepared in duplicate, was incubated at 34 °C for 30 min and the reaction was stopped by heating in a water bath at 100 °C for 2 min. The tubes were allowed to cool to room temperature and 0.2 ml of a neutral allumine suspension (1.35 g added to 3 ml 50 m*M* triethanolamine/HCl buffer, pH 7.6) was added, followed by vortexing and centrifugation at 3000 rpm for 5 min in an ALC Model 4226 centrifuge. Cyclic AMP production was evaluated on 50 μl of surnatant using a radioisotope dilution test with cyclic AMP binding protein, according to Gilman [6], with a cAMP kit purchased from Amersham Radiochemical Center (Amersham, UK). The radioactivity was measured in a Searle Nuclear Chicago Delta 300 Liquid Scintillation counter. An enzymatic unit of adenylate cyclase was expressed as the number of picomoles of 3′, 5′ cyclic AMP formed per ml of BCF per min.

EBV-transformed B lymphocytes (EBV-B cells) were used to see whether the toxin influenced the cellular replication in vitro. EBV-B cells were cultured for 24 h at 3×10^5 cells per well in 96-well flat-bottomed plates (Costar, Cambridge, Mass.) in 0.2 ml of RPMI 1640 (Gibco Laboratories, Paisly, Scotland) supplemented with L-glutamine (2 m*M*), 1% nonessential amino acids, 1% sodium pyruvate,

50 μg/ml gentamicin 5×10^{-5} *M* 2-ME, and 10% heat-inactivated FCS. BCF of one cytotoxic and one noncytotoxic HP organism, together with uninoculated broth, were added to the wells at the final dilutions of 1:5, 1:10, and 1:20 at the beginning of culture. Cells were pulsed for the last 18 h with 0.5 μCi of (^{3}H)thymidine (sp act, 185 GBq/mmol; Amersham International, Amersham, UK). Cells were then harvested on glass-fiber filters with a cell harvester (Skatron, Lier, Norway) and incorporated radioactivity was determined by liquid scintillation counting. Results were espressed as mean cpm $\times 10^{-3} \pm$ SE of duplicate cultures, and as proliferation indexes $\pm$SE.

Three cytotoxic and two noncytotoxic BCF, together 1 uninoculated broth, were tested to see whether they influenced phagocytosis: 100 μl of sample, 100 μl of polymorphonuclear (PMN) cell suspension containing 1×10^7 cells per ml, 100 μl of *Candida* spp. suspension containing 2×10^7 organisms per ml, and 100 μl of pooled human serum were mixed and incubated at 37 °C for 45 min. The percentage of PMN cells which had ingested yeasts was determined by microscopic examination (1000 magnifications). The evaluation of phagocyte oxygen metabolism in the presence of filtrates was undertaken using the Nitro Blue Tetrazolium (NBT) assay: 50 μl of sample, 50 μl of PMN cell suspension, and 50 μl of diluted NBT were mixed and incubated at 37 °C for 30 min with and without 5 μg of phorbol myristrate acetate. The NBT reduction rate was determined by microscopic examination (1000 magnifications).

In order to see whether subjects were simultaneously infected by cytotoxic and noncytotoxic strains, two to eight HP colonies for primary plate per patient were assayed for toxigenicity. A total of 40 patients who underwent diagnostic endoscopy were examined and 78 HP strains from the 22 infected patients were tested in all. Filtrates of culture in brucella broth with 10% FCS, 0.5% Vitox (Oxoid), 2.5 mg/l ferrum chloride, 2 mg/l amphotericin B, and the Skirrow mixture of chemotherapeutic agents incubated at 37 °C in a microaerobic environment, at 150 oscillations per min for 48 h were added to CHO and Vero cells, and in some cases to intestine 407 cells in vitro at 1:2, 1:3, 1:5, and 1:10 dilutions. Cells were examined for vacuolization and cytopathic effect after 24 h and 48 h of incubation.

To verify whether the toxin was produced in vivo, we assayed two mucosa samples from each of 40 patients with dyspepsia by putting one pice directly on agar which had been previously layered on CHO cells, and dipping the other in Eagles's medium which, after vortexing, was added to Vero cells. Cells were examined after 24 h and 48 h of incubation.

The searching for serum antibodies to the cytotoxin was carried out by the test of neutralization of in vitro toxic activity of filtrates and by immunoblotting test. Neutralization was performed on serum samples collected from 12 patients infected by cytotoxic HP organisms and from nine noninfected patients. Serum samples at various dilutions in Eagle's medium were mixed with BCF of HP strain G-32 which was cytotoxic on CHO cells up to a 1:10 dilution. Final dilution of the filtrate was 1:2. Mixtures were incubated at 37 °C for 60 min and at 4 °C for an additional 60 min, then added to CHO cells. Cells were observed for vacuolating effect after 24 h and 48 h of incubation. Four serum samples were also tested with BCF of the homologous and the heterologous strains.

Sixty serum samples (31 from patients infected by cytotoxic and 20 from patients infected by noncytotoxic HP organisms, and nine from noninfected subjects) were immunoblotted with a concentrated broth-culture filtrate (CBCF) of a cytotoxic HP strain. Ten serum samples which recognized cytotoxic-associated proteins were also blotted with CBCF of seven cytotoxic HP organisms. BCF were concentrated by precipitation with saturated ammonium-sulfate and dialysis in PBS.

CBCF proteins were separated on sodium dodecylsulfate (SDS) polyacrylamide gel electrophoresis (PAGE) and transferred to nitrocellulose. Protein-free sites were saturated with PBS supplemented with 3% defatted milk and 0.1% Triton (PBS-M-T) at room temperature for 30 min. Sheets were then dipped in serum samples diluted 1:100 in PBS-M-T and incubated on a shaker at room temperature o.n. After washing with PBS-M-T, antigen-coated sheets were shaken at room temperature for 90 min with peroxidase-labelled anti-human immunoglobulin G (IgG; Cappel, Cooper Biomedical, Malvern, Pa., USA) diluted in PBS-M-T. Anti-human IgA and IgM provided by Sigma Chemical (St. Louis, Mo., USA) were also used respectively in 20 cases of subjects of different groups and in eight cases of patients infected by cytotoxic HP strains. Strips were then washed in PBS-Triton, rinsed in Tris-BS, and stained with 4-chloro-1-naphtol/H_2O_2. CBCF were also blotted with serum samples from rabbits immunized with formalized whole-cell suspensions of cytotoxic and noncytotoxic HP strains.

For EM observation, biopsy samples were fixed at 4 °C with 2% glutaraldehyde in 0.2% cacodylate buffer, postfixed in 1% osmic acid, and dehydrated with graded alcohol. Ultrathin sections of samples embedded in Epon-araldite were stained with acetate-lead citrate and examined under a Philips TEM 400.

Results and Discussion

No strains proved to contain plasmids and mitomycin C-inducible phages. Together with the stability of the cytotoxin production [3], these data suggest that toxigenicity is under the control of chromosomic genes.

Certain filtrates were found to possess an adenylate cyclase activity, even though at very low degrees. For example, uninoculated broth induced the formation of 13.9 pmoles per ml of broth per min. One non-cytotoxic and three cytotoxic filtrates showed similar activity and were considered negative. The other filtrates stimulated the formation of a cAMP amount ranging from 20 to 53 pmoles. The cultivation of three strains for 96 h led to results similar to those obtained with a 48-h incubation culture. We do not know whether the stimulation also occurs in vivo and what the cellular target is. In any case, the adenylate cyclase activity did not seem associated with the toxin production.

Compared with the noncytotoxic broth culture filtrate G-50, which behaved like uninoculated broth, the cytotoxic G-32 filtrate caused a strong reduction of the proliferation index of B lymphocytes, especially at low dilutions (Table 1). The highest dilution of cytotoxic filtrate which caused vacuolization was 1:10. At 1:20 the proliferation index was still reduced by nearly three times (Table 1), indicating

Table 1. Proliferation assay of EBV-transformed B lymphocytes in the presence of cytotoxic and noncytotoxic filtrates and of uninoculated broth

Sample	Dilution	Cpm ± SE	Proliferation index ± SE
B-Cell medium		24.6 ± 1.39	1.00 ± 0.06
Uninoculated broth	1: 5	16.5 ± 2.68	0.67 ± 0.11
	1:10	21.4 ± 0.22	0.87 ± 0.01
	1:20	24.9 ± 0.36	1.02 ± 0.01
Noncytotoxic broth culture G-50	1: 5	19.6 ± 4.67	0.80 ± 0.19
	1:10	21.3 ± 0.53	0.87 ± 0.02
	1:20	27.1 ± 2.95	1.10 ± 0.12
Cytotoxic broth culture G-32	1: 5	1.1 ± 0.03	0.04 ± 0.00
	1:10	3.0 ± 0.34	0.12 ± 0.01
	1:20	10.3 ± 0.80	0.42 ± 0.03

that the proliferation assay was more sensitive than the determination of cytopathic effect in revealing cytotoxigenicity.

The vacuolating toxin has been defined as nonlethal because, even though vacuolated, cells in culture could exclude Tripan Blue [3]. The proliferation assay demonstrated that, at least as far as B lymphocytes are concerned, cells were negatively influenced by the toxin in their replication. In effect, CHO cells exposed to the toxin, after becoming vacuolated (Fig. 1 A), lose their morphology

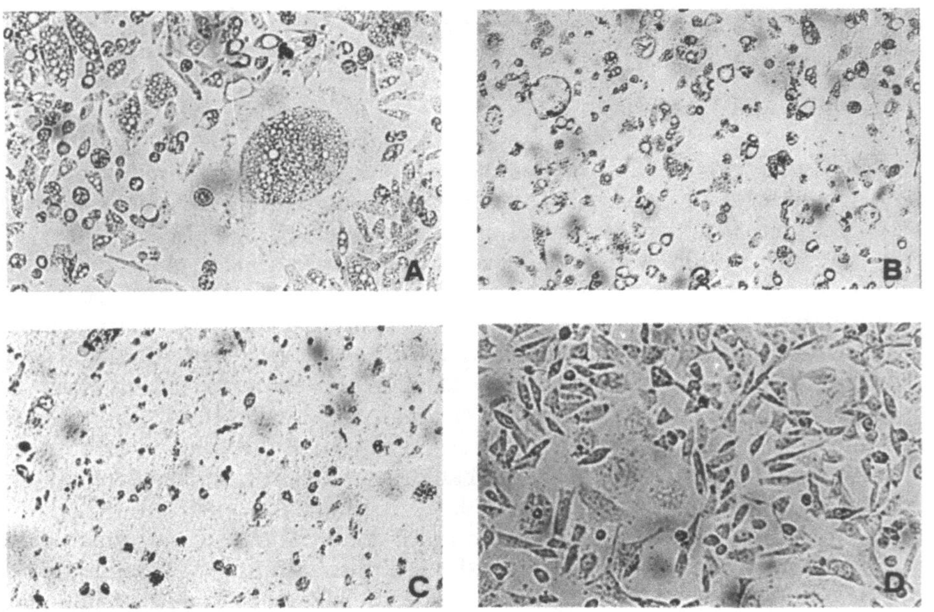

Fig. 1 A–D. CHO cells exposed to *H. pylori* vacuolating toxic filtrate for 24 h (**A**), 48 h (**B**), and 96 h (**C**). (**D**) Control CHO cells with uninoculated broth after 96 h of incubation

(Fig. 1 B), wrinkle up, and die (Fig. 1 C) in a few days – in any case far sooner than do cells with uninoculated broth (Fig. 1 D). Thus, even if vacuolated cells exposed to the toxin are still vital, they are nonetheless bound to die rapidly.

The three toxic BCF were found to afflict the phagocytic function of PMN cells by about 20%, while the two nontoxic BCF and uninoculated broth did not influence it at all. The intracellular oxidative metabolism was maintained in all cases, suggesting that the toxin exerted a generic toxic effect on PMN cells or damaged structures necessary to the movement of membrane (microtubules, for instance).

The vacuolating effect on cells in vitro directly exerted by biopsy specimens was observed in only one out of 15 cases in which in vitro cytotoxic HP organisms were isolated. Eaton et al. [7], in gnotobiotic piglets experimentally infected with HP strains, saw that one cytotoxic organism was more virulent than a noncytotoxic one. However, they could not demonstrate the presence of toxin in the gastric juice or the serological response to the toxin. It is reasonable to suppose that the in vivo production of toxin, if any, occurs at very low levels, hardly or not at all detectable by the cell in culture model. Searching for anticytotoxin antibodies in patient serum samples gave better results.

All serum samples from the 12 patients infected by cytotoxic HP strains and the one from the rabbit immunized with a cytotoxic HP organism neutralized the vacuolating effect of a toxic BCF at titers ranging from 1:4 to 1:32. The four sera tested neutralized the BCF toxicity of the homologous and also of the heterologous strains. This suggested that the toxin was also produced in vivo and was immunogenic and that the toxins in the four different BCF were immunologically related. However, four of the nine sera from noninfected subjects were able to neutralize the toxic effect. This could create confusion and impair the specificity of the in vitro toxicity neutralization tests. However, the immunoblotting assay clarified this matter. All the 31 subjects infected by cytotoxic HP strains, 14 of the 20 patients (70%) infected by noncytotoxic HP strains, and two of the nine subjects (22.2%) not infected had serum IgG which reacted with cytotoxin-associated proteins (CAP). Anti-CAP IgA were hardly detectable, mostly in serum from patients infected by cytotoxic HP organisms, while IgM were not detectable at all. These proteins were about 130 kDa, 95 kDa, and 80 kDa as shown in Fig. 2. The 130 kDa protein was the most frequently detected in the cytotoxic CBCF, and it was always recognized also in the other seven cytotoxic CBCF assayed by the ten serum samples tested (Fig. 3, lanes A, C, E, F). The serum sample from the rabbit immunized with a cytotoxic HP only reacted with the 80 kDa protein of the homologous CBCF. The serum from the rabbit injected with a noncytotoxic HP organism did not recognize any CAP. These observations suggested that the 95 kDa protein was a degradation product of 130 kDa CAP and that the 80 kDa protein constituted subunits of the major CAP. In any case these three CAP were lacking in CBCF of the two noncytotoxic HP strains tested (Fig. 3, lanes B and D).

These results confirmed the findings from the neutralization assays. The fact that some subjects considered noninfected on the basis of culture, microscopic examination, and rapid urease activity of gastric mucosa biopsy specimens [4], and that 70% of those infected by noncytotoxic HP organisms possessed serum antibodies to the toxin produced by HP suggested that the exposure to the toxin

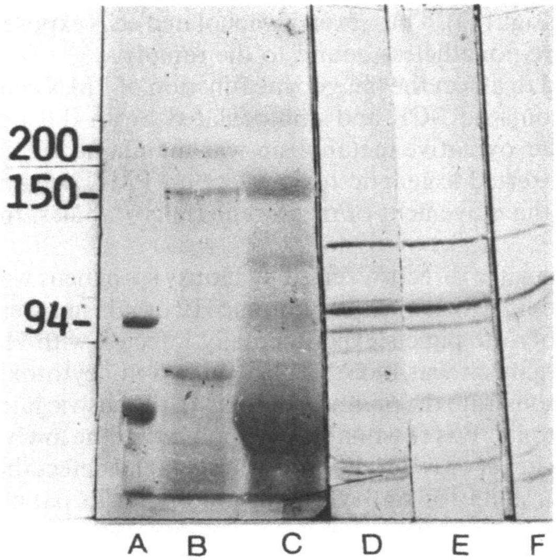

Fig. 2. Immunoblots of CBCF of a cytotoxic *H. pylori* strain with three serum samples from patients infected by cytotoxic organisms (*lanes D, E, F*). *A* Low standard molecular weights; *B*, high standard molecular weights; *C* CBCF stained with Ponceau red

Fig. 3. Immunoblots of four cytotoxic (*lanes A, C, E, F*) and two noncytotoxic CBCF (*lanes B* and *D*) with serum collected from a patient infected by both vacuolating and lytic *H. pylori* strains

was a common event, far more frequent than would have been expected on the basis of the results of in vitro toxigenicity tests. Perhaps testing only one HP colony per patient as we did in a previous work [4] was not indicative enough of the real situation. In fact, in only 15 of 22 patients infected by HP (among the 40 patients examined) were all strains from the same patient either cytotoxic or noncytotoxic. In the other seven cases (31.8%), cytotoxic strains were isolated

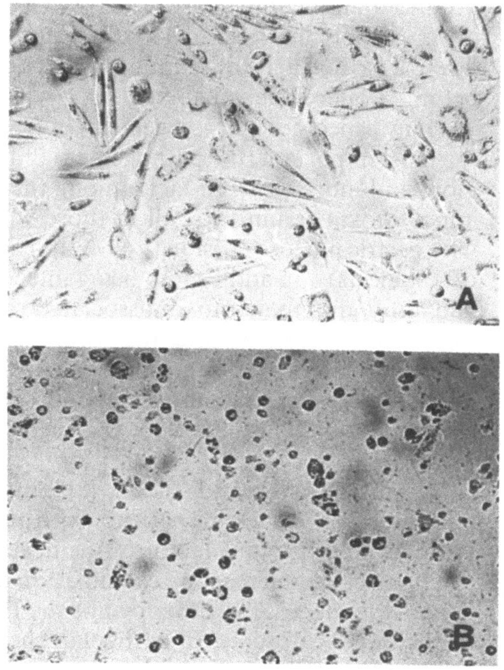

Fig. 4 A, B. CHO cells showing morphological responses to two HP broth-culture filtrates which were different from the vacuolating one. **A** Elongation effect; **B** Lytic effect

together with noncytotoxic HP organisms (78 HP strains were tested in all from the 22 patients). These results suggest that, for a better understanding of the pathogenic role of the cytotoxin, it is preferable to test several HP colonies per patient, even though some conclusions were similar to those obtained in a previous study in which a single HP colony per patient was tested for toxigenicity [4]. In fact, most strains isolated from the 11 patients with ulcer, of the 22 HP-infected patients, were cytotoxic (25 of 38 strains, or 65.7%), while only 15 of 40 strains (37.5%) isolated from the other 11 patients with chronic gastritis only produced the toxin ($P < 0.05$, chi-squared test, 1 df).

During the latter study we observed that certain filtrates caused morphological responses of cells in vitro that were different from the vacuolating response, an elongation of CHO cells (a cytotonic-like response), and a lytic effect on CHO and Vero cells (Fig. 4), suggesting that HP can produce more toxins. However, after 4 months of storage at −80 °C in Wilkins-Chalgren broth with 20% glycerol, the lytic HP strain (which had been isolated together with a vacuolating HP organism), became vacuolating. We do not know how to explain this phenomenon. It is reasonable to suppose that the lytic effect was due to a toxin different from the vacuolating one, and that the ability to produce it was lost due to the storage. However, the serum sample of the patient from whom the lytic strain had been isolated recognized some proteins in the already vacuolating CBCF of the homologous strain which were not visible in the CBCF of other cytotoxic or noncytotoxic strains (Fig. 3, lane F). This could mean that after storage the strain produced an incomplete lytic toxin. The fact that the lytic effect was originally observed only in an up to 1:5 dilution indicated that cytopathogenicity was not due to a particularly high amount of vacuolating toxin which caused cellular death before cells became vacuolated.

Many aspects of cytotoxigenicity in *H. pylori* are still obscure. However, one of the most important, the pathogenic significance of toxin production, is becoming more and more clear. In fact, among the 60 patients examined for possession of anticytotoxin serum IgG, all of those with duodenal ulcer ($n = 11$), all of those with gastric mucosa erosion ($n = 7$), and nearly all of those with gastric ulcer (7 out of 8 subjects) – in addition to being infected by mostly cytotoxic HP organisms – had antibodies to cytotoxin-associated proteins. Of the 31 patients with nonerosive chronic gastritis and the three with normal mucosa, 70.9% and none, respectively, had specific antibodies. In other words, 96.1% of patients with ulcers or gastric erosions had anti-CAP IgG vs 64.7% of patients without ulcerative or erosive lesions. This difference was highly significant ($P < 0.01$, chi-squared test, 1 df).

Recently, Cover and Blaser [8] obtained similar results on anticytotoxin antibody distribution in patients with peptic ulcer, and they, too, consider the toxin as an important factor in the development of peptic ulceration.

The electron microscopy examination did not reveal differences with regard to the gastric mucosa, whether or not the HP organisms isolated from the patients studied were cytotoxic. In particular, the mucosa samples of patients infected by HP strains which produced the vacuolating or the lytic toxin did not appear more damaged than that of subjects infected by noncytotoxic organisms. In addition, vacuols were never seen.

In conclusion, the toxin is produced in vivo, is immunogenic, and the exposure to it is quite common. Toxins produced by different *H. pylori* strains seem antigenically related. Toxins different from the vacuolating one probably exist. Strains isolated from patients with macroscopic lesions of gastric mucosa are mostly cytotoxic, and practically all individuals with ulcers have antibodies to the toxin. Thus, cytotoxicity is most probably important in the development of peptic ulcers.

Acknowledgement. We gratefully acknowledge Ms. D. Macchia, Ms. L. Tognelli, Mr. G. Mecacci, and Ms. C. Rinina, for their skilled technical assistance, Ms. F. Bartali for her secretarial assistance, and Ms. Zegelbone for correcting the English.

References

1. Blaser MJ, Brown WR (1989) Campylobacter and gastroduodenal inflammation. Adv Intern Med 34:21–42
2. Johnson WM, Lior H (1986) Production of heat-labile cytotoxin by Campylobacter pyloridis (Abstr B234).
3. Leunk RD, Johnson PT, David BC, Kraft WG, Morgan DR (1988) Cytotoxic activity in broth-culture filtrates of Campylobacter pylori. J Med Microbiol 26:93–99

4. Figura N, Guglielmetti P, Rossolini A, Barberi A, Cusi G, Musmanno RA, Russi M, Quaranta S (1989) Cytotoxin production by Campylobacter pylori strains isolated from patients with peptic ulcers and from patients with chronic gastritis only. J Clin Mircobiol 27:225–226
5. Davies RV, Botstein D, Roth JR (1980) Rapid isolation of plasmid and/or bacterial DNA from colonies or broth. In: Cold Spring Harbor Laboratory (ed) A manual for genetic engineering. Cold Spring Harbour, New York, p 120
6. Gilman GA (1970) A protein binding assay for adenosine 3′, 5′ cyclic monophosphate. Proc Natl Acad Sci USA 67:305–312
7. Eaton KA, Morgan DR, Krakowka S (1989) Campylobacter pylori virulence factors in gnotobiotic piglets. Infect Immun 57:1119–1125
8. Cover TL, Blaser MJ (1989) Human serologic response to Campylobacter pylori cytotoxin-associated proteins. Gastroenterology 96:A 101

Helicobacter pylori Hemagglutinins – Possible Gut Mucosa Adhesins

T. Wadström, J. L. Guruge, S. Wei, P. Aleljung, and A. Ljungh

Introduction

A number of recent studies indicate that *Helicobacter pylori* has unique cellular properties which allow this pathogen to colonize the human stomach mucosa [1, 2, 4, 7]. Despite this unique habitat for this microbe, very few studies have tried to identify surface proteins and other possible structures determining an association with the stomach mucosa. Emödy et al. [2] reported on surface hemagglutinins in *H. pylori* followed by reports from the USA, Ireland, and Japan. Studies by Emödy et al. [2] and Carlsson et al. [1] have defined a sialic acid specific hemagglutinin. More recently, a Canadian group of investigators defined a surface component of *H. pylori* interacting with a glycolipid from human erythrocyte membranes [5]. Furthermore, Huang et al. [4] described a hemagglutinin which caused neuraminidase and protease resistant hemagglutination. In view of these findings we decided to carry out a systematic study on *H. pylori* hemagglutinin profiles by screening clinical isolates and type culture collection strains for hemagglutination with a variety of erythrocytes.

Materials and Methods

Bacterial Strains

Forty-one strains of *H. pylori* isolated at hospitals in Sweden and France and one NCTC strain (11 637 = CCUG 17 874) were used in the study. The strains were stored in tryptic soy broth containing 15% (v/v) glycerol at −80 °C. Strains were cultivated on blood agar with 5% horse erythrocytes and incubated under microaerophilic conditions for 3–7 days.

Chemicals

All chemicals used were of analytical grade.

Department of Medical Microbiology, University of Lund. Sölvegatan 23, S-223 62 Lund, Sweden

Helicobacter pylori, Gastritis and Peptic Ulcer
Edited by P. Malfertheiner and H. Ditschuneit

Hemagglutination (HA) Assay

Bacteria were washed and resuspended in phosphate buffered saline (PBS, 0.15 *M* NaCl in 0.02 *M* sodium phosphate buffer pH 7.2) to a final concentration of approximately 10^{10} cells/ml). Horse, rabbit, human, bovine, guinea pig, pig, sheep, monkey, dove, cock, and mouse (citrated) blood were obtained fresh. Erythrocytes were washed two to three times and resuspended in PBS to a 2% (vol/vol) suspension. HA was performed on glass slides by mixing equal volumes (20 μl) of bacterial and erythrocyte suspensions. The reactions were read after 2 min.

HA Titer

Twofold dilutions of bacterial suspensions (25 μl) of 15 strains were made in V-bottomed 96-well plates (Greiner, Sohne, Nürtingen, FRG) and titrated against erythrocyte suspensions (25 μl) to obtain the minimal bacterial hemagglutinating concentrations. The plates were agitated to get even mixing, covered, and incubated at room temperature (20 °C) on an orbital shaker for 1 h. The plates were allowed to settle overnight before being read [4]. The endpoint was defined in principle as described by Nakazawa et al. [7] as the last dilution showing complete agglutination. Titers were expressed as reciprocals of endpoint dilutions. The specific activity of HA for each extract was calculated by dividing the HA titer by the protein concentration.

HA Inhibition

HA inhibition experiments were performed with titrated bacterial suspensions (25 μl) of two selected strains (17874 and 1139) according to Carlsson et al. [1]. After preincubation of bacteria with various potential monosaccharide and glycoconjugate inhibitors (25 μl) for 30 min at room temperature before the erythrocyte suspensions (25 μl) were added. For the control 25 μl PBS was used instead of the inhibitor. Determination of HA was done as described above.

Carbohydrates (Gal, NADGal, NADGlc, NANA, NANLac, Man, NGNA, and Glc) were dissolved at a final concentration of 0.1 *M* in PBS. Glycoconjugates (hog gastric mucin, FCS, bovine submaxillary mucin, orosomucoid and ganglioside type II, purchased from Sigma Chemical, St. Louis, USA) were dissolved at a final concentration of 0.1% in PBS.

Enzyme Treatment

Bacterial suspensions of strains 17874, 1139, and erythrocyte suspensions were pretreated separately with trypsin (50 μg/ml), pronase (100 μg/ml), and neuraminidase (1 U/ml) diluted in PBS, pH 7.2, for 30 min at 37 °C. The treated bacterial cells and the treated erythrocytes were washed three to four times and resuspended in PBS at final concentrations of 10^{10} cells/ml and 2% respectively before being used for HA. The untreated bacteria or the erythrocytes were used as the control.

Heat Treatment

Bacterial suspensions of strains 17874 and 1139 were incubated at 37 °C, 50 °C, and 100 °C for 30 min. Twofold dilutions of the treated bacterial suspensions were made on microtiter plates and titrated with erythrocytes to determine the HA titer.

Cell Surface Hydrophobicity Tests

Cell surface hydrophobicity was determined using the salt aggregation test (SAT) according to Emödy et al. [2]. Cells were washed with potassium phosphate buffer (0.02 *M*, pH 6.8) and serially diluted with ammonium sulfate solutions of varying molarity (0.2–2 *M*) on glass slides and observed for 30 sec. The highest dilution of ammonium sulfate in which visible aggregation occurred was the SAT value for surface hydrophobicity.

Preparation of Bacterial Cell Surface Extracts

Glycine Acid Extract. Glycine acid extracts were prepared according to Logan and Trust [6]. Cells were harvested and washed twice in distilled water and then suspended in 0.2 *M* glycine hydrochloride, pH 2.2 (4 g of cells per 100 ml). The suspension was stirred for 15 min at room temperature and cells were then removed by centrifugation. The supernatant was neutralized with NaOH and dialyzed overnight at 4 °C against PBS.

LiCl Extraction. Cells were washed twice in PBS and then suspended in 1 *M* LiCl. The suspension was heated at 45 °C for 2 h. Cells were removed by centrifugation and dialyzed as described above.

Guanidine Hydrochloride Extraction. Cells were washed in PBS, suspended in 5 *M* guanidine hydrochloride, and stirred for 10 min at 20 °C. Dialysis was done as described above.

Urea Extraction. Cells were washed in PBS, suspended in fresh 8 *M* urea, stirred at 20 °C for 10 min, and dialysed as described above.

Determination of Protein Concentration

Protein concentration of the extracts was determined using an acidic solution of Coomassie Brilliant Blue G-250 dye reagent (Bio-Rad, Richmond, UK). Standards were prepared with bovine serum albumin in PBS.

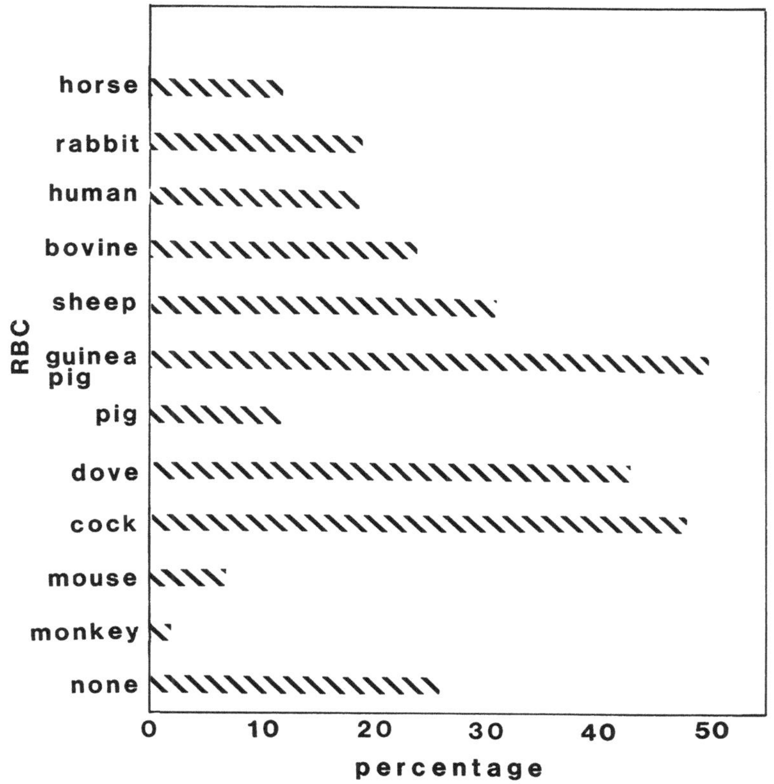

Fig. 1. Percentage of strains agglutinating various erythrocytes

Results

Hemagglutination (HA) activity of 42 *H. pylori* strains was tested with 11 different erythrocytes (Table 1). Nine different HA patterns were observed on the basis of number of erythrocytes agglutinated by each strain.

The percentage of strains agglutinating different erythrocytes is shown in Fig. 1. Guinea pig erythrocytes were agglutinated by the highest number of strains (50%), whereas monkey erythrocytes were agglutinated by only 2% of strains tested.

With various inhibitors tested, the HA of strains 17874 and 1139 was inhibited by NANLac only.

All strains showed low surface hydrophobicity by SAT except strain 72 (SAT value 1.0).

Two strains (17874 and 1139) agglutinating a number of erythrocytes were selected for further studies. The effect of heat treatment on HA titers is given in Table 2. The HA of strain 17874 was stable at 37 °C but reduced with human, bovine, sheep, monkey, dove, and mouse erythrocytes when treated at 50 °C for

Strain no.	Source of erythrocyte										
	Horse	Rabbit	Human	Bovine	Guinea pig	Pig	Sheep	Monkey	Dove	Cock	Mouse
Pattern 1											
17874	+	+	+	+	+	+	+	+	+	+	+
Pattern 2											
165	+	−	+	+	+	−	−	−	+	+	+
72	+	−	−	+	+	+	+	−	+	+	−
25/1139	−	+	+	+	+	−	+	−	+	+	−
Pattern 3											
12225	−	+	+	−	+	+	−	−	+	+	−
309/297	−	−	+	+	+	−	+	−	+	+	−
Pattern 4											
95	−	+	+	−	+	−	−	−	+	+	−
315	−	−	−	+	+	−	+	−	+	+	−
Pattern 5											
45	+	−	−	−	−	+	+	−	+	−	−
237	−	+	−	−	+	−	+	−	−	−	+
236	−	−	−	+	+	+	+	−	−	−	−
54/52	−	−	−	−	+	−	+	−	+	+	−
Pattern 6											
88	+	−	−	+	−	−	−	−	−	+	−
Pattern 7											
915	−	+	−	−	−	−	−	−	−	+	−
86	−	−	−	−	+	−	+	−	−	−	−
164	−	−	−	−	+	−	−	−	+	−	−
19106/154	−	−	−	−	+	−	−	−	−	+	−
94	−	−	−	−	−	−	+	−	−	+	−
73/249	−	−	−	−	−	−	−	−	+	+	−
Pattern 8											
19104	−	+	−	−	−	−	−	−	−	−	−
239/253/266	−	−	−	−	+	−	−	−	−	−	−
338/268	−	−	−	−	−	−	−	−	+	−	−
250	−	−	−	−	−	−	−	−	−	+	−
Pattern 9											

Table 2. Hemagglutination titer of strains 17874 and 1139 after treatment at 37 °C and 50 °C for 30 min with various erythrocytes

Source of RBCs	Untreated		Treated/30 min			
	20 °C		37 °C		50 °C	
	17874	1139	17874	1139	17874	1139
Horse	1	n.h.	1	n.h.	1	n.h.
Rabbit	2	4	2	2	2	2
Human	8	8	8	8	4	8
Bovine	8	8	8	8	4	8
Guinea pig	8	4	8	4	8	4
Pig	4	n.h.	4	n.h.	4	n.h.
Sheep	32	8	32	8	8	8
Monkey	4	n.h.	4	n.h.	1	n.h.
Dove	8	4	8	2	4	2
Cock	8	4	8	2	8	2
Mouse	64	n.h.	64	n.h.	32	n.h.

RBCs, red blood cells; n.h., no hemagglutination

30 min. The HA activity of 1139 was reduced with rabbit, dove, and cock erythrocytes at 37 °C and 50 °C. Both strains showed a weak activity with all erythrocytes when treated at 100 °C for 30 min.

Treatment of bacteria cells with trypsin, pronase, and neuraminidase partially inhibited the HA activity of strain 17874 whereas total inhibition by strain 1139 was observed with bovine, sheep, dove, and cock erythrocytes after treatment with neuraminidase (Table 3).

Table 3. Hemagglutination activity of strains 17874 and 1139 on various erythrocytes after treatment with trypsin, pronase, and neuraminidase for 30 min at 37 °C

Source of RBCs	Untreated		Treatment/30 min					
			Trypsin		Pronase		Neuraminidase	
	17874	1139	17874	1139	17874	1139	17874	1139
Horse	(+)	n.h.	–	n.h.	–	n.h.	–	n.h.
Rabbit	+	+	(+)	+	(+)	+	(+)	+
Human	+	+	(+)	+	(+)	+	(+)	+
Bovine	+	+	(+)	+	(+)	+	(+)	–
Guinea pig	+	+	(+)	+	(+)	+	(+)	+
Pig	+	n.h.	(+)	n.h.	(+)	n.h.	(+)	n.h.
Sheep	+	+	(+)	+	(+)	+	(+)	–
Monkey	+	n.h.	(+)	n.h.	(+)	n.h.	(+)	n.h.
Dove	+	+	(+)	+	(+)	+	(+)	–
Cock	+	+	(+)	+	(+)	+	(+)	–
Mouse	+	n.h.	(+)	n.h.	(+)	n.h.	(+)	n.h.

n.h., no hemagglutination; (+), weak hemagglutination

Table 4. Hemagglutination activity of strains 17874 and 1139 on various erythrocytes treated with trypsin, pronase, and neuraminidase for 30 min at 37 °C

Source of RBCs	Untreated		Treatment/30 min					
			Trypsin		Pronase		Neura-minidase	
	17874	1139	17874	1139	17874	1139	17874	1139
Horse	(+)	n.h.	–	n.h.	–	n.h.	–	n.h.
Rabbit	+	+	(+)	+	(+)	+	(+)	+
Human	+	+	(+)	+	(+)	+	(+)	+
Bovine	+	+	(+)	+	(+)	+	(+)	+
Guinea pig	+	+	(+)	+	(+)	+	(+)	+
Pig	+	n.h.	(+)	n.h.	(+)	n.h.	(+)	n.h.
Sheep	+	+	(+)	+	(+)	+	(+)	+
Monkey	+	n.h.	(+)	n.h.	(+)	n.h.	(+)	n.h.
Dove	+	+	(+)	+	(+)	+	(+)	+
Cock	+	+	(+)	+	(+)	+	(+)	+
Mouse	+	n.h.	(+)	n.h.	(+)	n.h.	(+)	n.h.

For symbols see Table 3

The HA activity of strains 17874 and 1139 on treated erythrocytes is shown in Table 4. Partial inhibition of HA by strain 17874 was observed with all erythrocytes whereas the HA of strain 1139 was not affected.

The results of the HA assay of different extracts is given in Table 5. The highest specific activity was observed with the LiCl extraction method.

Table 5. Hemagglutination assay of different extracts of *H. pylori*

Extract	Protein concentration (μg/ml)	HA titer	Specific activity/mg protein
Glycine HCl	126	1/4	31.7
LiCl	680	1/128	188.2
Guanidine HCl	1960	1/4	2.0
Urea	1760	1/64	36.4

HA titer was determined with rabbit RBCs

Discussion

Despite the fact that various mixtures of monosacharides do not inhibit hemagglutination (HA) of *H. pylori* strains with various HA profiles (Table 1), it seems most likely that *H. pylori* hemagglutinins recognize specific carbohydrate structures on erythrocyte membranes, i.e., various glycolipids and glycoproteins.

It seems most likely that *H. pylori* use such lectin-like interactions with mucosal surface glycoconjugates to colonize and penetrate the stomach mucosa into subepithelial tissues [1, 2]. However, since the microbe probably has to swim rapidly through the mucin layer it is unlikely that the lectins bind to mucin carbohydrate structures which could delay penetration through this barrier.

Hills et al. [3] have demonstrated the hydrophobic lining of the stomach lumen to be the gastric mucosal barrier. Our findings that *H. pylori* strains of different HA profiles express low cell surface hydrophobicity when grown under various conditions indicate that low hydrophobicity may also be important for rapid penetration of the gastric mucin layer. Interestingly, high surface hydrophobicity is very common among intestinal pathogens such as enterotoxigenic and enteropathogenic *E. coli* (ETEC and EPEC), *Yersinia enterocolitica*, and enteroinvasive pathogens such as congo red binding *Shigellae* [8].

The LiCl extraction method seems to be a mild method to extract surface hemagglutinins for further purification (Table 5). Purification of hemagglutinins and identification of various hemagglutinins among selected *H. pylori* strains of various patterns we describe in this communication are now in progress.

Acknowledgements. We would like to thank Dr. F. Megraud, Hopital des Enfants, Laboratoire de Bacteriologie, Bordeaux, France for kindly supplying us with *H. pylori* strains. This study was supported by a grant from the Swedish Medical Research Council (16x 04723).

References

1. Carlsson Å, Aleljung P, Emödy L, Ljungh A, Wadström T (1989) Carbohydrate receptor specificity of hemagglutination of Campylobacter pylori. In: Mégraud F, Lamouliatte H (eds) Gastroduodenal pathology and Campylobacter pylori. Excerpta Medica, Amsterdam, pp 375–378
2. Emödy L, Carlsson Å, Wadström T (1988) Mannose-resistant hemagglutination by Campylobacter pylori. Scand J Infect Dis 20:353–354
3. Hills BA, Butler BD, Lichtenberger LM (1983) Gastric mucosal barrier: hydrophobic lining to the lumen of the stomach. Am J Physiol 244:G 561–G 568
4. Huang J, Smyth CJ, Kennedy NP, Arbuthnott JP, Keeling PWN (1988) Hemagglutinating activity of Campylobacter pylori. FEMS letters 56:109–112
5. Lingwood CA, Law H, Pellizzari A, Sherman P, Drumm B (1989) Gastric glycerolipid as a receptor for Campylobacter pylori. Lancet 2:238–241
6. Logan SM, Trust TJ (1983) Molecular identification of surface protein antigens of Campylobacter jejuni. Infect Immun 42:675–682
7. Nakazawa T, Ishibashi M, Konishi H, Takemoto T, Shigeeda M, Kochiyama T (1989) Hemagglutination activity of Campylobacter pylori. Infect Immun 57:989–991
8. Qadri F, Hossain SA, Ciznar I, Haider K, Ljungh Ä, Wadström T, Sack DA (1988) Congo red binding and salt aggregation as indicators of virulence in Shigella species. J Clin Microbiol 26:1343–1348

Lectin Typing of *Helicobacter pylori*

F. Ascencio, J. L. Guruge, Å. Ljungh, F. Mégraud, S. Wei, and T. Wadström

Introduction

Typing of bacterial strains in epidemiological studies by using plant lectins has been found to be a powerful method [4] for differentiating strains of related species [1] and epidemiological studies of various infections (Table 1). Excellent reviews of lectin typing of bacteria have been published by Pistole [6] and Doyle and Keller.

The aim of this study was to make a preliminary evaluation of whether lectin typing can be useful means to discriminate strains within *Heliobacter pylori* since other typing systems such us serotyping and DNA restriction enzyme analysis are still in their infancy as epidemiological tools for studies on *H. pylori* infections [2, 3].

Table 1. Lectin typing of bacteria: some examples

Purpose	Bacterial specie	Reference
Taxonomic differentiation		
	Bacillus species	Cole et al. [1]
	Campylobacter species [a]	Wong et al. [7]
Epidemiology		
	Gonococcal infections	Korting and Abeck [4]
	Hemophilus ducreyi	Korting et al. [5]

[a] Strain characterization and grouping was with *C. jejuni* and *C. coli*. (No published studies of grouping *H. pylori* strains)

Based on standard methods for lectin typing of Bacillus sp. [1], we developed a system with 11 plant lectins for typing 50 *H. pylori* strains isolated in Sweden and France.

* This study was supported by grants from the Swedish Medical Council (A 6X 04723 to T. W.) and by grants from the Swedish Institute and Swedish Board for Technical Development to F. A. and T. W.

Department of Medical Microbiology, University of Lund, Sölvegatan 23, 223 62 Lund, Sweden

Helicobacter pylori, Gastritis and Peptic Ulcer
Edited by P. Malfertheiner and H. Ditschuneit

Materials and Methods

Bacteria

Most of the *H. pylori* strains were isolated at hospitals in Sweden and France. *H. pylori* strain NCTC 11637 was used as a reference strain. Bacterial strains were grown on blood agar with 5% horse erythrocytes and incubated under microaerophilic conditions at 37 °C for 3–5 days. Cells were harvested with a plastic loop and resuspended in 0.01 *M* potassium phosphate buffered saline, pH 7.2 (PBS) and washed once. Finally, bacterial cells were resuspended to approximately 10^{10} cells/ml in PBS.

Lectins

Highly purified lectins (Table 2) were kind gifts from B. Ersson, Biomedicum University of Uppsala, Sweden. They were dissolved in PBS containing 5 m*M* $CaCl_2$ to a final concentration of 200 µg/ml. Lectin solutions were stored frozen at −20 °C in working aliquots to avoid repetitive freezing and thawing.

Table 2. Lectins used to agglutinate *H. pylori* strains and their carbohydrate specificities

Lectin	Abbreviation	Specificity
Arachis hypogaea	PNA	D-Gal (1-3)-GalNAc
Canavalia ensiformis	Con A	D-Man〉 D-Glc〉 D-GlcNAc
Crotalaria juncea	CJA	D-Gal〉 D-Lac〉 D-Mel〉 L-Raf〉 D-GalNAc〉 D-Gal
Helix pomatia	HPA	D-GalNAC
Lens culinaris	LcH	D-Man
Lotus tetragonolobus	Lotus A	L-Fuc
Pisum sativum	PEA	D-Man
Triticum vulgaris	WGA	(D-GlcNAc), NeuNAc
Vicia ervilia	VEA	D-Man〉 D-Glc
Vicia faba	VFA	D-Man〉 D-Glc〉
Vicia villosa	VVA	D-GalNAc

Agglutination Assay

Binding of lectins to cells of various *H. pylori* strains was assayed in glass slides by mixing 20 µl of bacterial cell suspensions with an equal amount of a lectin solution. The two drops were mixed and agglutination reactions were read after 2 min. The agglutination reactions were scored from strong positive (3, strong agglutination reaction before 30 s; 2, strong agglutination reaction) to weakly positive (1, weak agglutination reaction) or negative (0, no agglutination reaction).

Human erythrocytes (resuspended in PBS to a final concentration of 2% v/v) and latex beads coated with various glycoproteins i.e., hog gastric mucin, bovine serum glycoprotein fraction VI, fetuin, hog submaxillary asialomucin, asialofetuin (prepared as in Ascencio et al., in preparation were used as positive controls for the lectin solutions) and BSA coated latex beads as negative controls.

Agglutination Inhibition Assays

To test whether the agglutination was inhibitable by specific glycoproteins and carbohydrates, lectin solutions (20 µl) were incubated at 20 °C for 1 h with an equal amount of the test substance, followed by addition of bacterial cell suspension (20 µl), and then the agglutination test was determined as described above.

Carbohydrates (Gal, NADGal, Fuc, NADGlc, NANA, Man, NGNA, and Glc) were dissolved at a final concentration of 0.1 *M* in PBS. Glycoproteins (hog gastric mucin, hog submaxillary asialomucin, submaxillary mucin, bovine serum glycoprotein fraction VI, orosomucoid, all purchased from Sigma Chemical, St Louis Mo.) and ganglioside type II were dissolved at a final concentration of 0.1% in PBS.

Treatment of Bacterial Cells

Cells of *H. pylori* strain 52 and 66 were treated with sodium periodate (0.075 *M*), sodium borohydride (0.075 *M*), sulfuric acid (0.025 *N*), and formalin (1% w/v) at 37 °C for 1 h. Bacterial cells were also heated at 60 °C for 1 h. Strains were also treated with proteinase K, trypsin, chymotrypsin, pepsin, and neuraminidase according to the specifications of the Worthington Enzyme Manual (Worthington Biochemical, Freehold, N.J. USA). After treatment, bacterial cells were washed three times with PBS and finally resuspended to 10^{10} cells/ml in PBS and used for the agglutination assays.

Results and Discussion

Preliminary studies showed that the method of Cole et al. [1] of heating cells before adding plant lectins (at 200 µg/ml) gave reproducible results for a small number of strains grown upon repeated subcultures under standard conditions. The distributions of lectin agglutination patterns are presented in Fig. 1. All 50 strains studied (except 6 strains, i.e., pattern 7) were agglutinated by one or two lectins in various combinations.

Two strains (52 and 66) were selected for further studies on the specificity of these lectin agglutinations (Tables 3 and 4). With our scoring system (from 0 to 3) for lectin-bacterial agglutinations we could demonstrate that specific sugars, glycoproteins, and mucins inhibited the reactions. While one or two monosaccharides were able to inhibit agglutination reactions with Lotus and Crotalaria

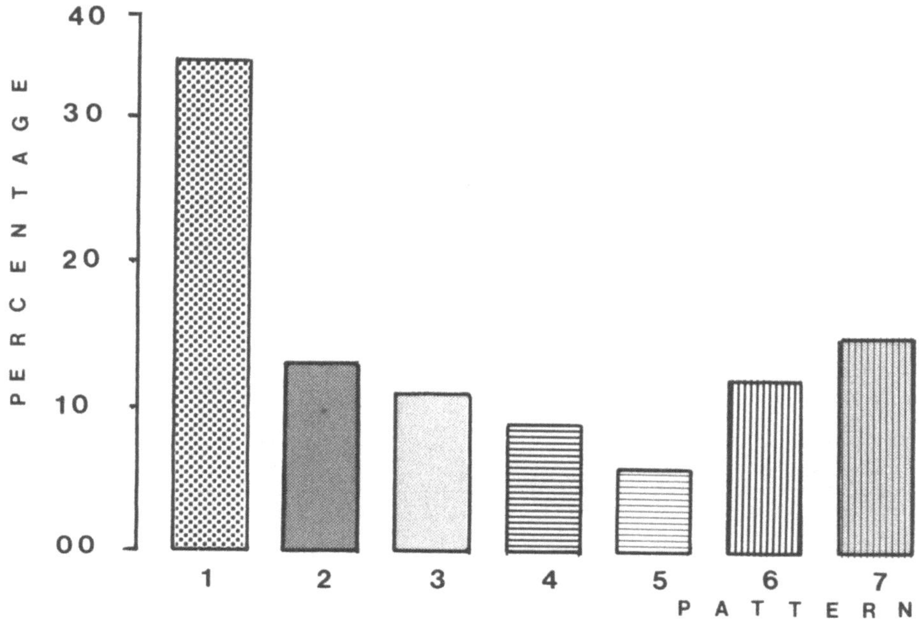

Fig. 1. Distribution of lectin-binding patterns among 50 *Helicobacter pylori* strains. *1*, Lotus A/CJA; *2*, CJA/WGA; *3*, lotus A/CJA/WGA; *4*, CJA; *5*, WGA; *6*, other lectin-binding patterns (includes the other lectins tested); *7*, none (no lectin binding)

Table 3. *Helicobacter pylori*-lectin binding inhibition

Inhibitor	*H. pylori* strain			
	52 (pattern 1)		66 (pattern 2)	
	Lotus	CJA	CJA	WGA
Control (PBS)	3	2	2	3
Glycoproteins				
Hog gastric mucin	1	0	0	0
Submaxillary asialomucin	3	0	0	0
Submaxillary mucin	3	1	1	3
Bovine serum glycoprotein	3	2	2	2
Fetuin	3	2	2	3
Orosomucoid	3	2	2	3
Gangliosides II	3	2	2	3
Carbohydrates				
Gal	3	0	1	3
GalNAc	3	0	0	3
Fuc	0	2	2	3
GlcNAc	3	2	2	3
NANA	3	2	2	3
Man	3	2	2	3
NGNA	3	2	2	3
Glc	3	2	2	3

PBS, phosphate buffered saline

Table 4. Effect of chemical and enzymatic cell treatments on *H. pylori*-lectin binding

Treatment		*H. pylori* strain			
		52 (pattern 1)		66 (pattern 2)	
		Lotus	CJA	CJA	WGA
Control (PBS)		3	3	3	3
$NaIO_4$	(0.075 *M*)	0	0	1	3
$Na(BH)_4$	(0.075 *M*)	3	2	2	3
H_2SO_4	(0.025 *N*)	3	2	3	3
Formalin	(1.0%)	3	3	3	3
60 °C	(1 h)	3	3	1	3
Neuraminidase		3	3	3	3
Trypsin		3	3	3	3
Pepsin		3	3	3	3
Proteinase K		3	3	3	3
Chymotrypsin		3	3	3	3

lectins, aggregation of *H. pylori* cells with wheat germ agglutinin was only inhibited with large molecular weight glycoconjugates. Furthermore, sodium periodate treatment of bacterial cells destroyed surface carbohydrate structures inhibiting agglutination reactions with each lectin except wheat germ agglutinin.

The reactions were shown to be resistant to protease treatment indicating that protease treatment could probably be a standard step in treating cells before typing. However, the reaction in the *H. pylori* strain 66 with the Crotalaria lectin was sensitive to heating of cells at 60 °C for 1 h. Suggesting that standard heat treatment in *H. pylori* typing should be avoided if specific protein-associated carbohydrate structures are important for the typing.

In summary, the results presented in this preliminary report seem promising for developing a lectin typing system of *H. pylori*. The simple procedure for lectin typing, compared with other possible typing methods, encouraged us to develop this system further. Studies by Korting et al. [5] showed that, by using 14 plant lectins, 29 lectin groups could be defined for epidemiological studies. These studies as well as studies by other investigators have generally shown that lectin agglutination types are stable and reproducible.

References

1. Cole HB, Ezzell JW, Keller KF, Doyle R (1984) Differentiation of Bacillus antracis and other Bacillus species by lectins. J Clin Microbiol 19:48–53
2. Delterne M, Burette A, Glupczinsky Y, Dekoster E, de Prez C, Jonas C (1987) Rapid identification of Campylobacter pylori in gastric biopsies. In: Menge H, Gregor M, Tytgat GNJ, Marshall BJ (eds) Campylobacter pylori. Springer, Berlin Heidelberg New York, pp 135–144
3. Kist M, Apel I, Jacobs E (1988) Protein antigens of Campylobacter pylori: the problem of species specificity. In: Menge H, Gregor M, Tytgat GNJ, Marshal BJ (eds) Campylobacter pylori. Springer, Berlin Heidelberg New York, pp 19–28

4. Korting HC, Abeck D (1988) Lektin-Typisierung als leistungsfähiges epidemiologisches Markersystem für Neisseria gonorrhoeae-Infectionen. Zentralbl Bakteriol Microbiol Hyg [A] 269:506–512
5. Korting HC, Abeck D, Johnson AP, Ballard RC, Taylor-Robinson D, Braun-Falco O (1988) Lectin typing of Haemophilus ducreyi. Eur J Clin Microbiol Infect Dis 7:678–670
6. Pistole TG (1981) Interaction of bacteria and fungi with lectins and lectin-like substances. Annu Rev Microbiol 35:85–112
7. Wong KH, Skelton SK, Feeley JC (1986) Strain characterization and grouping of Campylobacter jejuni and Campylobacter coli by interaction with lectins. J Clin Microbiol 23:407–410

Association of *Helicobacter pylori* with Epithelial Cells

J. L. Fauchère and M. J. Blaser

Introduction

The presence of *Helicobacter pylori* on gastric mucosa has been associated with type B gastritis and peptic ulcer disease. Despite its uncertain role in pathogenesis, gastric mucosa colonization by *H. pylori* is at least a good indicator for gastroduodenal inflammatory diseases [1, 14, 19].

The ecological niche of *H. pylori* seems to be restricted to the human stomach and duodenum, where it persists at least for several months and probably for years [1]. Histological studies show *H. pylori* to be closely associated with the luminal side of mucosal cells, particularly in the antrum and over gastric metaplasia of the duodenum where peptic ulceration also is preferentially located [1, 18]. The motility, shape, and strong urease activity of *H. pylori* allow it to reach the mucosa through the viscous mucus despite the high acidity of the gastric juice [8], but the ultimate interactions between *H. pylori* and mucosal cells have been poorly understood as yet. The high specificity of *H. pylori* for human gastric mucosa may be due to microecological factors or may result from specific binding of the organism to a particular cellular receptor. It is thus of interest to study the colonization factors of *H. pylori* and their cellular receptors.

In Vitro Assessment of *Helicobacter pylori* Association with Encaryotic Cells

Microscopic Method

To properly study and quantitate the association of *H. pylori* with epithelial cells, standardized in vitro assays are needed. The most common method entails mixing bacteria and cultivated cells and then enumerating by microscopy the bacteria remaining associated with the cells after washing. Using this method, Neman-Simha et al. [15, 16], Evans et al. [4], and Fauchère et al. [5], demonstrated that all *H. pylori* strains studied were adherent to Hep-2, INT 407, Y1, and HeLa cell lines. In vitro adherence of *H. pylori* on epithelial cells appears to be independent of the cell line. This method, allowing direct observation of the adherence

Laboratoire de Microbiologie, Faculté de Médecine Necker-Enfants Malades, 156, rue de Vaugirard, 75730 PARIS CEDEX 15, and Veterans Administration Medical Center, Denver, Colorado 80220, USA

Helicobacter pylori, Gastritis and Peptic Ulcer
Edited by P. Malfertheiner and H. Ditschuneit

phenomenon, is qualitative but may be considered as the "gold standard" for adherence of *H. pylori*. However, with this method, quantitation is difficult. We have demonstrated that the adherence expressed as the mean number of bacteria binding to each cell is dependent on the initial bacteria to cell ratio [5] (Fig. 1a, b).

a

b

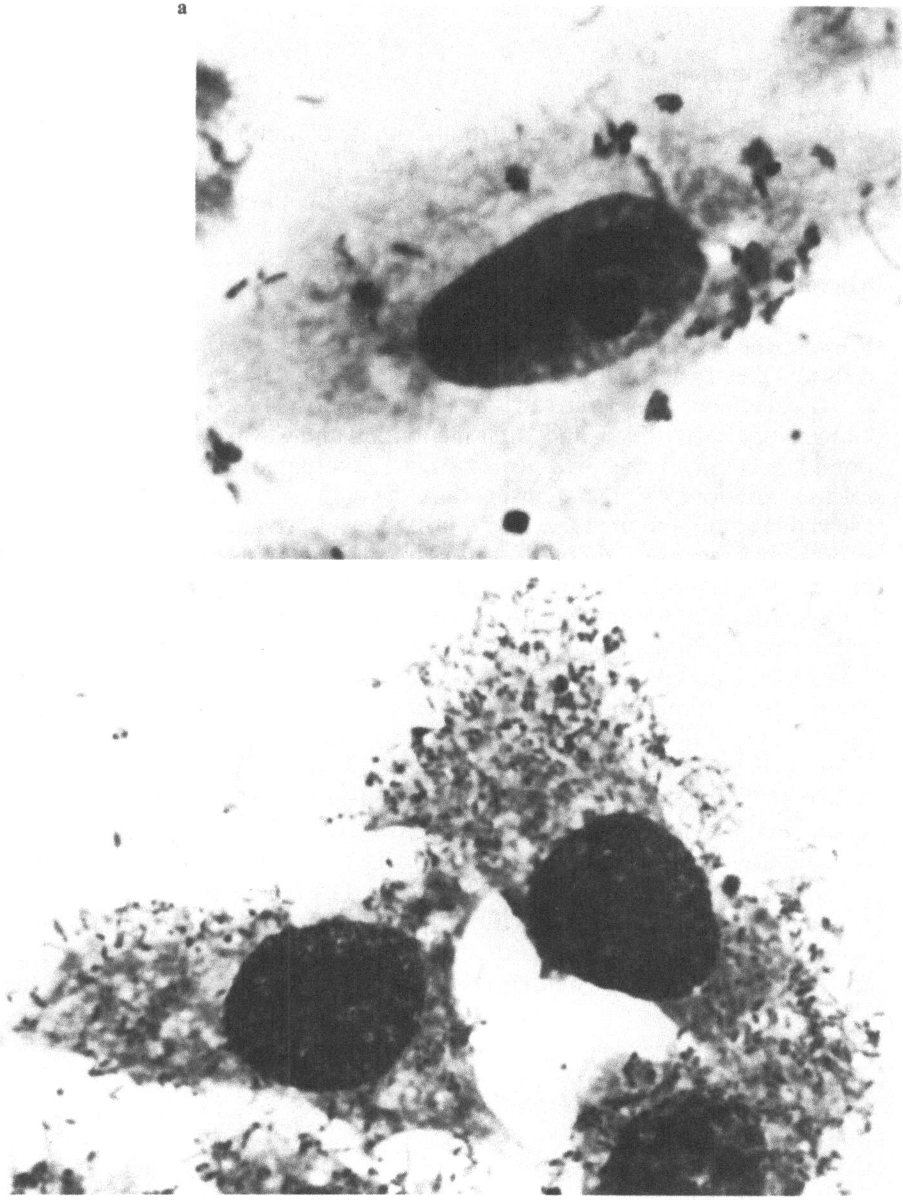

Fig. 1 a, b. Adherence of *H. pylori* to HeLa cells. **a** Initial ratio of 10^4 bacteria per cell. **b** Initial ratio of 10^6 bacteria per cell. ×1000

Moreover, this laborious method is not adapted to the screening of numerous strains and does not allow the study of adherence of subcellular components. It may be considered as a reasonable qualitative reference method for assessing binding of *H. pylori* cells to encaryotic cells, but less time-consuming and more quantitative methods needed to be developed.

Hemagglutination

H. pylori agglutin red blood cells from a variety of animals and hemagglutination assays based on methods similar to those used for *E. coli* have been developed as alternative methods for *H. pylori* colonization factors [2, 10].

Microtiter Enzyme Assays

We described two microtiter enzyme assays to assess *H. pylori* adherence to HeLa cells [5]. The urease method detects adherent *H. pylori* using their strong urease activity. HeLa cells are cultivated on 96-well microplates, a standard inoculum of viable *H. pylori* is added, and, after incubation and washing, urease activity of the bound bacteria in the presence of urea is evaluated by determination of the released ammonia. For standardization, the initial inoculum is evaluated by both the viable count and urease activity, providing a standard curve which allows the expression of urease activity in terms of number of viable bacteria. The adherence is expressed as the percentage of the initial inoculum remaining in the well after the nonspecific adherence to a cell-free well had been subtracted. This expression of adherence allows the results to be independent of the initial bacteria to cell ratio. The ELISA method is based on the imunological detection of bacterial material bound onto cellular ligands. Microtiter-plate wells are coated with a binding ligand consisting of either a HeLa cell fraction or bovine serum albumin (BSA) used as a nonspecific-binding ligand. Residual polystyrene binding sites are blocked, the bacterial ligand is added, and after incubation and washing the bound material is revealed by a classic ELISA using an antiwhole-cell serum. The specific adherence optical density (OD) is calculated by subtracting the OD corresponding to the nonspecific binding to BSA from the OD corresponding to the binding on HeLa. This specific adherence OD can be transformed into μg of adhering antigen using standard curves [5]. The urease method and the ELISA method produce results well correlated with the microscopic reference method (Table 1).

Colonization Factors of *Helicobacter pylori*

Colonization factors (CF) are defined as bacterial components or subcellular fractions, that facilitate binding onto encaryotic target cells. To identify CF, a genetic approach may be the most efficient, but until recently any genetic tools to generate nonadherent isogenic variants from an adherent strain did not exist, nor

Table 1. Adherence of four *H. pylori* and control strains onto HeLa cells, assessed using three different methods

Strain	*Microscopic method* (adhesion score)[a]	*Urease method* (% bound bacteria)[b]	*ELISA method* (specific OD × 100)
H. pylori 87–338	2.0±0.1	10.5±2.4	43±3
H. pylori 87–456	3.1±0.3	8.15±0.19	36±5
H. pylori 87–263	2.4±0.4	4.5±0.7	56±3
H. pylori 84–182	2.8±0.0	8.6±2.8	39±3
H. fetus 84–107	0.0	ND[b]	0
Providencia sp.[c]	0.3±0.2	0.0(0)	ND

[a] Calculated on the basis of number of bound bacteria per cell
[b] Proportion of the initial inoculum binding onto the cells
[c] Nonadherent urease-producing control strain
OD, optical density

did any model of pathogenesis that was adapted for screening of variants. Thus, previous work was based on the phenotypic identification of CF using wild type adherent strains.

Fibrillar Hemagglutinin

Evans et al. identified a fibrillar hemagglutinin using inhibition of *H. pylori* hemagglutination activity by treating red blood cells with various enzymes or using various putative inhibitors of hemagglutination. This hemagglutinin is extracted from the bacterial cell by N-octylglucose and appears to be an external protein as indicated by immunogold staining [2, 3]. The fibrillar hemagglutinin preferentially binds to the Neu-Ac [2, 3] galactose isomer of Neu-Ac-lactose. This kind of receptor is widespread and not restricted to the gastric epithelial cells which contrasts with the specificity of *H. pylori* for gastric mucosa. Moreover, a glycerolipid unrelated to Neu-Ac-lactose has been found to be another adherence receptor for *H. pylori* [12]. Lelwala Guruge et al. defined three different patterns of *H. pylori* hemagglutinin, according to the type of erythrocytes agglutinated. Their pattern 1 is likely due to the neuraminyl-lactose fibrillar hemagglutinin whereas the other patterns to be of different specificity [11].

Superficial Adhering Material

Using the ELISA adherence assay, we tried to identify the adherent bacterial material extractable from *H. pylori* and to determine the specificity of binding [6]. The adherent material initially contained in untreated bacteria cultivated on blood agar was largely recovered in the saline or water first extracts, whereas further extracts including that with glycine (pH 2.2) contained proportionately less. Thus, some of the adherent material seems to be surface exposed and easily extractable.

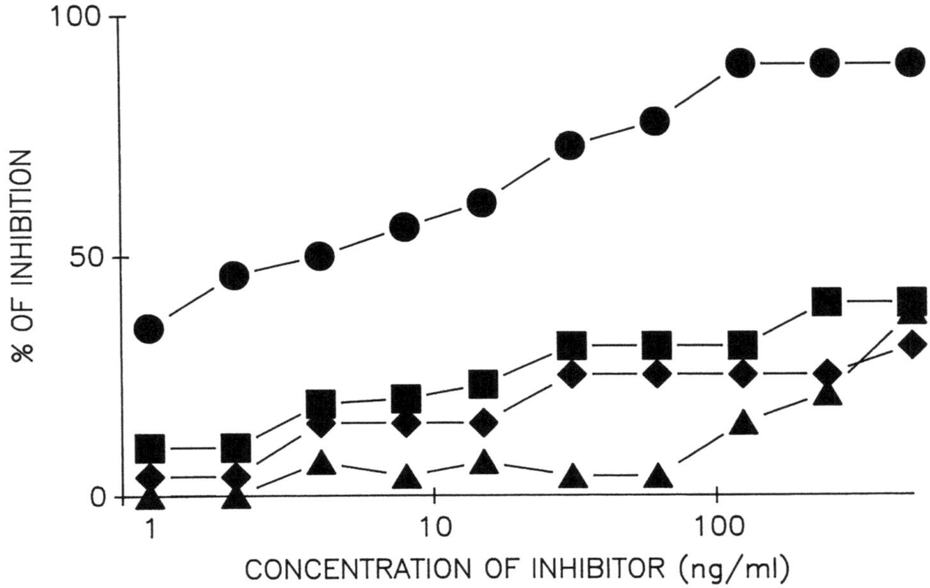

Fig. 2. Inhibition of binding of the superficial adhesin of *H. pylori* by various components: ●, HeLa membranes; ▲, calf fetal fetuin; ◆, N-acetyl-neuraminic acid; ■, bovine serum albumin

This material was called "superficial adhering material" (SAM). SAM bound more on a HeLa cell membrane-enriched fraction (100% binding) as the binding ligand than on cytosolic protein (33%), polystyrene (30%), human serum albumin (25%), calf fetal fetuin (10%), or BSA (5%). Even at low concentrations, HeLa cell membranes inhibit SAM binding, while calf fetal fetuin and N-acetyl neuraminic acid are not better inhibitors than was BSA (Fig. 2). Pretreatment of the HeLa-cell membranes with *C. perfringens* neuraminidase does not prevent the binding of SAM. These data indicate that the SAM of *H. pylori* binds on a membrane receptor different from the N-acetyl-neu-lactose hemagglutinin receptor.

The first saline or water extracts containing a large amount of SAM were used to identify this adhering component. The first extracts were fractionated by gel exclusion chromatography and each fraction was assesses for protein concentration, proportion of SAM as measured by ELISA, and urease activity (Fig. 3). Most of the bacterial antigenic material was found between fractions 14 and 40. The adherence activity was found between fractions 15 and 30 and was maximum in fractions 17 and 18. Urease activity was concentrated in fraction 18. We then analysed by immunoblot using an antiwhole-cell serum the fractions exhibiting the maximum adherence and compared them with those lacking adherence (Fig. 4). A major antigen of approximately 60 kDa was present in the fraction exhibiting the maximum adherence activity and was absent from the others. This antigen was also revealed by an anti-urease serum. Thus, SAM appears to copurify with urease.

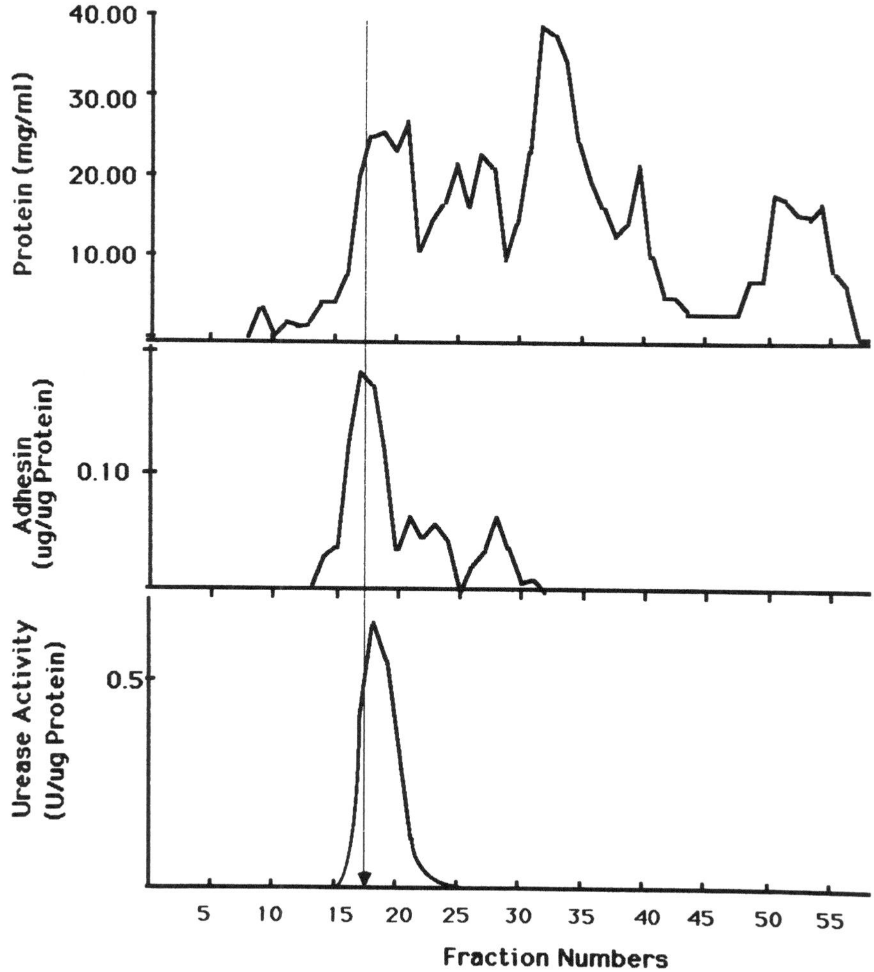

Fig. 3. Protein concentration, adherence, and urease activity of gel filtration fractions of water extract from *H. pylori*. Protein concentration determined using the bicinchoninic acid method. Adherence determined by ELISA; urease activity determined by NADH-dependent coupled enzymes assay

Conclusions

After reaching its natural ecological niche beneath the gastric mucus layer, *H. pylori* appears to bind to the gastric epithelial cells. This phenomenon has been studied in vitro, with models assessing interaction between viable *H. pylori* and cultivated cells and further by hemagglutination and by microtiter enzyme assays. With all these methods, all studied strains have been found to be strongly adherent

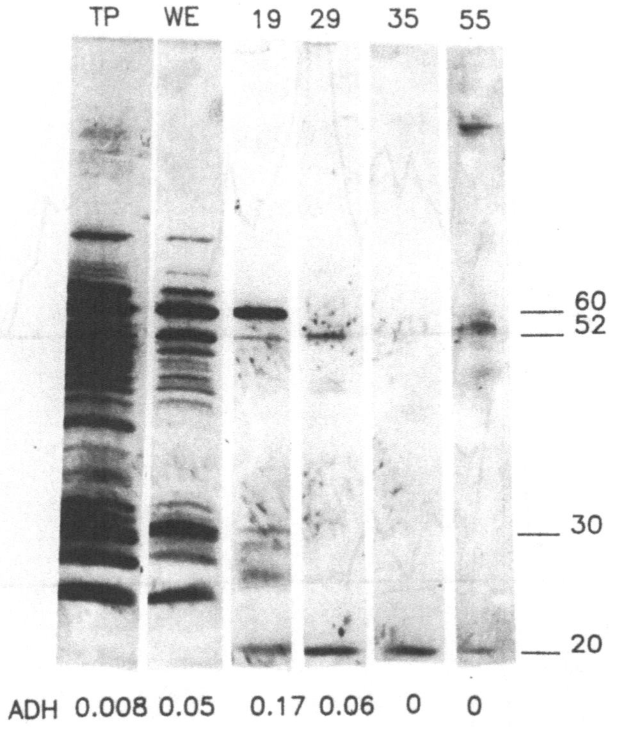

Fig. 4. Western blot of gel filtration fractions of water extract from *H. pylori*, as revealed by antiwhole bacterial cells serum. *TP*, total protein from bacterial cells; *WE*, water extract; *19–55*, fraction numbers; *ADH*, adherence activity (µg adhesin per µg bacterial antigen)

to epithelial cells. Two types of colonization factors have been described: external fibrillar hemagglutinins binding to N-acetyl-neuraminyl lactose and superficial adhering components binding to another unidentified membrane receptor. This superficial adhering component copurifies with urease. These two CF may be relevant in *H. pylori* colonization of the human stomach.

It would be of great interest to determine the relevance of *H. pylori*-gastric cell interactions in the pathogenesis of gastritis. For this purpose, both in vivo and in vitro models are needed. These models should allow a simple and accurate quantitation of bacteria cell interactions, they should be flexible for extensive screening and should allow testing of target cells from various origins, particularly those involved in natural infections. The microtiter enzyme assays allow the study of interactions between the bacterial component of cellular subunits and should be useful for identification of the CF and their cellular receptors.

References

1. Blaser MJ (1990) Gastric. Helicobacter pylori and the pathogenesis of gastroduodenal inflammation. J Infect Dis (in press)
2. Evans DG, Doyle JE Jr, Moulds JJ, Graham DY (1988) N-acetylneuraminyllactose-binding fibrillar hemagglutinin of Campylobacter pylori: a putative colonization factor antigen. Infect Immun 56:2896–2906

3. Evans DJ, Evans EG, Smith KE, Graham DY (1989) Serum antibody responses to the N-acetylneuraminyllactose-binding hemagglutinin of Campylobacter pylori. Infect Immun 57:664–667
4. Evans DG, Evans DJ Jr, Graham DY (1989) Receptor-mediated adherence of Campylobacter pylori to mouse Y-1 adrenal cell monolayers. Infect Immun 57:2272–2278
5. Fauchère JL, Blaser MJ (1989) Simplified determination of Camylobacter pylori adherence to epithelial cells by microtiter enzyme assays (Abstr) 89th Annual Meeting of the American Society for Microbiology, 14–18 May, New Orleans
6. Fauchère JL, Zhiheng P, Blaser MJ (1989) Adherence of superficial components of Campylobacter pylori to epithelial cell membranes. Klin Wochenschr 67 [Suppl 18]:19
7. Goodwin CS, Armstrong JA, Marshall BJ (1986) Campylobacter pyloridis, gastritis and peptic ulceration. J Clin Pathol 39:353–365
8. Hazell SL, Lee A, Brady L, Hennessy W (1986) Campylobacter pyloridis and gastritis: association with intercellular spaces and adaption to an environment of mucus as important factors in colonization of the gastric epithelium. J Infect Dis 153:658–663
9. Hemalatha SG, Drums B, Shuman PM (1989) Adherence of Campylobacter pylori to human gastric epithelial cells in vitro. Klin Wochenschr 67 [Suppl 18]:40
10. Huang J, Smyth CJ, Kennedy NP, Arbuthnott JP, Napolean Keeling PW (1988) Haemagglutinating activity of Campylobacter pylori. FEMS Microbiol Lett 56:109–112
11. Lelwala Guruge J, Ljungh A, Megraud F, Wei S, Wadströn T (1989) Hemagglutinin profiles of Campylobacter pylori strains isolated in Sweden and France. Klin Wochenschr 62 [Suppl 18]:40
12. Lingwood CA, Pellizzari A, Law H, Sherman P, Drumm B (1989) Gastric glycerolipid as a receptor from Campylobacter pylori. Lancet 1:238–240
13. Marshall BJ, McGechie DB, Rogers PA, Glancy RJ (1985) Pyloric Campylobacter infection and gastroduodenal disease. Med J Aust 142:439–444
14. McNulty CAM, Watson DM (1984) Spiral bacteria of the gastric antrum. Lancet 1:1068–1069
15. Neman-Simha V, Brugmann D, Megraud F (1989) Interaction of Campylobacter pylori with epithelial cells: role of the urease. Klin Wochenschr 67 [Suppl 18]:48
16. Neman-Simha V, Megraud F (1988). In vitro model for Campylobacter pylori adherence properties. Infect Immun 56:3329–3333
17. Rathbone BJ, Wyatt JI, Heatley RV (1986) Campylobacter pylori – a new factor in peptic ulcer disease. Gut 27:635–641
18. Van Spreeuwel JP, Duursma GC, Meijer CJ, Bax R, Rosekrans PC, et al. (1985) Campylobacter colitis: histological, immunohistochemical, and ultrastructural findings. Gut 26:945–951
19. Warren JR, Marshall BJ (1983) Unidentified curved bacilli on the gastric epithelium in active chronic gastritis. Lancet 1:1273–1275

Influence of *Helicobacter pylori* on Intragastric Environment: Bile Acids and Biliary Lipids

R. RAEDSCH, S. POHL*, A. STIEHL, and J. PLACHKY

Introduction

In recent years *Helicobacter pylori* (HP) was identified as an important factor in the pathogenesis of gastritis and ulcer disease. A metabolic activity of HP may influence the intragastric environment and represent a link to the pathogenicity of these bacteria. The intragastric milieu is determined by aggresive factors like acid formation, pepsinogen, bile acids, and lysolecithin. In the present study we investigated the interactions of HP with bile acids and biliary lipids.

Bile Acids

Bile acids are synthesized in the liver cell and excreted with bile into the duodenum. Bile acids reach the stomach via duodenogastric reflux. Also in physiological conditions reflex episodes are found. Quantification of duodenogastric reflux may be performed using scintigraphic methods.

Some bile acids are strong detergents. Bile acids are amphoteric molecules and consist of the steroid nucleus, one carboxyl group, and one to three hydroxyl groups in 3a, 7a, and 12a position. The physicochemical properties of the bile acids are greatly determined by the position of these hydroxyl groups and additional conjugation of the carboxyl group with glycine or taurine (amidation; Fig. 1).

In several studies a toxic effect of bile acids on gastric mucosa could be demonstrated with disruption of gastric mucosal barrier [1–3]. The detergent effects of bile acids decrease with increasing numbers of hydroxyl groups. Thus, mono- and dihydroxylated bile acids litho-, deoxy-, and chenodeoxycholate are more toxic than cholic acid. Amidation of bile acids decreases toxicity [4]. The dihydroxylated (3a, 7β) ursodeoxycholic acid differs greatly from the a-hydroxylated bile acids and exerts no detergent effect on gastric mucosa or liver cells [4, 5].

We investigated the concentrations of individual bile acids in gastric juice of 90 patients with gastritis [6]. Gastric juice was aspirated during routine endoscopy.

University of Heidelberg, Dept. of Medicine and Dept. of Microbiology*,
Bergheimerstr. 58, 6900 Heidelberg, FRG

Helicobacter pylori, Gastritis and Peptic Ulcer
Edited by P. Malfertheiner and H. Ditschuneit

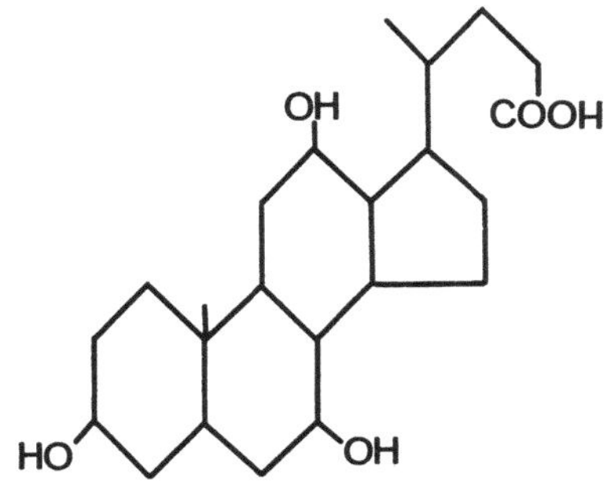

Fig. 1. Chemical structure of the bile acid molecule (cholic acid)

Results

Primary bile acids cheno- and cholic acid were the predominant bile acids in both HP-positive and HP-negative patient groups. In gastric juice of HP-positive patients higher amounts of toxic secondary bile acids, litho- and deoxycholic acid, and of not glycine or taurine conjugated bile acids were found ($P < 0.001$). As bile acids are almost exclusively excreted by the liver cell as glycine or taurine conjugates, deamidation of bile acids must have occurred within the gastric juice. As high quantities of not amidated bile acids only have been found in HP-positive patients a metabolic activity of HP to deamidate bile acids was suggested. A possible hydroxylase activity of HP remains to be established.

Total bile acid concentrations in gastric juice were significantly ($P < 0.001$) higher in HP-negative patients (0.52 ± 0.2 mmol) compared with HP-positive patients (0.25 ± 0.06 mmol). These findings suggested that high duodenogastric reflux rates diminish the growth of HP in the antrum of the stomach. To evaluate the effects of individual bile acids in physiological concentration ranges were incubated HP in bile acid containing agar and determined minimum inhibition concentrations [7]. An effective inhibition of HP was demonstrated by bile acids in concentration ranges found in gastric juice. A correlation of minimum inhibitory concentrations of individual bile acids for HP could be shown with polarity and detergent activity of the bile acid molecule.

The inhibitory effect of bile acids on the growth of HP may explain the decreased HP colonization rates in patients with high duodenogastric reflux rates as also found in patients after gastric surgery [8]. Recolonization of HP in gastric remnant was reported after biliary diversion in peptid ulcer surgery [9]. The inhibition of HP by refluxed bile acids represents a local effect mainly restricted to the distal part of the stomach; HP colonization of corpus and fundic area are not supposed to be affected.

Biliary Lipids

The main constituents of biliary lipids are cholesterol and phospholipids. Biliary lipids reach the stomach by duodenogastric reflux. Biliary lecithin may be metabolized to lysolecithin by phospholipase A2. Lysolecithin represents a well-known cytotoxin. Its strong detergent activity disrupts the gastric mucosal barrier and membranes of gastric epithelium [10].

We investigated the concentrations of phospholipid metabolites and lysolecithin in gastric juice of 90 patients with gastritis undergoing routine endoscopy [11]. We found substantial concentrations of lysolecithin in gastric juice of HP-positive patients. Concentrations of lysolecithin were significantly higher in HP-positive patients with 0.11 ± 0.02 mmol than in HP-negative patients with 0.07 ± 0.02 mmol. Therefore, we investigated a possible phospholipase A2 activity of HP, which may be responsible for increased formation of lysolecithin in HP-positive patient groups. HP strains obtained from 27 patients with chronic gastritis were kept by subcultivation and a phospholipase A2 activity was examined. Colonies of HP were scraped off agar plates and suspended in buffered saline. Quantification of HP in suspension was performed using the turbidity method of McFarland. After sonication phospholipase A2 activity was assayed using phosphatidyl-choline as substrate. Standard curves were prepared from phospholipase A2 derived from Crotalus durissus (Boehringer Mannheim). All measurements were carried out in triplicate.

Results

Suspension of HP colonies in saline yielded up to 650×10^8 bacteria/l. In all HP strains tested phospholipase A2 activity could be demonstrated. Quantitatively the activity of phospholipase A2 was 224 ± 29 U/l per 100×10^8 bacteria (Fig. 2).

These results demonstrate that HP produces substantial amounts of phospholipase A2, which may metabolize lysolecithin from biliary lecithin refluxed into the stomach. Lipolytic activities of HP were described recently by another author [12]. This metabolic activity of HP may explain the increased concentrations of cytotoxic lysolecithin found in gastric juice of HP-positive patients.

Summary and Conclusions

The pathomechanisms by which HP contributes to gastritis and ulcer disease include the production of urease, proteases, adhesive factors, cytotoxins, and phospholipase. We could demonstrate interactions of HP with bile acids and biliary lipids. Metabolic activities of HP may cause increased formation of detergent bile acid metabolites in the stomach. HP strains produce phospholipase A2 in quantities which may metabolize considerable amounts of lysolecithin from biliary lecithin refluxed into the stomach. This metabolic activity of PH leads to increased concentrations of the cytotoxic lysolecithin in the stomach of HP-positive patients and may represent a link to the pathogenicity of HP.

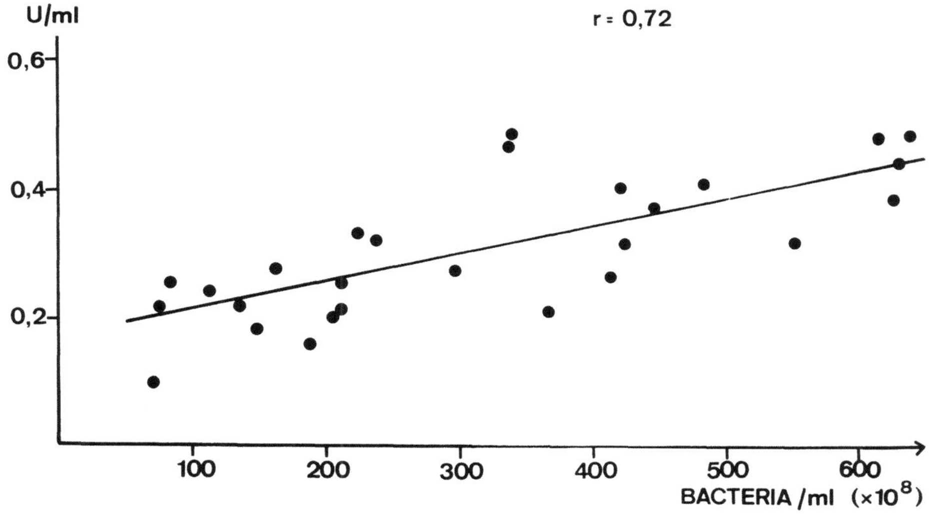

Fig. 2. Phospholipase A2 activity of HP strains

References

1. Duane WC, Wiegand DM (1980) Mechanisms by which bile salt disrupts the gastric mucosal barrier in the dog. J Clin Invest 66:1044–1049
2. Ritchie WP, Felger TS (1981) Differing ulcerogenic potential of dihydroxy and trihydroxy bile acids in canine gastric mucosa. Surgery 89:342–347
3. Duane WC, Wiegand DM, Sievert CE (1982) Bile acid and bile salt disrupt gastric mucosal barrier in the dog by different mechanisms. Am J Physiol 242:G95–99
4. Lillemoe KD, Kidder GW, Harmon JW, Gadacz TR, Johnson LF, Bunte RM, Hofmann AF (1983) Tauroursodeoxycholic acid is less damaging than taurochenodeoxycholic acid to the gastric and esophageal mucosa. Dig Dis Sci 28:359–364
5. Galle P, Theilmann L, Kohl B, Raedsch R, Stiehl A (1989) Ursodeoxycholate reduces toxicity of different bile acids in human primary hepatocytes cultures. Gastroenterology 96:A599
6. Raedsch R, Stiehl A, Waldherr R, Pasch B, Plachky J, Kommerell B (1988) Intragastrale Konzentrationen individueller Gallensäuren und Lysolezithin bei Gastritis-Patienten mit und ohne Campylobacter pylori Besiedelung des Magens. Inn Med 15:40–45
7. Raedsch R, Pohl S, Plachky J, Stiehl A, Kommerell B (1989) The growth of Campylobacter pylori is inhibited by intragastric bile acids. In: Mégraud F, Lamouliatte H (eds) Gastroduodenal pathology and Campylobacter pylori. Excerpta Medica, Amsterdam, pp 409–412
8. O'Connor HJ, Dixon MF, Wyatt JI (1986) Effect of duodenal ulcer surgery and enterogastric reflux on Campylobacter pylori. Lancet 2:1178–1181
9. O'Connor HJ, Newbold MK, Alexander-Williams J, Thompson H, Drumm J, Donovan IA (1989) Effect of Roux-en-Y biliary diversion on Campylobacter pylori. Gastroenterology 97:958–964
10. Orchard R, Reynolds K, Fox B, Andrews R, Parkins RA, Johnson AG (1977) Effect of lysolecithin on gastric mucosal structure and potential difference. Gut 18:457–461
11. Raedsch R, Stiehl A, Waldherr R, Pasch B, Kommerell B (1988) Elevated concentrations of not amidated and secondary bile acids and lysolecithin in gastric juice in the presence of Campylobacter pylori. Gastroenterology 94:A267
12. Slomiany BL, Nishikawa H, Piotrowski J, Okazaki K, Slomiany A (1989) Lipolytic activity of Campylobacter pylori: effect of sofalcone. Digestion 43:33–40

Animal Models of *Helicobacter pylori* Gastritis

L. ENGSTRAND

Introduction

A large number of investigators have agreed that *H. pylori* is an emerging candidate as an agent for the genesis of gastritis and peptic ulcers. There have been rapid advances in our knowledge of *H. pylori* infection, but as with many human diseases, progress in the understanding of *H. pylori* gastritis would be increased if there were available for experimental use an animal model of the infection. For future research high priority should be given to the development of a challenge model in an animal that is raised conventionally. The development of such a model may help to fulfil Koch's postulate and to clarify the pathogenetic mechanisms as well as therapeutic regimens.

In order to assess the suitability of an animal as model regarding *H. pylori* infection some factors need to be taken into consideration: Are there similarities to the human gastrointestinal system (anatomical, histological, physiological, hormonal, acids, etc.)? Is naturally occurring gastritis or ulcer present in the animal? Does the animal harbor gastric campylobacter-like organisms in the stomach and how do the normal gastric flora appear? Cost, size, age, and details of animal breeding are other factors that might determine whether the animal is attractive or not and if the model will be convenient.

Gastric ulcer and gastritis are well known in some animals like cats, dogs, and pigs. Endoscopic investigations of the esophagus, stomach, and duodenum in dogs and even cats with symptoms are routine in some veterinary hospitals. Ulceration in pigs causes economic consequences to animal breeders worldwide but because of the difficulties of diagnosis, the early stages of ulceration are usually undetected. In rhesus monkeys chronic gastritis has been observed in combination with *H. pylori* and in ferrets gastritis has been observed in combination with gastric spiral bacteria but the association with gastritis is inconsistent.

Naturally Occurring Ulcers/Gastritis in Dogs and Pigs

In a recent paper van der Gaag [1] described the histopathological appearance of endoscopic gastric biopsy specimens in dogs with or without symptoms. Of the

Dept. of Clinical Bacteriology, Uppsala University Hospital, Sweden

Helicobacter pylori, Gastritis and Peptic Ulcer
Edited by P. Malfertheiner and H. Ditschuneit

dogs included in the study 35% showed slight to severe gastritis but gastric erosions and ulcers were seldom found. The classification of dog gastritis is nearly identical to the conventional human classification of gastritis.

In pigs there is a difference between ulcers occurring in the glandular and the nonglandular area of the stomach [2]. The nonglandular area surrounds the cardia and is frequently the site of ulceration. The peptic ulcer of the glandular area is primarily seen in the fundic region but is on rare occasions accompanied by ulceration of the glandless esophageal area. These ulcers are often found at slaughter. The etiology of the esophageal ulceration still needs clarification but some causative agents like infections, psychosomatic and other stress factors, gastric acidity, dietary factors, and food preservatives have been established [3].

Gastritis of the glandular region in pigs is most pronounced on the greater curvature. The pathogenesis is poorly understood. Physiologic hyperemia, hemorrhage, infarction, and erosion may be observed but are nonspecific. The lesion merely reflects an enteric disease or septicemic process [4].

Naturally Occurring *H. pylori* Infection in Animals

When *H. pylori* has been observed in animals it was been in association with gastritis. *Macaca mulatta*, *M. nemestrina* [5], and probably baboons seem to be the only animals so far that naturally harbor *H. pylori*. Baskerville and Newell [6] found histologically confirmed gastritis in combination with *H. pylori* infection. Antibodies to *H. pylori* were detected in serum by ELISA. Colonization with *H. pylori* in monkeys results in lesions in the antrum identical to *H. pylori*-associated gastritis in man.

Experimental Animal Models of *H. pylori* Infection

Oral challenge with *H. pylori* of numerous small laboratory animals like mice, rats, rabbits, guinea pigs, and germ-free rats have, so far, all been unsuccessful [7]. The gastrointestinal systems of these animals are all different from that of humans and may explain why these animals seem to be resistant to the infection.

However, attempts to infect small laboratory animals will continue since there is the need for a convenient animal model that can easily be reproduced by other investigators. To this end, oral challenge of only a few animals has been successful, i.e., gnotobiotic and barrier-born pigs and recently even germ-free dogs (Krakowka 1989, personal communication).

H. pylori and the Gnotobiotic Pig

Krakowka et al. [8] and Lambert et al. [9] describe the successful colonization of *H. pylori* in gnotobiotic piglets. After pretreatment with H_2-blockers, *H. pylori* was given orally and the pigs were killed at intervals over 4 weeks. Bacteria colonized the gastric and proximal duodenal mucosa, development of gastritis was observed (infiltration of mononuclear leukocytes), and *H. pylori* antibodies were found in serum samples collected at necropsy. The use of this model may be limited because of the short observation period permitted, difficulties with endoscopic examination, treatment regimens, and the method of animal breeding (the pigs must be kept in separated isolation units).

H. pylori and the Barrier-Born Pig

The barrier-born pig (specific pathogen free) is of a race originating in Sweden, is not delivered by hysterectomy, and is kept isolated from common infectious agents in pigs. These pigs could be used to assess the therapeutic possibilities once a *H. pylori* infection is established.

Gastroscopy is technically easy to perform in anesthetized pigs and long-term observation (at least 12 weeks) can be done without difficulty.

The experimental design is comparable with a human inoculation experiment [10] and with the gnotobiotic piglet experiment [8, 9]. Endoscopy with biopsy was repeated weekly for the first 3 weeks and then every 2nd week. Serum samples were taken at the same intervals. After 10 weeks four infected pigs were treated with a combination of bismuth, metronidazole, and ampicillin for 2 weeks.

H. pylori could be detected by growth as shown in Fig. 1. Once infected, the pigs remained *H. pylori* positive during the observation period and the *H. pylori* negative pigs remained negative. After therapy the bacteria disappeared from the gastric mucosa.

A stained cryostat section with *H. pylori* is shown in Fig. 2. Mononuclear cells were present in the lamina propria. In the *H. pylori* positive pigs the intensity of

Pig no	Week 0	1	2	3	4	6	8	10	12
1	-	-	+	+	+	+	+	+	-
2	-	-	+	+	+	+	+	+	-
3	-	-	-	+	+	+	+	+	-
4	-	-	+	+	+	+	+	+	-
5	-	-	-	-	-	-	-		
6	-	-	-	-	-	-	-		
7	-	+	+	+	+	+	+	+	+
8	-	-	-	-	-	-	-		
9	-	-	-	-	-	-	-		

Treatment

Fig. 1. Detection of *H. pylori* from gastric biopsy specimens. Growth of *H. pylori* in antral biopsy specimens from pigs inoculated with *H. pylori* and from uninfected controls. Treatment was performed in all pigs in litter one after 10 weeks observation

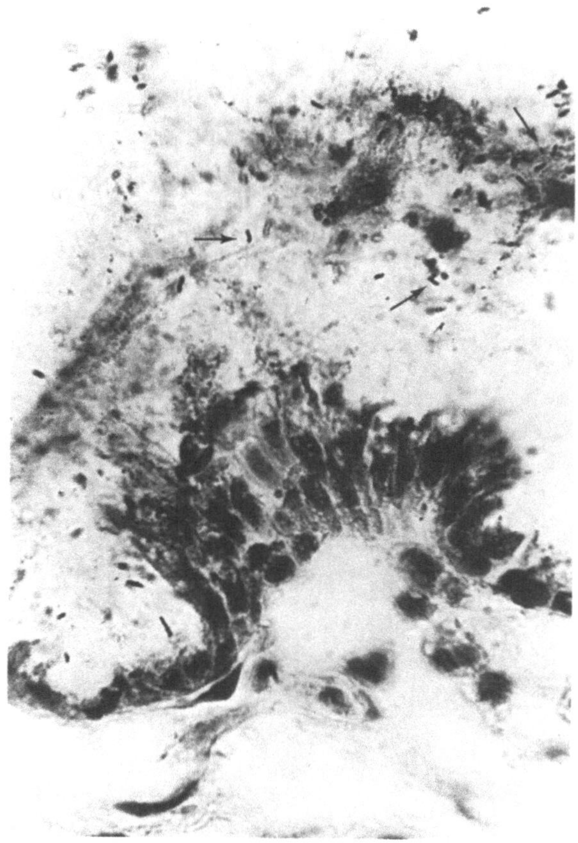

Fig. 2. Cryostat section of gastric pig mucosa stained by monoclonal antibodies and peroxidase-antiperoxidase. Bacteria are indicated by arrows: section is counterstained using hematoxylin

these cells increased 4–6 weeks after inoculation which is comparable to a superficial focal gastritis in man (Fig. 3a, b). These findings in the *H. pylori* positive pigs were varied but with a clear difference from the uninfected pigs. Raised antibody levels to *H. pylori* were detected in the serum from all infected pigs 2–4 weeks after challenge and remained at high levels throughout the observation period. Controls and pigs inoculated but not colonized had no antibody response specific to *H. pylori*.

Future Strategies in Animal Models of *H. pylori* Infection

An in expensive, easily reproducible, and convenient animal model of *H. pylori* infection is urgently needed. The rhesus monkey naturally harbors the organism and could potentially be a candidate as an in vivo experimental animal. However, the cost, difficult breeding, and problems with availability make this animal less attractive.

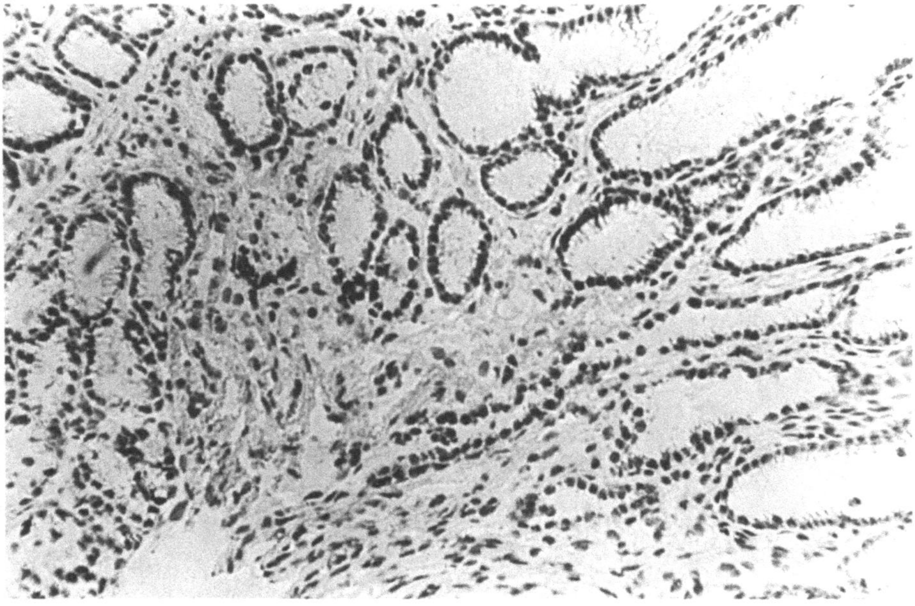

a

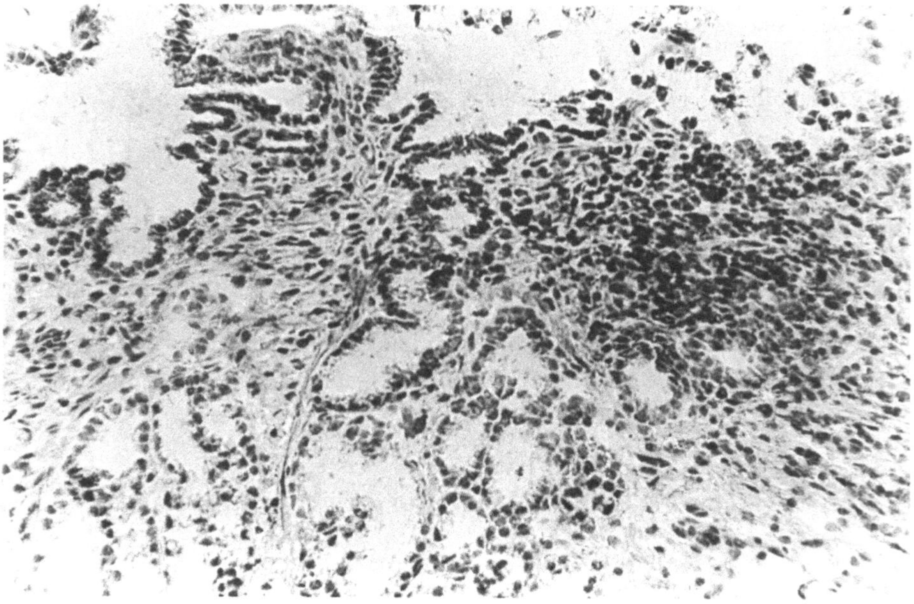

b

Fig. 3a, b. Cryostat sections of gastric pig mucosa before (**a**) and after (**b**) inoculation with *H. pylori*. Note mononuclear cell infiltration in **b**; sections are counterstained using hematoxylin

It seems that the animal fulfilling the criteria for a good experimental animal model must be protected in some way against common infectious agents before challenge with *H. pylori*. Gnotobiotic pigs and germ-free dogs are susceptible to *H. pylori* and advances have been made with the barrier-born pig model. The pig is a well-known laboratory animal and its gastric physiology has similarities to that of the human. Recently an interesting study by Duriex et al. [11] has described the gastric mucosa concentrations of antibiotics in a pig model. This technique could be valuable in different pharmacokinetic studies of anti- *H. pylori* drugs. Furthermore, a recent study describes the successful in vitro infection of pig mucosa with *H. pylori* [12] and another describes a gastric glycerolipid detected both in the human and the pig stomach tissue that is specifically recognized by isolates of *H. pylori* [13].

In summary, the best current models for experimental *H. pylori* infection today are the gnotobiotic piglet and the barrier-born pig models. The facts that these models are expensive and cumbersome with respect to animal breeding and size of the animals after 12 weeks make them less attractive. However, the pig fulfils many of the above-mentioned criteria for a suitable animal model of *H. pylori* infection. Work on experimental animal models of *H. pylori* infection in humans is just beginning.

References

1. Van der Gaag I (1988) The histological appearance of peroral gastric biopsies in clinically healthy and vomiting dogs. Can J Vet Res 52:67–74
2. O'Brien JJ (1969) Gastric ulceration (of the pars oesophagea) in the pig. Vet Bull 39:75–82
3. Kowalczyk T (1975) Gastric ulcers. In: Dunne HW, Leman AG (eds) Diseases of swine, 4th edn. Iowa State University Press, Ames, pp 978–1010
4. Jubb KVF, Kennedy PC, Palmer N (1985) Pathology of domestic animals, 3rd edn. Academic, New York
5. Bronsdon MA, Schoenknecht FD (1988) Campylobacter pylori isolated from the stomach of the monkey Macaca nemestrina. J Clin Microbiol 26:1725–1728
6. Baskerville A, Newell DG (1988) Naturally occurring chronic gastritis and Campylobacter pylori infection in the rhesus monkey: a potential model for gastritis in man. Gut 29:465–472
7. Ehlers S, Warrelmann M, Hahn H (1988) In search of an animal model for experimental Campylobacter pylori infection: administration of Campylobacter pylori to rodents. Zentralbl Bacteriol Microbiol Hyg 268:341–346
8. Krakowka S, Morgan DR, Kraft WG, Leund RD (1987) Establishment of gastric Campylobacter pylori infection in the neonatal gnotobiotic piglet. Infect Immun 55:2789–2796
9. Lambert JR, Borromeo M, Pinkard KJ, Turner H, Chapman CB, Smith ML (1987) Colonization of gnotobiotic piglets with Campylobacter pyloridis an animal model? J Infect Dis 155:1344
10. Marshall BJ, Armstrong JA, McGechie DB, Glancy RJ (1985) Attempt to fulfil Koch's postulates for pyloric campylobacter. Med J Aust 142:436–439
11. Duriex D, Westblom TU, Madan E, Midkiff BR, Subik M (1990) Gastric concentrations of ciprofloxacin in a guinea pig model: a comparison to in vitro susceptibilities in human isolates of Campylobacter pylori. In: Marshall B, McCallum R, Peterson W (eds) Gastroduodenal disease and Campylobacter pylori. Blackwell, Cambridge (in press)
12. Rosberg K, Hubinette R, Berglindh T, Rolfsen W (1989) Studies of Campylobacter pylori in a gastric mucosa in vitro model. Klin Wochenschr [Suppl] 67:59–60
13. Lingwood CA, Law H, Pellizzari A, Sherman P, Drumm B (1989) Gastric glyceriolipid as a receptor for Campylobacter pylori. Lancet 2:238–241

The Pathogenic Mechanisms of *H. pylori* – a Comment

D. G. Newell

At the First European Symposium, in Bordeaux, much of the discussion centred on whether *H. pylori* was pathogenic or not. Many argued in favour of a commensal role for the organism. The controversy now appears to be resolved, at least in part, and *H. pylori* colonisation of human gastric mucosa is now thought to be aetiological in histological gastritis and, probably, in peptic ulcer. In addition, a role for *H. pylori* in gastric cancer is now being seriously considered. It seems that in less than 1 year the question has changed from one of "is *H. pylori* pathogenic?" to "what are the pathogenic mechanisms involved in *H. pylori*-associated disease?".

If *H. pylori* causes such diverse pathological entities as histological gastritis and peptic ulcer, it is likely that the disease mechanisms will be both complex and multiple. In the workshop on pathomechanisms of *H. pylori* we invited speakers to review and discuss some of the presumed virulence factors already identified and to speculate on those bacterial features, which could contribute to *H. pylori* pathogenicity and which are, as yet, relatively uninvestigated.

When considering the pathogenic mechanisms of a bacterial infection it may be advantageous to distinguish between those bacterial properties which allow the establishment and maintenance of colonisation – the colonisation factors – and those properties which cause disease – the disease–associated factors. Such distinctions are, of course, not absolute and some factors which allow colonisation, for example urease, may also cause disease.

Bacterial properties of *H. pylori* allowing colonisation and causing disease:

- Colonisation factors
  - Adherence
  - Protection from acid
  - Protection from host immune response
- Disease-causing factors
  - Direct
    - Toxin production
    - Immunopathological effects
  - Indirect
    - Disruption of gastric mucosal barrier
    - Bacterial metabolism increasing bile acid toxicity

PHLS – Centre for Applied Microbiology and Research – Pathology, Porton Down, Salisbury, Wiltshire SP4 OJG, UK

Helicobacter pylori, Gastritis and Peptic Ulcer
Edited by P. Malfertheiner and H. Ditschuneit

Properties of *H. pylori* Allowing Colonisation

In *H. pylori* infection the organism colonises, chronically and probably life-long, the gastric mucosa, which is a uniquely hostile microenvironment. The organism has to survive not only exposure to the acidic milieu and the unremitting flow of the viscous mucous overlying the gastric epithelium, but also the attacks of a well-established local immune response.

Adhesion

Ultrastructurally, *H. pylori* is seen to adhere intimately to the gastric epithelium. This adherence is so strong as to distort the surface of the epithelial cell into an adherence pedestal with a concomitant loss of microvilli. Adherence to the gastric mucosa is apparently essential for the maintenance of the infection, protecting against bacterial loss during the passage of food and in the continuous mucous flow. However, some bacteria appear to be free-floating, particularly down in the crypts where there is little turbulence.

The adherence of *H. pylori* is totally tissue specific, to the extent that even gastric metaplasia occurring in the duodenum or rectum can be colonised. Conversely, in those areas of the stomach where there is intestinal metaplasia colonisation terminates abruptly. There is also considerable host specificity. Only humans, non-human primates and possibly the pig are naturally colonised. Such specificity suggests the expression of complex adhesins by the bacterium.

Several adhesins have been identified to date, though none, as yet, account for the significant tissue specificity observed. Multiple lectin-type adhesins detectable by haemagglutination reactions were described by Dr. Wadstrom (see p. 96), and it is evident that such reactions are complex with at least nine patterns distinguishable. Both glycoproteins and glycolipids appear to be involved as cell surface receptors of these haemagglutinins (see pp. 96) [6]. At least one type of haemagglutinin is sialic acid specific and neuraminidase (see pp. 96). A similar adhesin has been partly characterised [4] as a surface expressed fibrillar material extractable by *N*-octylglucose. Other adhesins, of which there are several, are neuraminidase insensitive. One of these is detectable in a tissue culture cell adhesion model and has been partly purified and characterised (see pp. 96). This adhesin is also surface associated and co-purifies with urease activity by gel filtration chromatography. However, this relationship needs clarification as other bacterial components also co-purify with urease using chromatographic techniques.

It is significant that none of the adhesion studies to date adequately account for the absolute tissue specificity. Such studies may well require the development of in vitro systems using suitable cellular material from gastric sources.

Motility

In order to reach the epithelial cell surface the organism must first penetrate the viscous mucin layer. Motility is a notable feature of *H. pylori* which has between three and seven sheathed flagella sited at one pole. Aflagellate strains poorly colonise the stomach of the gnotobiotic piglet model [3] suggesting that motility is an essential virulence factor. *H. pylori* flagellin has been identified recently as a 51 kDa polypeptide [5]. This protein has been particularly difficult to isolate and purify considering the number of flagella expressed by the organism. As expected, treatment of organisms with acid conditions causes the dissolution of the flagella into flagellin subunits. It seems likely that the flagella sheath material protects the flagella, to some extent, from the acidic microenvironment. The sheath material has not yet been identified but it may well contain urease activity and is certainly highly antigenic.

Even with extremely active flagella, penetration of the viscous mucin layer overlying the gastric epithelium would be difficult, but the spiral morphology appears to contribute to penetration and allows at least a proportion of the organisms to survive in the mucus flow without attachment.

Acid Protection

It is generally believed that the pH at the gastric epithelial surface, where most of the bacteria colonise, is close to neutral. However, bacteria evidently closely approach acid secreting cells without damage and presumably must survive the gastric lumen during transmission. This resistance from the acidic microenvironment must, at least in part, be due to the presence of urease activity. The importance of urease as a colonisation factor has been confirmed in the gnotobiotic piglet model using mutants of *H. pylori* which poorly synthesis urease and which cannot colonise the gastric epithelium [3]. The urease of *H. pylori* has been investigated widely and several investigators provided the most up to date information. It now seems that urease has a molecular weight of about 500 kDa and comprises two subunits essential for enzymic activity (61 kDa and 28 kDa). This enzyme can be purified to homogeneity by chromatographic techniques [2] and by affinity techniques with monoclonal antibodies (see pp. 81). Both subunits have been cloned into *E. coli* and sequenced (see pp. 74). The importance of urease in the colonisation of the gastric mucosa has been illustrated by the demonstration of antigenically cross-reactive enzymes in other gastric colonising bacteria from humans and animals, but interestingly the ureases produced by *H. mustelae* do not cross-react antigenically (see pp. 81) and show a poor degree of DNA homology (see pp. 74) suggesting a different evolutionary origin.

The location of the urease in the bacterium is essential to its activity. By cytochemical staining and immunolabelling at the ultrastructural level, the urease is located primarily on the bacterial surface probably lossely associated with the outer membrane.

Polymorphism

The transmission of bacterial infections is largely dependent on the organism's ability to survive between hosts. Coccal forms of *H. pylori* are produced *in vitro* in a process during which spiral organisms become spherical. These coccal forms are more common in older cultures. Although the cocci are metabolically quiescent, they are apparently viable (see pp. 63). Such morphological pleomorphism could protect the organism in hostile environmental situations, such as dehydration. There is evidence that these cocci are present *in situ* and it is possible that such forms are more resistant to the effects of antibiotics, thereby allowing the recrudescence of infection after therapy.

Avoidance of Host Immune Responses

H. pylori colonises an apparently immunologically hostile microenvironment. During infection the host mobilises both humoral and cellular immunity at the infected site. The specificity of the local antibody response towards *H. pylori* surface antigens has been experimentally demonstrated. There is no evidence to suggest that the gastric mucosa is an immunologically privileged site. On the contrary, it would appear that host immunoglobulin both reaches and binds to the bacterium *in situ*. Therefore the organism must possess mechanisms to avoid this response. As yet, little effort has been made to investigate this crucial research area, but a number of potential mechanisms can be suggested:

Antigenic variation
Induction of ineffective antibodies
Loss of antigen-antibody complexes from the bacterial surface
Antibodies inactivated by soluble antigens
Non-specific binding of host immunoglobulin masking antigens
Immunosuppression
Non-specific stimulation of host immune response
Production of enzymes which destroy antibodies

Properties of *H. pylori* Causing Disease

There are a limited number of mechanisms by which bacteria cause disease including invasion of host cells with concomitant local tissue damage, toxin production affecting local or distal sites, and immunopathological reactions, which are host-mediated events. In the gastric mucosa there is also an indirect potential mechanism for causing disease. In this case disruption of the mucin layer overlying and protecting the gastric epithelium would result in local tissue damage involving inflammation, possibly leading to peptic ulceration.

Toxin Production

As *H. pylori* is not invasive, potential toxic mechanisms have been investigated by several laboratories. It was initially thought that *H. pylori* produced proteases which allowed mobility through the mucin layer but which disrupted the gastric mucosal barrier. *H. pylori* is surrounded *in situ* by a cave-like structure. Moreover, during colonisation the ultrastructure of the gastric mucus changes from net-like to globular (see pp. 63). Such changes are supportive evidence of proteolytic activity. However, recent investigations [1], confirmed by others at the workshop, could find no significant proteolytic activity by *H. pylori*.

The gastric mucosal barrier may also be susceptible to the detergent properties of products of bacterial metabolism, of bile acids and of biliary lipids (see page 118).

The production of specific cytotoxins by *H. pylori* would also cause cell damage. About 50% of strains produce a non-lethal, heat-labile protein factor which causes vacuolation in tissue culture cells (see pp. 86). This cytotoxin affects a number of physiological processes including the rate of cell death. It has been partly characterised, and several cytotoxin-associated polypeptides have been detected (130 kDa, 95 kDa and 80 kDa) which are not present in non-toxic strains. The significance of this cytotoxin *in vivo* is still debatable. However, specific antibodies against the cytotoxin-associated polypeptides are produced during infection.

Almost certainly many more bacterial factors which interfere with host cell biochemistry and physiology will be detected in the future. Nevertheless, the relevance of such activities to the pathological processes needs to be carefully considered before a toxicological affector mechanism can be attributed to *H. pylori*-associated gastroduodenal diseases.

Immunopathological Effects

The local immune response to *H. pylori* is well-established. This response is directed primarily against antigenic material, including urease, which is continuously eluted from the bacterial surface. Such antigenic insult of an immunologically virgin mucosal surface would cause a rapid local recruitment of antigen presenting cells and antigen-specific T helper lymphocytes to stimulate the proliferation and differentiation of antibody-producing B cells. This massive immune response requires down regulation by antigen-specific T suppressor/cytotoxic cells. The mucosal surface then becomes a reservoir of immunocompetent cells. Such cells are the components of the histological gastritis associated with *H. pylori* infection. In this way an immunological over-responsiveness by host-mediated defence mechanisms can account for the bacteria-associated disease without recourse to other pathomechanisms. Considerably more work will be required on the characterisation of surface antigens and the identification of antigen-specific T and B cells within the gastric mucosa before the significance of this mechanism in the disease can be assessed.

In Vivo Models for Investigation of *H. pylori*

Pathogenic Mechanisms

Investigation of the virulence factors of *H. pylori* has been seriously hampered by the lack of a suitable animal model. It now seems unlikely that small laboratory animals can be colonised with *H. pylori*, even when genetically or immunologically manipulated. To date only the primate and gnotobiotic piglet models have seemed scientifically useful. However, both are restricted to a few specialised laboratories. The description of the barrier-born piglet model (see pp. 122) indicates that, at least in susceptible animals, bacterial competition does not necessarily prevent colonisation.

The accessibility of *in vivo* models will be the limiting factor in defining and characterising pathogenic mechanisms. Therefore, the development of suitable *in vitro* models becomes essential, especially to assess the virulence of the genetically engineered bacteria now being produced. It should also be remembered that the expression of virulence factors is frequently dependent on growth conditions. Without doubt current culture conditions cannot reflect the natural growth environment. The development of continuous culture techniques with controlled microenvironments which can simulate the gastric mucosa could provide complex models more suitable for detecting virulence factors.

H. pylori infection has uniquely combined the research resources of many disciplines with successful collaborations between microbiology, immunology, biochemistry, molecular biology, gastroenterology, and pathology. The enthusiastic, and sometimes heated, discussion initiated during this workshop illustrated the willingness of workers in this field to both communicate and share their ideas. Such enthusiasm will undoubtedly unravel the pathogenic mechanisms of *H. pylori*-associated diseases in the future.

References

1. Baxter A, Campbell CJ, Cox DM, Grinham CJ, Pendlebury JE (1989) Proteolytic activities of human Campylobacter pylori and ferret gastric campylobacter-like organism. Biochem Biophys Res Commun 163:1–7
2. Dunn B (1989) oral communication
3. Eaton KA, Morgan DR and Krakowka S (1989) Campylobacter pylori virulence factors in gnotobiotic piglets. Infect Immun 57:1119–1125
4. Evans DG, Evans DJ, Moulds JJ and Graham DY (1988) N-acetylneurminyllactose-binding fibrillar hemagglutinin of Campylobacter pylori: a putative colonization factor antigen. Infect Immun 56:2896–2906
5. Geis C, Leying H, Snerbaum S, Mai U and Opferhuch W (1989) Ultrastructure and chemical analysis of Campylobacter pylori flagella. Infect Immun 27:436–11
6. Lingwood CA, Pellizzari A, Law H, Sherman P and Drumm B (1989) Gastric glycerolipid as a receptor for Campylobacter pylori. Lancet i:238–240

Local and Systemic Immune Response

Methods of Studying the Immune Response

H. von Wulffen

The successful culturing of *Helicobacter pylori* (formerly *Campylobacter pylori*) by Marshall [6] opened the way for conducting serological studies using this organism as antigen. Various serologic test methods have been employed since, including the complement fixation test [3, 9], agglutination test [3], passive hemagglutination assay [5], enzyme-linked immunosorbent assay [1, 2], and immunoblot techniques [4, 9]. In principle, all of these methods work to show that colonization or infection with *H. pylori* leads to a detectable humoral immune response. However, studies comparing the performance of older and more recent methods, e.g., complement fixation with immunoblot [9], usually reveal the superiority of the ELISA and the immunoblot method over the other techniques. For this reason, I shall be limiting my discussion to these two methods.
The advantages of the immunoblot method are:

1. High sensitivity
2. Detection of class-specific antibodies
3. Assessment of antigen preparation and variation
4. Differentiation of immune response to distinct proteins
5. Recognition of cross-reactions

This technique yields a very high sensitivity, one reason for this being that any particular immunogenic protein is being concentrated in a very thin line on the nitrocellulose membrane. Also, by using class-specific detecting antibodies in the conjugate step, we can differentiate the immune response with regard to the different immunoglobulin classes, something we are very much interested in when studying a chronic mucosal inflammatory state.

In studying the immune response to *H. pylori*, it has become evident that it is not so much the method used as the antigen preparation employed that decides on the quality of the serological assay. The immunoblot technique provides the tools that are necessary to study the antigen preparation and assess its variations. The comparative analysis of the protein profiles of different *H. pylori* isolates on the stained gel or the developed blot may help to decide, for example, whether to use one or more strains as antigen. The immunoblot analysis may also reveal the presence or absence of cross-reacting proteins in different preparations of the antigen. Thus, no matter what method is used for serology, the immunoblot technique is an excellent instrument for quality control of the antigen preparation.

Institut für Medizinische Mikrobiologie und Immunologie, Universitätskrankenhaus Eppendorf, Martinistr. 52, D-2000 Hamburg 20

Helicobacter pylori, Gastritis and Peptic Ulcer
Edited by P. Malfertheiner and H. Ditschuneit

When interpreting an individual blot strip developed with a patient's serum, one may have difficulties in deciding whether the reactive bands are specific for *H. pylori* or not. In a previous study [10] the analysis of a large number of sera led us to the selection of a set of four apparently *H. pylori*-specific protein bands. On the basis of the evaluation of these protein bands we constructed a scoring system whereby one point was assigned to each reactive band and half a point to a weakly reactive band, and we were thus able to establish semiquantitative correlations between the number of reactive bands and various groups of patients.

Although this immunoblot method is a very efficient tool for studying the immune response, it does have a number of drawbacks. It is:

1. Laborious
2. Apt to trouble-shooting
3. Only semiquantitative
4. Not easily able to identify a particular protein band in a complex antigen mixture
5. Not always able to transfer all proteins of interest transferred during a blot
6. Capable of denaturing immunodominant epitopes

Especially because it is quite labor-intensive it is not suited for testing large numbers of sera. Like any other sophisticated method it is apt to trouble-shooting, and once it is not working correctly, it may be very time-consuming to find the fault. As mentioned above, the method only allows a semiquantitative interpretation of the results. In complex antigen mixtures such as a whole cell lysate or a sonicate where one gets a very dense set of reactive bands, it may become very difficult to identify a particular protein band. Also, depending on the quality and percentage of the polyacrylamide and the method of transfer used, not all proteins of interest may become transferred to the nitrocellulose during the blotting procedure. Since the antigen samples are boiled in sodium dodecyl sulfate (SDS) and other denaturing agents, some of the immunodominant epitopes may no longer be recognized.

Let us now turn to the ELISA method. The advantages of this procedure are:

1. Ease of handling large numbers of samples
2. High sensitivity
3. Detection of class-specific antibodies
4. Quantitation

In contrast to the immunoblot method, there is no problem with screening large numbers of serum samples in parallel. Its sensitivity is comparable, and one can also differentiate the immune response with regard to the immunoglobulin classes. As the ELISA results are printed out as absorbance readings, quantitative interpretations may be obtained.

There are several ways that ELISA results may be quantitatively expressed. Of course, the classical way is to use serial dilutions and titrate the end-point. This, however, would mean forgoing the advantage of being able to process large numbers of sera in parallel.

Some investigators simply take the raw optical density readings, a method that may present problems in as much as you cannot avoid day-to-day variations. In a previous study [8] we decided to express our results as ELISA units (U/ml) based on serial dilutions of an internal standard serum. The absorbance readings of this calibration serum at a dilution of 1:3200 for IgG were arbitrarily set at 100 U/ml. For each test serum done in duplicate the mean absorbance value was recorded, and the result was calculated from the standard curve and expressed in U/ml. Another way that has been suggested [7] is to measure the concentration of the anti-*H. pylori* antibodies by using a standard curve of human immunoglobulin mass. That way one can express the results as μg bound immunoglobulin per ml.

Serological assays are usually judged on the basis of their performance values such as sensitivity, specificity, positive and negative predictive values. To derive these data, one needs a set of clearly defined positive and negative reference sera. What should be the "golden standard" for defining the terms positive and negative in relation to *H. pylori*? For the purist microbiologist, of course, there is no doubt: culture of the organism must be the golden standard. On the other hand, we know of the patchy distribution of *H. pylori* and the related type B gastritis in the stomach mucosa. So, how many biopsies and cultures do we need to get 100% of our proposed golden standard? What about a biopsy urease test, ^{14}C-urea breath test, microscopic detection of the typical organism, or even the histopathologist's diagnosis of type B gastritis as a reference method for serology? The choice one makes probably depends on the quality of the reference methods available in a given situation. Table 1 illustrates this: using the same acid glycine extract as antigen for an IgG ELISA, we arrived at quite different performance values depending on the reference method employed. Overall, we found the best values using the breath test as "golden standard."

Table 1. Sensitivity and specificity of *H. pylori* acid extract IgG ELISA in relation to reference method used

	Reference method		
	Histological grading of gastritis %	Detection of *H. pylori* by Gram stain or culture %	^{14}C-urea breath test %
Sensitivity	85	94	89
Specificity	62	57	84
Positive predictive value	89	67	85
Negative predictive value	54	84	89

References

1. Booth L, Holdstock G, MacBride H, Hawtin P, Gibson JR, Ireland A, Bamforth J, DuBoulay CE, Lloyd RS, Pearson A (1986) Clinical importance of Campylobacter pyloridis and associated serum IgG and IgA antibody responses in patients undergoing upper gastrointestinal endoscopy. J Clin Pathol 39:215–219
2. Goodwin CS, Blincow E, Peterson G, Sanderson C, Cheng W, Marshall BJ, Warren JR, McCulloch R (1987) Enzyme-linked immunosorbent assay for Campylobacter pyloridis: correlation with presence of C. pyloridis in the gastric mucosa. J Infect Dis 155:488–494

3. Jones DM, Lessells AM, Eldridge J (1984) Campylobacter like organisms on the gastric mucosa: culture, histological and serological studies. J Clin Pathol 37:1002–1006
4. Kaldor J, Tee W, Nicolacopolous L, Demirtzoglou K, Noonan D, Dwyer B (1986) Immunoblot confirmation of immune response to Campylobacter pyloridis in patients with duodenal ulcers. Med J Aust 145:133–135
5. Marshall BJ, McGechie DB, Francis GJ, Utley PG (1984) Pyloric campylobacter serology. Lancet II:281
6. Marshall BJ, Royce H, Annear DJ, Goodwin CS, Pearman JW, Warren JR, Armstrong JA (1984) Original isolation of Campylobacter pyloridis from human gastric mucosa. Microbios 25:83–88
7. Newell DG, Tathborie BJ (1989) The serodiagnosis of Campylobacter pylori infection. Serodiagn Immunother Infect Dis 3:1–6
8. von Wulffen H, Grote HJ (1988) Enzyme-linked immunosorbent assay for detection of immunoglobulin A and G-antibodies to Campylobacter pylori. Eur J Clin Microbiol Infect Dis 7:559–565
9. von Wulffen H, Heesemann J, Bützow GH, Löning T, Laufs R (1986) Detection of Campylobacter pyloridis in patients with antrum gastritis and peptic ulcers by culture, complement fixation test, and immunoblot. J Clin Microbiol 24:716–720
10. von Wulffen H, Grote HJ, Gatermann S, Löning T, Berger B, Buhl C (1988) Immunoblot analysis of immune response to Campylobacter pylori and its clinical associations. J Clin Pathol 41:653–659

Serodiagnosis of *Helicobacter pylori* Infections: Suitability of Various Antigen Preparations

A. M. Hirschl and M. L. Rotter

Introduction

The detection of antibodies specific for *Helicobacter (Campylobacter) pylori* is gaining importance for the diagnosis of chronic gastritis associated with this pathogen [12, 14]. Compared with invasive diagnostic methods or breath tests, serodiagnosis is much easier to perform, cheaper, and, above all, less stressful to the patient. Furthermore, for the purpose of studying the efficacy of therapeutic regimens, it appears well suited for monitoring the colonization of the gastric mucosa with *H. pylori* (Fig. 1). This is advantageous since in studies such as these frequent endoscopies are necessary over a long period, which is expensive and not a factor conducive to patient compliance. Of course, biopsy speciriens

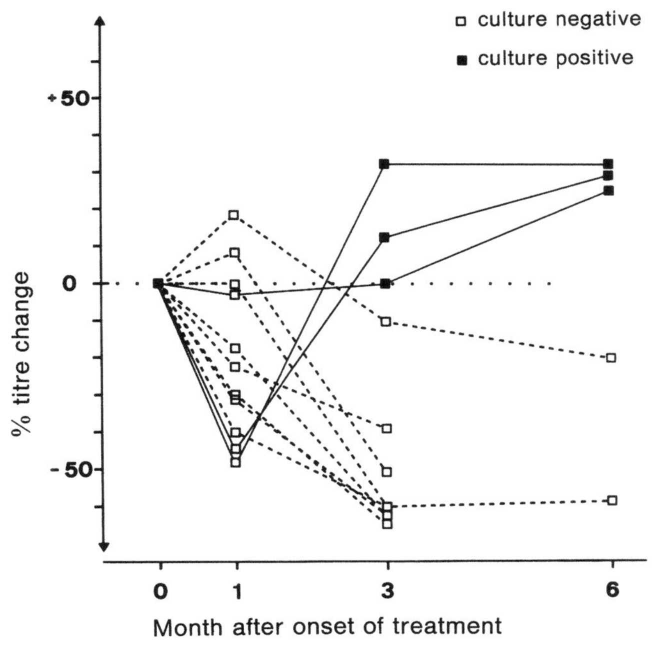

Fig. 1. Serum IgG titers (ELISA) to *H. pylori* in patients with duodenal ulcer before and after treatment with amoxycillin (750 mg t.i.d./12 days) + metronidazole (500 mg t.i.d./12 days). ELISA was performed as already described [7]; centrifuged sonicate of patient's own strain served as antigen. ELISA values expressed as % of the titer at time 0 (= before onset of treatment) for each individual patient

Hygiene-Institute of the University, Vienna, Austria

Helicobacter pylori, Gastritis and Peptic Ulcer
Edited by P. Malfertheiner and H. Ditschuneit

would still be indispensable for cultivation of the pathogen, typing, and sensitivity testing, as well as for histological check-up, but the frequency of taking them could possibly be significantly reduced.

Of all serological techniques employed, today the IgG-ELISA is used most frequently because it is relatively easy to perform and highly flexible [1–7, 9–11, 13, 15–18].

H. pylori Antigens for ELISA

These antigens may be classified into three categories: (a) whole-cell antigens and ultrasonicates of them, (b) partially purified antigens, and (c) highly purified antigens.

Whole-Cell Antigens and Ultrasonicates

These antigens contain a multitude of epitopes which are not very well defined. Therefore, nonspecific binding of immunoglobulins and cross-reactions (mainly with other *Campylobacter* spec.) occur frequently. This results in a relatively high background of false-positive reactions and consequently in unsufficient specificity, examples of which are given in Table 1. The data listed there and also in the subsequent tables are not directly comparable with each other as technical details of the test, interpretation of results (e.g., threshold), and, above all, choice of diagnostic methods for the detection of *H. pylori* are not uniform. Generally, it can be stated that the specificity can be enhanced by changing the interpretative threshold but only together with a nonproportional loss of sensitivity. Furthermore, an "artificial" fixing of cut-off limits does not take into account the distributions of positive and negative sera which usually is bimodal. This in turn, can create problems since a high number of readings may happen to lie within the cut-off region, thus giving rise to a purely arbitrary allocation of patients' sera to the interpretative categories "negative" and "positive." Also fixing of two cut-off values, one for high specificity and one for high sensitivity, as it has been suggested [5, 9], does not solve the problem in the situation of routine testing.

Table 1. Sensitivity and specificity of *H. pylori* IgG-ELISA using whole cells or sonicates of them as antigen

Antigen	Sensitivity (%)	Specificity (%)	Reference
Whole cells	91.0	75.0	Morris et al. [10]
Whole cells	97.0	78.0	Schaber et al. [18]
Whole cell supernatant	98.7 [a]	73.1 [a]	Musgrove et al. [11]
Sonicate	96.5	90.0	Perez-Perez et al. [16]
Sonicate	98.9	70.6	Hirschl [6]

[a] Numbers calculated from data shown in cited paper

When using whole cells or ultrasonicates of them as an antigen it also seems important to know which strain(s) and a mixture of how many strains should be employed. In any case, in some of our own studies [6] significantly higher IgG titers were observed with ELISA testing when the patient's own strain was used as an antigen rather than a type strain.

Partially Purified Antigens

Among these antigens centrifuged cell sonicates and acid-glycine extracts produce test results of markedly higher specificity than sensitivity (Table 2). This is true, at least, if the interpretative cut-off is chosen as the value calculated from the mean titer of negative sera +3 SD (own results). In contrast, crude urease preparations give rise to test results with a distribution pattern of readings similar to that found with nonpurified antigen preparations, consequently offering relatively low specificity only. Dent et al. [2] believe that the high proportion of so-called false-positive results of this serodiagnostic approach is due to failure of the other diagnostic methods such as culture or histology to detect the presence of *H. pylori* rather than low specificity of the ELISA. This hypothesis is supported by results of Evans et al. [3] who obtained very good results with an antigen preparation they describe as "high molecular weight cell associated proteins exerting strong urease activity," thus with an antigen which is probably very similar to the above-mentioned urease preparation though obtained in a different way. In this study, the decision as to whether a serum stems from a patient infected with *H. pylori* or not was based on the results of the ^{13}C-urea breath test. Also very good results were reported by Stacey and Newell [19] who separated a cell ultrasonicate by FPLC, pooled certain fractions, and used this mixture as an antigen.

Highly purified Antigens

Results obtained with antigens of this group are distinguished by very high specificity but suboptimal sensitivity; thus, antigens of this group produce a rather

Table 2. Sensitivity and specificity of *H. pylori* IgG-ELISA using partially purified antigens

Antigen	Sensitivity (%)	Specificity (%)	Reference
Centrifuged sonicate	76.0	97.6	Own results
Ultracentrifuged sonicate	83.8	95.1	Own results
Acid-glycine extract	70.3	97.6	Own results
	85.7[a]	97.0[a]	Newell and Stacey [13]
Crude urease	94.3[a]	72.7[a]	Newell and Stacey [13]
	100.0[a]	73.2[a]	Dent et al. [2]
High molecular weight cell-associated proteins	98.7	100.0	Evans et al. [3]

[a] Numbers calculated from data shown in the cited paper

Table 3. Sensitivity and specificity of *H. pylori* IgG-ELISA using highly purified antigens

Antigen	Sensitivity (%)	Specificity (%)	Reference
Purified urease	65.7 [a]	100.0 [a]	Newell and Stacey [13]
Protein (approx. 120 kDa)	83.8	97.6	Own results
N-Acetylneuraminyllactose-binding hemagglutinin	74.8	98.3	Evans et al. [4]

[a] Numbers calculated from data shown in the cited paper

large proportion of false-negative results (Table 3). This is not surprising as it is well known that the immune response to *H. pylori* can be very variable [8, 13]. Consequently, it is unlikely that a highly purified antigen with its relatively low number of epitopes will recognize products of the immune response to *H. pylori* in each individual patient. But combining antigen preparations such as these with only partially purified antigens leads to interesting results. And, indeed, in our hands a combination of an acid-glycine extract or an ultracentrifuged cell sonicate with the 120 kDa protein of *H. pylori* [7] led to nearly optimal values for specificity and sensitivity (Table 4).

Table 4. Sensitivity and specificity of *H. pylori* IgG-ELISA using mixtures of antigen preparations

Antigen	Sensitivity (%)	Specificity (%)	Reference
Ultracentrifuged sonicate + protein 120 kDa	97.3	100.0	Own results
Acid glycine extract + protein 120 kDa	97.3	100.0	Own results

Conclusions

Whole cells and nonpurified cell sonicates of *H. pylori* are in general not (well) suited for serodiagnosis of an infection with this pathogen because the specificity of the test system is insufficiently low. In contrast, results obtained with "unicomponent antigens" are highly specific but the test systems are of suboptimal sensitivity; with regard to the latter feature the 120 kDa protein comes off comparatively best.

Among the group of "multicomponent but partially purified antigens" the acid-glycine extract though, as yet, frequently regarded as a standard is inferior to antigen preparations containing components of urease. Mixtures of various antigen preparations also led to very satisfying results.

Summarizing, it can be stated that several excellent test systems are available for the serodiagnosis of *H. pylori* infection; thus, invasive and/or expensive diagnostic procedures may at least partially be replaced by these methods. Which of these test systems (or others) will be regarded as a reference system in future will probably also depend on the suitability of the technique of antigen preparation for large-scale production.

References

1. Booth L, Holdstock G, MacBridge H, Hawtin P, Gibson JR, Ireland A, Bamforth J, du Boulay C, Lloyd RS, Pearson AD (1986) Clinical importance of Campylobacter pyloridis and associated serum IgG and IgA antibody responses in patients undergoing upper gastrointestinal endoscopy. J Clin Pathol 39:215–219
2. Dent JC, McNulty CAM, Uff JS, Gear MWL, Wilkinson SP (1988) Campylobacter pylori urease: a new serological test. Lancet 1:1002
3. Evans DJ, Evan DG, Graham DY, Klein PD (1989) A sensitive and specific serological test for detection of Campylobacter pylori infection. Gastroenterology 96:1004–1008
4. Evans DJ, Evans DG, Smith KE, Graham DY (1989) Serum antibody responses to the N-acetylneuraminyllactose-binding hemagglutinin of Campylobacter pylori. Infect Immun 57:664–667
5. Goodwin CS, Blincow E, Peterson G, Sanderson C, Cheng W, Marshall B, Warren JR, McCulloch R (1987) Enzyme-linked immunosorbent assay for Campylobacter pyloridis: correlation with presence of Campylobacter pyloridis in the gastric mucosa. J Infect Dis 155:488–494
6. Hirschl AM (1987) Frequency of occurrence of Campylobacter pylori and analysis of the systemic and local immune response. Zentralbl Bakteriol Microbiol Hyg [A] 266:526–542
7. Hirschl AM, Pletschette M, Hirschl MH, Berger J, Stanek G, Rotter ML (1988) Comparison of different antigen preparations in an evaluation of the immune response to Campylobacter pylori. Eur J Clin Microbiol Infect Dis 7:570–575
8. Jacobs E, Apel I, Kist M, Bredt W (1988) Antibody response of patients against a 120 kDa surface protein of Campylobacter pylori. Zentralbl Bakteriol Microbiol Hyg [A] 268:271–276
9. Loffeld RJLF, Stobberingh E, Flendrig JA, van Spreeuwel JP, Arends JW (1989) Diagnostic value of an immunoassay to detect anti Campylobacter pylori antibodies in nonulcer dyspepsia. Lancet 1:1182–1185
10. Morris A, Nicholson G, Lloyd G, Haines D, Rogers A, Taylor D (1986) Seroepidemiology of Campylobacter pyloridis. N Z Med J 99:657–659
11. Musgrove C, Bolton FJ, Krypczyk AM, Temperley JM, Cairns SA, Owen WG, Hutchinson DN (1988) Campylobacter pylori: clinical histological, and serological studies. J Clin Pathol 41:1316–1321
12. Newell DG, Rathbone BJ (1989) The serodiagnosis of Campylobacter pylori infection. Serodiagn Immunother 3:1–6
13. Newell DG, Stacey AR (1989) Antigens for the serodiagnosis of Campylobacter pylori infection. Gastroenterol Clin Biol 13:37B–41B
14. Newell DG, Stacey AR (1989) The serology of Campylobacter pylori infection. In: Rathbone BJ, Heatley RV (eds) Campylobacter pylori and gastroduodenal disease. Blackwell, Oxford, pp 74–82
15. Newell DG, Johnston BJ, Ali MH, Reed PI (1988) An enzyme-linked immunsorbent assay for the serodiagnosis of Campylobacter pylori-associated gastritis. Scand J Gastroenterol 23:53–57
16. Perez-Perez GI, Dworkin BM, Chodos JE, Blaser MJ (1988) Campylobacter pylori antibodies in humans. Ann Intern Med 109:11–17
17. Rathbone BJ, Wyatt JI, Worsley BW, Shires SE, Trejdosiewicz LK, Heatley RV, Losowsky MS (1986) Systemic and local antibody responses to gastric Campylobacter pyloridis in non-ulcer dyspesia. Gut 27:642–647

18. Schaber E, Umlauft F, Stöffler G, Aigner F, Paulweber B, Sandhofer F (1989) Indirect immunofluorescence test and enzyme-linked immunosorbent assay for detection of Campylobacter pylori. J Clin Microbiol 27:327–330
19. Stacey AR, Newell DG (1989) The development of antigens for improved serodiagnostic tests of Campylobacter pylori associated gastritis. In: Mégraud F, Lamouliatte H (eds) Gastroduodenal pathology and Campylobacter pylori. Excerpta Medica, Amsterdam pp 193–195

Possible Clinical Uses of Serology of *Helicobacter pylori*

G. M. SOBALA

Introduction

The highly specific and sensitive serological tests for *H. pylori* which have recently been developed clearly have many uses in epidemiology and other research [12], but their role in clinical practice is unclear. There have been suggestions that serology could be used in the diagnosis of dyspepsia and also in the monitoring of anti-*Helicobacter* therapy. This paper will discuss the problems underlying such approaches.

Diagnosis of Dyspepsia

H. pylori serology may find a role in the diagnosis of dyspepsia because the current situation is unsatisfactory. Dyspepsia is a very common symptom in the community [6], and its investigation constitutes a major workload for both radiology and endoscopy services. Most centres have experienced a steady increase in the yearly number of upper gastrointestinal (GI) endoscopies over the past decade (Table 1). The majority of these investigations yield normal results or show minor pathology of doubtful clinical significance. Only about 20% of

Table 1. Yearly numbers of upper GI endoscopies performed at Leeds General Infirmary, 1975–1988

Increasing endoscopy workload			
Year	Upper GI endoscopies performed	Year	Upper GI endoscopies performed
1975	247	1982	2105
1976	1025	1983	2549
1977	1280	1984	2399
1978	1506	1985	2364
1979	1835	1986	2306
1980	1917	1987	3282
1981	2031	1988	3191

Gastroenterology Unit, The General Infirmary Great George Street, Leeds, West Yorkshire LS1 3EX, United Kingdom

Helicobacter pylori, Gastritis and Peptic Ulcer
Edited by P. Malfertheiner and H. Ditschuneit

Table 2. Diagnoses in 1119 patients attending a dyspepsia clinic at Leeds General Infirmary,1987–1989

Causes of Dyspepsia		
Diagnosis	No. of patients	Percentage of total
Neoplasm	17	1.5
Peptic Ulcer	236	21.1
Oesophagitis	193	17.2
Erosions	31	2.8
Minor abnormality	282	25.2
Normal	360	32.2
Total	1119	

investigated patients prove to have peptic ulcer disease, and 1%–2% have gastric cancer (Table 2).

The discovery of the very close association between *H. pylori* and peptic ulcer disease, and duodenal ulcer disease in particular, has led to suggestions that noninvasive tests for the presence of this organism could be used as an adjunct or an alternative to endoscopy in diagnosing dyspepsia [10, 12, 14]. There are two methods by which *H. pylori* can be detected noninvasively: serology and carbon-labelled urea breath testing. In the best hands both are very sensitive and specific, but serology would undoubtedly be cheaper to use on a mass scale and does not have the logistic problems associated with the breath tests. Serology can be performed on a blood sample drawn from the patient by this primary care practitioner, whereas breath testing requires patients to fast (which they dislike), at least 0.5 h of their time, and probably a visit to a central hospital.

The main aims for introducing *H. pylori* serology into the diagnostic process would be to endoscopy and radiology workloads and to reduce the overall cost of investigating dyspepsia. However, clinicians hold widely divergent views as to which sorts of patients should be investigated, why they should be investigated, and as to the optimal therapy for each of the causes of dyspepsia [1]: these issues are more pertinent to devising a suitable strategy for serology in the diagnosis of dyspepsia than are technical issues such as the sensitivity and specificity of the assays. The key point is that a strategy should actually influence management. Three different strategies and the assumptions behind them will be discussed below.

Serology as Replacement for Endoscopy

The most radical suggestion is that serology should replace endoscopy/radiology. This strategy is based on the assumption that *H. pylori* is a key factor in dyspepsia and that in essence only two forms of dyspepsia exist: *H. pylori*-positive dyspepsia (whether due to ulcers or not) which should be treated by Helicobactericidal therapy and *H. pylori*-negative dyspepsia which should be managed in other ways. It does not need to be stressed that these assumptions are currently highly controversial. Although there is increasing evidence for a role for *H. pylori* in duodenal ulcer disease, its role in gastric ulcer disease is less clear and its role in "nonulcer dyspepsia" (NUD) is completely unproven. Such a strategy could only sensibly by implemented if it were eventually proved that nonulcer dyspepsia in *H. pylori*-positive patients does respond to the same Helicobactericidal therapy as peptic ulcers, and that such therapy is safe when used relatively indiscriminately. The results of many further clinical studies are awaited to answer these questions.

The "success" of this strategy would be measured by the number of endoscopies it saved, and also by the accuracy with which it categorised patients as *H. pylori* positive or negative: the latter clearly depends on the sensitivity and specificity of the serological method used.

Serology as Screening Procedure Prior to Endoscopy

A more widely suggested strategy is that of using serology as a screening procedure [12, 14]. Patients with dyspepsia would have their *H. pylori* status determined by serology. Positives would go on to endoscopy to determine whether they had peptic ulcer disease, while negatives would be reassured and treated for "nunulcer dyspepsia". This policy is basedon the assumption that it is important to make a positive diagnosis of peptic ulcer disease: in other words, that peptic ulcer patients should be managed differently from other patients with dyspepsia. Indeed, the assumption is that the chief aim of endoscopy in most patients is to detect ulcer disease rather than minor upper GI tract pathologies (e.g. oesophagitis) which are less closely associated with *H. pylori*. For physicians who use H2 receptor antagonists for all forms of dyspepsia such a strategy would be irrelevant, while for those using Helicobactericidal therapy only for ulcer patients it would be ideal.

The "success" of this strategy would be measured by the number of endoscopies it saved and also by its accuracy in detecting peptic ulcer disease.

Screening for *H. pylori*-Negative Patients

A third suggestion is a curious blend of of the first two strategies [9]. It is that patients with dyspepsia should first have serology performed. Patients with positive results should be treated with Helicobactericidal drugs; patients with negative results should go on to upper GI endoscopy to determine the cause of their symptoms – which ipso facto must be a pathology not associated with *H. pylori*. This strategy, like the first, assumes that treatment for peptic ulcer disease and *H. pylori* associated nonulcer dyspepsia should be the same, and yet the authors themselves believe in "a non-existent direct relationship between gastritis and dyspepsia" [8]. It is also based on the odd belief that the "causes" of *H. pylori*-negative dyspepsia are likely to be visible endoscopically. This is not in line with general experience.

How Well Would Such Strategies Work?

No studies of an adequate size have yet been published on the actual use of *H. pylori* serology to screen patients for endoscopy, but we have extrapolated the likely results from biopsy determined *H. pylori* status in 842 dyspepsia clinic patients, and these data are presented elsewhere in this volume.

Problems with Screening Strategies

H. pylori-Positive Nonulcer Dyspepsia

As has already been pointed out, it is not really possible to choose a rational strategy until the problem of the interrelationship (if any) between *H. pylori* and nonulcer dyspepsia has been worked out.

Gastric Cancer

A problem with all three strategies outlined above is that in their "raw forms" they overlook the issue of gastric cancer. It is increasingly being recognised that successful therapy of gastric cancer hinges on early diagnosis and thus on early investigation of dyspepsia. Early diagnosis has to be by endoscopy and biopsy, and this is clearly at odds with any policy which screens patients away from endoscopy. However, it has been pointed out that (in the UK at least) gastric cancer is extremely rare in patients under 45 years of age, constituting only 1.8% of all cases [16], and thus a modification could be to apply the strategies only to patients under this age. Older patients, in whom gastric cancer is a real risk, would be investigated anyway. It is still not clear how many of the rare gastric cancers in young patients would actually evade detection using such modified strategies: the precise relationship between *H. pylori* and gastric cancer has still to be determined. Although it has been known for many years that chronic gastritis is usually present in cases of gastric cancer [4, 11], and Correa and Ruiz [3] consider that *H. pylori* may be a promoter of gastric carcinogenesis, the organism has not universally been found in cases of gastric cancer. However, even assuming that serology would be positive in only 70% of young patients with gastric cancer, a "missed" gastric cancer in a young patient (under 45) would occur very rarely indeed: perhaps once for every 7500 patients of all ages seen. Although each individual missed gastric cancer in a younger patient is a tragedy, the effect of a screening strategy could well be to free endoscopic resources for the earlier detection of cancers in patients over 45: thus for every one missed cancer in a patient under 45, several patients over 45 may have lesions detected at an earlier and curative stage.

H. pylori-Negative Peptic Ulcers

If the serological test in use has a high sensitivity for detecting *H. pylori* infection, then the peptic ulcers missed by the strategy will be *H. pylori* negative. If an age limit is in force then these will mostly be duodenal in site. We do not know enough about these ulcers to assess the clinical importance of missing such a diagnosis. It is not known whether they follow a different clinical course to *H. pylori*-positive peptic ulcers, and especially whether they are as likely to relapse, bleed or perforate. Before a serology strategy can be "judged", this knowledge has to be obtained.

Oesophagitis and Other Upper GI Pathology

The discussion so far has centred around peptic ulcer disease and gastric cancer. Clearly there are other pathologies which can be detected endoscopically. Even if one assumes that gastric and duodenal erosions can be subsumed into the broader category of ulcer disease, this leaves diagnoses such as hiatus hernia and oesophagitis. However, with these two diagnoses one has to question the value of the endoscopy in affecting the management of the patient. A patient with heartburn is likely to be treated for gastro-oesophageal reflux whether or not endoscopy shows active oesophagitis: endoscopy is not the gold standard for the diagnosis of this condition [5, 7], and if serology can act as a tool in the diagnostic process by excluding concurrent peptic ulcer disease then the indications for endoscopy are diminished. The relationship between *H. pylori* and such conditions in young patients is also not completely clear: while no close association exists as for dudenal ulcer, weaker associations have not been excluded [2].

Commercial vs Research Assays

Before any screening strategy can be implemented, prospective studies must be done on large numbers of patients looking closely at precisely how many cases of each relevant diagnosis would be missed by implementing the strategy. Such studies must be performed with the same commercial kits which would be used in the implementation of the strategy, and ideally the assays themselves should be performed in routine diagnostic laboratories, rather than the research laboratories in which the assays were developed. Kits tend to perform rather better in the hands of their creators than in the "real world"!

Epidemiology of H. pylori

Clearly there is little point in implementing a strategy if in fact the prevalence of *H. pylori* in young patients in the local population is very high: few endoscopies would be saved. Similarly it may be dangerous to implement a strategy in a population with a high incidence of gastric cancer in the young.

What Would Happen to the "Screened" Patient Anyway?

The final problem is that a strategy may fail in its goal of reducing the total number of endoscopies being performed if the patients who are screened away from endoscopy actually end up having one anyway – perhaps after a brief therapeutic tussle in the out-patient clinic. For a strategy to work, one requires a clear management plan for *H. pylori*-negative patients that does not include endoscopy at an early stage. The clinicians must be able to trust the serological test's exclusion powers.

Serology as Form of Monitoring Therapy

The questions surrounding the possible clinical uses of serology as a tool in monitoring the effectiveness of therapy directed against *H. pylori* are a lot more straightforward. Such therapy is being used increasingly widely, and methods of determining its success or failure are required. Endoscopy and biopsy is accurate but elaborate, unpleasant and otherwise probably unnecessary. Carbon-laballed urea breath tests are very useful in this role, having high sensitivities and specificities, but they still require the patient to fast and attend a central unit, and either require very expensive equipment (C^{13} urea), or involve a small radiation dose (C^{14} urea). If serology could achieve comparable sensitivity and specificity (above 90%), then it would answer a real need. There is certainly evidence that titres to *H. pylori* do fall within a few months of eradicating the organism [13, 15], but the immunologists' problem is no longer that of defining cut-off values for past or current infection. What are required are accurate guidelines as to how great a fall in titres over what time period indicate eradication of *H. pylori*, and the sensitivity and specificity of such predictions. It is possible that such criteria may emerge from further work in this field, but much remains to be done.

References

1. Anon Ymous (1988) Management of dyspepsia: report of a working party. Lancet 1:576–578
2. Cheng EH, Bermanski P, Silversmith M, Valenstein P, Kawanishi H (1989) Prevalence of Campylobacter pylori in esophagitis, gastritis and duodenal disease. Arch Intern Med 149:1373–1375
3. Correa P, Ruiz B (1989) Campylobacter pylori and gastric cancer. In: Rathbone BJ, Heatley RV (eds) Campylobacter pylori and gastroduodenal disease. Blackwell, Oxford, pp 139–145
4. Du Plessis DJ (1974) The distribution of gastritis in carcinoma of the stomach. Br J Surg 61:521–523
5. Johansson KE, Ask P, Boeryd B, Fransson S-G, Tibbling L (1986) Oesophagitis, signs of reflux, and gastric secretion in patients with symptoms of gastrooesophageal reflux disease. Scand J Gastroenterol 21:837–847
6. Jones R, Lydeard S (1989) Prevalence of symptoms of dyspepsia in the community. Lancet 1:47–51
7. Knill-Jones RP, Card WI, Crean GP, James WB, Spiegelhalter DJ (1984) The symptoms of gastro-oesophageal reflux and oesophagitis. Scand J Gastroenterol 19 [Suppl 106] 19:72–76
8. Loffeld RJLF, Potters HVJP, Stobberingh E, Flendrig JA, van Spreeuwel JP, Arends JW (1989) Campylobacter associated gastritis in patients with non-ulcer dyspepsia: a double blind placebo controlled trial with colloidal bismuth subcitrate. Gut 30:1206–1212
9. Loffeld RJLF, Stobberingh E, Flendrig JA, van Spreeuwel JP, Arends JW (1989) Diagnostic value of an immunoassay to detect anti Campylobacter pylori antibodies in non-ulcer dyspepsia. Lancet 1:1182–1185
10. Martin DF, Montgomery E, Dobek AS, Patrissi GA, Peura DA (1988) Rule out peptic ulcer disease with a question and a blood test. Gastroenterology 94:A285
11. Meister H, Holubarsch C, Haferkamp O, Schlag P, Herfath C (1979) Significance and location of atrophic gastritis and glandular dysplasia in benign and malignant gastric disease.

In: Herfath C, Schlag P (eds) Gastric cancer. Springer, Berlin Heidelberg New York, pp 105–107
12. Newell DG, Rathbone BJ (1989) The serodiagnosis of Campylobacter pylori infection. Serodiagn Immunother Infect Dis 3:1–6
13. Oderda G, Vaira D, Holton J, Ainley C, Altare F, Ansaldi N (1989) Amoxycillin plus tinidazole for Campylobacter pylori gastritis in children: assessment by serum IgG antibody, pepsinogen I and gastrin levels. Lancet 1:690–692
14. Sobala GM, Rathbone BJ, Wyatt JI, Dixon MF, Heatley RV, Axon ATR (1989) Investigating young patients with dyspepsia (Letter). Lancet 1:50–51
15. Vaira D, Holton J, Cairns SR, Falzon M, Polydorou A, Dowsett SF, Salman PR (1988) Antibody titres to Campylobacter pylori after treatment for gastritis. Br Med J 297:297–397
16. Williams B, Luckas M, Ellingham JHM, Dain A, Wicks ACB (1988) Do young patients with dyspepsia need investigation? Lancet 2:1349–1351

Evaluation of a Clinical Role for Serology to *Helicobacter pylori*

G. M. Sobala, B. J. Rathone, T. Shallcross, J. I. Wyatt, M. F. Dixon, R. V. Heatley, and A. T. R. Axon

Introduction

Demand for endoscopy services is growing inexorably and diverse strategies have been proposed to use them more effectively [1, 8]. One suggestion has been centred around the close association between *Helicobacter pylori* and peptic ulcers and the development of accurate serological tests for the presence of this organism. This could allow endoscopy to be targetted to patients serologically positive for *H. pylori*, among whom nearly all peptic ulcers should occur. We have examined the likely results of implementing three different strategies for easing endoscopic workload, using retrospective data from dyspepsia clinics.

Method

Retrospective data was available on 842 patients either randomly or consecutively selected into four different studies from three separate dyspepsia clinics in two hospitals in Leeds over 4 years. All patients had been referred for investigation of dyspepsia by their general practioners. The following data were available for these patients: age, presence/absence of peptic ulcer disease on endoscopy, and *H. pylori* status on the basis of modified Giemsa staining of at least two gastric biopsy specimens. In 696 patients data were available on the recent usage of nonsteroidal anti-inflammatory agents (NSAIDs). The data had been stored in computer files, using three separate proprietary databases. The files were combined and the final dataset was analysed using SAS 5.16 running on an Amdahl mainframe. Due to differences between the original studies, the final dataset could not contain data on patients with gastric cancer, details of the site of the peptic ulcers, or information on other endoscopic findings.

The likely effects of three strategies were examined.

Strategy 1

The chief aim of this strategy to diagnose gastric cancer. The endoscopic diagnosis of peptic ulcer disease, while useful, is not held to be essential [8]. Accordingly

Dr. G. M. Sobala, Gastroenterology Unit, The General Infirmary, Great George Street, Leeds, West Yorkshire LS1 3EX, United Kingdom

Helicobacter pylori, Gastritis and Peptic Ulcer
Edited by P. Malfertheiner and H. Ditschuneit

endoscopy is restricted to patients over a certain cut-off age, set at around 40–45 years. Dyspeptic patients under this age would not be endoscoped as gastric cancer is very rare in these younger patients.

Strategy 2

The aims of this strategy are to use endoscopy both to detect gastric cancer and to make a positive diagnosis of peptic ulcer disease [6, 7]. Accordingly to catch the gastric cancers, as in strategy I, all patients over a cut-off age would be endoscoped. Patients under this age would be screened with serology to *H. pylori*. Those with positive results would be endoscoped. A negative result would be deemed to have excluded peptic ulcer disease, and endoscopy would not be necessary.

Strategy 3

This is a modification of strategy 2, and has the same aims. The modification is based on the belief that a substantial proportion of *H. pylori*-negative peptic ulcers occur in patients using nonsteroidal anti-inflammatory drugs. Therefore all patients over the cut-off age or using NSAIDs would be endoscoped. Other patients would be screened with *H. pylori* serology as outlined above.

The model was used to assess the number of endoscopies that would be saved, and the proportion of all peptic ulcers that would be "missed" by each strategy at a selection of cut-off ages. The assumption was made that serological testing would have produced the same result as to *H. pylori* status as the biopsy Giemsa technique actually used. In practice agreement is likely to be very good (perhaps 95% or better) but not perfect, and so the results must be considered to be slightly "fuzzy". Of course, it is perfectly possible that serodiagnosis of *H. pylori* could yield "better" results than biopsy in these circumstances.

Results

In our study 192 patients (22.8%) had peptic ulcers, while 356 (42.3%) had *H. pylori* present in their gastric biopsy specimens. The results of applying the three strategies are shown in Tables 1 and 2. As the NSAID consumption data were incomplete, strategy III has been assessed in two ways. The first analysis is on all the patients in the dataset: those patients whose NSAID consumption was not recorded are treated as nonconsumers. The second analysis is on the subset of 696 patients in whom NSAID usage was fully documented.

Discussion

An implementation of the first strategy certainly would succeed in reducing the number of endoscopies performed, but at the cost of failing to diagnose a

Table 1. Consequences of implementing the three strategies at different cut-off ages: percentage reduction in total number of endoscopies performed. NSAID data were incomplete so two sets of figures are given for strategy III: (a) analysis on all patients and (b) analysis only for patients with NSAID data available

Endoscopies saved			
Cut-off age	Strategy I	Strategy II	Strategy III (a) (b)
30	22.4	13.3	12.5/12.1
40	40.5	21.9	20.2/19.7
50	60.5	29.0	26.2/26.3

Table 2. Consequences of implementing the three strategies at different cut-off ages: percentage of all peptic ulcers missed. NSAID data were incomplete so two sets of figures are given for strategy III: (a) analysis on all patients and (b) analysis only on patients with NSAID data available

Peptic ulcers missed			
Cut-off age	Strategy I	Strategy II	Strategy III (a) (b)
30	13.5	2.1	1.6/2.0
40	28.1	3.1	2.6/2.0
50	51.6	6.8	4.7/4.0

substantial number of patients with peptic ulcer disease. While it could be argued that these patients' symptoms would reliably respond to empirical H2 antagonists it does seem desirable to have a positive diagnosis to aid in the management of a condition characterised by a relapsing course extending over decades. It is also unclear as to what would happen to those young patients whose symptoms did not respond to H2 antagonists. They would be more likely to have nonulcer dyspepsia than ulcer disease, and this may lead to a paradoxical situation in which endoscopy in the young is performed most often in those least likely to have organic pathology.

An implementation of the second strategy would not save as many endoscopies as the first, but a saving of 21.9% (at a cut-off age of 40 years) would still be worth having, and would free stretched resources significantly. This strategy succeeded very well in not missing peptic ulcer patients: at a cut-off age of 40 years the negative predictive value for ulcer disease was 96.7%, certainly good enough to base mangement plans upon.

The third strategy offered further improvement on these figures, but perhaps not as much as was hoped. This is because although approximately one-half of all *H. pylori*-negative peptic ulcer patients were NSAID users, they were mostly elderly and would have been endoscoped anyway because of their age. *H. pylori* ulcers in the young were rare to begin with, and few of these were secondary to NSAIDs. However, although this strategy was not much better than the second in detecting peptic ulcers, equally it was not much worse at saving endoscopies, and is thus probably preferable. Indeed, the negative predictive value for ulcer disease at a cut-off age of 40 years was 97.8%.

Comparison with Symptom Screening Strategies

An alternative approach to screening patients for endoscopy is based upon the differences in symptomatology between patients with ulcer disease and nonulcer dyspepsia. Several symptom scoring systems have been derived, varying widely in their complexity [1, 3, 4]. How does an approach based on ulcer aetiology (*H. pylori* and NSAIDs) compare with these descriptive systems, based on the clinical manifestations of disease? A comparison between strategy III and two published symptom-scoring systems is shown in Table 3. The serology-based strategy gives considerably superior results to that of Davenport et al. [1] and also to the system of Mann et al. [4] as reanalysed by Davenport et al. [1]. Holdstock et al. [3] reported a system giving only slightly inferior results to ours. However, this study was performed on a more affluent southern UK population with a lower incidence of peptic ulcer disease in young persons (and presumably also of *H. pylori* infection) and indeed was heavily weighted towards not investigating the young. This is why the predecessor of this strategy [4] fared so much worse when transplanted to the poorer North by Davenport et al. [1]: only 9.7% of Mann's patients under 50 years of age had ulcers, compared with 21.6% of Davenport's patients and 19.5% of our own patients in this age group. Our serology-based strategy would actually perform better if applied in a region with a low incidence of *H. pylori* infection and peptic ulcer disease in the young.

We conclude that screening strategies based on *H. pylori* derology may be able to play a major role in reducing endoscopy workload, and are superior to screening strategies based on symptomatology. The problems of implementing such strategies are discussed elsewhere in this volume.

Table 3. Comparison between serology-based screening strategy and three symptom-based screening strategies

Serological screening vs symptom-based screening		
Strategy	Endoscopies saved	Ulcers missed
Strategy III (b) *H. pylori*/NSAID (cut-off age, 40 years)	19.7%	2.0%
Davenport et al. 1985	28.9%	12.5%
Mann according to Davenport et al. 1985	33.2%	14.1%
Holdstock et al. 1986	22.8%	4.1%

References

1. Davenport PM, Morgan AG, Darnborough A, de Dombal FT (1985) Can preliminary screening of dyspeptic patients allow more effective use of investigational techniques? Br Med J 290:217–219
2. Heatley RV, Rathbone BJ (1987) Dyspepsia: a dilemma for doctors? Lancet 2:779–782
3. Holdstock G, Harman M, Machin D, Patel C, Lloyd RS (1986) Prospective testing of a scoring system designed to improve case selection for upper gastrointestinal investigation. Gastroenterology 90:1164–1169
4. Mann J, Holdstock G, Harman M, Machin D, Loehry CA (1983) Scoring system to improve effectiveness of open access endoscopy. Br Med J 287:937–940
5. Martin DF, Montgomery E, Dobek AS, Patrissi GA, Peura DA (1988) Rule out peptic ulcer disease with a question and a blood test. Gastroenterology 94:A285
6. Newell DG, Rathbone BJ (1989) The serodiagnosis of Campylobacter pylori infection. Serodiagn Immunother Infect Dis 3:1–6
7. Sobala GM, Rathbone BJ, Wyatt JI, Dixon MF, Heatley RV, Axon ATR (1989) Investigating young patients with dyspepsia (Letter). Lancet 1:50–51
8. Williams B, Luckas M, Ellingham JHM, Dain A, Wicks ACB (1988) Do young patients with dyspepsia need investigation? Lancet 2:1349–1351

Serum Antibodies to the Vacuolating Toxin Produced by *Helicobacter pylori*

N. FIGURA, O. FANTI, P. GUGLIELMETTI, A. BARBERI, A. ROSSOLINI, A. ZANCHI, M. G. CUSI, R. A. MUSMANNO, M. RUSSI, and S. QUARANTA

Introduction

Cytotoxin is one of the candidate factors of virulence of *Helicobacter pylori* (HP) [1]. Toxigenicity occurs frequently in strains isolated from patients with peptic ulceration [2], and there is serological evidence that cytotoxin production also occurs in vivo. In fact, in immunoblotting tests we saw that all the 31 individuals infected by cytotoxic (CT^+) HP strains tested and 14 of 20 patients (70%) infected by noncytotoxic (CT^-) organisms had serum immunoglobulin G (IgG) which recognized HP cytotoxin-associated proteins (CAP). IgA to CAP were hardly detected, mostly in patients infected by CT^+ strains, while IgM were not detected at all (N. Figura et al., in preparation).

However, although the difference in the detection frequencies of anti-CAP in the two groups of patients was significant ($P < 0.01$, chi-squared test, 1 df), the immunoblotting technique could not differentiate, in certain cases, between an ongoing and a previous CT^+ HP infection. In fact, some patients, considered not infected on the basis of culture, microscopic examination, and rapid urease test of gastric mucosa specimens, also had anti-CAP serum IgG, suggesting that these antibodies were still detectable after bacteriological healing of the infection. In addition, the method was not sensitive enough to clearly detect anti-DAP IgA.

Thus we used the enzyme-linked immunosorbent assay (ELISA), which is more quantitative than immunoblotting, to see whether there were differences in the levels of IgG and IgA to HP cytotoxin in serum of patients infected by CT^+ and CT^- HP strains, and to see whether it was possible to individualize, with a noninvasive method, subjects infected by CT^+ organisms.

Materials and Methods

To compare levels of anticytotoxin antibodies 36 serum samples – 21 from patients with CT^+ HP strains and 15 from patients with CT^- organisms – were tested. Specificity and reliability of the ELISA test procedure were assayed with 23 serum samples, eight from patients with CT^+, eight from patients with CT^- HP strains, and seven from subjects not infected by *H. pylori*.

Istituto di Malattie Infettive, Università di Siena, via P. A. Mattioli, 10, I-53100 Siena[1], Istituto di Microbiologia, Università di Siena[2], Servizio di Gastroenterologia, Divisione Medica II, Ospedale di Grosseto[3], Italy

Helicobacter pylori, Gastritis and Peptic Ulcer
Edited by P. Malfertheiner and H. Ditschuneit

One cytotoxic and one noncytotoxic HP strain were grown in brucella broth with 10% fetal calf serum (FCS) at 37 °C, in a microaerobic environment, at 150 oscillations per min, for 48 h. Broth cultures were then centrifuged and filtered through 0.22 μm-filters.

Microtiter plates were left to sensitize with filtrates and uninoculated broth diluted 1:2 in carbonate buffer pH 9.6 at 4 °C o.n. (200 μl of antigen per well). Wells were then washed three times with 0.1 *M* phosphate-buffered saline (PBS; pH 7.2), and 150 μl of serum diluted in PBS was added to each well. Sera were assayed at 1:40 and 1:160 dilutions. In tests carried out to determine specificity of ELISA, sera were diluted 1:100. Plates were incubated at 37 °C for 90 min. After three washes with PBS-0.5% Tween 20, 150 μl of conjugate (alkaline phosphatase-conjugated (APC) anti-human IgG antibody developed in rabbit, and peroxidase-conjugated (PC) anti-human IgA antibody developed in goat), diluted at the dilutions suggested by the suppliers (Behringwerke, and Sigma Chemical, respectively) were added to each well. After incubation at 37 °C for 90 min, the plates were washed three times with PBS-0.5% Tween 20 and 100 μl of substrate were added to each well (p-nitrophenyl phosphate, and o-phenylenediamine · 2 HCl, respectively, for APC and PC). The reaction was stopped after 30 min with 50 μl 1 *N* sulfuric acid per well for IgA determination, and after 45 min with 50 μl 2 *N* sodium hydroxide per well for IgG determination. Absorbance was read at 492 nm for IgG determination and at 405 nm for IgA determination on a Titertek Uniskan reader (Flow Laboratories), and results were reported as optical density ±SD × 1000. All tests were performed in duplicate. The precision of the ELISA test procedure was determined as differences of intra- and interassay readings obtained with all 23 serum samples on wells sensitized with CT^+ filtrate.

Results and Discussion

Mean serum IgG and IgA levels to uninoculated broth were similar or not significantly different in the three groups of patients (Table 1). Levels of IgA to the CT^- filtrate were similar in the three groups, and levels of IgG in infected patients

Table 1. Mean serum IgG and IgA levels to cytotoxic and noncytotoxic filtrates and to uninoculated broth

Antigen	Antibody class	Patients infected by *H. pylori*		Noninfected patients
		Cytotoxic	Noncytotoxic	
Uninoculated broth	IgG	143±102	123±103	83±50
	IgA	90±53	96±89	91±91
Noncytotoxic broth culture filtrate	IgG	214±142	141±63	97±48
	IgA	127±64	150±60	178±154
Cytotoxic broth culture filtrate	IgG	387±227	357±249	105±68
	IgA	614±359	393±145	155±61

Table 2. Mean IgG and IgA levels to cytotoxic filtrate in sera of patients infected by CT^+ and CT^- organisms

	Serum diluted 1:40		Serum diluted 1:160	
	CT^+	CT^-	CT^+	CT^-
IgG	166 ± 111	154 ± 133	115 ± 77	96 ± 81
IgA	327 ± 179 [a]	142 ± 145 [a]	175 ± 116 [b]	47 ± 60 [b]

[a] $P = 0.0019$
[b] $P = 0.0003$ (Student's t test)

were not significantly different. IgG levels to CT^+ filtrate were similar in HP-infected patients, while IgA levels in serum samples of subjects infected by CT^+ HP strains were significantly higher (Table 2).

The intraassay variations of the ELISA tests were 9.0% for IgG and 0.8% for IgA. The interassay variations were 34.4% for IgG and 13.5% for IgA.

Results referring to the significant difference between IgA levels to cytotoxic filtrate in CT^+ and CT^- HP-infected patients were confirmed by further tests in which a higher number of sera were assayed: 21 were collected from patients infected by CT^+ HP strains and 15 from patients infected by CT^- organisms. In these tests, wells sensitized with the filtrate of a noncytotoxic strain were used as "blanks" in order to avoid the interference exerted by the antigens present in the CT^- filtrate. Patients infected by CT^+ HP strains showed serum IgA to the toxic filtrate at levels significantly higher than subjects infected by CT^- organisms (Table 2). The IgG levels were similar in the two groups of patients, while IgM, looked for in eight serum samples of patients infected by CT^+ organisms, were not detectable at all.

These results confirm in part those obtained with the immunoblotting technique. In particular, both techniques (immunoblotting and ELISA) indicated that the toxin was also produced in vivo since the infected patients had antibodies to it. The anticytotoxin IgA measurement, together with the determination of antibodies to HP cellular antigens, can contribute to a better definition of the immune status of the patients. In addition, the anti-CAP IgA measurement may be helpful in the diagnosis of HP-associated ulcer. In fact, the IgA levels to the toxic filtrate in our patients with peptic ulcer were significantly higher than in patients without ulcer (201 ± 176 vs 84 ± 95, $P = 0.014$, with 1:160 serum dilution), indicating that toxigenicity may be relevant to the pathogenesis of the mucosal alterations caused by *H. pylori*.

References

1. Leunk RD, Johnson PT, David BC, Kraft WG, Morgan DR (1988) Cytotoxic activity in broth-culture filtrates of Campylobacter pylori. J Med Microbiol 26:93–99
2. Figura N, Guglielmetti P, Rossolini A, Barberi A, Cusi G, Musmanno RA, Russi M, Quaranta S (1989) Cytotoxin production by Campylobacter pylori strains isolated from patients with peptic ulcers and from patients with chronic gastritis only. J Clin Microbiol 27:225–226

Local Immune Responses to *Helicobacter pylori* Infections

A. R. Stacey, P. R. Hawtin, and D. G. Newell

It is widely accepted that the majority of people with *H. pylori* infections amount a large circulating antibody response against that organism [1]. This response has proved to be so predictable that it forms the basis of serodiagnostic tests such as ELISA [2, 3]. The sensitivity and specificity of such tests depends largely on the antigen preparations used, although ELISAs using a variety of crude and more purified antigens are reported to be greater than 90% sensitive and specific for detecting *H. pylori* infection [4, 5].

Immunogold labelling techniques clearly demonstrate that the antibodies of this circulating immune response are directed against the surface of the organism including the body, the flagella and the loosely bound material. These antibodies are mainly of the IgG class although specific IgA may also be raised but rarely IgM. The IgG1 subclass predominates with some IgG2 and IgG4 but no IgG3 [6].

Despite this antibody response the organism continues to colonise the gastric epithelium and will persist unless eliminated by treatment. Hence, the circulating antibody response appears to be unable to eradicate the infection. Moreover, it is unable to prevent recrudescence following temporary clearance of the organism [7].

The persistence of this infection forces us to address the question why is this immune response apparently ineffective?

It is of course possible that the circulating immune response does not reflect responses at the site of infection. We must therefore establish whether the gastric mucosa is an immunocompetent surface capable of mounting a local immune response.

Immunohistological techniques demonstrate the presence of immunoglobulin secreting B cells within the gastric mucosa of *H. pylori* colonised patients. Moreover, specific anti-*H. pylori* antibodies are present in gastric juice [8] and in tissue culture supernatants of gastric mucosal [9] and duodenal [10] biopsy specimens taken from *H. pylori*-positive patients. The majority of these locally produced antibodies, unlike the circulating antibodies, are of the IgA class. Some IgG antibodies mainly of the IgG1 subclass are also present. Small amounts of specific IgG2 and IgG4 antibodies can also be detected (Fig. 1). Such antibody classes provide appropriate responses for effective elimination of bacterial infections at mucosal surfaces. In particular the production of IgA class antibodies

Immunodiagnostics Group, PHLS-CAMR Porton Down, Salisbury, Wilts., U.K.

Helicobacter pylori, Gastritis and Peptic Ulcer
Edited by P. Malfertheiner and H. Ditschuneit

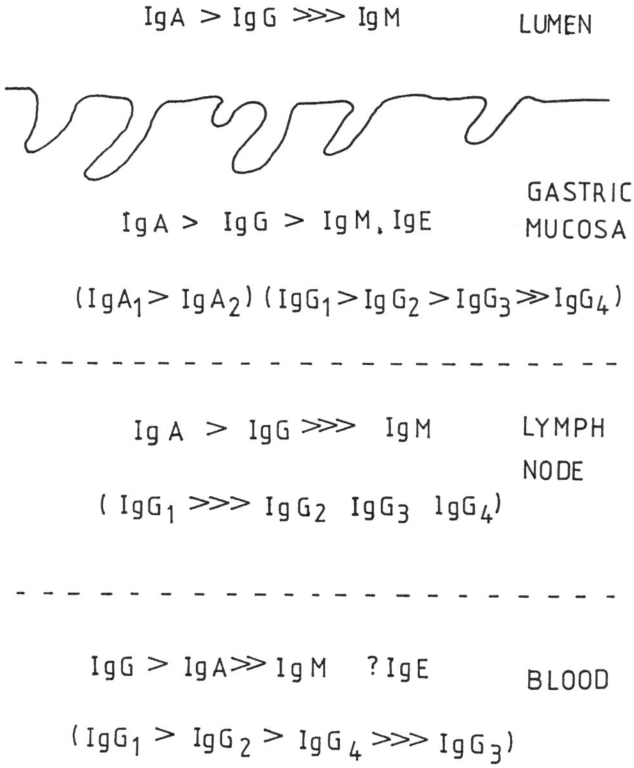

Fig. 1. The distribution of antibody class in systemic and locally produced anti-*H. pylori* antibodies. (Note the major class of antibody present in the serum is IgG whereas, in the locally produced antibodies it is IgA)

prevent the adherence of the organism to the gastric epithelium. Equally IgG1 antibodies may be involved in the activation and fixation of complement.

At the cellular level immunohistological studies of gastric mucosal biopsy specimens also demonstrate the presence of increased numbers of T cells, of both the CD4 positive and CD8 positive subsets and also antigen presenting cells in areas of *H. pylori*-associated gastritis [11].

This evidence indicates that the gastric mucosa is an immunocompetent surface and that *H. pylori* colonised patients mount a specific mucosal immune response against the organism which is, nevertheless, ineffective in clearing the organism.

Several reasons for this ineffectiveness can be postulated. These fall into two main categories (Table 1) related either to the ability of the organism to avoid the host immune response or to the hosts' ability to respond to relevant antigens.

Table 1. Potential reasons for the apparent ineffectiveness of the immune response

Bacterial factors	Host factors
Antigenic drift	No response to relevant antigens
Loss of antigen antibody complexes	Incorrect response to antigens
Nonspecific absorption of Ig	
Nonspecific activation of host	

Little is known about the immunocompetence of the gastric mucosa during chronic bacterial infections. Investigating the role of the local immune response in *H. pylori*-associated gastritis presents the investigator with several problems. First, the quantity of material available is a limiting factor; namely the size and number of the biopsy specimens obtainable. Second, the techniques currently used impose limitations. For example, immunohistological techniques using monoclonal antibodies have enabled identification of the infiltrating cell types and study of their patterns of distribution. However, techniques such as this do not allow the antigenic specificity of these infiltrating cells to be established. Such information requires the isolation and expansion of the T and B cell populations from gastric mucosal biopsy material or mucosally associated lymphoid tissue by cellular cloning techniques. We have used the technique of EBV transformation to immortalise B cells isolated from lymph nodes draining the stomach taken at vagotomy.

The antigenic specificity of these antibodies produced by the host has been studied by ELISA using FPLC fractionated *H. pylori* whole cell sonicates [12] as antigen (Fig. 2). Using this method of preparation antigens are obtained in their native three-dimensional state. The cloned and immortalised cells produced antibodies which were found to share some of the antigenic specificity seen in systemic and locally-produced antibodies from gastric mucosal biopsy material. These antibodies are probably directed against the flagella. Antibodies produced by the nontransformed lymph-node cells fully reflect the patterns of antigenic specificity seen with the systemic antibodies [9].

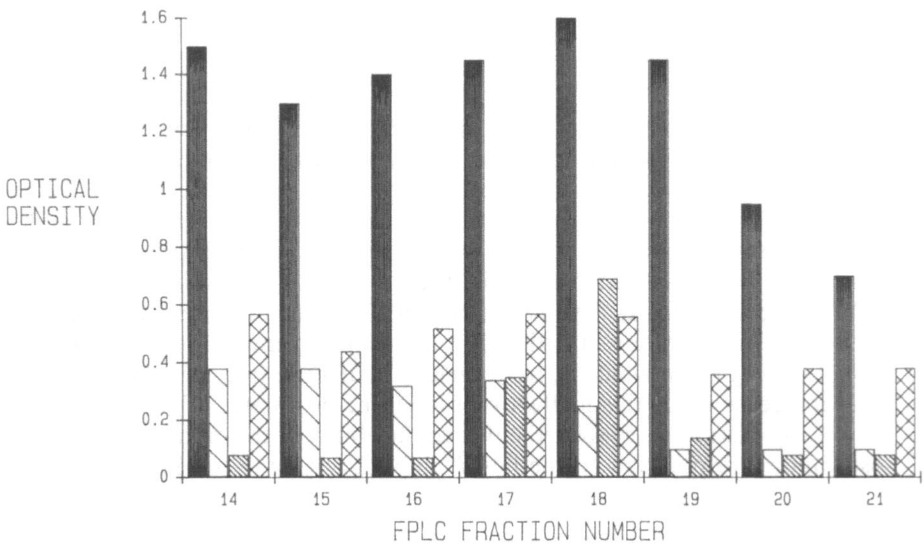

Fig. 2. The antigenic specificity of systemic and locally produced antibodies as determined using FPLC fractionated *H. pylori* sonicates as antigen. The systemic antibodies ■, and antibodies from the gastric biopsy specimens ⊠ and nontransformed lymph-node cells ⧅ reacted with all the FPLC antigens including the urease (fractions 14/15) and the putative flagella (fractions 17/18) whereas, antibodies produced by the transformed lymph-node cells ▧ reacted solely with the putative flagella

The host therefore clearly mounts a local antibody response against *H. pylori*. The presence of infiltrating CD 4 positive and CD 8 positive T cells within the lamina propria indicates that there is also a large cellular immune response occurring. Antigen is evidently presented to antigen specific CD 4 positive T helper cells by antigen presenting cells including infiltrating monocytes or possibly directly by the gastric epithelial cells. T cell help derived from the CD4 positive T cells would then lead to B cell differentiation and ultimately antibody synthesis and secretion. The release of T cell growth factors from the CD 4 positive T cells would lead to increased T and B cell infiltration and increased antibody secretion. Cellular infiltration and proliferation such as this would require considerable immunoregulation. This would involve the stimulation and proliferation of the other T cell subset, the CD 8 positive cytotoxic cells. *H. pylori*-associated gastritis may, therefore, be an immunopathological disease subsequent to the regulation of the local immune response.

In conclusion colonisation of the gastric mucosa by *H. pylori* continues despite an immunologically hostile microenvironment. It would, therefore, appear that the persistence of the infection is largely dependent upon the mechanisms by which *H. pylori* evades the host immune response.

Since *H. pylori* is the only organism to commonly colonise the human gastric epithelium we can assume that the immunocompetent cells present during *H. pylori* associated gastritis are there as a result of that infection. This provides us with a unique opportunity to investigate the mechanisms by which gastric mucosal immune responses during chronic bacterial infections are initiated and controlled.

References

1. Newell DG, Rathbone BJ (1989) The serodiagnosis of Campylobacter pylori infection – a review. Serodiagn Immunother 3:1–6
2. Newell DG, Johnston BJ, Ali MH, Reed PI (1988) An enzyme-linked immunosorbent assay for the serodiagnosis of Campylobacter pylori-associated gastritis. Scand J Gastroenterol 23 [Suppl 142]:53–57
3. Jones DM, Eldridge J, Fox AJ, Sethi P, Whorwell PJ (1986) Antibody to the gastric campylobacter like organism ("Campylobacter pyloridis") – clinical correlations and distribution in the normal population. J Med Microbiol 22:57–62
4. Newell DG, Stacey AR (1989) Antigens for the serodiagnosis of Campylobacter pylori infections. Gastroenterol Clin Biol 13:37B–41B
5. Goodwin CS, Blincow E, Peterson G, Saunderson C, Cheng W, Marshall B, Warren JR, McCullouch R (1987) Enzyme linked immunosorbent assay for Campylobacter pyloridis: correlation with the presence of C. pyloridis in the gastric mucosa. J Infect Dis 155:388–494
6. Steer HW, Hawtin PR, Newell DG (1987) An ELISA technique for the serodiagnosis of Campylobacter pyloridis infection in patients with gastritis and benign duodenal ulceration. Serodiagn Immunother 1:253–259
7. Langenberg W, Rauws EAJ, Widjokusumo A, Tygat GNJ, Zanen HC (1986) Identification of Campylobacter pylori isolates by restriction endonuclease DNA analysis. J Clin Microbiol 24:414–417
8. Rathbone BJ, Wyatt JI, Worsley BW, Shires SE, Trejdosiewicz LK, Heatley RV, Losowsky MS (1986) Systemic and local antibody responses to Campylobacter pyloridis in non-ulcer dyspepsia. Gut 27:642–647
9. Newell DG, Stacey AR (1989) The isotype and specificity of local and systemic anti-Campylobacter pylori antibodies. Proceedings of the second international symposium on Campylobacter pylori. Bad Nauheim, Springer, Berlin Heidelberg New York (in press)

10. Crabtree JE, Rathbone BJ, Shallcross T, Wyatt JL, Heatley RV (1989) Local humoral responses to Campylobacter pylori in patients with duodenitis. Poster presentation, 2nd Meeting of the European Campylobacter pylori Study Group, Ulm
11. Engstrand L, Schneyus A, Phalson C, Grimelius L, Schawn A, Gustavasson S (1989) Association of Campylobacter pylori with induced expression of class II transplantation antigens on gastric epithelial cells. Infect Immun 57:827–832
12. Stacey AR, Newell DG (1989) The development of antigens for improved serodiagnostic tests of Campylobacter pylori-associated gastritis. In: Mégraud F, Lamouliatte H (eds) Proceedings of the workshop of gastroduodenal pathology and Campylobacter pylori, Bordeaux. Elsevier, Amsterdam, p 193

Molecular Cloning of Immunodominant Antigen Epitopes of *Helicobacter pylori* for Development of a Specific ELISA to Diagnose *H. pylori* Infection*

C. L. Clayton, H. Kleanthous, J. Herrmann, E. Jacobs, M. Kist, and S. Tabaqchali

Helicobacter (*Campylobacter*) *pylori* has been implicated in the etiology of gastritis and peptic ulcer disease. *H. pylori* infection is normally diagnosed by microscopy and culture of biopsy samples. However, this is not only invasive but is time consuming and expensive. Various noninvasive serological techniques have been developed for diagnosing infection and ELISA is the most commonly used test. ELISAs that are based on *H. pylori* whole cells, sonicates or crude extracts have levels of sensitivity and specificity of 70%–90% [4]. A slight improvement has generally been reported by using acid cell extracts [7]. However, all these preparations contain antigens that cross-react with other bacterial species and cause false-positive reactions [5]. Adjusting the cut-off value only increases the number of false negatives. Second-generation ELISA systems have been developed which are based on high molecular weight conserved *H. pylori* specific purified antigens [4, 5]. They have high sensitivity and specificity and contain urease as the major antigenic component. High molecular weight fractionated antigens of *H. pylori* have also been isolated from SDS-PAGE gels [6] and these preparations may also be used for a *H. pylori* specific ELISA system. Molecular cloning of *H. pylori* immunodominant polypeptides can provide a useful, quick, high yield of *H. pylori* antigens that can be standardised and used for a specific diagnostic ELISA. We have previously reported the cloning of genes encoding 66 and 31 kDa immunodominant polypeptides of *H. pylori* in *E. coli* [3]. These antigens were demonstrated to be conserved, *H. pylori*-specific subunits of the *H. pylori* urease enzyme [2].

Immunoblot studies with human sera from patients infected with *H. pylori* have shown that the majority of them have antibodies to the 66 and 31 kDa urease polypeptides [8]. Another *H. pylori* immunodominant polypeptide is a 120 kDa surface exposed antigen that has been found to be reactive with over 80% of *H. pylori* antibody positive sera [1]. We therefore set out to try and clone the 120 kDa polypeptide gene in *E. coli*.

---

* This work was supported by grants from the Wellcome Trust and the British Medical Research Council

Department of Medical Mircobiology, St. Bartholomew's Hospital Medical College, West Smithfield, London, EC1A 7BE, UK
Institute for Medical Microbiology and Hygiene, University of Freiburg, Freiburg, F.R.G.

Helicobacter pylori, Gastritis and Peptic Ulcer
Edited by P. Malfertheiner and H. Ditschuneit

Methods

The 120 kDa membrane protein was purified from *H. pylori* by preparative SDS-PAGE and electroelution. A gene bank of *H. pylori* DNA in *E. coli* was constructed by cloning *EcoR1* cleaved DNA fragments (0–7 kb) into the bacteriophage λgt11 expression vector. The DNA is inserted within the beta-galactosidase gene and is expressed as a fusion protein with beta-galactosidase (mol. wt. 114 kDa). The bacteriophage library was plated on *E. coli* Y 1090 and Lac Z directed gene expression was induced by the addition of IPTG. Rabbit antisera was raised to the purified 120 kDa antigen, adsorbed with *E. coli* Y 1090 cells and used to screen the λgt11 library for antigen recombinants by in situ ELISA of plaques replicated onto nitrocellulose filtres.

Results

The 120 kDa protein of *H. pylori* was demonstrated to be immunodominant by immunoblot analysis of *H. pylori* total proteins with both *H. pylori* whole-cell rabbit antisera and seropositive human sera (Fig. 1). Two positive antigen-producing recombinants (120λ1, 120λ2) were detected on screening the gene

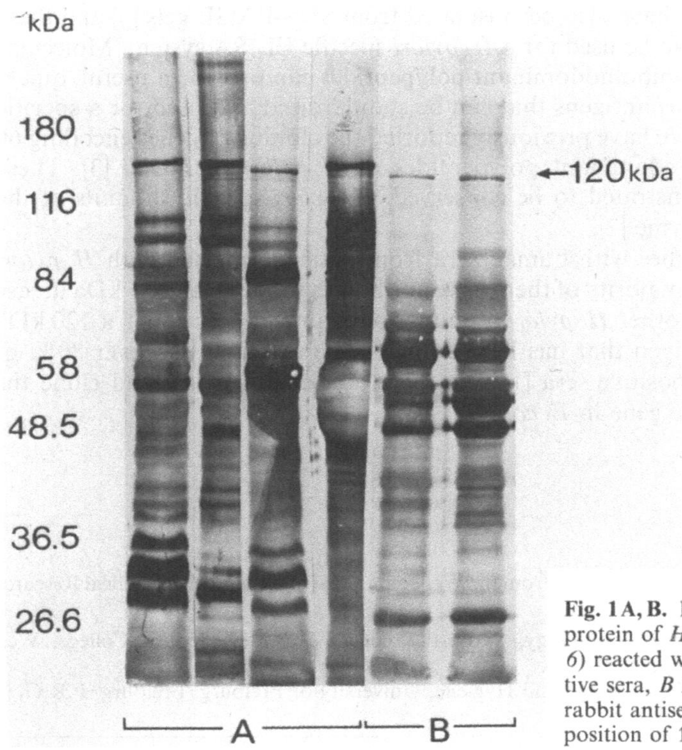

Fig. 1A, B. Immunoblots of total protein of *H. pylori* strains (*lanes 1–6*) reacted with: *A* human seropositive sera, *B H. pylori* whole cell rabbit antisera. *Arrow* indicates position of 120 kDa antigen

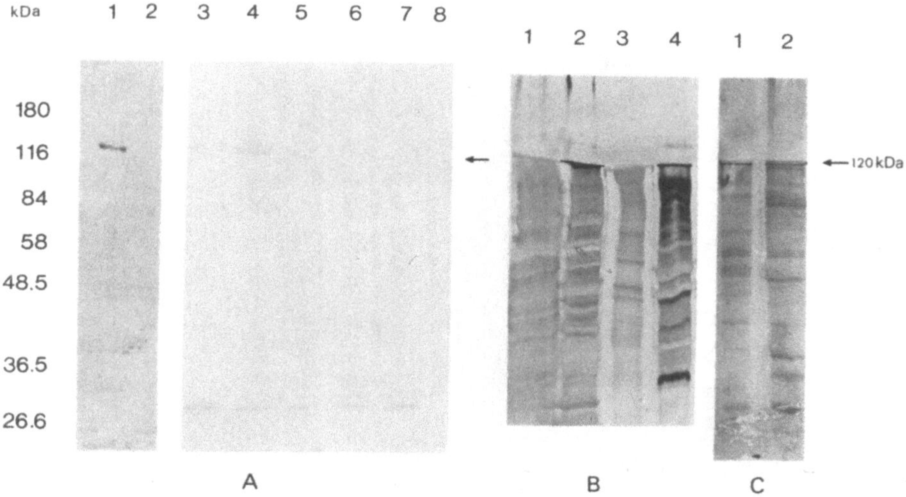

Fig. 2A–C. Immunoblots of proteins of: **A** Positive phage recombinant 120λl lysate (*lane 1*), negative control 120λ-lysate (*lane 2*), recombinant lysogens 120λL5 (*lanes 3, 4*), 120λL13 (*lane 5*), 120λL5 with no IPTG (*lane 6*), 120λL48 (*lane 7*), and beta-galactosidase (*lane 8*) reacted with adsorbed 120 kDa antiserum I. *Arrow* indicates position of fusion proteins. **B** recombinant lysogens 120λL48 (*lane 1*), 120λL5 (*lane 2*), beta-galactosidase (*lane 3*), and *H. pylori* CP 246 total protein (*lane 4*) reacted with absorbed 120 kDa antiserum II. **C** *H. pylori* CP49, CP178 total protein (*lanes 1, 2*) reacted with adsorbed antiserum developed against 120λl positive recombinant lysate. *Arrow* indicates position of 120 kDa antigen

library and were then purified. The recombinant phage 120λ2 was found to encode an IPTG inducible B-galactosidase fusion protein of approximate mol. wt. of 130 kDa. Recombinant lysogens were prepared by lysogenising *E. coli* Y 1089 with the recombinant phage. The lysogen derived from 120λ1 (120λL5) was also found to be induced by IPTG and had a slightly higher molecular weight of approximately 140 kDa. Immunoblot analysis showed that the phage recombinant and the derived lysogens produced fusion proteins which reacted with the adsorbed 120 kDa specific antiserum (Fig. 2). This reaction was due to the presence of a *H. pylori* antigen epitope in the fusion protein as there was no reaction with beta-galactosidase alone (Fig. 2). Antisera raised to the lysate of the positive recombinant phage 120λ1 was found to react with the recombinant lysogen fusion protein (120λL5) and with the 120 kDa polypeptide of *H. pylori* strains (Fig. 2B, C) suggesting that the cloned epitope was part of the 120 kDa polypeptide of *H. pylori*.

DNA was extracted from the recombinant phage 120λ1 but the insert size could not be determined as the phage DNA would not digest with *EcoR1*. Digestion of 120λ1 DNA with the enzymes *Kpn1* and *Sst1* whose sites flank the *EcoR1* cloning site suggested that the recombinant phage contained 0.3 kb of *H. pylori* DNA as part of a 2.4 kb *Kpn1-Sst1* fragment. Southern hybridisation confirmed that 120λ1 contained *H. pylori* DNA sequences and that they were

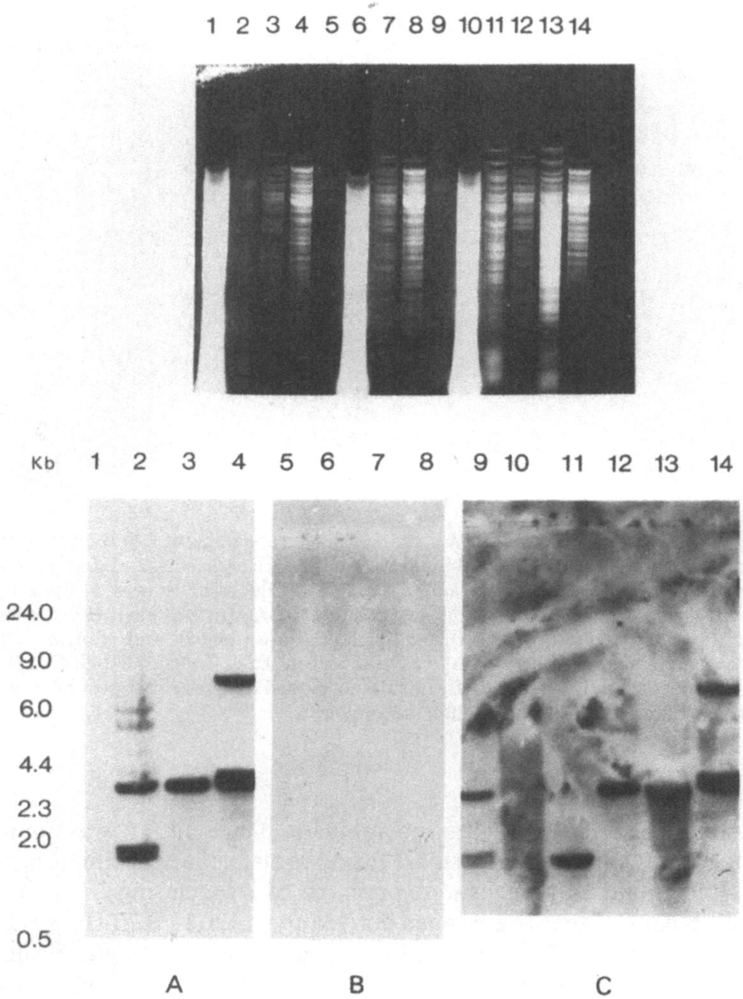

Fig. 3A–C. Southern hybridisation of total DNA *Hae* III digests of *H. pylori* strains CP 290 (*landes 1, 6, 10*), *CP 195* (*lanes 2, 5, 9*), CP 178 (*lanes 3, 7, 12*), CP 126 (*lanes 4, 13*), CP 150 (*lane 11*), and CP 296 (*lane 13*) with ^{32}P labelled. **A** The 2.4 kb *Kpn* 1-*Sst 1* fragment of positive recombinant 120λl. **B** Total DNA of control λgtgapll. **C** Total DNA of positive recombinant phage 120λl

located on the 2.4 kb *Kpn 1-Sst 1* fragment (Fig. 3). The DNA insert was found not to be homologous with the digested total DNA of *Campylobacter jejuni* and *Campylobacter coli*. The 2.4 kb *Kpn 1-Sst 1* probe hybridised to a number of *H. pylori HaeIII* DNA fragments of between 1.8 to 8.5 kb (Fig. 3). The probe detected homologous sequences in 7/9 (78%) of *H. pylori* strains.

Discussion

The 120 kDa antigen was found to be a major immunodominant polypeptide of *H. pylori*. This antigen is probably the same as the 110 and 128 kDa polypeptides that have been reported to be highly immunogenic components of *H. pylori* [6, 9]. *H. pylori* specific epitopes of the 120 kDa antigen were expressed as *Lac* promoter dependent beta-galactosidase fusion proteins. The DNA encoding this epitope will be used as a radiolabelled probe to screen plasmid or cosmid *H. pylori* gene libraries in order to try and clone the entire 120 kDa gene. The epitope DNA sequence was found to be absent from some *H. pylori* strains. These strains could represent the *H. pylori* isolates that have been found not to express this antigen [1] or the DNA encoding for the cloned epitope may represent a divergent region of the 120 kDa gene. The majority of sera from infected patients have been found to react with 66 and 31 kDa [7] and 63 and 28 kDa polypeptides [9] which probably represent the cloned urease antigens. However, it has been reported that circulating anti-urease antibodies detected by FPLC purified urease ELISA are not produced by all infected patients [8] and it seems a more complex antigenic preparation will be necessary to detect all patients infected with *H. pylori*. A combination of the cloned 120 kDa antigen and the cloned urease polypeptides should give a more standardised specific ELISA for diagnosing *H. pylori* infection.

Acknowledgements. Thanks to Ann Carey and Sandy Gale for typing of the manuscript.

References

1. Apel I, Jacobs E, Kist M, Bredt W (1988) Antibody response of patients against a 120 kDa surface protein of Campylobacter pylori. Zentralbl Bakteriol Microbiol Hyp 268:271–276
2. Clayton CL, Topping AJ, Kleanthous H, Wren BW, Mullany P, Tabaqchali S (1989) Molecular cloning and characterisation of Campylobacter pylori species-specific antigens in Escherichia coli K12. In: Mégraud F, Lamoulliatte P (eds) Gastroduodenal pathology and Campylobacter pylori. Elsevier, Amsterdam, pp 115–118
3. Clayton CL, Wren BW, Mullany P, Topping A, Tabaqchali S (1989) Molecular cloning and expression of Campylobacter pylori species-specific antigens in Escherichia coli K-12. Infect Immun 57:623–629
4. Dent JC, McNulty CAM, Uff JS, Gear MLW, Wilkinson SP (1988) Campylobacter pylori urease: a new serological test. Lancet 1:1002
5. Graham DY, Evans DJ, Evans DG (1989) Detection of Campylobacter pylori infection. Lancet 2:569
6. Hirschl AM, Pletschette M, Hirschl MH, Berger J, Stanek G, Rotter ML (1988) Comparison of different antigen preparations in an evaluation of the immune response to Campylobacter pylori. Eur J Clin Microbiol Infect Dis 7:100–105
7. Newell DG (1987) Human serum antibody responses to the surface protein antigens of Campylobacter pylori Serodiagn Immunother 1:209–213
8. Newell DG, Stacey AR, Hawtin PR, Clayton CL (1989) The antigenicity of Campylobacter pylori urease. 5th International Workshop on Campylobacter Infections, Puerto-Vallarti
9. Von Wulffen H, Grote HJ, Gatermann S, Loning T, Berger B, Buhl C (1988) Immunoblot analysis of immune response to Campylobacter pylori and its clinical associations. J Clin Pathol 41:653–659

The Effect of Treatment on Circulating Anti-*Helicobacter pylori* Antibodies – a Two-Year Follow-Up Study

D. G. NEWELL, G. D. BELL, J. WEIL, P. JONES, P. GRANT, and G. HARRISON

The majority of patients colonised with *Helicobacter pylori* elicit a detectable, systemic antibody response against this organism. The isotypes involved in this response are consistent with a chronic bacterial infection of a mucosal surface, with IgG and IgA predominating and IgM rarely seen [5]. This response forms the basis of a noninvasive, ELISA diagnostic test for *H. pylori* infection [4], which is suitable for screening patients prior to endoscopy [3]. However, the value of this test for monitoring treatment is questionable [2] because chronic bacterial infections tend to induce predominantly long-lived antibody responses. Nevertheless, serology is the only rapid, cheap, noninvasive and readily accessible diagnostic test available. We have, therefore, attempted to quantitatively assess the persistence of circulating anti-*H. pylori* antibodies after eradication of the organism, as determined using the ^{14}C-urea breath test.

Materials and Methods

To date 101 patients have been investigated as part of ongoing treatment studies. All patients were diagnosed as being colonised with *H. pylori* on the basis of culture, rapid urease tests and/or histology of endoscopic biopsy specimens. All patients were given an antimicrobial treatment regimen.

^{14}C-Urea breath tests [1] were performed before treatment, at the end of treatment, and 1, 3, 6, 12, and 24 months post-treatment, to monitor *H. pylori* status. The threshold of positivity for the breath test was taken as an area under the curve of less than 40. Those patients in whom the ^{14}C breath test was still positive at 1 month post-treatment were retreated.

Antibody levels were determined before treatment ($n = 42$), at the end of treatment ($n = 19$) and at 1 month ($n = 30$), 3 months ($n = 8$), 6 months ($n = 21$), 12 months ($n = 19$) and 24 months ($n = 9$) post-treatment. The serology was performed using a ELISA technique as described by Newell et al. [4], with acid extractable antigenic material, excepting that the serum dilution was 1:200. IgG and IgA were detected by direct ELISA with class-specific antibody peroxidase conjugates. IgG subclasses were determined by indirect ELISA with mouse monoclonal subclass-specific antibodies. Bound mouse IgG was then detected with anti-mouse IgG antibody peroxidase conjugate [5]. Antibody concentrations

PHLS, Centre for Applied Mircobiology and Research, Porton Down, Salisbury, SP4 OJG, Department of Medicine and PHLS, The Ipswich Hospital, Ipswich, IP4 5PD

Helicobacter pylori, Gastritis and Peptic Ulcer
Edited by P. Malfertheiner and H. Ditschuneit

were determined using a standard curve of immunoglobulin mass bound against optical density for each of the detection systems used.

Using this assay system for IgG antibodies, and a threshold of 10 μg/ml, on 45 unrelated patients of accurately defined *H. pylori* status (24 positive and 23 negative) as detected by culture, histology, rapid urease test and ^{14}C breath test the sensitivity and specificity of the test for detecting *H. pylori* infection was 95% and 100% respectively.

Results

When the specific IgG levels of all patients were analyzed on the basis of treatment period there were no significant decreases in IgG antibody levels either during treatment or in those patients who were unsuccessfully treated (Fig. 1). However, the specific antibody levels in patients successfully treated fell progressively throughout the period of study. The observed differences were statistically significant, from the pretreatment levels, at 1, 3, 6, 12 and 24 months post-eradication ($P > 0.05$, $P > 0.05$, $P > 0.001$, $P > 0.001$, and $P > 0.001$ respectively). Additionally, there was a progressive fall in specific IgA levels, starting by the end of treatment and continuing throughout the period studied except for the 6-month group (Fig. 2). This anomaly was due to two patients who were breath test negative but were inexplicably still IgA antibody positive, although their IgG was low. Interestingly, unsuccessfully treated patients, who were still breath test positive at 1 month, nevertheless showed a relative fall in specific IgA.

The IgG class and subclass antibody response to successful treatment has been followed in some of the patients during the early stages of eradication. About 70% of these patients showed a fall in specific IgG by 1 month and 90% by 3 months but 10% did not demonstrate any falling antibody levels even 3 months post-eradication. In these patients the specific IgG1 response largely mirrored the total

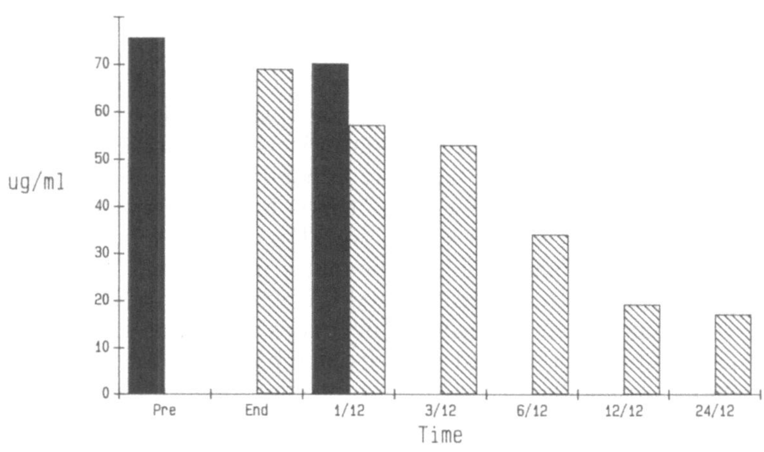

Fig. 1. Anti-*H. pylori* acid extract IgG antibodies in treated patients. ■, ^{14}C breath test positive; ▧, ^{14}C breath test negative

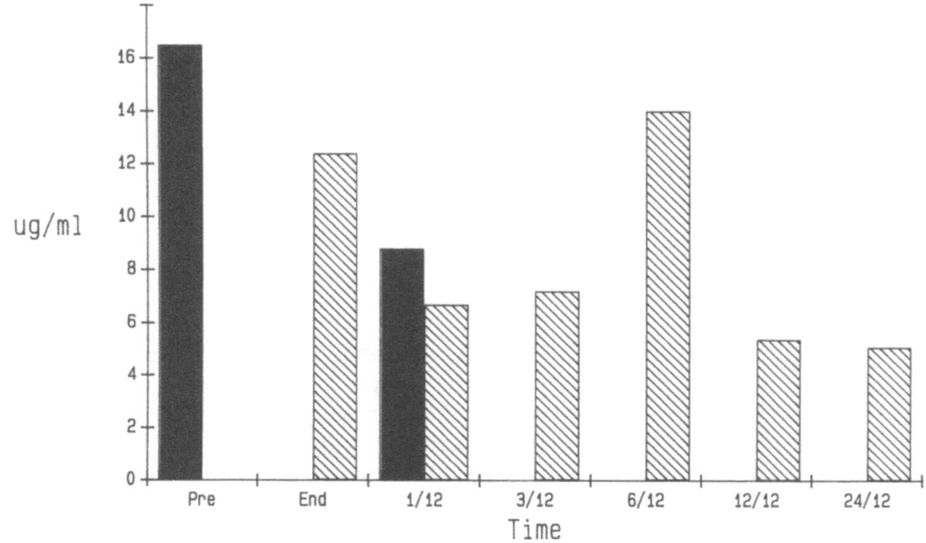

Fig. 2. Anti-*H. pylori* acid extract IgA antibodies in treated patients. ■, ^{14}C breath test positive; ▧, ^{14}C breath test negative

IgG response with the majority of patients showing a fall by 1 month but a few remaining unchanged at 3 months. Most of these patients showed a more rapid fall in specific IgG2. Moreover those patients in whom the IgG1 remained unchanged at 3 months showed a fall in IgG2 by 1 month. IgG4 antibodies also fell significantly in these patients, notably in several cases by the end of treatment.

Conclusions

During successful treatment there is a steady fall in antibodies. Conversely, patients who are unsuccessfully treated show little or no fall in IgG antibodies. This fall occurs in both IgG and IgA class antibodies. IgG subclass antibodies were analyzed because they may respond more rapidly to clearance of infection. The fall in IgG is significant by 1 month post-treatment though IgA, IgG2 and IgG4 may fall a little sooner. IgG1 is the most prevalent of the circulating antibody responses and predominates in the local IgG response so it is not surprising that it reflects the total IgG response. An IgG2 response is a frequent occurrence in chronic bacterial infections at mucosal surfaces but IgG2, like IgG4 is a relatively short-lived antibody, and both responses though small appear to be more predictable of the current infection than total IgG. IgG3 antibodies are rarely raised in *H. pylori* infections [5].

IgA is also a potentially interesting antibody for monitoring current infection as it is the predominant response at the gastric mucosal surface. However, not all patients produce a detectable IgA circulating response. Nevertheless, the significant fall in IgA following treatment, albeit unsuccessful, suggests that the IgA antibody level generally reflects the reduced antigenic load.

It would appear from these preliminary studies that serology is useful in determining the long-term effectiveness of treatment. A word of caution, however: even by 24 months post-eradication, as determined by the ^{14}C-urea breath test, some patients continued to have a high antibody level, significantly above the 10 μg/ml threshold we would usually apply to antibody seronegative patients. It has yet to be established if this is a reflection of a low-level persistent infection.

References

1. Bell GD, Weil J, Harrison G, Morden A, Jones PH, Gant PW, Trowell JE, Yoong AK, Daneshmend TK, Logen RFA (1987) ^{14}C-urea breath test analysis, a non-invasive test for Campylobacter pylori in the stomach. Lancet 1:1367–1368
2. Langenberg W, Rauws EAJ, Houthoff HJ, Oudbier JH, van Bohemen CG, Tytgat GNJ, Rietra PJGM (1988) Follow-up study of individuals with untreated Campylobacter pylori-associated gastritis and of noninfected persons with non-ulcer dyspepsia. J Infect Dis 157:1245–1249
3. Newell DG, Rathbone BJ (1989) The serodiagnosis of Campylobacter pylori infection. Serodiagn Immunother Infect Dis 3:1–6
4. Newell DG, Johnston BJ, Ali MH, Reed PI (1988) An enzyme-linked immunosorbent assays for the serodiagnosis of Campylobacter pylori-associated gastritis. Scand J Gastroenterol (Suppl 142) 23:53–57
5. Steer HW, Hawtin PR, Newell DG (1987) An ELISA technique for the serodiagnosis of Campylobacter pyloridis infection in patients with gastritis and benign duodenal ulceration. Serodiagn Immunother Infect Dis 1:254–259
6. Weil J, Bell GD, Harrison G, Trowell J, Gant P, Jones P (1989) Use of ^{14}C-urea breath test to monitor treatment response in C. pylori positive patients – two-year follow-up figures. Klin Wochenschr [Suppl 18] 67:70

Local and Systemic Immune Response – a Comment

B. Rathbone

Soon after the original isolation of *H. pylori* it became obvious that there was a systemic immune response in colonized subjects. Since that time groups world wide have been involved in studying these responses and the workshop on "Local and Systemic Immune Responses" held at the second meeting of the European *Helicobacter pylori* Study Group enabled different aspects of the immune response to *H. pylori* to be discussed, the contributions above representing some of the topics discussed.

Humoral responses to *H. pylori* as with many other microbial infections can be diagnosed by a number of different techniques. Although the protein profiles of different *H. pylori* strains appear similar, the antigen specificity of the systemic humoral response shows considerable variation between different subjects. As a result of this no single protein antigen is suitable for optimally identifying the humoral response. Different workers have employed diverse antigen preparations in ELISA assays and, although the ELISA technique is widely used and very flexible, there is no universal method for presenting results or defining the positive/negative cut-off, which makes comparisons between different studies difficult. Direct comparisons of different antigens are discussed (see p. 141) and these studies demonstrate that the sensitivity and specificity of two commonly used antigen preparations can be improved by the addition of the 120-kda protein. Such antigen preparations have a very good predictive value in diagnosing *H. pylori* colonization, but in the future a commercially available combination of cloned antigens should provide a more easily standardized specific antigen (see p. 167).

With certain exceptions (such as recent bismuth or antibiotic therapy), current serological assays are excellent at diagnosing *H. pylori* colonization. As serology is cheap and non-invesive, the technique certainly lends itself to epidemiological studies, but does it have a role in everyday clinical practice? This complex issue is discussed in detail, with an assessment of how different screening strategies might work (see p. 147). In terms of screening dyspeptic patients it is important to clearly define the conditions you wish to diagnose and treat. A knowledge of their prevalence in the local population and their exact association with *H. pylori* is also required. With these details suitable strategies can be devised and then evaluated.

Department of Gastroenterology St. James's University Hospital, Leeds LS9 7TF, UK

Helicobacter pylori, Gastritis and Peptic Ulcer
Edited by P. Malfertheiner and H. Ditschuneit

Theoretically more straight forward than screening is the use of serology in treatment follow up (see pp. 172). The data available certainly demonstrate significant decreases in antibody titres in the months following *H. pylori* eradication. However, for serology to be clinically useful in individual patients post treatment, the sensitivity, specificity and predictive value of any decrease in terms of clearance needs to be known. These data are not currently available for any serological assay.

H. pylori is unique in the human gut in that it colonises a niche long term, in association with an acute and chronic inflammatory response in isolation from other bacteria. The available evidence strongly supports *H. pylori* as being the cause of chronic gastritis. The immunological studies on the local response (see p. 137, 162) demonstrate a specific local humoral response to *H. pylori* and marked changes in T-cell populations. This raises the question: Why is the immune response ineffective in clearing colonization? A number of host or bacterial factors may be important in this and further study is clearly important. The continuing local immune response in association with colonization is also a potentially damaging factor and must be considered as a possible pathogenic mechanism.

Study of the immune responses to *H. pylori* has relevance to the diagnosis, follow up and pathogenesis of *H. pylori*-associated conditions. The infection also provides a unibacterial model of epithelial colonization in association with a specific long-term immune response. Detailed study of this model may provide important knowledge concerning bacterial pathogenicity and host mucosal defences.

Morphological Aspects of Gastritis

Helicobacter pylori Induced Gastritis in Childhood*

S. J. Czinn

Introduction

In 1984, Marshall and Warren [10] reported the presence of spiral organisms in the gastric mucosa of adults with gastritis. Since then, this organism has been isolated and characterized as *Helicobacter pylori* (*H. pylori*). A number of investigators have reported an association between *H. pylori* and gastritis or gastric and duodenal ulcers in adults [6, 7, 13]. With the advent of small flexible fiber-optic endoscopes, several pediatric studies have investigated the relationship of *H. pylori* infection and gastritis [2–4]. In industrialized countries, *H. pylori* infection is a rare finding in the gastric mucosa of children. The low prevalence as well as the lack of confounding variables such as smoking, alcohol intake, or use of nonsteroidal antiinflammatory drugs (NSAIDS) in children may allow such pediatric studies of clarify the role of *H. pylori* in the pathogenesis of gastritis and gastric or duodenal ulcers. The purpose of the present study was: (a) to determine the relationship between *H. pylori* and gastritis or peptic ulcer disease in children and (b) to determine the natural history of *H. pylori* in children with gastritis.

Using pediatric gastric antrum biopsy material from 1979–1988, we studied the prevalence of gastric *H. pylori* and evaluated the association between this organism and the clinical, endoscopic, radiologic, and pathologic presentation [8].

Endoscopic antral biopsy specimens obtained from children aged 5 though 18 years were examined. In all patients endoscopy and gastric biopsy were performed for assessment of possible acid peptic disease. Children with secondary gastritis due to Crohn's disease, graft versus host disease, or eosinophilic gastritis were excluded from this study. All biopsy specimens had been fixed in Zenker's solution and stained with hematoxylin and eosin. The severity of inflammation was graded 0 to 3+. Inflammation was assessed as chronic, active, or chronic active. The numbers of organisms were graded from 0 to 3+.

Over the 9-year period 179 children within the study age group had gastric antral biopsy specimens taken. Studies using the aforementioned techniques have determined at 21% of children undergoing endoscopy and gastric biopsy at our institution have evidence of *H. pylori* infection. In the 81 children with histologic evidence of gastritis, 47% were positive for *H. pylori*. None of the children

* This work was supported in part, by Grant AI 25818 from the National Institute of Allergy and Infectious Diseases.

From the Department of Pediatrics, School of Medicine, Case Western Reserve University, and the Division of Gastroenterology, Rainbow Babies and Childrens Hospital, Cleveland, Ohio

Helicobacter pylori, Gastritis and Peptic Ulcer
Edited by P. Malfertheiner and H. Ditschuneit

without gastritis had evidence of *H. pylori* infection. The mean age of children with *H. pylori* gastritis was 12 years. The presenting signs and symptoms included epigastric abdominal pain (50%), vomiting (40%), and hematemesis (16%). Unfortunately, the clinical presentation was not helpful in discriminating children with *H. pylori* gastritis from those with *H. pylori*-negative gastritis.

Results

Endoscopic Findings

In children the endoscopic appearance of *H. pylori* gastritis can vary from normal to the presence of gross ulcerations. In this study 80% of the children had *H. pylori* gastritis, 30% had duodenal ulcers, 6% had a gastric ulcer, and 15% had normal appearing gastric mucosa. Interestingly, the majority of these children had a nodular appearing gastric antrum (Fig. 1). This finding is pathognomonic of *H. pylori* infection and has not been reported in any of the adult studies. Presently the etiology of this finding is unclear.

Radiologic Findings

Unlike adults with *H. pylori* gastroduodenal disease, a subset of infected children present with chronic vomiting or hematemesis. Sixteen such patients underwent radiologic examination of the upper gastrointestinal tract at the time of

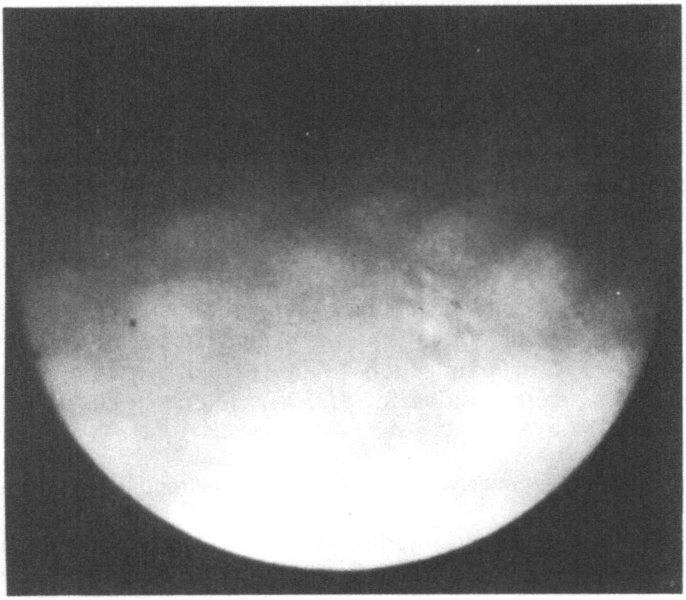

Fig. 1. Antral nodularity present in children with *H. pylori* gastritis

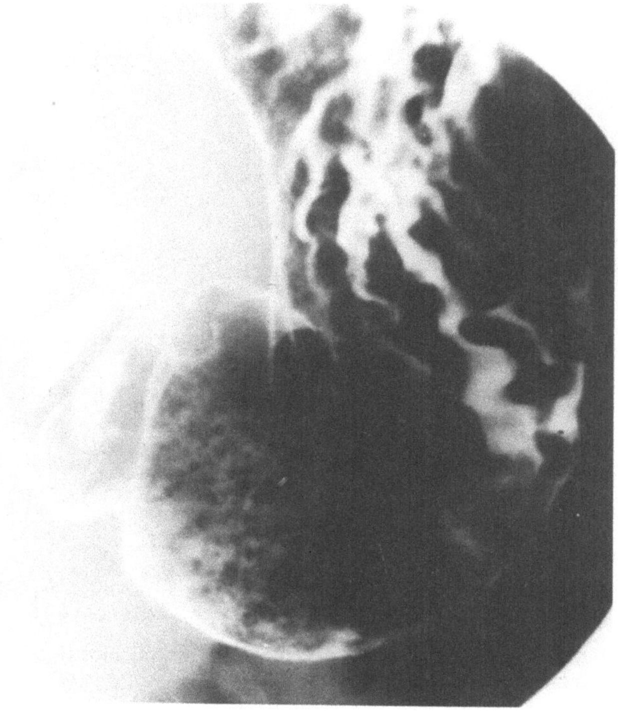

Fig. 2. Air-contrast examination of a child with *H. pylori* gastritis. Note the thickened folds in the body and antrum of the stomach. [From 11]

presentation. Seven of these children (40%) were noted to have hypertrophic or enlarged gastric folds (Fig. 2). Therefore, infection with *H. pylori* should be added to the list of causes for radiologic evidence of hypertrophic gastric folds in children [11].

Histologic Findings

The children with antral gastritis were analyzed according to the type of gastritis. Of the 38 children with *H. pylori* gastritis 82% had diffuse chronic gastritis highlighted by the presence of lymphocytes and plasma cells within the lamina propria, with only occasional foci of active inflammation within the glandular epithelium (Fig. 3). Eighteen percent had only diffuse chronic inflammation and no children presented with active inflammation alone. Thus, most children with *H. pylori* have lymphocytes and plasma cells as the predominant inflammatory cell within the lamina propria. This differs significantly from the descriptions of *H. pylori* gastritis in adults [1]: the predominant inflammatory cell within the lamina propria as well as the glands is the polymorphonuclear neutrophil. In our

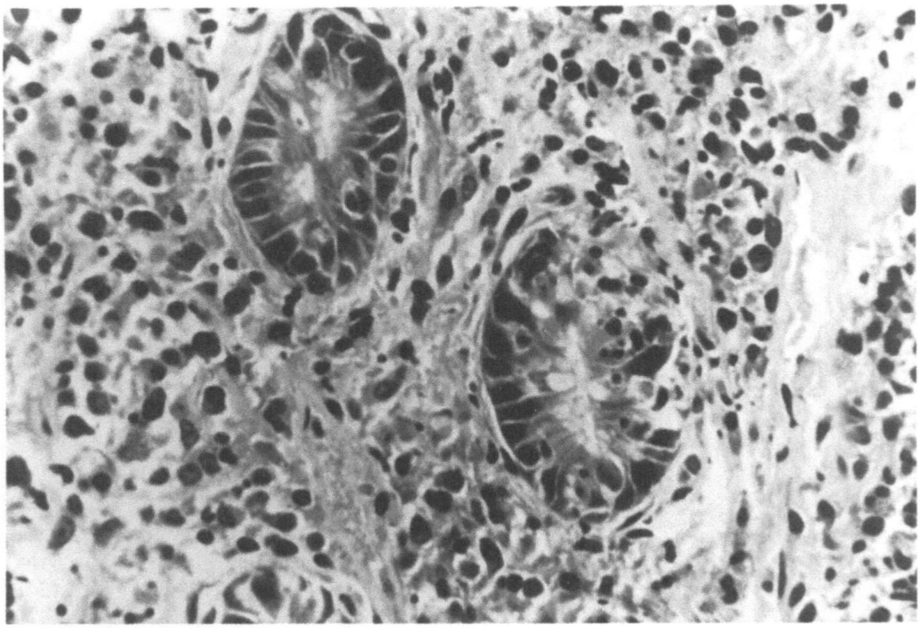

Fig. 3. High magnification of a pediatric antral biopsy specimen. Note the focus of active inflammation within the gland

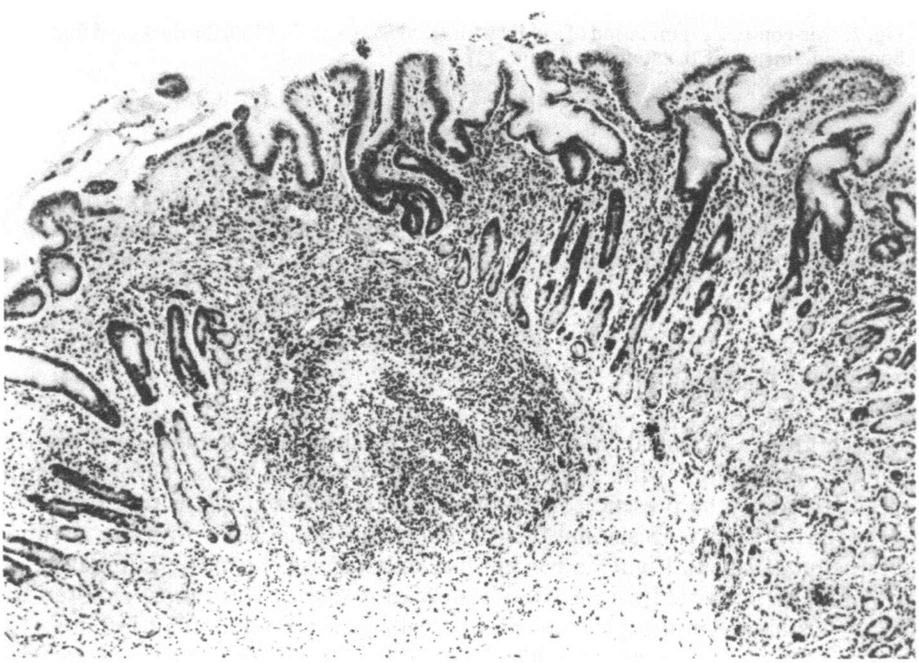

Fig. 4. Gastric biopsy of a child with *H. pylori* gastritis. Note the diffuse chronic gastritis with lymphoid follicle at center. [From 3]

experience, lymphoid follicles are only rarely present in the pediatric population (Fig. 4) and may not explain the gross antral nodularity seen at the time of endoscopy.

Additionally, the degree of gastritis was assessed and correlated with the presence of *H. pylori*. As reported in adult studies, most children with *H. pylori* have moderate or severe gastritis. Of the 81 children with histologic antral gastritis, 46 had mild gastritis, and 9 of these (19%) had *H. pylori* infection. Twenty of 81 had moderate gastritis, and 16 (80%) had *H. pylori* infection. Fifteen of 81 children had severe gastritis, and 13 (87%) had *H. pylori* infections. Thus, *H. pylori* gastritis is responsible for the majority of cases of moderate and severe gastritis. The degree of gastritis did not correlate with the number of *H. pylori* organisms.

Seventeen *H. pylori*-positive children, treated with 6–8 weeks of a standard H2-antagonist therapy, were followed over a 2-year period to evaluate the course of gastritis (Table 1). Eight of the 17 children had a decrease in the level of gastritis. Six remained unchanged and in only three children were we able to document a progression of the gastritis.

Table 1. IgG Antibody response to *H. pylori* gastritis

Degree of gastritis	Serum IgG (OD_{405})
Mild	0.274 ± 0.126
Moderate	0.778 ± 0.55
Severe	1.32 ± 0.60

Serologic Findings

In addition to the histologic and microbiologic techniques for identification of *H. pylori* infection, we and others [5, 12] have demonstrated the utility of serologic techniques to identify infected individuals. Thus, utilizing sonicated whole-cell proteins as the detecting antigen in the enzyme-linked immunosorbent assay, the serum IgG antibody response was compared with the degree of gastritis (Table 2). The serum IgG immune response in children directly correlates with the degree of gastritis.

Table 2. Course of *H. pylori* gastritis in children treated with H_2-antagonists

Decreased	Unchanged	Increased
Severe → moderate (3)	Severe (2)	Moderate → severe (2)
Severe → mild (2)	Moderate (3)	Mild → moderate (1)
Moderate → mild (3)	Mild (1)	

Discussion

H. pylori gastritis is uncommon in the pediatric population. Unlike adults, pediatric patients with *H. pylori* gastritis present with abdominal pain, vomiting, and hematemesis, and comprise the majority of children with moderate to severe gastritis. Of these children 50% present with the acute onset of symptoms (symptomatic for less than 3 months) and therefore present an opportunity to study the early phase of *H. pylori* infection. Endoscopically, the gastric antrum is very nodular. Additional mucosal abnormalities are noted on upper gastrointestinal radiologic examination. Specifically, 47% of children with *H. pylori* gastric disease have radiologic evidence of hypertrophic gastric folds. Histologically, a more chronic and less active inflammatory response is noted. Interestingly, these endoscopic and histologic findings have been reproduced in the gnotobiotic piglet model of *H. pylori* disease where the initial inflammatory response to *H. pylori* is the presence of chronic inflammatory cells [9]. Evidence for a pathogenic role of *H. pylori* is seen by the development of serum anti-*H. pylori* antibodies. In children, the serum immune response correlates directly with the degree of gastritis. This association has not been noted in adults and may only be obvious during the early phases of the infection. To date, no consistent progression of gastritis from mild to severe or to frank ulcerations has been noted. However, long-term prospective studies are needed to elucidate the natural history of *H. pylori* gastroduodenal disease.

Acknowledgements. I am grateful to Katie Chipps for her assistance in preparing this manuscript.

References

1. Bode G, Malfertheiner P, Ditschuneit H (1988) Pathogenic implications of ultrastructural findings in Campylobacter pylori related gastroduodenal disease. Scand J Gastroenterol 23(142):25–39
2. Cadranel S, Hennequin Y, Souayah H, et al. (1988) Campylobacter pylori (CP) in children is it frequent and what does it mean? Workshop on Gastroduodenal Pathology and Campylobacter pylori Bordeaux
3. Czinn SJ, Dahms BB, Jacobs GH, et al. (1986) Campylobacter-like organisms in association with symptomatic gastritis in children. J Pediatr 109:80–83
4. Drumm B, Sherman P, Cutz E, et al. (1987) Association of Campylobacter pylori on the gastric mucosa with antral gastritis in children. N Engl J Med 316:1557–1561
5. Evans DJ Jr, Evans DG, Graham DY, Klein PD (1989) A sensitive and specific serologic test for detection of Campylobacter pylori infection. Gastroenterology 96(4):1004–1008
6. Jones DM, Lessells AM, Eldridge J (1984) Campylobacter-like organisms on the gastric mucosa: culture, histological and serological studies. J Clin Pathol 37:1002–1006
7. Kasper G, Dickgiesser N (1984) Isolation of Campylobacter-like bacteria from gastric epithelium. Infection 12:179–180
8. Kilbridge PM, Dahms BB, Czinn SJ (1988) Campylobacter pylori-associated gastritis and peptic ulcer disease in children. Am J Dis Child 142:1149–1152
9. Krakowka S, Morgan DR, Kraft W et al. (1987) Establishment of gastric Campylobacter pylori infection in the neonatal gnotobiotic piglet. Infect Immunol 55:2789–2796
10. Marshall B, Warren JR (1984) Unidentified curved bacilli in the stomach of patients with gastritis and peptic ulceration. Lancet 1:1311–1314

11. Morrison S, Hoffenberg E, Dahms BB, Czinn SJ (1978) Enlarged gastric folds in association with Campylobacter pylori gastritis. Radiology 171:819–821
12. Rathbone BJ, Wyatt JI, Worsley BW et al. (1986) Systemic and local antibody responses to gastric Campylobacter pyloridis in non-ulcer dyspepsia. Gut 27:642– 647
13. Rollason TP, Stone J, Rhondes JM (1984) Spiral organisms in endoscopic biopsies of human stomach. J Clin Pathol 7(3):23–26

Comparison of Several Methods of Detecting *Helicobacter (Campylobacter) pylori*

R. J. L. F. LOFFELD, E. STOBBERINGH, J. A. FLENDRIG, and J. W. ARENDS

Introduction

There is increasing evidence that *Helicobacter (Campylobacter) pylori* is a causative factor in type B (antral) gastritis [1–8]. The diagnosis of a Helicobacter-associated gastritis in patients with nonulcer dyspepsia (NUD) may be of therapeutic importance, since drugs active against *H. pylori* are able to reduce colonization of this microorganism with improvement of gastritis [9, 10].

Precise identification of *H. pylori*, therefore, is an important step in the detection of Helicobacter-associated gastritis, especially during follow-up after therapeutical intervention.

So far, microbiological, histological, and serological methods have been widely applied for detection of *H. pylori*. Culture is generally accepted to be very specific. However, its sensitivity is low due to adverse influences resulting in false-negative culture results [11].

H. pylori has been studied in tissue sections stained using conventional nonspecific methods, like the silver stain according to Warthin and Starry [12], Gram's [13], modified Giemsa [14, 15], and acridine orange stain [16]. The specificity of these staining methods is questionable since they solely rely on the detection of microorganisms with morphological features analogous to those of *H. pylori*. Recognition of antigens of *H. pylori* is not possible. Immunohistochemical identification of *H. pylori*, therefore, would be a more specific histological method of detection if a specific antibody is used.

We used an immunoperoxidase technique employing a specific anti-*H. pylori* antiserum to detect *H. pylori* in antral biopsy specimens and compared its diagnostic yield with that of culture and modified Giemsa stain in a group of patients with nonulcer dyspepsia.

Patients and Methods

Gastric biopsy specimens of 200 consecutive patients in whom a nunulcer dyspepsia was diagnosed were studied. These patients included a group of 50 patients with nonulcer dyspepsia who had given informed consent for a

Departments of Internal Medicine, Microbiology and Pathology, University Hospital Maastricht, PO BOX 1918 6201 BX Maastricht, The Netherlands

Helicobacter pylori, Gastritis and Peptic Ulcer
Edited by P. Malfertheiner and H. Ditschuneit

therapeutic trial, as reported elsewhere [10]. These trial patients underwent two or three endoscopic investigations with antral biopsy specimens taken with a 1-month interval. All these biopsy specimens were included in the study. The group was comprised of 85 men (mean age 41.4 years, range 19–76) and 115 women (mean age 48.7 years, range 16–82).

At upper gastrointestinal endoscopy using the Olympus GIF Q gastroscope four antral biopsy specimens were taken, one for culture of *H. pylori* and three for histological assessment of gastritis and Helicobacter-like organisms. The total number of endoscopic investigations was 302. All histological and microbiological detection methods could be applied in 244 cases.

Histological Examination

Biopsy specimens were stained with haematoxylin and eosin for the assessment of gastritis according to Whitehead [17] (grade 0 and 1, normal antral histology; grades 2 and 3, showing features consistent with gastritis).

Scoring of *H. pylori* was performed on the basis of positive modified Giemsa [14] and/or immunoperoxidase stain. The presence of Helicobacter-like organisms was noted on a semiquantitative scale: grade 0, no bacteria seen; grade 1, sporadically some bacteria seen; grade 2, bacteria seen in almost all high power fields (×400); grade 3, bacteria seen in all high power fields, also lying in clusters. All sections were examined blindly without knowledge of culture results. The whole biopsy specimen was examined at high power magnification (×400).

Culture

One biopsy specimen was cultured. It was put in sterile saline 0.9% and transported to the microbiology department. The biopsy specimen was incubated on blood agar containing 6% sheep blood, under microaerophilic conditions for 5–7 days at 37 °C.

Antiserum Preparation

Five different strains of *H. pylori* were washed two to three times with saline 0.9%. A bacterial suspension was made which contained at least 100 million bacteria per ml. Two rabbits were inoculated with 1 ml suspension (bacterial suspension with Freud's adjuvant in a 1:1 solution). The first injection was administrated intracutaneously, after 4 weeks 1 ml bacterial suspension without Freud's adjuvant was given intramuscularly followed by a final intravenous injection of the same suspension. After this injection the rabbits were bled and the antiserum was harvested.

Immunoperoxidase Stain

All available biopsy material was recut and deparaffinised with xylol and alcohol 90% and treated with a hydrogen peroxidase methanol 0.5% solution during 15 min. The sections were pretreated with a pepsine solution (0.1% pepsine in 0.1 *N* hydrogen chloride) during 30 min. After cleaning with Tris buffered saline the sections were preincubated with normal swine serum in a solution of 1:5 during 10 min. After incubation with primary antiserum (polyclonal rabbit antiserum, 1:10 solution, 45 min), the sections were washed three times with Tris buffer. The biopsy specimens were incubated with swine anti-rabbit serum in a solution of 1:150 for 45 min. After cleaning with Tris buffer the sections were stained (4 min) with 3.3 aminobenzidine tetrahydrochloride with the use of 10% imidazol to intensify the reaction. Counterstaining was performed with haematoxylin. All sections were examined semiquantitatively for the presence of *H. pylori*.

Cross-Reactivity

The polyclonal antiserum was tested for cross-reactivity with *Campylobacter jejuni*, *Campylobacter fetus*, *Campylobacter coli*, *Escherichia coli*, and *Enterobacter*. Cultures of *H. pylori* and the other microorganisms were washed with phosphate buffered saline (three times). A suspension was made and fixed in buffered formaline solution. After centrifugation for 10 min at 3000 rpm, this material was embedded in warm agar (0.4 g agar in 10 ml phosphate buffered saline, at 37°–40 °C). After cooling down the bacteria were embedded in paraffin. The blocks were cut with the microtome and processed with the same technique as used for normal biopsy specimens. Gram stains of the sections were made in order to be sure that the sections contained bacteria. All sections containing embedded bacteria were stained according to the immunoperoxidase staining technique as described in Methods. The preparations were studied blindly without knowledge of the embedded bacteria.

Statistical analysis

The chi-squared test was employed for statistical analysis.

Results

Cross-Reactivity of the Antiserum

Cross-reactivity with Campylobacter species was negligible. It was absent with respect to other microorganisms. Hence the polyclonal antiserum proved to be rather specific for *H. pylori*.

Diagnostic Yield of the Three Detection Methods

As shown in Table 1 immunoperoxidase stain revealed the highest diagnostic yield, 218 (89%) of biopsy specimens being positive for *H. pylori*. Modified Giemsa stain was positive in 191 (78%) of the cases, whereas 108 (44%) biopsy specimens yielded a positive culture. Culture, moreover, due to technical difficulties remained inconclusive in 18 biopsy specimens.

Table 1. Diagnostic yield of different methods of detecting *H. pylori*

	Positive	Negative	Inconclusive [a]	Total
Culture	108 (44%)	118 (48%)	18 (8%)	244
Modified Giemsa	191 (78%)	53 (22%)	–	244
Immunoperoxidase	218 (89%)	26 (11%)	–	244

[a] In these cases culture failed because of technical difficulties

Immunoperoxidase Stain and Culture

One hundred eight biopsy specimens (44%) revealed a positive culture. *H. pylori* was seen in the immunoperoxidase stain in all these specimens. No false-negative immunoperoxidase stain was seen. Of 136 biopsy specimens with a negative or inconclusive culture 110 showed *H. pylori* in the immunoperoxidase stain, in the remainder (26 cases) no bacteria were seen.

Immunoperoxidase Stain and Modified Giemsa Stain

One hundred eighty-six biopsy specimens (76%) showed microorganisms in the immunoperoxidase stain as well as in the modified Giemsa stain. In 32 cases (13%) no bacteria were detected in the modified Giemsa stain, whereas the immunoperoxidase stain definitely showed *H. pylori*. In these cases *H. pylori* invariably appeared to be present in very low numbers (grade 1), always confined to the deep layers of the gastric pits.

In 21 cases (8%) both detection methods did not show microorganisms, whereas 5 (3%) biopsy specimens with a positive modified Giemsa stain revealed no bacteria in the immunoperoxidase stain.

Modified Giemsa Stain and Culture

Out of 108 positive cultures 104 (96%) revealed *H. pylori* in the modified Giemsa stain, only 4 (4%) negative stains were found, hence false-negative Giemsa stains. One hundred eighteen biopsy specimens did not show bacterial growth, 87 (74%) of these, however, demonstrated Helicobacter-like microorganisms in the modified Giemsa stain.

Culture and Semiquantitative *H. pylori* Score

Out of 108 positive cultures 40 (37%) showed grade 1 presence of *H. pylori*; 28 (26%) grade 2 presence, whereas the remainder, 40 cases (37%), scored grade 3. One hundred eighteen biopsy specimens with a negative culture showed grade 0/1 presence in 87 (73%); grade 2 in 19 (16%) cases; whereas only 12 cases (11%) scored grade 3. The relation between culture success rate and Helicobacter score is shown in Table 2 and appeared to be statistically significant, also when the results of inconclusive cultures were assessed to be either positive or negative.

Table 2. Correlation of culture and *H. pylori* score

	H. pylori score		
Culture	Grade 0/1	Grade 2	Grade 3
Positive $n = 108$	40 (37%)	28 (26%)	40 (37%)
Negative $n = 118$	87 (73%)	19 (16%)	12 (11%)

$P < 0.0001$

Discussion

Immunoperoxidase and immunofluorescence techniques with polyclonal and monoclonal antibodies, so far, have been only employed in small groups of patients [18–21]. The results of this study show that the immunoperoxidase stain can be applied in a large number of biopsy specimens. In all cases of microbiologically confirmed presence of the bacterium immunoperoxidase staining appeared to be positive, indicating that the technique is sensitive. In addition cross-reactivity with Campylobacter species and a panel of other bacteria was absent or negligible indicating a high specificity of the method.

Many workers in the field accept culture as the gold standard for detection of *H. pylori*. However, many use a combination of two or three different detection methods, including culture. A "gold standard" by definition has a sensitivity and a specificity of 100%. In an earlier study we reported that culture, although specific, lacks sensitivity [22]. This can be explained by our observation that a significant correlation exists between culture results and the bacterial load of the biopsy specimens. It is conceivable that false-negative cultures could be due to low numbers of bacteria present in the biopsy specimen.

Although after anti-*H. pylori* treatment improvement of gastritis has been described and the bacterium generally cannot be detected with culture [9, 23, 24], recrudescence of gastritis occurs in almost all cases. Restriction DNA analysis showed that the bacteria responsible for this recurrence are the same as before the start of therapy [25], implying endogenic reinfection. Hence, false-negative culture results occur after therapy. Since our study nicely showed that the success rate of culture in cases with low bacterial load, which can be expected after anti-*H. pylori* therapy, is low, culture is not reliable for excluding the persistence of the infection after therapeutic intervention. The immunoperoxidase stain can show very low numbers of bacteria in these cases.

H. pylori can exhibit a patchy distribution in the human antrum, and the number of microorganisms seen in several biopsy specimens from the same patient can vary widely. Hence it is possible that false-negative histological and/or microbiological detection can occur if only one biopsy specimen is examined. This situation especially occurs in the group of patients showing normal histological features and low numbers of *H. pylori* after a course of colloidal bismuth subcitrate (CBS) [10].

We already demonstrated that the modified Giemsa stain is superior to the silver stain according to Warthin and Starry [12]. In this study we showed that the most specific staining method is the immunoperoxidase staining technique. The modified Giemsa stain, however, still remains a reliable means of detecting *H. pylori* in histological sections. Only 5 (3%) modified Giemsa stains were found in which the immunoperoxidase stain failed to show *H. pylori*. There are two possibilities to explain this discrepancy.

First, the discrepant cases occurred when only low numbers of *H. pylori* were present. In such cases it is conceivable that the immunoperoxidase stain may be negative due to the different level of the section which was stained. Second, the possibility exists that the curved bacilli seen in these cases belong to other spiral organisms in the human stomach as recently described [26–30]. In all cases of positive immunoperoxidase stain and negative modified Giemsa stain, the microorganisms were found lying deep in the gastric pits, an area which is difficult to inspect in nonspecific staining methods, because of high background staining and lack of contrast. In the immunoperoxidase stain background staining is negligible and *H. pylori* can be detected easily deep with in the gastric pits.

In conclusion, immunohistology provides a sensitive and specific means of studying Helicobacter-associated gastritis. Since almost all patients with type B gastritis are positive for *H. pylori*, detection of the microorganism may not absolutely be required if gastritis is demonstrated on routine histological investigation. However, in cases of therapeutical intervention it is necessary, especially if normal antral mucosa is found during follow-up, not to rely on culture, but to use a specific or nonspecific staining method in order to differentiate between suppression or eradication of *H. pylori*. Because of its rather high sensitivity and its ease the modified Giemsa stain is applicable in daily clinical practice. For more refined detection of *H. pylori*, however, and for research purposes the immunoperoxidase stain is preferable.

References

1. Marshall BJ, Warren JR (1984) Unidentified curved bacilli in the stomach of patients with gastritis and peptic ulceration. Lancet 1:1311–1314
2. Rathbone BJ, Wyatt JI, Heatley RV (1986) Campylobacter pyloridis: a new factor in peptic ulcer disease? Gut 27:635–641
3. Goodwin CS, Armstrong JA, Marshall BJ (1986) Campylobacter pyloridis, gastritis and peptic ulceration. J Clin Pathol 39:353–365
4. McNulty CAM (1986) Campylobacter pyloridis-associated gastritis. J Infect 13:107–113

5. Axon ATR (1986) Campylobacter pyloridis; what role in gastritis and peptic ulcer? Br Med J 293:772
6. Hornick H (1987) Peptic ulcer disease, a bacterial infection? N Engl J Med 316:1598–1600
7. Marshall BJ, Armstrong JA, McGechie DB, Glancy RJ (1985) Attempts to fulfil Koch's postulates for pyloric Campylobacter. Med J Aust 142:436–439
8. Morris A, Nicholson G (1987) Ingestion of Campylobacter pyloridis causes gastritis and raises fasting gastric pH. Am J Gastroenterol 82:192–199
9. McNulty CAM, Geary JC, Crump B, et al. (1986) Campylobacter pyloridis and associated gastritis; investigator blind, placebo controlled trial of bismuth salicylate and erythromycin ethylsuccinate. Br Med J 293:645–649
10. Loffeld RJLF, Potters HVPJ, Stobberingh E, van Spreeuwel JP, Flendrig JA, Arends JW (1989) Campylobacter associated gastritis in patients with non-ulcer dyspepsia: a placebo controlled trial with colloidal bismuth subcitrate. Gut 30:1206–1212
11. Marshall BJ, McGechie DB, Rogers PA, Glancy RJ (1985) Pyloric Campylobacter infection and gastroduodenal disease. Med J Aust 142:439–444
12. Warthin AS, Starry AC (1920) A more rapid and improved method of demonstrating spirochaetes in tissues. Am J Syph 4:97–99
13. Montgomery EA, Martin DF, Peura DA (1988) Rapid diagnosis of Campylobacter pylori by Gram's stain. Am J Clin Pathol 90:606–609
14. Potters HVPJ, Loffeld RJLF, Stobberingh E, van Spreeuwel JP, Arends JW (1987) Rapid staining of Campylobacter pyloridis. Histopathology 11:1223
15. Gray SF, Wyatt JI, Rathbone BJ (1986) Simplified techniques of identifying Campylobacter pyloridis. J Clin Pathol 39:1279–1280
16. Walters LL, Budin RE, Paull G (1986) Acridine orange to identify Campylobacter pyloridis in formalin fixed paraffin-embedded gastric biopsies. Lancet 1:42
17. Whitehead R (1979) Mucosal biopsy of the gastrointestinal tract. In: Bennington JL (ed) Major problems in pathology, 2nd edn, vol 3. Saunders, Philadelphia, p 15
18. Anderson LP, Holck S, Povlson CO (1988) Campylobacter pylori detected by indirect immunohistochemical techniques. APMIS 96:559–564
19. Barbosa AJA, Quiroz DMM, Mendes NE, Rocha GA, Lima GF, Oliveira CA (1988) Immunocytochemical identification of Campylobacter pylori in gastritis and correlation with culture. Arch Pathol Lab Med 112:523–525
20. Engstrand L, Pahlson C, Gustavsson S, Schwan A (1986) Monoclonal antibodies for rapid identification of Campylobacter pyloridis. Lancet 2:1402–1403
21. Steer HW, Newell DG (1985) Immunological identification of Campylobacter pyloridis in gastric biopsy tissue. Lancet 2:38
22. Loffeld RJLF, Potters HVPJ, Stobberingh E, Arends JW, Flendrig JA, van Spreeuwel JP (1988) Campylobacter associated gastritis in patients with non-ulcer dyspepsia. J Clin Pathol 41:85–88
23. Glupczynski Y, Burette A, Labbe M, Deprez C, Reuck M, Deltenre M (1988) Campylobacter pylori associated gastritis: a double blind placebo controlled trial with amoxycillin. Am J Gastroenterol 83:365–372
24. Rauws EAJ, Langenberg W, Houthof HJ, Zanen HC, Tytgat GNJ (1988) Campylobacter pyloridis associated chronic active gastritis. Gastroenterology 94:33–40
25. Langenberg W, Rauws EAJ, Widjojokusumo A, Tytgat GNJ, Zanen HC (1986) Identification of Campylobacter pyloridis isolates by restriction endonuclease DNA analysis. J Clin Microbiol 24:414–417
26. Kasper G, Dickgiesser N (1985) Isolation from gastric epithilium of Campylobacter-like bacteria that are distinct from "Campylobacter pyloridis". Lancet 1:111–112
27. Goodwin S, Blincow E, Armstrong J, McCulooch R (1985) Campylobacter pyloridis is unique: GCLO-2 is an ordinary Campylobacter. Lancet 2:38–39
28. Morris A, Ali M, Thomson L, Vanderwee M, Hollis B (1989) Tightly spiral shaped bacteria in gastric antrum: are they urease positive? J Clin Pathol 42:216
29. Dent JC, McNulty CAM, Uff JC, Wilkinson SP, Gear MWL (1987) Spiral organisms in the gastric antrum. Lancet 2:96
30. Lee A, Dent J, Hazell S, McNulty C (1988) Origin of spiral organisms in human gastric antrum. Lancet 1:300–301

Endoscopic and Histologic Aspects of Gastritis

G. N. J. Tytgat, S. Hofer, and E. A. J. Rauws

One of the most difficult areas in clinical gastroenterology today is a precise description of the inflammatory changes of the stomach and the endoscopic parameters upon which a macroscopic endoscopic and microscopic diagnosis of inflammation of the gastric mucosal lining may be based. Reading the literature one may readily become confused because of the lack of precise terminology and discrepancies in interpretation of macroscopic and microscopic changes. A compilation of unequivocally visible and microscopic alterations should allow a proper diagnosis of any gastric abnormality. An ongoing debate concerns the question whether any macroscopic changes justify the use of the term "gastritis" or whether this term should be reserved only for histologically ascertained inflammatory changes. One way to solve this problem could be to use the broad term of "gastropathy" to indicate any macroscopically visible change from the normal. By doing so one may avoid the criticism that occasionally there is poor correlation between endoscopically observable alterations and the histologically seen inflammation.

The Normal Appearance of the Gastric Mucosa

There is a great deal of ignorance with respect to the "normal" macroscopic appearance of the gastric lining. In all probability many of the endoscopic appearances which endoscopists call "normal" are presumably not normal [1–8]. We are so used to see some alterations of the mucosa in middle-aged or elderly people in comparison with the "normal" appearance in young healthy volunteers, that the common slightly altered findings in the adult and elderly population are considered to represent the normal spectrum.

The mucosal appearance in the stomach is called normal when there is an even shade of pink color, smoothness, and luster throughout that area of mucosa. Usually the antrum appears flat or reveals only slightly elevated prepyloric folds with adequate but not excessive insufflation. Less commonly one may see one or two roofing folds, forming arches over the pyloric channel or a few more prominent antropyloric folds. The fold or rugal pattern in the corpus-fundus area is regular and even, in cross diameter not exceeding 5 mm. The folds are pliant and distend readily upon insufflation. In a normal stomach there is no adherent mucous-fibrinous exudate.

Academic Medical Centre, Meibergdreef 9, 1105 AZ Amsterdam, The Netherlands

Helicobacter pylori, Gastritis and Peptic Ulcer
Edited by P. Malfertheiner and H. Ditschuneit

The Endoscopic Spectrum of Gastropathy

Gastropathy is defined as the presence of presumably inflammatory alterations in the mucosal membrane. Endoscopically or macroscopically gastropathy may be diagnosed when some or all of the following abnormalities are unequivocally present and visible either focally or diffusely in the stomach. When one or more of these macroscopic parameters appear to be dominant, then this is used to indicate the most appropriate subcategory within the gastropathy spectrum.

Swelling or Edema

Edema is readily diagnosed when it is pronounced or severe but is rather difficult to detect reliably when it is mild or moderate in intensity. Edema may occasionally give a somewhat whiter or slightly opalescent tinge or may accentuate the areae gastricae pattern with accentuation of the hexagonal lineae gastricae.

Erythema or Redness (Hyperemia)

Erythema is diagnosed when there are patchy or diffuse areas of mucosa unequivocally discernible as redder in shade than the adjacent mucosa in that area. Erythema may be mild, representing minimal but obvious change, moderate, or severe or marked when the color change is beefy-red in intensity.

On close inspection erythema appears to be discontinuous, composed of innumerable tiny (1–3 mm) minimally raised dots of redness. Separating these slightly raised dots of erythema are yellow-white lines, the lineae gastricae. Histologic superficial gastritis has been associated with this appearance in 40% or more of cases [4, 9]. The distribution of erythema may be variable, focal, segmental, or diffuse. Presumably a combination of capillary dilatation or engorgement and mucin depletion alters the light transmission and reflection from the mucosal surface, giving a redder appearance.

Loss of Shininess or Luster

In the presence of inflammation the mucosa may become dull or luster less, instead of having the normal smooth and shiny appearance.

Exudate

Mucus-like or gray-yellowish or sometimes brownish or greenish material may adhere to the mucosal surface, called exudate. Occasionally the adherence of such punctate or strands of exudate is tenacious and resists vigorous rinsing with a water-jet. Not uncommonly tiny whitish punctate spots of exudate may be seen, especially in the antral area in *Helicobacter pylori*-associated inflammation, to be discovered only upon close inspection. The latter need differentiation from spots of highlighting which are in general more brilliant and less sharply demarcated.

Erosion

Per definition, an erosion is a visible break in the mucosal integrity. When the erosion is present within the niveau of the mucosa, the latter is called flat. When the erosion is elevated, the erosion is called raised (varioliform). Flat erosions are superficial whitish necrotic abnormalities, which in principle do not disrupt or destroy the muscularis mucosae. They may appear with or without a surrounding red halo. They vary in size from pinpoint up to 1 cm in diameter. They may be solitary or multiple. They may predominate on prominent prepyloric folds. Endoscopic detection is relatively easy, especially when the base of the erosive defect is covered with a layer of fibrinopurulent exudate. Characteristic for raised erosions (varioliform erosions) are discrete mounds of elevated mucosa, capped by a central defect, the so-called volcano-like erosions.

Fold Enlargement (Hyperrugosity)

Folds are considered prominent if they have a diameter between 5 and 10 mm; folds are considered large if they exceed 10 mm in diameter. Genuinely enlarged folds do not flatten or flatten only partially during insufflation. Enlarged folds may reveal superimposed caliber changes or frankly nodular irregularities.

Fold Atrophy or Thinning or Disappearance

Fold atrophy is characterized by flattening or thinning and ultimately disappearance (bald corpus fundus). This process is usually most obvious in the corpus fundus area. Occasionally tiny mucosal elevations or excressences may remain in areas where the folds have disappeared.

Visibility of the Vascular Pattern

The thinning of the mucosa allows the endoscopic appearance or visibility of small and larger ramifying vessels in the wall in the stomach. Often these vascular structures are rather unsharply delineated and irregular and tortuous in appearance. To avoid confusion with vascular transparency, excessive distension with air of the stomach should be avoided.

Intramural Bleeding Spots

Disruption of vascular integrity leads to extravasation of erythrocytes. The latter may be seen as punctate red spots in a congested mucosa (petechiae) or as somewhat larger red-brownish or dark ecchymotic spots or flecks. Often such lesions are multiple and are not in the same stage of development causing inequalities in color due to various stages of decomposition of the blood. Occasionally extravasation into the lumen may be seen originating from the intramural bleeding spots.

Granularity

The mucosa is called finely or coarsely granular or nodular when the evenness of the lining has disappeared. Then the granularity is rather regular, the term "mamillation" is sometimes used.

Endoscopic Abnormalities in *Helicobacter pylori* Gastritis

The endoscopic appearance of the stomach in the acute infectious phase of *Helicobacter pylori*-induced inflammation is poorly delineated. A rare patient has been described with a strikingly red, friable, and slightly microeroded appearance, covered with abundant fibrinopurulent exudate, in the antrum [10]. In many patients with chronic *Helicobacter pylori* colonization usually in the distal part of the stomach, no or at the most only discrete changes in the endoscopic appearance of the stomach are detectable [11–13]. If endoscopic abnormalities are seen, they usually consist of patches of erythema, contrasting and alternating with paler areas, slight unevenness of the mucosal lining, and some loss of shininess. In a variable proportion of patients, close inspection, especially of the antral area, reveals tiny punctate whitish spots of exudate together with small erosions, often on a standing prepyloric fold. Rather exceptionally more obvious erosive gastritis is visible in the distal part of the stomach.

More characteristic is the combination of discrete endoscopic abnormalities in the antrum combined with endoscopic evidence of inflammation in the bulb. Endoscopic bulbitis is usually characterized by foci of striking punctate erythema and areas of punctate or confluent erosions. Additional features are slight edema of the mucosa and accentuation of the fold pattern in the bulb and the conspicuous bulbar motor activity.

Not uncommonly both the abnormalities in the antrum and in the bulb may develop or progress during acid suppressive therapy.

Preliminary data have been obtained from the literature that successful eradication of *Helicobacter pylori* may lead to regression if not disappearance of the endoscopic abnormalities [14].

To document the endoscopic abnormalities in *Helicobacter pylori* infection, a series of patients with chronic dyspepsia were photographed in a standard way. The endoscopic slides were blindly reviewed and scored without knowledge of the *Helicobacter pylori* status. The results are summarized in Tables 1 and 2. It is readily obvious that several of the inflammatory parameters are more commonly present in *Helicobacter pylori*-related gastropathy.

Table 1. Endoscopic appearances of the antral mucosa in 96 nonulcer dyspepsia patients in relation to absence or presence of *H. pylori* on culture and associated antral gastritis [26]

	H. pylori positive ($n = 55$)	*H. pylori* negative ($n = 41$)	p^a
Normal	3	8	0.05
Edema	41	18	< 0.003
Erythema	49	29	< 0.03
Atrophy	3	9	< 0.02
Mamillation	3	2	
< 5 Erosions	20	5	< 0.01
> 5 Erosions	2	1	
Punctate exudate	14	5	0.10

[a] Fisher's exact test (two-tailed)

Table 2. Endoscopic appearances of the duodenal bulb in 50 nonulcer dyspepsia patients in relation to the *H. pylori* status of the antral mucosa [26]

	H. pylori positive ($n = 30$)	*H. pylori* negative ($n = 20$)	p^a
Normal	1	4	0.14
Edema	26	12	0.04
Erythema	28	15	0.10
Less than 5 tiny erosions	8	5	N.S.
More than 5 tiny erosions	9	4	N.S.
Punctate exudate	3	2	N.S.

[a] Fisher's exact test (two tailed)

Histologic Classification of Gastritis

Over the years many classifications have been developed for the assessment and staging of the severity of inflammatory changes in the gastric mucosa [15–24]. Obviously in the light of the discovery of *Helicobacter pylori* gastritis classification needs revision.

The criteria of Whitehead et al. [16] have been used by many investigators. In this classification the following parameters are used:

1. Chronic inflammatory infiltration of the superficial mucosal layers
2. Inflammatory infiltration of the deep mucosal layers
3. Elongation of gastric pits
4. Atrophy of gastric glands
5. Intestinal metaplasia

Superficial gastritis is diagnosed when only superficial epithelium, gastric pit region, and related lamina propria are affected. Chronic gastritis is diagnosed when, additionally, a deep extension of the inflammatory infiltrate around the tubules is found. The criterion of atrophic gastritis is based on the detection of atrophy of the tubules.

The three histological diagnoses (superficial gastritis, chronic gastritis, and atrophic gastritis) can be subdivided into mild, moderate, and severe.

Sipponen et al. [24] from Finland, classify gastritis as:

1. Acute (erosive) gastritis
2. Chronic gastritis
3. Special types (eosinophilic, granulomatous, and lymphocytic) and a miscellaneous group including the hyperplastic gastropathies and gastritis in gastric resection stomachs

Chronic gastritis is graded as chronic gastritis without atrophy or superficial gastritis with slight, moderate, or severe degree of inflammation, and atrophic gastritis with slight, moderate, or severe degrees of atrophy (Table 3).

Table 3. Grading of chronic gastritis (Sipponen)

Normal (N)	Score 0
Superficial chronic gastritis without atrophy (S)	0.5, slight 1, moderate 1.5, severe
Atrophic gastritis (A)	2, slight 3, moderate 3, severe

Wyatt and Dixon [23] outlined a new approach to the interpretation of inflammatory gastric disease based on characteristic patterns of mucosal response to the underlying mechanisms of injury. Their classification utilizes the previous categories of type A and type B gastritis because they feel that they are conveniently linked to their autoimmune and bacterial causation but also acknowledge the recently identified categories of reflux and lymphocytic gastritis. They subdivide histologic inflammation into:

1. Type A chronic gastritis (autoimmune pathogenesis)
2. Type B chronic gastritis (mainly bacterial infection due to *Helicobacter pylori*)
3. Reflux gastritis (due to chemical injury)
4. Lymphocytic gastritis (presumably an atypical reaction to *Helicobacter pylori*)
5. Eosinophylic gastritis (presumably allergic in nature)
6. Isolated granulomatous gastritis (of which the pathogenesis remains unknown)

Recently Stolte and Heilmann [25] launched a new German etiopathogenetic classification for gastritis. The following subdivisions were proposed: autoimmune gastritis (type A); bacterially induced gastritis (type B); combination of type A and type B gastritis; chemically-toxically induced gastritis (type C); lymphocytic gastritis, and various other forms such as acute infectious gastritis, granulomatous gastritis, eosinophylic gastritis, and Crohn's disease of the stomach.

For the *Helicobacter pylori* studies in Amsterdam a scoring system was used based on the following parameters: the degree of mononuclear inflammatory cell density (grade 0–2); the density of polymorphs in the lamina propria (grade 0–3); the degree of intraepithelial polymorphonuclear leukocytes (grade 0–3); and the degree of epithelial necrosis and microerosions (grade 0–2) [26].

Gastric Inflammation in *Helicobacter pylori* Colonization

In its most characteristic expression, gastric mucosal biopsy specimens reveal the combination of chronic inflammation with large numbers of mononuclear inflammatory cells in the lamina propria and an acute component characterized by the presence of polymorphleukocytes within the lamina propria but especially permeating between the epithelial cells (Fig. 1). Occasionally the polymorphonuclear infiltration can be rather dense, sometimes leading to a glandular crypt abscess. Especially in the more severe forms of inflammation there is a conspicuous lack of mucous granules within the epithelial cells. Focally there may be evidence of microerosions at the luminal surface with small defects where the epithelial layer is completely necrosed and exchanged for a layer of inflammatory cells (Fig. 2). In most of the biopsy specimens the inflammatory infiltration is limited to the superficial layers of the lamina propria. Presumably only in severe long-standing cases there is also atrophy of the glandular area.

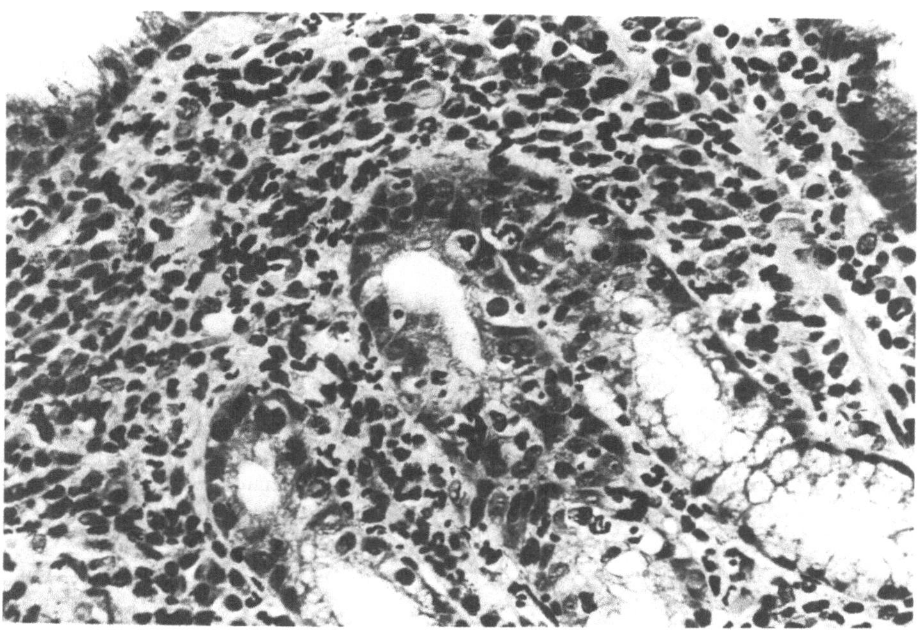

Fig. 1. Chronic active gastritis with a dense infiltrate in the lamina propria and invasion of granulocytes in the epithelial layer

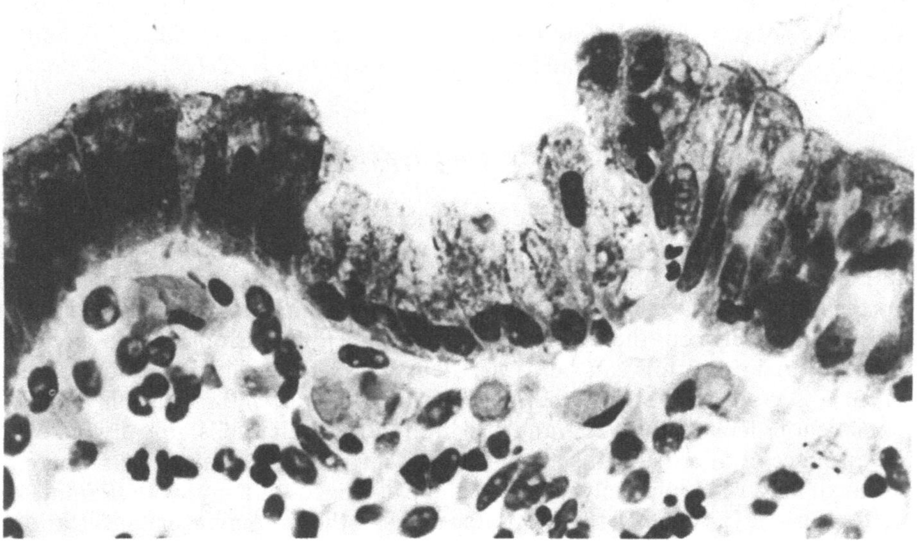

Fig. 2. Foveolar epithelium with focal erosion

Usually the inflammatory changes are more pronounced in the antral biopsy specimens with the biopsy specimens compared from the corpus fundus area.

The gastritis score in many biopsy specimens taken sequentially in our patients with *Helicobacter pylori*-related dyspepsia is shown in Table 4. As can be seen in those patients in whom no eradication was achieved the gastritis score remains more or less constant during the course of 1 year. In contrast in those patients in whom eradication could be achieved there is a gradual improvement in the overall gastritis score. Rather rapidly the acute polynuclear component of the inflammatory reaction disappears whereas the mononuclear component takes longer to diminish and disappear.

Table 4. Mean gastritis scores in patients with chronic active gastritis and positive culture for *H. pylori* before (B) and immediately after (A) treatment. Mean gastritis scores are given for those patients that remained *H. pylori* positive and those that became and remained *H. pylori* negative after treatment with amoxicillin, colloidal bismuth subcitrate, or the combination of these two agents. Note the ultimate complete disappearance of the inflammatory changes of the gastric mucosa (gastritis score, 0) after persistent eradication of *H. pylori* during follow-up

			Follow-up months			
	B	A	1	3	6	12
H. pylori, positive, ($n = 51$)	5.8	1.7	4.3	5.1	5.1	5.5
H. pylori, negative, ($n = 34$)	5.1	1.1	0.8	0.5	0.3	0

Correlation Between Endoscopic Abnormalities and Histologic Findings

In several studies it has been stated that the endoscopic appearances do not reliably predict the presence or absence of histologic gastritis [1–8]. A rather poor correlation is indeed usually found in the presence of discrete or equivocal abnormalities. When the endoscopic changes are unequivocal or more pronounced and certainly in the presence of erosions, there is almost always histologic evidence of inflammation. Therefore, the more severe the gastric endoscopic abnormalities, the better the correlation with histologic evidence of inflammation. The same holds true for endoscopic bulbitis [27–28]. Furthermore, it is well possible that the overall poor correlation is related to a large extent to inaccuracies of endoscopic interpretation, to ignorance with respect to the optimal site for targeted biopsy, to the optimal number of biopsies, and generally accepted criteria for histologic diagnosis of chronic superficial inflammation.

Concluding Remarks

To fully document the role of endoscopy, in parallel to histology, in the various forms of gastric inflammation, new studies have to be designed with careful scoring of any endoscopic abnormality and adequate sampling of the gastric mucosa for histological analysis. Both endoscopy and biopsy have to be performed before and after therapeutic attempts to correct the cause of the gastric inflammation. Perhaps the standardized use of videoendoscopy with the possibilities for more objective analysis of the macroscopic image (degree and distribution of redness, unevenness of erythema, presence and number of spots of punctate exudate, or erosions, etc.) will allow a more objective and accurate endoscopic interpretation of the gastric mucosal lining.

It is not to be expected that the discrete chronic variants of superficial inflammation without alteration of the overall mucosal thickness or of the mucus-secreting epithelium will induce readily obvious endoscopic changes. In contrast, active inflammation with mucin depletion and focal necrosis and disruption of the epithelial layer should be readily detectable by endoscopic means.

If ultimately endoscopic biopsies prove to be the key diagnostic modality, where should biopsies then be taken? Currently it is not known whether at random biopsies or targeted biopsies are equally informative in *Helicobacter pylori* gastritis. This will of course depend to a large extent upon the equal and diffuse character of the inflammation. If inflammation is of equal severity over a certain segment of the stomach, then random biopsies would seem to be adequate. If, on the other hand, inflammation is rather patchy, then more precise targeting may be more appropriate. It may be wise to sample in particular those areas where there is distortion or disruption of the epithelial lining or where there is evidence of punctate exudate.

References

1. Wenger J, Petty AM, Mendelhoff J (1970) Simultaneous gastroscopic, photographic and histological study of common types of gastritis. Gastrointest Endosc 16:234–236
2. Cronstedt JL, Simpson JW (1973) Correlation between gastroscopic and direct vision biopsy findings. Gastrointest Endosc 19:174–195
3. Myren J, Serck-Hanssen A (1974) The gastroscopic diagnosis of gastritis with particular reference to mucosal reddening and mucus covering. Scand J Gastroenterol 9:457–462
4. Taor RE, Fox B, Ware J, Johnson AG (1975) Gastritis – gastroscopic and microscopic. Endoscopy 7:209–215
5. Fung WP, Papadimitriou JM, Matz LR (1979) Endoscopic, histological and ultrastrutural correlations in chronic gastritis. Am J Gastroenterol 71:269–279
6. Sauerbruch T, Schreiber MA, Schussler P, et al. (1984) Endoscopy in the diagnosis of gastritis. Diagnostic values of endoscopic criteria in relation to histological diagnosis. Endoscopy 16:101–104
7. Stolte M (1988) "Normal" endoscopic findings in the gastrointestinal tract. When should a biopsy be taken? Endoscopy 20:111–113
8. Vaira D, Holton J, Osborn J, et al. (1989) Is endoscopy without biopsies worthless when gastritis is considered? Gut 30:A710
9. Mackintosh CE, Kreeb L (1977) Anatomy and radiology of the areae gasticae. Gut 18:855–864
10. Salmeron M, Desplaces N, Lavergne A, Houdart R (1986) Campylobacter-like organisms and acute purulent gastritis. Lancet 2:975–976
11. McNulty CAM, Gearty JC, Grump B, et al. (1986) Campylobacter pyloridis and associated gastritis: investigator blind, placebo controlled trial of bismuth salicylate and erythromycin ethylsuccinate. Br Med J 293:645–649
12. Jones DM, Eldridge J, Fox AJ, Sethi P, Whorwell PL (1986) Antibody to the gastric campylobacter-like organism ("Campylobacter pyloridis") – clinical correlations and distribution in the normal population. J Med Microbiol 22:57–62
13. Pettross CW, Appleman MD, Cohen H, et al. (1988) Prevalence of Campylobacter pylori and association with antral mucosal histology in subjects with and without upper gastrointestinal symptoms. Dig Dis 33:649–653
14. Rauws EAJ, Tytgat GNJ (1989) Campylobacter pylori. Thesis, University of Amsterdam
15. Motteram R (1951) A biopsy study of chronic gastritis and gastric atrophy. J Pathol Bacteriol 63:389–394
16. Whitehead R, Truelove SC, Gear MWL (1972) The histological diagnosis of chronic gastritis in fibre optic gastroscope biopsy specimens. J Clin Pathol 25:1–11
17. Strickland RG, MacKay IR (1973) A reappraisal of the nature and significance of chronic atrophic gastritis. Am J Dig Dis 18:426–440
18. Correa P (1980) The epidemiology and pathogenesis of chronic gastritis: three etiologic entities. Front Gastrointest Res 6:98–108
19. Weinstein WM (1981) The diagnosis and classification of gastritis and duodenitis. J Clin Gastroenterol [Suppl 2] 3:7–16
20. Whitehead R (1985) Mucosal biopsy of the gastrointestinal tract, 3rd edn. Philadelphia, Saunders
21. Kekki M, Siurala M, Varis K, et al. (1987) Scand J Gastroenterol [Suppl 141] 22:1–28
22. Correa P (1988) Chronic gastritis. A clinico-pathologic classification. Am J Gastroenterol 83:504–509
23. Wyatt JI, Dixon MF (1988) Chronic gastritis – a pathogenetic approach. J Pathol 154:113–124
24. Sipponen P, Seppälä K, Äärynen M, et al. (1989) Chronic gastritis and gastroduodenal ulcer: a case control study on risk of coexisting duodenal or gastric ulcer in patients with gastritis. Gut 30:922–929
25. Stolte M, Heilmann KL (1989) Neue Klassifikation und Gradierung der Gastritis. Leber Magen Darm 5:220–226

26. Rauws EAJ, Langenberg W, Houthoff HJ, Zanen HC, Tytgat GNJ (1988) Campylobacter pyloridis-associated chronic antral gastritis. Gastroenterology 94:33–40
27. Fontan AW, Rapaport M, Celener D, et al. (1978) Chronic nonspecific duodenitis (bulbitis). Endoscopy 10:94–98
28. Joffe SN (1982) Relevance of duodenitis to non-ulcer dyspepsia and peptic ulceration. Scand J Gastroenterol [Suppl 79] 17:88–97

Correlation Between the Grade of Activity of Type B Gastritis and Synthesis of Glycoproteins

K. Baczako*, A. Stanescu, P. Fabritius, and P. Malfertheiner

The histological pattern of gastritis is characterized by the degree of inflammatory cellular infiltration of the mucosa, the grade of activity, and the change of mucosal architecture, such as atrophy and intestinal metaplasia [1–4]. The actual activity of inflammation is indicated by the grade of tissue invasion by polymorphonuclear leukocytes. While the histological changes of gastritis are well defined, our knowledge of the functional importance of the mucosal inflammatory process is still inadequate. The mucus production of the gastric epithelium seems to be a suitable indicator of physiological or pathological mucosal activity and of the active defensive barrier function [5–8].

Histologically, especially in *Helicobacter pylori* (HP) infected samples, an increased mucus production can be detected. Sometimes we can see an explosive release and often a mucus depletion [9, 10]. In a previous study we found biochemically an increased glycoprotein metabolism by the inflamed gastric mucosa, which is in contrast to histologically normal tissue [11]. In the present morphofunctional study we compared the glycoprotein metabolism of gastric antral mucosa in vitro with the grade of inflammatory activity in HP-infected and HP-free biopsy samples.

Patients, Materials and Methods

We examined 58 patients with uncharacteristic upper abdominal complaints by gastroduodonoscopy. Patients taking steroidal or nonsteroidal antiinflammatory drugs were excluded from the study. Mucosal biopsy specimens were taken from the gastric antrum for histology and for tissue culture. The control group included 14 young and healthy volunteers.

Histological Examination

Two gastric mucosal biopsy specimens for routine diagnostic microscopy were fixed in 4% buffered formaldehyde solution and embedded in Paraplast. These samples were cut in 5-µm sections and stained with hematoxylin and eosin. We estimated the granulocytic activity of gastritis histologically, according to the criteria summarized in Table 1.

* Pathologisches Institut, Heidenheimer Str. 59, Postfach 3767, 7900 Ulm
Dept. of Internal Medicine, University of Ulm, R. Kochstr. 8, 7900 Ulm

Helicobacter pylori, Gastritis and Peptic Ulcer
Edited by P. Malfertheiner and H. Ditschuneit

Table 1. Histological grading of activity of chronic gastritis

Grade 0:	Lymphocytes and plasma cells in the tunica propria. No or few PMNs
Grade I:	Increased number of PMNs in the tunica propria mucosae without infiltration of foveolar epithelium
Grade II:	Strongly increased number of PMNs in the tunica propria with definite infiltration of foveolar epithelium (leukopedesis of neck cells and superficial epithelium)
Grade III:	Extremely dense infiltration of PMNs in the tunica propria with infiltration of foveolar epithelium, often with glandular microabscesses

PMNs, polymorphonuclear leukocytes

Stromal infiltrates by lymphocytes and plasma cells were always present. Semiquantitatively we graduated the presence and behavior of polymorphonuclear leukocytes. Samples without detectable granulocytic activity were evaluated as grade 0. Those with excessive dense infiltration and invasion of foveolar epithelium, microabscesses, and superficial granulocytic plugs were evaluated as grade 3. This grading is as was suggested by Stolte [4].

The histological sections were stained with silver impregnation by Warthin-Starry [12] for the detection of HP. Table 2 shows the criteria for the semiquantitative histological grading of HP colonization of gastric mucosa.

Table 2. Histological grading of *Helicobacter pylori* colonization (with silver impregnation by Warthin-Starry)

Grade 0:	Histologically no bacteria detectable
Grade I:	Few bacteria, mainly in the superficial mucus layer
Grade II:	Moderate amount of *H. pylori*, focally or evenly distributed in the superficial mucus layer and/or colonization of the gastric pits and glands
Grade III:	Dense *H. pylori* colonization of the mucus layer and gastric pits and obvious tendency to inter-cellular infiltration between the mucus cells

Organ Culture

Three biopsy specimens were transferred immediately onto a stainless steel organ culture grid and cultured in plastic humid chambers following the method as described elsewhere [11, 13, 14].

Figure 1 shows a plastic culture dish. In the middle of the grid is the mucosal explant (a) floating on the surface of the culture medium (c) in the central well of the plastic dish. The biopsy specimens were incubated for 2, 4, and 6 h in a Trowell culture medium. As glycoprotein precursor we used D-6-^{3}H-glucosamine in 5 µCi/ml concentration in the medium.

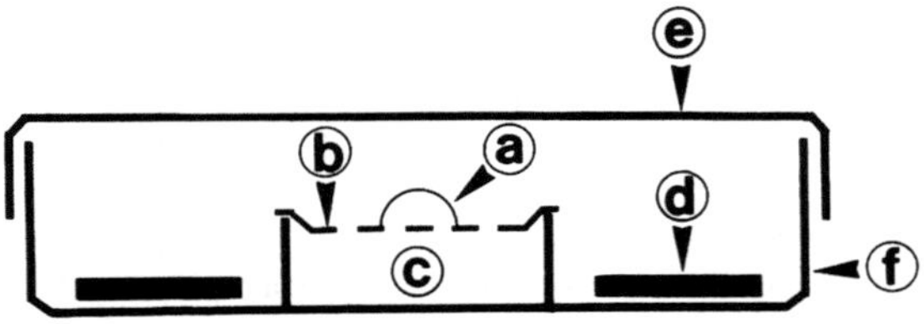

Fig. 1. Plastic organ culture dish: *a*, gastric antral mucosa (explant); *b*, stainless steel grid; *c*, culture medium; *d*, paper ring moistened with distilled water; *e*, f, plastic dish

The explants were harvested and washed. The tracer incorporation was estimated after ultrasound homogenization of the tissue. The proteins were precipitated, washed, and resolved in scintillation solution. The activity was counted in a Beckmann liquid scintillation system. The protein-bound radioactivity in the culture medium was used as a parameter for mucus secretion. It was measured after removing the free precursor by dialysis.

Figure 2 demonstrates the characteristic curves of time-dependent incorporation and secretion of ^{3}H-labelled precursor in short-term culture. For further statistical evaluations we used the areas under the curve for each patient as parameter for incorporation, as well as for secretion. The statistical analysis was carried out using Student's t test for paired samples and the Wilcoxon test.

Histologically normal human and animal mucosal tissues can be cultured for a long time. However, the gastric mucosa with dense inflammatory infiltration is very vulnerable and the results are often not reproducible after longer incubation periods.

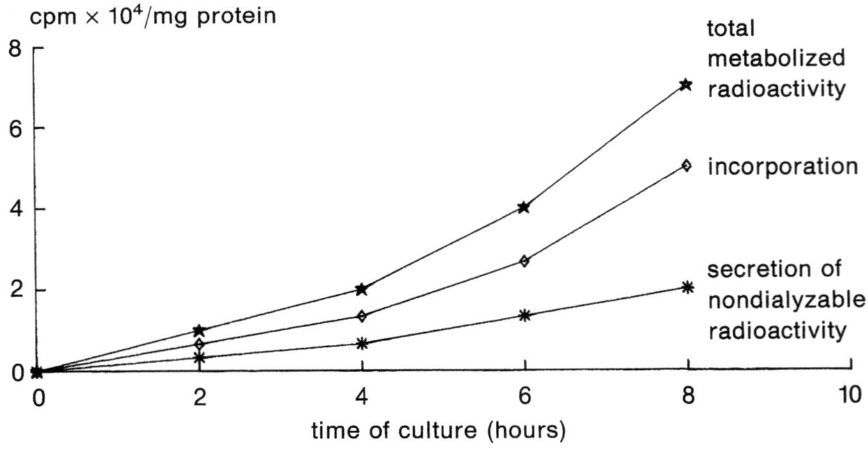

Fig. 2. Time-dependent incorporation and secretion of ^{3}H-labelled glucosamine by human gastric antrum tissue cultures

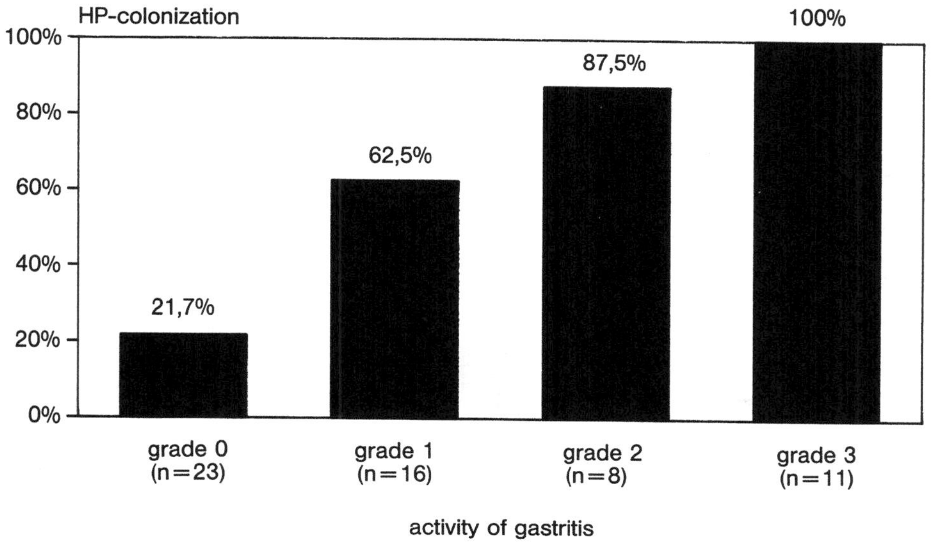

Fig. 3. Histologically verified activity of gastritis and the grade of HP-colonization; n = number of patients

Results

The results of the histological analysis of biopsy specimens demonstrate a positive correlation between the grade of activity of antral gastritis and the grade of HP colonization (Fig. 3).

The measured glycoprotein metabolism showed in the gastritis group without histologically detectable HP infection, as well as with HP colonization, significantly increased values of incorporation and secretion, when compared with the biopsy specimens from healthy control persons. Glycoprotein metabolism was significantly increased in HP-infected mucosal samples compared with those without HP infection (Fig. 4).

We found an increased glycoprotein metabolism of mucosal samples with HP colonization grade 1 and grade 2, but no further increase at HP grade 3 (Fig. 5). When the various grades of activity of gastritis were compared, a similar pattern of increased 3H-glucosamine metabolism was evident (Fig. 6). The tracer incorporation and the total metabolized radioactivity showed obviously increasing values in the groups with activity grades from 0 to 2. With activity grade 3 a fall of the tracer incorporation, as well as a decreased total metabolism could be observed.

Discussion

The tissue culture method enables us to compare the glycoprotein metabolism of diseased gastric antral mucosa with the metabolism of the histologically normal

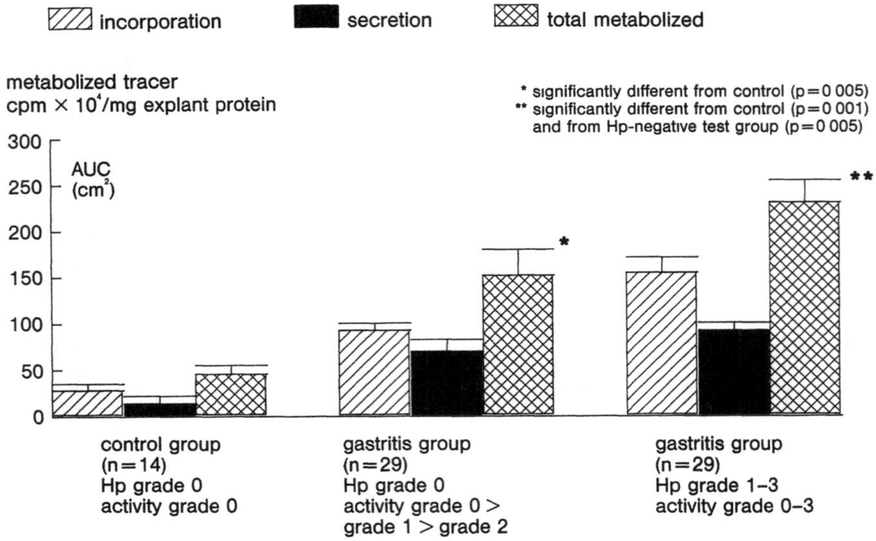

Fig. 4. ^{3}H-Glucosamine incorporation and secretion of radiolabelled glycoproteins by inflamed gastric antral mucosa with and without HP-colonization

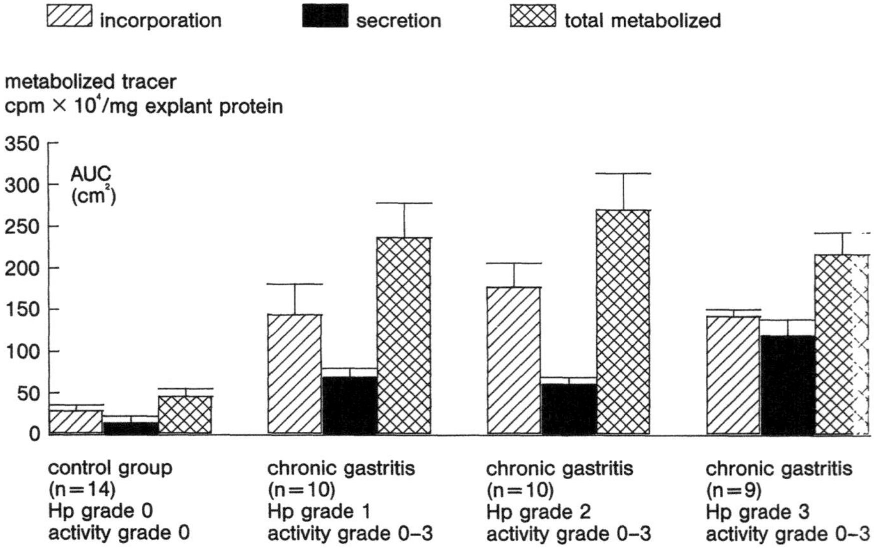

Fig. 5. ^{3}H-Glucosamine incorporation and secretion of radiolabelled glycoproteins by HP-infected gastric antral mucosa (grade 0–3)

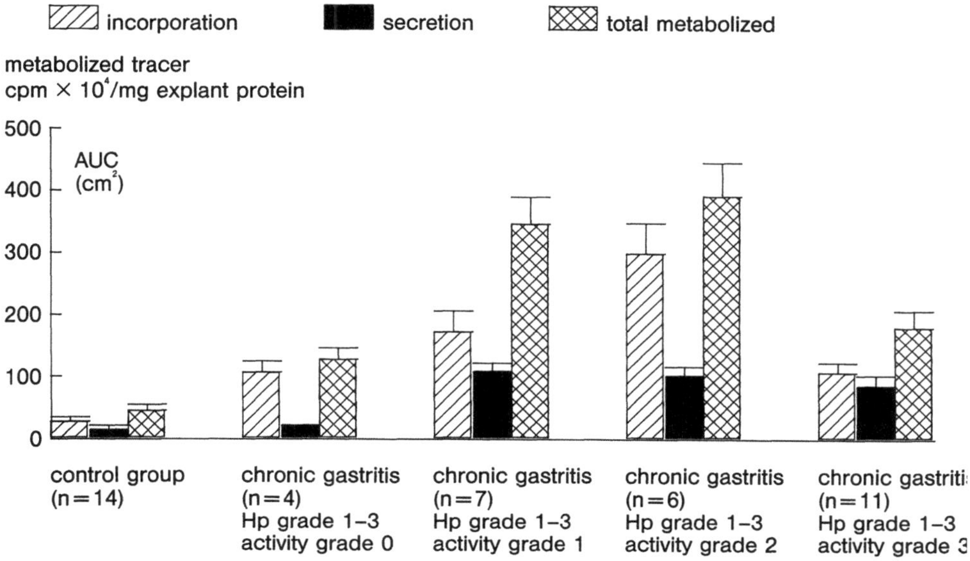

Fig. 6. ^{3}H-Glucosamine incorporation and secretion of radiolabelled glycoproteins by HP-infected antral mucosa with chronic gastritis (grade of activity 0–3)

mucosa. With this method we were able to demonstrate an increased ^{3}H-glucosamine turnover in all inflamed mucosal samples. However, the in vitro ^{3}H-glucosamine metabolism was significantly higher in HP-positive than in HP-negative biopsy specimens. Furthermore, the slightly and the moderately colonized samples, as well as biopsy specimens with slight and with moderate granulocytic activity metabolized significantly more radioactivity than did controls. Samples with severe HP colonization and high activity however have rather a supressed metabolic function. The latter may be a sign of imbalance in the barrier function and of mucosal breakdown.

The influence of inflammation on the gastrointestinal mucus production is still obscure. Presumably the inflammatory cells and endogenous factors derived from macrophages play an important role in the stimulation of mucus production and secretion [13–15]. Biologically active factors derived from macrophages have been described as mucus secretagogues acting on human respiratory epithelium [16, 17].

Investigations to characterize the action of the various biological factors of gastritis on the gastric glycoprotein metabolism may be important for clarifying the pathogenesis of mucosal damage.

Some observations must be made concerning the method. First, the results must be interpreted with caution and in close relationship with the morphological aspects of gastritis. The method is expensive and time-consuming. Second, individual variations of the values measured within a test group are relatively wide. Third, the mucosal samples altered by inflammation survive short culture periods only. Nevertheless, at present we have no valid methods for studying the mucosal function in its natural environment. One cannot, however, simply apply these in vitro findings to the conditions in vivo. On the other hand, the dynamics of the precursor's time-dependent metabolism give us some information about the actual functional state of the mucosal tissue and also about its ability to be stimulated.

References

1. Whitehead R (1979) Mucosal biopsy of the gastrointestinal tract. In: Bennington JL (ed) Major problems in pathology, vol III, 2nd edn. Saunders, Philadelphia
2. Whitehead R (1982) Gastritis – histopathological background. Scand J Gastroenterol [Suppl] 79:40–43
3. Siurala M, Sipponen P, Kekki M (1986) Campylobacter pylori in a sample of Finnish population: relations to morphology and functions of the gastric mucosa. Gut 29:909–915
4. Stolte M, Eidt S, Ritter M, Bethke B (1989) Campylobacter pylori und gastritis. Pathologe 10:21–26
5. Hollander F (1954) The two-component mucus barrier. Arch Intern Med 93:107–120
6. Menguy R (1969) Gastric mucus and gastric mucous barrier. Am J Surg 117:806–812
7. Allen A (1981) Structure and function of gastrointestinal mucus. In: Johnson LR (ed) Physiology of gastrointestinal tract. Raven, New York, pp 617–639
8. Flemström G, Knutson L, Kivilaakso E (1988) The mucus bicarbonate barrier and its role in gastroduodenal mucosal protection. In: Menge H, Gregor M, Tytgat GNJ, Marschall BJ (eds) Campylobacter pylori. Springer, Berlin Heidelberg New York pp 69–75
9. Madan E, Westblom TU, Subik M (1988) Mucus depletion in Campylobacter pylori mediated gastritis. J Clin Pathol 90:512
10. Kraft WG, Morgan DR, Leunk RD, Krakowka S (1988) An animal model of gastritis induced by Campylobacter pylori. In: Menge H, Gregor M, Tytgat GNJ, Marschall BJ (eds) Campylobacter pylori. Springer, Berlin Heidelberg New York pp 47–54
11. Baczako K, Melzner I, Malfertheiner P, Pietrzyk C, Brass B (1986) In vitro glycoprotein synthesis and secretion by gastric mucosal tissue of patients with gastritis. Am J Gastroenterol 81:346–353
12. Luna LG (ed) (1968) Manual of histologic staining methods of the Armed Forces Institute of Pathology, 3rd edn. McGraw-Hill, New York, pp 238–240
13. Shelhamer JH, Marom Z, Kaliner M (1980) Immunologic and neuropharmacologic stimulation of mucus glycoprotein release from human airways in vitro. J Clin Invest 66:1400–1408
14. Marom Z, Shelhamer JH, Kaliner M (1981) Effects of arachidonic acid, monohydroxyeicosatetraenoic acid and prostaglandins on the release of mucous glycoproteins from human airways in vitro. J Clin Invest 67:1695–1703
15. Peatfield AC, Rich B, Richardson PS, Williams IP (1983) The action of prostaglandins on secretion from human airways in vitro. J Physiol (Lond) 340:30P
16. Marom Z, Shelhamer JH, Kaliner M (1984) Human pulmonary macrophage-derived mucus secretagogue. J Exp Med 159:844–860
17. Marom Z, Shelhamer JH, Kaliner M (1985) Human monocyte-derived mucus secretagogue. J Clin Invest 75:191–198

Immunohistological Patterns of the Local Immune Response in *Helicobacter pylori* Gastritis

T. Kirchner, A. Melber, W. Fischbach, K. L. Heilmann, and H. K. Müller-Hermelink

Introduction

For a long time studies of gut-associated immunity concentrated on the small and large bowel, but almost neglected the stomach as an organ without significant immunocompetence. Attention increasingly turned to the gastric immune system, since the common B cell gastric lymphoma was reassessed to be an archetype of lymphoproliferative disease in mucosa-associated lymphoid tissue (MALT) [11, 20]. Moreover, the rediscovery of spiral bacteria on the gastric mucosa [31], which were defined as a new bacterial genus called *Helicobacter pylori* (HP) [9], decisively stimulated the interest in gastric immunity. Up to now a growing body of evidence emphasizes that the most common type B gastritis is induced by HP and that it may reflect a mucosal immune response to the exogenous bacterial antigens [4, 8, 19, 26, 32]. Therefore, histologic evaluations of gastritis require an immunologic basis, and the immunohistochemistry could provide some insight into the local immune response induced by the HP colonization.

Material and Method

Endoscopic biopsy specimens were obtained from 23 patients with the clinical diagnosis of nonulcer dyspepsia (NUD). None of the patients had been treated with H2 receptor blockers, antiinflammatory drugs, colloidal bismuth, or antibiotics within 2 months before the endoscopy. Two biopsy particles of each antral and body mucosa were fixed with formalin and routinely processed to paraffin sections. They were stained with hematoxylin and eosin for the histologic diagnosis and with the Wharthin-Starry method for the detection of HP. Normal mucosa without HP (antrum, 6 cases; body, 8 cases), inactive type B gastritis with HP (antrum, 2 cases; body, 3 cases), active type B gastritis with HP (antrum, 11 cases; body, 8 cases); lymphocytic gastritis with HP (2 cases); and type A gastritis without HP (2 cases) were distinguished.

Parallel to the routine procedure two fresh biopsy specimens of each antral and body mucosa were snap-frozen in isopentane-liquid nitrogen using OCT (Tissue Tek) as embedding medium. Immunohistochemistry was performed on air-dried, acetone-fixed cryostat sections (5 μm) by applying the indirect immunoperoxidase

Institute of Pathology, University of Würzburg, Josef-Schneider-Str. 2, 8700 Würzburg, FRG

Helicobacter pylori, Gastritis and Peptic Ulcer
Edited by P. Malfertheiner and H. Ditschuneit

technique in a three-stage procedure. A similar indirect immunoperoxidase method was used on dewaxed paraffin sections for the demonstration of immunoglobulins and the secretory component. Staining of the following antibodies was studied in each case: TCRδ1 (T$\gamma\delta$-receptor) and TβF1 (T$\alpha\beta$-receptor) from T Cell Sciences, Cambridge Mass., USA; Leu4 (CD3), Leu3a (CD4), Leu2a (CD8), Leu7 (CD57), and anti-HLA-DR from Becton-Dickinson, Oxnard Calif., USA; Dako-CD22, anti-HLA-A, B, C, and anti-SC from Dakopatts, Hamburg, FRG; anti-IgM, -IgG, and -IgA from Diagnostica Merck, Darmstadt, FRG; Ki-M4 and Ki-M8 from Radzun, Kiel, FRG; ViT6 (CD1) from Knapp, Vienna, Austria; HML1 from Dianova, Hamburg, FRG.

The number of intraepithelial lymphocytes (IEL) was determined as lymphocytes per 100 epithelial cells (ECs) after counting at least 500 epithelial nuclei. The density of lymphocyte subsets in the lamina propria (LP) was evaluated by applying an ocular square (area, 0.014 mm^2 at $\times 400$ magnification). The determination of the average number of lymphocytes per square unit was based on counts of at least 5 square units in each section.

Immune System of the Normal Gastric Mucosa

The normal gastric mucosa showed a highly organized immune system (Table 1). Its main component was IEL homing in the foveolar and glandular epithelium of the antral and body mucosa. The IEL showed T cell markers (Fig. 1a) and expressed the HML1-antigen of mucosal lymphocytes. The average number HM1-positive IEL was 6 per 100 ECs. The IEL predominantly exhibited the CD8-antigen, and the ratio of CD8-positive cells to CD4-positive cells was 3 to 1. Approximately 45% of the IEL expressed T cell receptors of the $\gamma\delta$-type, which were not detected on lymphocytes of the LP.

Compared with the intraepithelial compartment fewer lymphocytes were seen in the LP (Fig. 1a). CD4-positive T cells predominated at this site, and the ratio of CD4-positive cells to CD8-positive cells in the LP was 2.5 to 1. B cells and plasma cells were extremely rare in the normal mucosa, and occured exclusively in the LP (Fig. 1b). NK/K cells were not detected.

Table 1. Mucosal immune system of the normal stomach

Cell types	Intraepithelial		Lamina propria
	Foveolar	Glandular	
T cell CD8+	+++	+++	+
T cell CD4+	+	+	++
T cell $\gamma\delta$-receptor	++	+	–
NK/K cells CD57+	–	–	–
B cells CD22+	–	–	(+)
Plasma cells	–	–	(+)
Macrophages Ki-M8+	–	–	+
IDCs CD1+	–	–	(+)

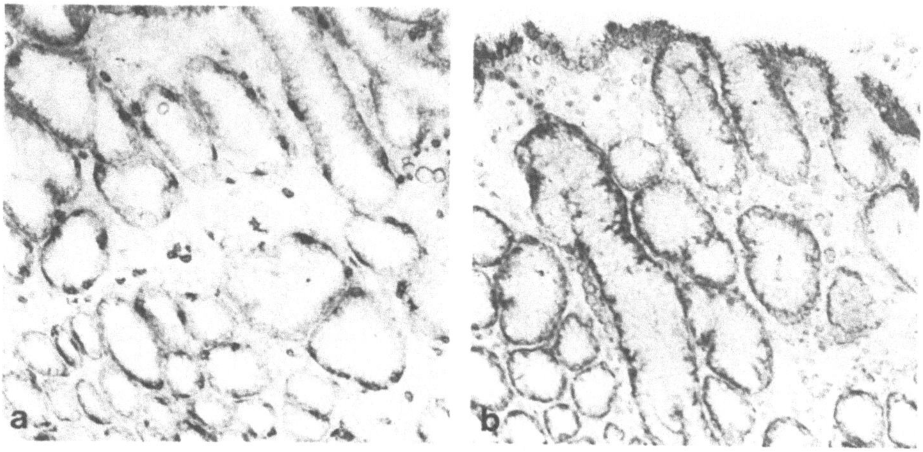

Fig. 1a, b. Serial sections of a normal antral mucosa. **a** CD3-positive T cells are seen in the gastric epithelium, fewer of them occur in the lamina propria (LP), × 190. **b** CD22-positive B cells are almost absent, × 190

A possible zone for antigen presentation were the basal pits of the normal mucosa. They showed a weak epithelial expression of class II major histocompatibility complex (MHC) antigens (HLA-DR) and of class I MHC antigens (HLA-A, B, C), whereas the other epithelial cells were negative. A few CD1-positive interdigitating dendritic cells (IDCs) serving as potent antigen presenters were sometimes found near the basal pits, especially in the antral region.

Patterns of the Local Immune Reaction in HP Gastritis

Presence of IDCs and Epithelial Expression of MHC Antigens

The prerequisites for antigen presentation were enhanced in HP gastritis. This was indicated by an increase of CD1-positive IDCs around the basal pits. Furthermore, the epithelial expression of HLA-A, B, C and HLA-DR antigens was continuously expanded to the upper foveolar and surface epithelium and to the glandular neck region in inactive gastritis. In active stages of gastritis all gastric epithelial cells exhibited a strong positive staining for HLA-A, B, C and HLA-DR antigens.

Perifoveolar Lymphocyte Clusters in the LP

Perifoveolar LP lymphocytes were constantly increased in HP infection, and distinct clusters of lymphocytes occurred around the basal pits. These clusters consisted of CD3-positive T cells (Fig. 2a). Counts showed an accumulation of

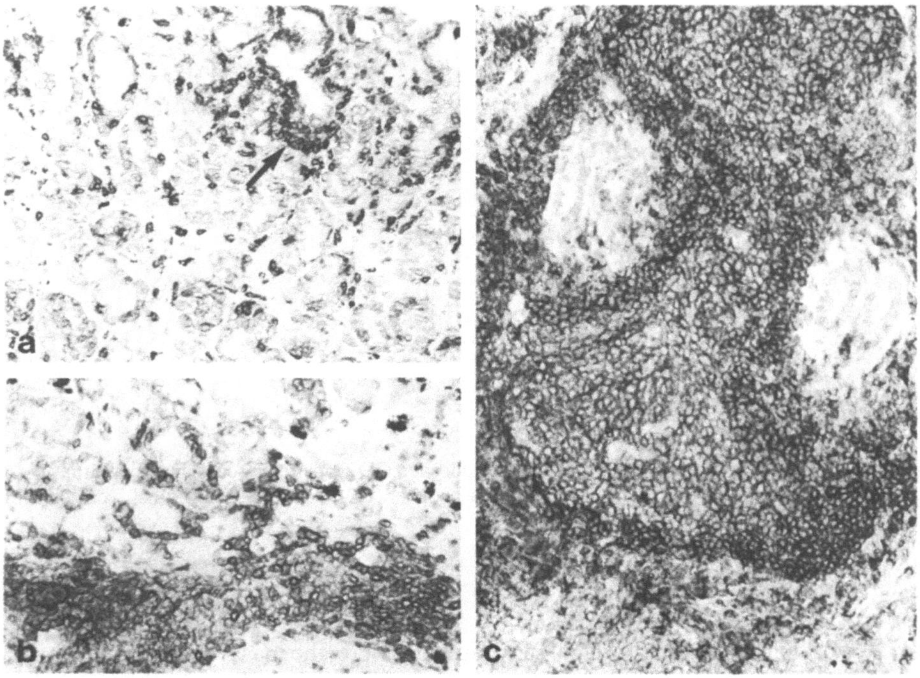

Fig. 2a–c. Immunohistological patterns of *H. pylori* (HP) gastritis. **a** Perifoveolar cluster (*arrow*) of CD3-positive T cells infiltrating the LP, ×150. **b** Basal lymphocyte aggregate formed by CD3-positive T cells, ×240. **c** Large intramucosal lymphoid follicle comprising CD22-positive B cells, ×190

both CD4- and CD8-positive T cells in inactive gastritis (CD4:CD8 = 1:1) and a further rise of CD4-positive cells in active gastritis (CD4:CD8 = 2:1). B cells did not participate in the perifoveolar clustering, but rather showed a diffuse increase being twofold in inactive gastritis and threefold in active gastritis compared with the perifoveolar LP of the normal mucosa.

Basal Lymphocyte Aggregates and Lymphoid Follicles

Another focal point for the gastric immune response were basal lymphocyte aggregates above the mucosal muscular layer. The aggregates comprised mainly T cells (Fig. 2b), with a preponderance of the CD4-positive subset, and some B cells. They could be precursors of intramucosal lymphoid follicles which were also localized in the basal part of the mucosa and were distinguished from the lymphocyte aggregates by the predominance of B cells (Fig. 2c) and the presence of Ki-M4-positive follicular dendritic cells (FDCs).

Basal lymphocyte aggregates rarely occurred in the normal mucosa. They were mostly seen in inactive HP gastritis, whereas true lymphoid follicles were mainly observed in the active stages of inflammation.

Plasmocytosis and Epithelial Expression of the Secretory Component

Plasma cells predominantly accumulated in the superficial part of the mucosa colonized by HP. There was a clear preponderance of IgA-producing plasma cells over IgG-positive and IgM-positive plasma cells (IgA $\gg$ IgG > IgM). This IgA plasmocytosis was associated with a de novo expression of the secretory component (SC) in foveolar epithelial cells. In the complex system of transepithelial immunoglobulin secretion, the SC synthesized by mucosal ECs becomes attached to the dimeric IgA secreted by mucosal plasma cells. After combining the completed secretory immunoglobulin is transported through the ECs to the mucosal surface [18]. No staining for SC was detected in the ECs of the normal gastric mucosa, whereas an expression of SC was a constant finding in active stages of HP gastritis. Furthermore, all foci of intestinal metaplasia exhibited an intense epithelial staining for SC.

IEL and Lymphocytic Gastritis in HP Infection

The number of foveolar IEL was increased approximately twofold in common type B gastritis with HP colonization, and slightly elevated in type A gastritis without HP. The expression of the HML 1 antigen, the staining for T cell markers, and the ratio of CD 8-positive cells to CD 4-positive cells (3:1) was similar to that in the normal mucosa. However, the percentage of T cells bearing $\gamma\delta$-receptors was reduced to approximately 15% of the IEL in active HP-induced gastritis.

Two cases of lymphocytic gastritis exhibiting a tenfold increase of HML 1-positive IEL were observed in HP infection (Fig. 3a, b). Both cases also showed an occurrence of lymphoid follicles and LP plasmocytosis predominated by IgA-producing plasma cells. The intense accumulation of IEL was only seen in the

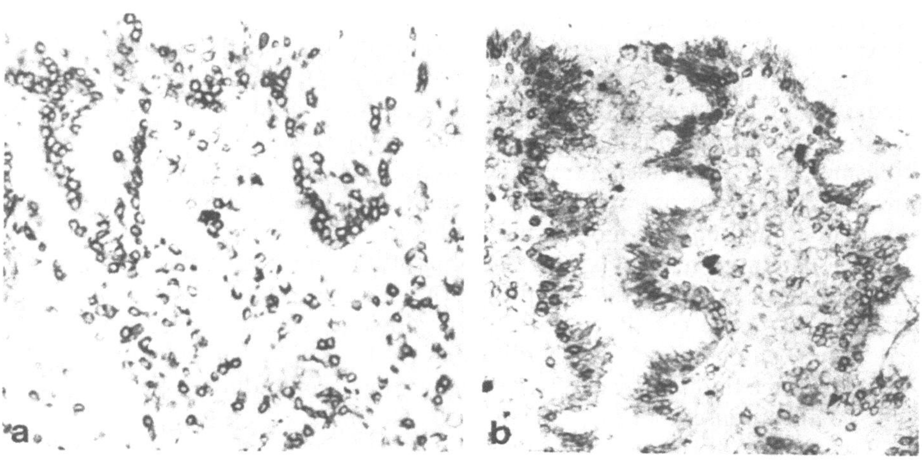

Fig. 3a, b. Lymphocytic gastritis in HP infection. **a** Intense accumulation of CD 3-positive T cells in the epithelium of the surface and foveolae, × 240. **b** Demonstration of the HML 1-antigen on the increased IEL, × 240

body mucosa. It was restricted to the epithelium of the mucosal surface and the pits, but did not affect the glands.

Different immunophenotypes of IEL were observed in this lymphocytic gastritis. One case mainly showed an accumulation of intraepithelial T cells with $\gamma\delta$-receptors, part of them being CD8- and CD4-negative. The second case exhibited no increase of the $\gamma\delta$-T cells, but an intraepithelial aggregation of $\alpha\beta$-T cells with CD8 expression.

Periglandular LP Lymphocytes

An emphasis of periglandular T cell infiltrates and a focal aggregation of T cells immediately around the glands of the body mucosa were features in both cases of type A gastritis without HP. Counts demonstrated a periglandular accumulation of both CD4- and CD8-positive T cells. This periglandular clustering of T cells was not evident in type B gastritis without HP. Here the lymphocyte infiltrate of the glandular region showed either a random distribution or a formation of lymphocyte "streets" connecting basal and perifoveolar lymphocyte accumulations.

Discussion

Two major components of the MALT are generally distinguished [2]. On the one hand cell-mediated immunity is apparently provided by special subsets of T cells which are able to home in the epithelium. These IEL predominantly show CD8 antigens, and some of them belong to the recently described T cell subpopulation expressing $\gamma\delta$-receptors [6, 12, 14]. On the other hand humoral immunity is related to intramucosal lymphoid follicles and to plasma cells producing secretory immunoglobulins that can inhibit bacterial adherence, neutralize viruses and toxins, and prevent uptake of antigens impinging on mucous membranes [2, 18].

The coexistence of both components of the MALT characterizes the physiologic immune system in the small and large intestine. In contrast, the normal gastric mucosa contains mainly T cells with a majority of IEL, whereas B cells and plasma cells are extremely rare, and lymphoid follicles are absent. Therefore, the normal gastric mucosa apparently remains in a naive immune status with a cell-mediated immunologic monitoring of the mucosal integrity, but without active humoral immune responses. Mechanismus of a humoral mucosal immunoreactivity are only seen in gastritis, and the rise of these processes seems to be an essential finding in type B gastritis. Some immunologic aspects of type B gastritis have already been described by other authors [7, 10, 21, 28–30]. The present study extends these findings by demonstrating distinct immunohistological patterns of the local response in HP colonization (Table 2).

One feature of HP gastritis is the clustering of T cells and the increase of IDCs around the basal pits. These events take place in both inactive and active gastritis and may indicate an enhanced IDC-dependent antigen presentation and stimulation of T cells at this zone. Furthermore, basal lymphocyte aggregates consisting

Table 2. Immunophenotype and HP association of major immunohistological patterns in gastritis

Immunohistological pattern	Immunophenotype	HP-Associated
Perifoveolar lymphocyte clusters	T cells CD4 > CD8	Yes
Basal lymphocyte aggregates	T cells > B cells CD4 > CD8	Yes
Lymphoid follicles	B cells ≫ T cells Ki-M4+ FDCs	Yes
Plasmocytosis	IgA ≫ IgG > IgM	Yes
Accumulation of intraepithelial lymphocytes	$T\gamma\delta+/CD8-$ or $T\alpha\beta+/CD8+$	Rarely
Periglandular lymphocyte aggregates	T cells CD4 = CD8	No

of T and B cells with a preponderance of CD4-positive T cells are only rarely seen in the normal mucosa, but are constantly detected in inactive and active HP gastritis. Decisive interactions of T and B cells and a stimulation of B cells by T helper cells might happen at this basal site of the mucosa. The consequence could be a local B cell proliferation and rise of lymphoid follicles, which are also localized in the basal part of the mucosa and mostly occur in active stages of gastritis. In contrast, the accumulation of plasma cells is confined to the perifoveolar mucosal area. This distribution and the predominance of IgA-producing plasma cells signals a secretory humoral response to luminal antigens, the effectiveness of which is further underlined by the de novo epithelial expression of SC in HP gastritis.

Experimental studies demonstrated that recombinant interferon-γ secreted by lymphocytes and tumor necrosis factor-α produced by macrophages can upregulate the epithelial expression of functional SC in a dose-dependent manner [16, 24]. Interferon-γ can also induce an epithelial expression of HLA-DR antigens [23], which is restricted to the basal pits in the normal gastric mucosa but is strongly positive on all gastric epithelial cells in active gastritis. In vitro studies supported an important role of HLA-DR molecules in the induction of IgA and IgM production [15], and it was also shown that the magnitude of immune responses is related to the density of the HLA-DR expression [13, 17]. Furthermore, IgA-producing plasma cells tend to accumulate around HLA-DR-positive ducts in human salivary glands [1, 27].

Therefore, complex mechanisms of lymphoepithelial interaction and local immune regulation are involved in the mucosal immune responses [2]. The immunohistological patterns described in this study may provide some framework for the dynamic immune process taking place in HP gastritis. However, the sequence of the reactions outlined in Fig. 4 is still hypothetic and has to be supported by a functional analysis of progressing and resolving phases in the local response. Problems concerning the gastric immunity in HP infection are stressed, for example, by the rare occurrence of lymphocytic gastritis [5]. The intense

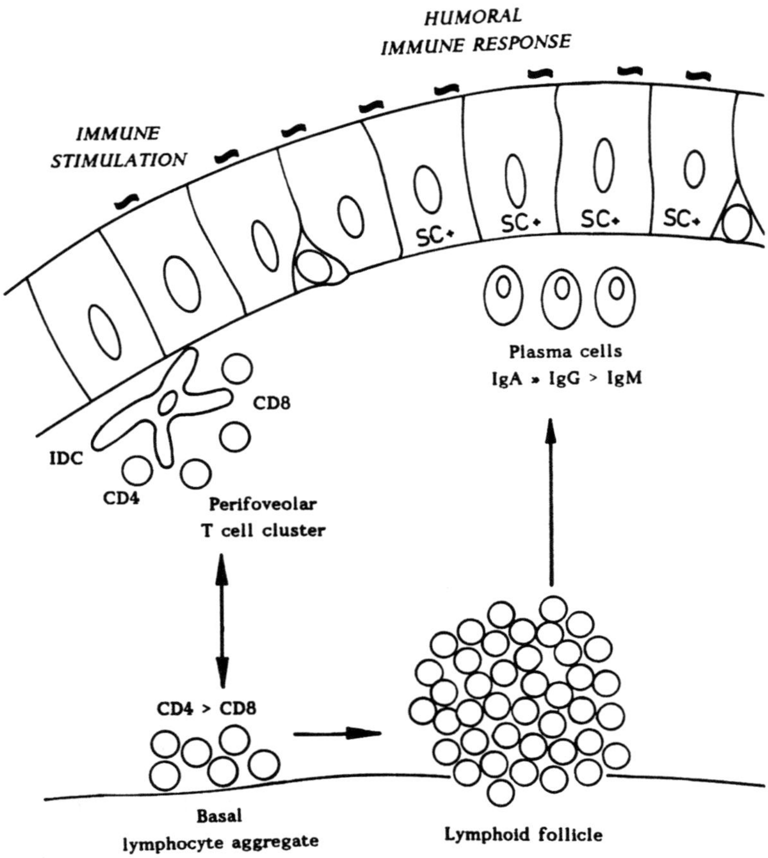

Fig. 4. Focal points and hypothetic sequence of the local immune reaction in HP colonization of the gastric mucosa

accumulation of IEL in this type of HP-associated gastritis is similar to findings in intestinal celiac disease [25] and might indicate a modulation of the local immune response due to host factors such as individual HLA determinants. On the other hand, special virulence factors of HP or other environmental immunogens could also be important. Finally, it has to be shown whether the different immunophenotypes in lymphocytic gastritis characterize transient or stable events of the local response.

Since the immunohistological findings indicate an essential switch from the naive status of immunologic monitoring in the normal gastric mucosa to processes of an active local immune defense in HP colonization of the mucous membrane, two crucial questions have still to be addressed. One concerns the specificity of the local response to HP antigens. The finding of HP-specific immunoglobulins produced in the gastric mucosa [3, 22], and the in situ demonstration of a HP-coating by immunoglobulins [32] certainly provide strong evidence that secretory mucosal antibodies are specifically directed to the bacteria. However, activated T cells clones specific for HP-derived antigenic determinants have still to be isolated from the gastric mucosa, and it has to be clarified whether other potential pathogens or environmental factors – like nutritional antigens – contribute to the mucosal immune stimulation.

The other question concerns the pathogenic relevance and effect of the local immune defense in HP colonization. Periglandular T cell aggregates as in autoimmune gastritis were not observed in this study. Specific immunologic processes causing mucosal damage and/or glandular atrophy have not yet been demonstrated in HP infection. At least the inactive HP gastritis is reminiscent of the physiologic immunity in the large bowel colonized by various commensal bacteria. Therefore, some cases of inactive HP gastritis might characterize neither a starting nor a resolving phase of inflammation, but rather a persisting carrier stage of HP colonization with a balanced local immunity and without a risk of continuing mucosal damage. Precise clinicopathological definitions of initial, spontaneously regressing, persisting, and progressing stages in the natural course of the HP infection and the HP-induced immune reaction are still missing to a large extent and are urgently required with regard to formulating therapeutic strategies.

References

1. Brandtzaeg P (1985) The human secretory immune system: general review. In: Revillard JP, Voisin C, Wierzbicki N (eds) Mucosal immunity. IgA and polymorphonuclear neurophils. Journee Immunologiques – Fondation Franco-Allemande, Paris, pp 11–43
2. Brandtzaeg P, Sollid LM, Thrane PS, Kvale D, Bjerke K, Scott H, Kett K, Rognum TO (1988) Lymphoepithelial interactions in the mucosal immune system. Gut 29:1116–1130
3. Brassens-Rabbe MP, Megraud F, Teyssou R, Vialette G (1989) Specific IgA response to Campylobacter pylori in gastric mucosa. Klin Wochenschr [Suppl 18] 67:7
4. Correa P, Munoz N, Cuello C, Fox J, Zavala D, Ruiz B (1989) The role of Campylobacter pylori in gastro-duodenal disease. In: Fenoglio-Preiser C, Wolff M, Rilke F (eds) Progress in surgical pathology, vol 10. Field and Woods, Philadelphia, pp 191–210
5. Dixon FM, Wyatt JI, Burke DA, Rathbone BJ (1988) Lymphocytic gastritis – relationship to Campylobacter pylori infection. J Pathol 154:125–132
6. Dobbins WO III (1986) Human intestinal intraepithelial lymphocytes. Gut 27:972–985
7. Engstrand L, Scheynius A, Pahlson C, Grimelius L, Schwan A, Gustavsson S (1989) Association of Campylobacter pylori with induced expression of class II transpalantation antigens on gastric epithelial cells. Infect Immun 57:827–832
8. Goodwin CS, Armstrong JA, Marshall BJ (1986) Campylobacter pyloridis, gastritis, and peptic ulceration. J Clin Pathol 39:353–365
9. Goodwin CS, Armstrong JA, Chilvers T, Peters M, Collins MD, Sly L, McConnell W, Harper WES (1989) Transfer of Campylobacter pylori and Campylobacter mustelae to Helicobacter gen. nov. as Helicobacter pylori comb. nov. and Helicobacter mustelae comb. nov., respectively. Int J Syst Bacteriol 39:397–405
10. Isaacson P (1982) Immunoperoxidase study of the secretory immunoglobulin system and lysozyme in normal and diseased gastric mucosa. Gut 23:578–588
11. Isaacson PG, Spencer J (1987) Malignant lymphoma of mucosa-associated lymphoid tissue. Histopathology 11:445–462
12. Janeway CA Jr (1988) Frontiers of the immune system. Nature 333:804–806

13. Janeway CA Jr, Bottomly K, Babich J (1984) Quantitative variations in Ia antigen expression plays a central role in immune regulation. Immunol Today 5:99–105
14. Janeway CA Jr, Jones B, Hayday A (1988) Specificity and function of T cells bearing $\gamma\delta$ receptors. Immunol Today 9:73–76
15. Kawanishi H (1987) Immunocompetence of normal human appendiceal lymphoid cells: in vitro studies. Immunology 60:19–28
16. Kvale D, Lövhaug D, Sollid LM, Brandtzaeg P (1988) Tumor necrosis factor-α upregulates expression of secretory component, the epithelial receptor for polymeric Ig. J Immunol 140:3086–3089
17. Matis LA, Glimcher LH, Paul WE, Schwartz RH (1983) Magnitude of response of histocompatibility-restricted T-cell clones is a function of the product of the concentrations of antigen and Ia molecules. Proc Natl Acad Sci USA 80:6019–6023
18. Mestecky J, McGhee J (1987) Immunoglobulin A (IgA): molecular and cellular interactions involved in IgA biosynthesis and immune response. Adv Immunol 40:153–245
19. Musgrove C, Bolton FJ, Krypczyk AM, Temperly JM, Cairns SA, Owen WG, Hutchinson DN (1988) Campylobacter pylori: clinical, histological, and serological studies. J Clin Pathol 41:1316–1321
20. Myhre MJ, Isaacson PG (1987) Primary B-cell gastric lymphoma – a reassessment of its histogenesis. J Pathol 152:1–11
21. Papadimitriou CS, Ioachim-Velogianni EE, Tsianos EB, Moutsopoulos HM (1988) Epithelial HLA-DR expression and lymphocyte subsets in gastric mucosa in type B chronic gastritis. Virchows Arch [A] 413:197–204
22. Rathbone BJ, Wyatt JI, Tompkins D, Heatley RV, Losowsky MS (1986) In vitro production of Campylobacter pyloridis specific antibodies by gastric mucosal biopsies. Gut 27:A607
23. Sollid LM, Gaudernack G, Markussen G, Kvale D, Brandtzaeg P, Thorsby E (1987) Induction of various HLA class II molecules in a human colonic adenocarcinoma cell line. Scand J Immunol 25:175–180
24. Sollid LM, Kvale D, Brandtzaeg P, Markussen G, Thorsby E (1987) Interferon-γ enhances expression of secretory component, the epithelial receptor for polymeric immunoglobulins. J Immunol 138:4303–4306
25. Spencer J, MacDonald TT, Diss TC, Walker-Smith JA, Ciclitira PJ, Isaacson PG (1989) Changes in intraepithelial lymphocyte subpopulations in coeliac disease and enteropathy associated T cell lymphoma (malignant histiocytosis of the intestine). Gut 30:339–346
26. Stolte M, Eidt S, Ritter M, Bethke B (1989) Campylobacter pylori und Gastritis. Assoziation oder Induktion? Pathologe 10:21–26
27. Thrane PS, Brandtzaeg P, Huitfeldt H (1986) Spatial relationship between IgA-positive cells and epithelium expressing HLA class II determinants in salivary glands (Abstr 1.14.13) International Congress on Immunology, Toronto
28. Tsutsumi Y, Nagura H, Watanabe K (1984) Immune aspects of intestinal metaplasia of the stomach: an immunohistochemical study. Virchows Arch [A] 403:345–359
29. Valnes K, Brandtzaeg P, Elgjo I, Stave R (1984) Specific and non-specific humoral defense factors in the epithelium of normal and inflamed gastric mucosa. Immunohistochemical localization of immunoglobulins, secretory component, lysozyme, and lactoferrin. Gastroenterology 86:402–412
30. Vecchi M, Berti E, Primignani M, Monti M, Agape D, Torgano G, Arosio E, de Franchis R (1985) In situ identification of immune competent cells in gastrointestinal mucosa: an evaluation by immunoelectron microscopy. Virchows Arch [A] 406:407–415
31. Warren JR, Marshall BJ (1983) Unidentified curved bacilli on gastric epithelium in active chronic gastritis. Lancet 1:1273–1275
32. Wyatt JI, Rathbone BJ, Heatley RV (1986) Local immune response to gastric Campylobacter in non-ulcer dyspepsia. J Clin Pathol 39:863–870

Risk of Peptic Ulcer in Gastritis

P. Sipponen

In the multifactorial pathogenesis of peptic gastroduodenal ulcer, chronic gastritis may influence both aggressive and defence mechanisms. Depending upon the grade and type of gastritis, chronic inflammation and coexistent atrophy of the underlying mucosa may decrease the mucosal resistance [16], but they may also lower the effect of aggressive factors, that is, acid and pepsin secretions from the gastric body (corpus) mucosa [23].

Chronic gastritis is a progressive lesion. It begins as a chronic inflammation and may slowly progress to an atrophy, i.e., to a loss of normal mucosal glands, and to a growth of new metaplastic epithelium and glands [19, 20, 24]. The progression of gastritis to atrophic endstages affects often dissimilarly the antrum and the corpus [20, 22]. The atrophy may occur either in the antrum (B type of atrophic gastritis), in the corpus (A type), or in both (AB type). Correspondingly, the different atrophic endstages are dissimilarly linked to the impairments in gastric function and physiology [21, 23]. The A and AB types of gastritis typically indicate hypochlorhydria or even achlorhydria, whereas the acid output in the B gastritis is usually normal or even high. On the other hand, an impairment in the gastrin response is a typical phenomenon in the B or AB gastritis. Thus, it is understandable that the different grades and topographic types of gastritis may dissimilarly associate with the different gastric diseases such as duodenal (DU) or gastric ulcer (GU) [21].

Based on convincing evidence [47], a microbial origin and cause is suggested in the pathogenesis of gastritis, particularly in the types which tend to progress to the B or AB types of atrophic gastritis. In the early phases of the progression of gastritis (superficial gastritis) over 90% of subjects show a colonization of *Helicobacter* (*Campylobacter*) *pylori* (HP), or related bacteria, at the gastric surface epithelium [8, 12, 13].

HP certainly has noxious metabolic and immunologic properties which are capable of distorting the integrity of gastric or duodenal mucous membrane [4, 7]. Because of the common occurrence of the bacterium in both gastritis and ulcer patients, the cause and the pathogenesis of ulcer is often suggested to be causally related to these bacteria. It could be more plausible, however, that HP is a common cause of gastritis and that the gastritis itself also exerts influences in the pathogenesis of DU and GU [3, 8, 15]. The recent epidemiologic data support this view. The risk and probability of DU and GU can be shown to be strongly related to the presence, to the particular types and grades of gastritis [16, 17]. In addition,

Dept. of Pathology, Jorvi Hospital, SF-02740 Espoo, Finland

Helicobacter pylori, Gastritis and Peptic Ulcer
Edited by P. Malfertheiner and H. Ditschuneit

the studies suggest a "dose-response-like" relationship between the grade of gastritis and the ulcer risk, supporting the possibility that gastritis (and previous HP infection) has a causative role in the pathogenesis of DU and GU, instead of the role and influence of HP alone.

Association Between Gastritis and Peptic Ulcer

Although gastritis is extremely common in the general population [18], it is yet more common in patients with DU or GU. A great number of epidemiologic studies have convincingly shown already decades ago that a normal, nongastritis stomach is exceptional in the ulcer patient, if those patients are excluded in whom the ulcer is probably induced by some drugs (acetosalisylic acid or other NSAIDs) [1, 2, 5, 6, 9–11, 14].

In Tables 1 and 2, the age-adjusted relative risks (RR) of DU or GU in different grades of antral and corpus gastritis (superficial gastritis, slight, moderate, or severe atrophic gastritis) are presented when the risk of ulcer in the presence of normal mucosa is considered as a baseline reference (RR = 1). These calculations are epidemiologic estimates and are based on case-control approaches involving ulcer patients and nonulcer controls [16]. They suggest an extremely strong association between gastritis and peptic ulcer.

A simple generalization is that the grade of antral gastritis correlates positively and linearly with the risk of peptic ulcer, whereas an opposite relationship exists in the correlation between the corpus gastritis and the ulcer risk. In patients with gastritis of the B type (antral mucosa is markedly atrophic but corpus mucosa is nonatrophic), the RR of DU or GU can be estimated to be 30 times higher than the risk of ulcer in patients with normal stomach. On the other hand, the ulcer risk is again very low in subjects with gastritis of the A or AB type [16].

Table 1. Age-adjusted relative risk (RR) of duodenal (DU) or gastric ulcer (GU) in different grades of antral gastritis. [From 16]

Mucosa	Patients/Controls	RR	CI95
Males			
N	15/184	1.0	
S	197/196	12	7–20
A1	63/33	22	12–42
A2–A3	42/19	19	9–38
Females			
N	17/271	1.0	
S	134/246	7	5–12
A1	58/61	10	6–17
A2–A3	22/28	9	4–18

N, normal mucosa; S, superficial chronic gastritis; A1–A3, slight, moderate or severe gastritis; CI95 = 95% confidence interval

Table 2. Age-adjusted relative risk (RR) of duodenal (DU) or gastric ulcer (GU) in different grades of body gastritis. [From 16]

Mucosa	Patients/Controls	RR	CI95
Males			
N	69/221	1.0	
S	211/155	4	3–6
A1	15/30	1	0.5–2
A2–A3	3/23	0.2	0.1–0.6
Females			
N	57/289	1.0	
S	138/218	3	2–4
A1	16/47	1	0.6–2
A2–A3	7/49	0.4	0.1–1.2

N, normal mucosa; S, superficial chronic gastritis; A1–A3, slight, moderate or severe gastritis; CI95 = 95% confidence interval

The gastritis-ulcer relationship is easily explained by alterations in the physiology of the underlying mucosa. In AB gastritis, the gastritis- (and HP-) related negative effects on the mucosal resistance are opposed by low acid and pepsin secretion from the corpus mucosa, resulting in a marked decrease of ulcer risk. Correspondingly, an imbalance between the aggressive and defence mechanisms can be seen to be highest in patients with B gastritis, a gastritis in which the epidemiologic estimates of the ulcer risk are highest indeed [16].

The Preceding Role of Gastritis in Peptic Ulcer Disease

The concept of the role of gastritis as a risk factor for peptic ulcer implies that the inflammation (and the HP infection) precedes the ulcer formation. An association between gastritis and ulcer may mean that the gastritis is a simple result of ulcer, or that the gastritis and the ulcer are only independent parallel phenomena instead of forming a causative sequence. Some available epidemiologic data support the view of the causative sequence from gastritis to ulcer formation [16]. However, direct follow-up data on this matter are scanty, and our knowledge is nearly solely based on cross-sectional analyses of gastritis in ulcer patients and in varying control populations.

In regard to the cumulative risk of DU or GU, a short summary of the results from a recent 10-year follow-up of patients with or without gastritis (unpublished data) is presented in Table 3. In this follow-up, new symptomatic DU or GU cases were found nearly entirely in the patients who had gastritis in the initial endoscopy. Only one patient had contracted ulcer within the follow-up period without evidence of gastritis at the beginning of follow-up. She developed pyloric GU 10 years after the first endoscopy (there was gastritis in both the antrum and body at the time of diagnosis of ulcer).

In the follow-up, the cumulative risk of peptic ulcer was particularly accentuated in the middle-aged males (40–60 years of age) who had superficial antral

Table 3. Cumulative prevalence of new duodenal (DU) or gastric ulcer (GU) cases in 10-year follow-up of patients with or without gastritis in antrum and body (males and females as a common group)

Antrum/Body	Total no.	DU or GU in follow-up	
		no.	Percentage
N/N	133	1	0.8
S–A3/N–S	233	29	12.4
S–A3/A2–A3	22	1	4.5

N, normal; S, superficial chronic gastritis; A1–A3, slight, moderate, or severe atrophic gastritis

gastritis or atrophic gastritis of the B type in the initial endoscopy. Approximately every fourth male was recorded to have contracted symptomatic DU or GU within the follow-up period.

The available follow-up data are in line with the data from epidemiologic cross-sectional case-control studies: there is an approximately tenfold increase in ulcer risk in the presence of antral gastritis in general, this risk increasing further to up to some tenfol in the presence of type B gastritis. Correspondingly, the cumulative risk and probability of DU or GU is low in subjects with normal stomach or in those who show marked atrophic alterations (moderate or severe atrophic gastritis) in the gastric body.

Conclusions

Gastritis is strongly associated with peptic ulcer. Gastritis precedes the formation of ulcer, and it can be seen as a risk factor for peptic ulcer, for both DU and GU. The risk of DU or GU is increased up to 10 to 30-fold in the presence of antral gastritis compared with ulcer risk in subjects with a normal stomach. In middle-aged males, the 10-year cumulative risk and probability of symptomatic peptic ulcer are highest and can be approximated to be even 20%–30% in the occurrence of antral gastritis. On the other hand, the presence of atrophic gastritis in the body mucosa markedly decreases this risk, both in males and females, to the level which is nearly indistinguishable from the risk in a normal stomach. In the presence of a normal stomach, the risk and cumulative propability of symptotic DU and GU can be estimated to be very low (at most a few percent) in both males and females.

References

1. Earlam RJ, Amerigo J, Kakavoulis T, Pollock DJ (1985) Histological appearances of oesophagus, antrum and duodenum and their correlation with symptoms in patients with a duodenal ulcer. Gut 26:95–100
2. Gear MWL, Truelove SC, Whitehead R (1971) Gastric ulcer and gastritis. Gut 12:639–645
3. Goodwin CS (1988) Duodenal ulcer, Campylobacter pylori, and the "leaking roof" concept. Lancet 24/31:1467–1469

4. Graham DY (1989) Campylobacter pylori and peptic ulcer disease. Gastroenterology 96:615–625
5. Hui W-M, Lam A-K, Ho J, et al. (1986) Chronic antral gastritis in duodenal ulcer. Natural history and treatment with prostaglandin E1. Gastroenterology 91:1095–1101
6. MacKay IR, Hislop IG (1966) Chronic gastritis and gastric ulcer. Gut 7:228–233
7. Marshall BJ (1988) The Campylobacter pylori story. Scand J Gastroenterol [Suppl 146] 23:58–66
8. Marshall BJ, McGechie DB, Rogers PA, Glancy RJ (1985) Pyloric Campylobacter infection and gastroduodenal disease. Med J Aust 142:439–444
9. Meikle DD, Taylor KB, Truelove SC, Whitehead R (1976) Gastritis, duodenitis, and circulating levels of gastrin in duodenal ulcer before and after vagotomy. Gut 17:719–728
10. Meister H, Holubarsch C, Haferkamp O, Schlag P, Herfarth C (1979) Gastritis, intestinal metaplasia, and dysplasia versus ulcer in stomach and duodenum and gastric carcinoma. A histotopographic study. Pathol Res Pract 164:259–269
11. Overgaard Nielsen H, Munoz JD, Kronborg O, Andersen D (1981) The antrum in duodenal ulcer patients. Scand J Gastroenterol 16:491–493
12. Price AB, Levi J, Oolby JM, et al. (1985) Campylobacter pyloridis in peptic ulcer disease: microbiology, pathology and scanning electron microscopy. Gut 26:1183–1188
13. Rauws EAJ, Langenberg W, Houthoff HJ, Zanen HC, Tytgat GNJ (1988) Campylobacter pyloridis-associated chronic active gastritis. A prospective study of its prevalence and the effects of antibacterial and antiulcer treatment. Gastroenterology 94:33–40
14. Schrager J, Spink R, Mitra S (1967) The antrum in patients with duodenal and gastric ulcers. Gut 8:497–508
15. Sipponen P, Varis K, Cederberg A, et al. (1988) Campylobacter pylori is associated with chronic gastritis but not with active peptic ulcer disease. APMIS 96:84–88
16. Sipponen P, Seppälä K, Äärynen M, Helske T, Kettunen P (1989) Chronic gastritis and gastroduodenal ulcer: a case control study on risk of coexisting duodenal or gastric ulcer in patients with gastritis. Gut 30:922–929
17. Sipponen P, Äärynen M, Kääriäinen I, Kettunen P, Helske T, Seppälä K (1989) Chronic antral gastritis, Lewisa+phenotype, and male sex as factors in predicting coexisting duodenal ulcer. Scand J Gastroenterol 24:581–588
18. Siurala M, Isokoski M, Varis K, Kekki M (1968) Prevalence of gastritis in a rural population. Scand J Gastroenterol 3:211–223
19. Siurala M, Varis K, Kekki M (1980) New aspects on epidemiology, genetics, and dynamics of chronic gastritis. Front Gastrointest Res 6:148–166
20. Siurala M, Kivilaakso E, Sipponen P (1984) Gastritis. In: Demling L, Domschke S (eds) Klinische Gastroenterologie, vol 1. Thieme, Stuttgart, pp 321–337
21. Stadelmann O, Elster K, Stolte M, et al. (1971) The peptic gastric ulcer – histotopographic and functional investigations. Scand J Gastroenterol 6:613–623
22. Strickland RG, MacKay IR (1973) A reappraisal of the nature and significance of chronic atrophic gastritis. Dig Dis Sci 18:426–440
23. Varis K, Ihamäki T, Härkönen M, Samloff IM, Siurala M (1979) Gastric morphology, function and immunology in first-degree relatives of probands with pernicious anaemia and controls. Scand J Gastroenterol 14:129–139
24. Whitehead R, Truelove SC, Gear MWL (1972) The histological diagnosis of chronic gastritis in fibre optic biopsy specimens. J Clin Pathol 25:1–11

Differences Between *Helicobacter pylori* Associated Gastritis in Patients with Duodenal Ulcer, Pyloric Ulcer, Other Gastric Ulcer, and Gastritis Without Ulcer

S. EIDT and M. STOLTE

Introduction

Helicobacter (Campylobacter) pylori (HP) is the main causal agent of type B gastritis [1–3]. The spectrum of HP-associated pathology extends from gastritis in the absence of lesions, duodenal ulcer, and pyloric ulcer, to gastric ulcer [4]. We undertook a study to establish whether, among these various groups, differences in the density of colonization by HP or in the degree or the activity of gastritis can be found.

Material and Methods

Our investigations covered 1258 cases in which at least two biopsy specimens each from the antrum and body were available (450 cases of duodenal ulcer, 78 cases of pyloric ulcer, 70 cases of gastric ulcer in other regions, and 660 cases of gastritis without lesions). The degree of HP colonization and the degree and the activity of gastritis were scored semiquantitatively according to Stolte et al. [3] in sections stained with hematoxylin and eosin, and Warthin-Starry. In addition, the HP colonization in each group was estimated by adding the results of each case and a final division by the number of patients. In the same way a mean gastritis score uniting the lymphoplasmocellular and the neutrophil polymorph infiltrations was evaluated. The results were separately considered for the antrum and body. Statistical evaluation was done using the chi-squared test.

Results

The age distribution revealed the well-known predominance of younger patients in the case of duodenal ulcer (Table 1), while the sex distribution demonstrated the dominance of men in the cases of all peptic ulcers (Table 2).

An analysis of the differing density of HP colonization in the antrum and body revealed that colonization by HP in duodenal and pyloric ulcer is virtually identical and high-grade in more than 60%. The groups of patients with type B

Institut für Pathologie (Leiter: Prof. Dr. M. Stolte), Klinikum Bayreuth, Preuschwitzer Strasse 101, D-8580 Bayreuth

Helicobacter pylori, Gastritis and Peptic Ulcer
Edited by P. Malfertheiner and H. Ditschuneit

Table 1. Distribution of ages in the different groups of gastritis

	Mean age (years)
Duodenal ulcer	49.2+14.6
Pyloric ulcer	54.0+12.7
Other gastric ulcer	57.9+13.5
Without ulcer	53.2+14.8

Table 2. Sex distribution in the different groups of gastritis

	Men : women
Duodenal ulcer	2 :1
Pyloric ulcer	2.3:1
Other gastric ulcer	1.7:1
Without ulcer	1.1:1

gastritis but no lesions and with gastric ulcer showed a low level of colonization of the antral mucosa by HP (Fig. 1). These differences were statistically significant. In contrast, an analysis of HP colonization in the body revealed, vis-à-vis the antrum, a smaller density of HP colonization and no significant differences between the groups (Fig. 2).

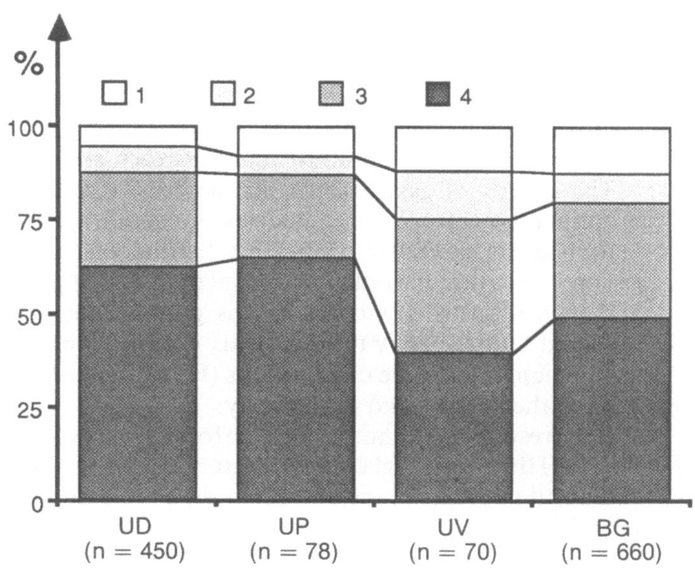

Fig. 1. *H. pylori* (HP) colonization in the antrum (UD, duodenal ulcer; UP, pyloric ulcer; UV, other gastric ulcer; BG, gastritis without ulcer)

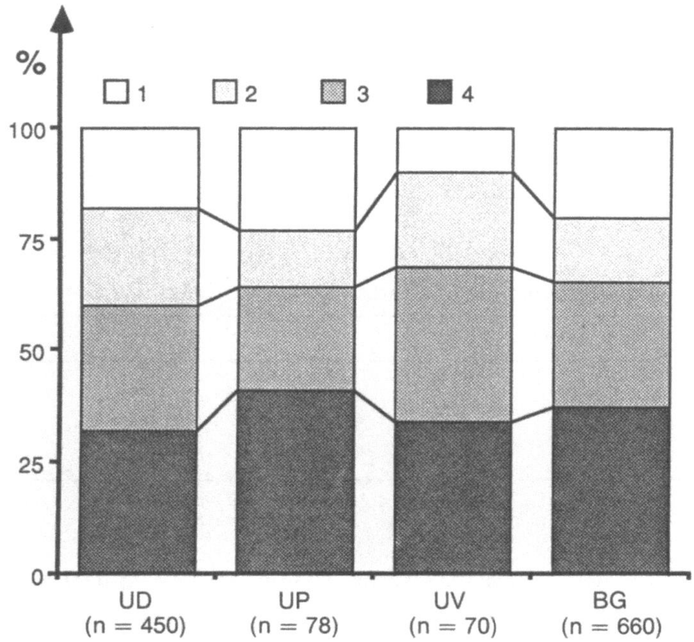

Fig. 2. HP colonization in the body

The degree of gastritis, that is, the density of infiltration of the lamina propria with lymphocytes and plasma cells, is, overall, greater in the antrum than in the body. In contrast to the three ulcer groups, only in the group of gastritis without lesions a lower grade of gastritis could be found (Fig. 3). In the body, gastritis is, for the most part, only low-grade. A higher grade is observed only in the case of gastric ulcer (Fig. 4). Both distributions reach statistical significance.

The activity of gastritis, that is, the density of the infiltration of the lamina propria with neutrophil granulocytes, is, overall, again greater in the antrum than in the body, in particular in the case of the duodenal ulcer and pyloric ulcer groups. In contrast to this, the inflammatory activity was appreciably lower in gastric ulcer and type B gastritis without lesions groups (Fig. 5). Within the body, it was observed that the activity of gastritis is appreciably greater in the gastric ulcer group than in the three other groups (Fig. 6). Statistical significance was reached both in the antrum and in the body.

If the results are summarized in the form of mean score values, it is first observed that the HP score is higher in the antrum than in the body and that the most dense colonization occurs in the antrum in the case of duodenal ulcer and pyloric ulcer (Fig. 7). Within the body, in contrast, there are no differences in the HP scores (Fig. 8).

In contrast, an analysis of the gastritis score reveals a reversal of this picture: in the antrum, the score computed on the basis of the summation of lymphocytes,

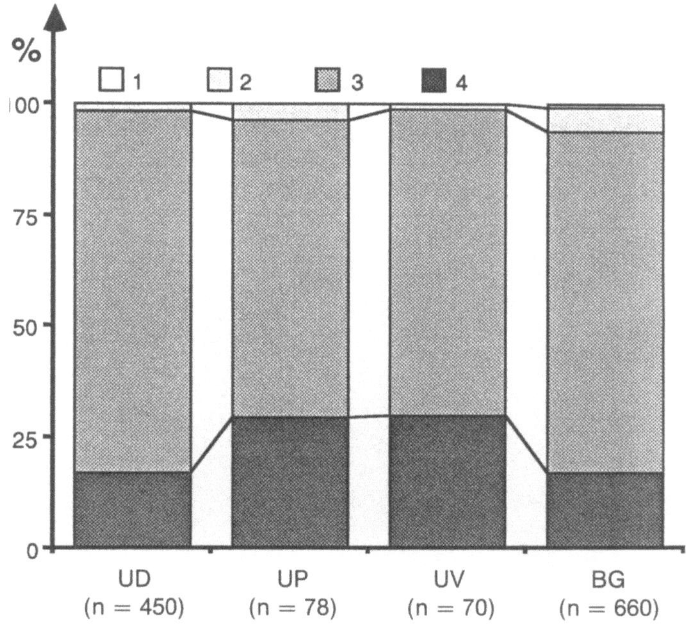

Fig. 3. Degree of antral gastritis

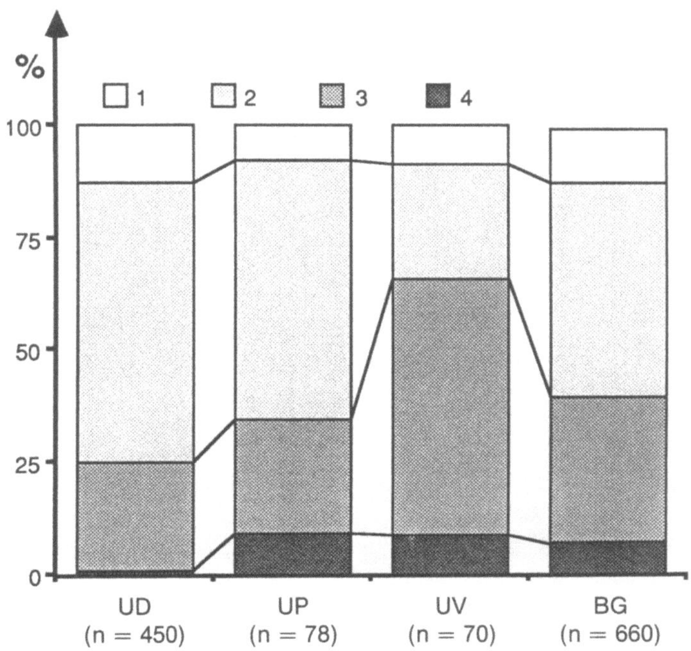

Fig. 4. Degree of body gastritis

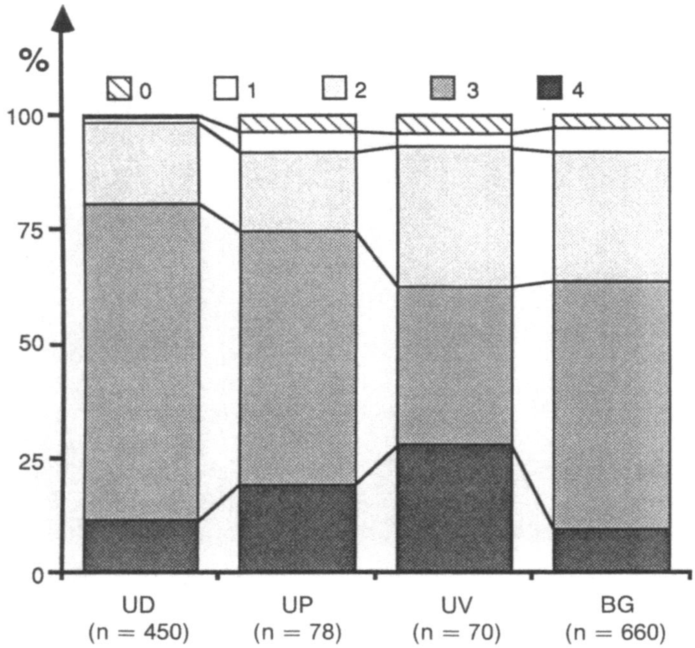

Fig. 5. Activity of antral gastritis

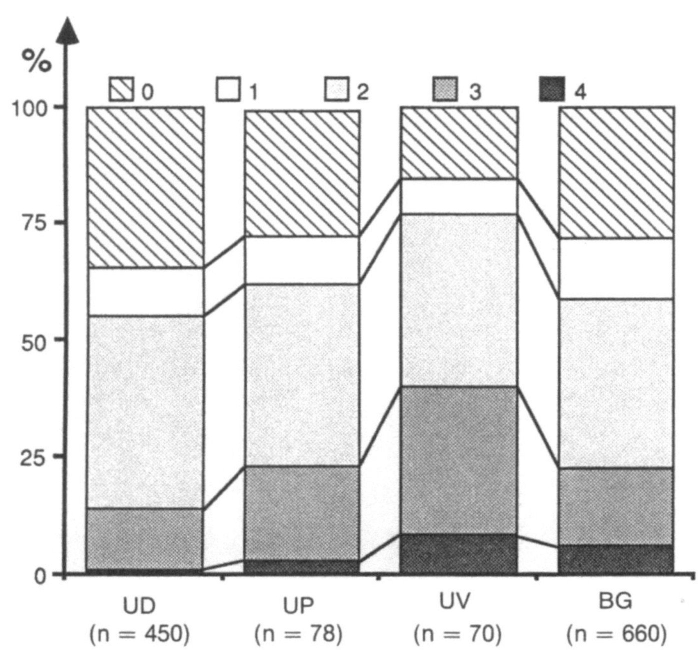

Fig. 6. Activity of body gastritis

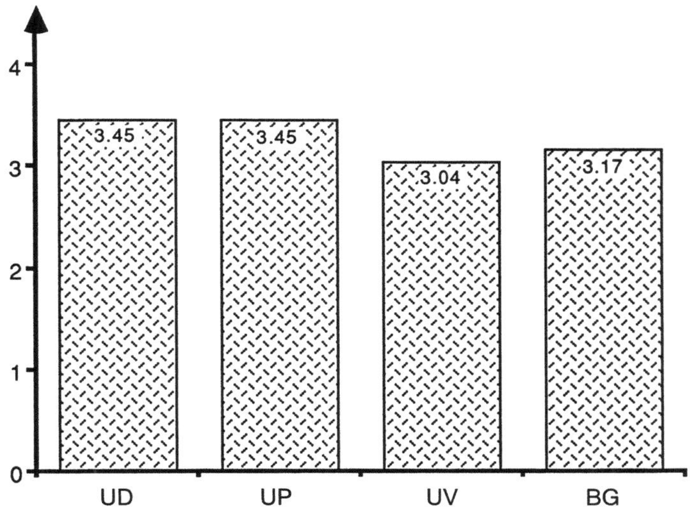

Fig. 7. Score of antral HP colonization

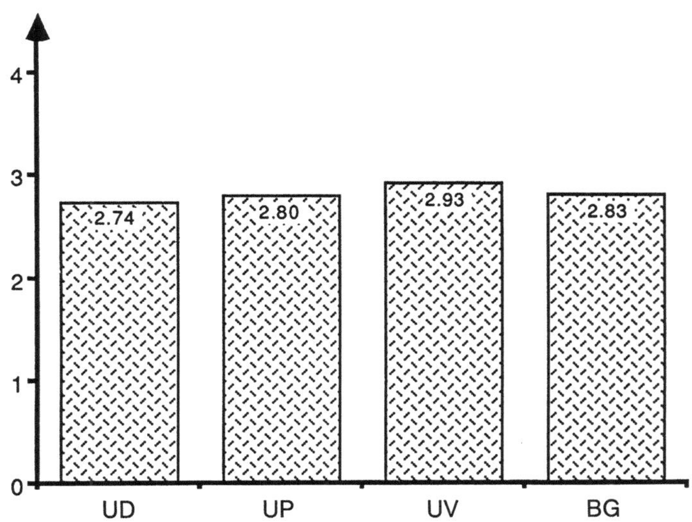

Fig. 8. Score of body HP colonization

plasma cells, and neutrophil granulocytes is high for all the ulcer groups; the gastritis is least pronounced in patients with no ulcers (Fig. 9). In the body, however, there are obvious differences. Here, gastritis is least marked in the case of duodenal ulcer and most pronounced in the case of gastric ulcer (Fig. 10).

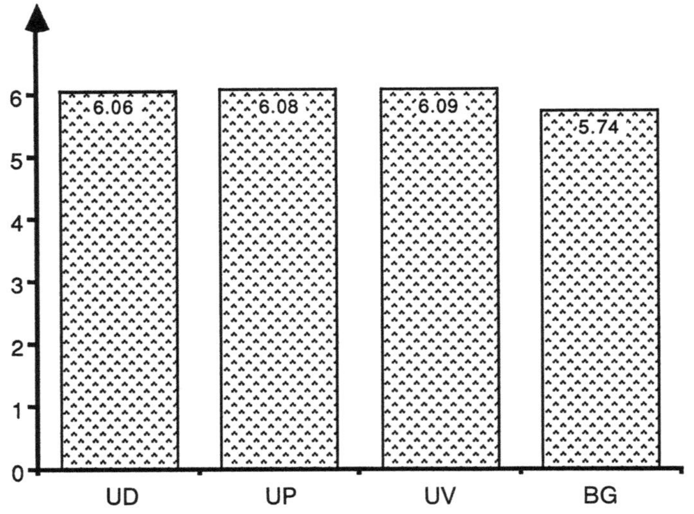

Fig. 9. Score of antral gastritis

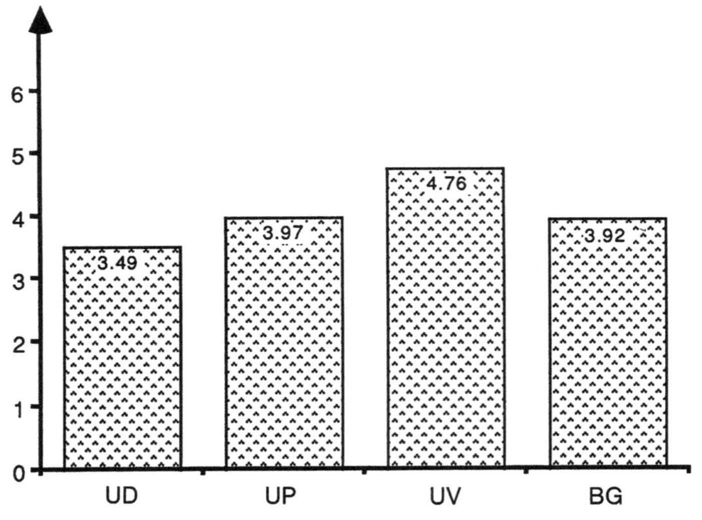

Fig. 10. Score of body gastritis

Discussion

The results in the groups of duodenal and pyloric ulcers are very similar especially concerning the HP colonization of the antral mucosa (see Fig. 1), the activity of the antral gastritis (see Fig. 5), and the antral HP score (Fig. 7). On the other hand, the results of the group of gastric ulcer patients concerning the degree and activity of body gastritis are clearly higher than in all the other groups (Fig. 4, 6) – a fact well represented in the gastritis mean score (Fig. 10). In conclusion, we propose

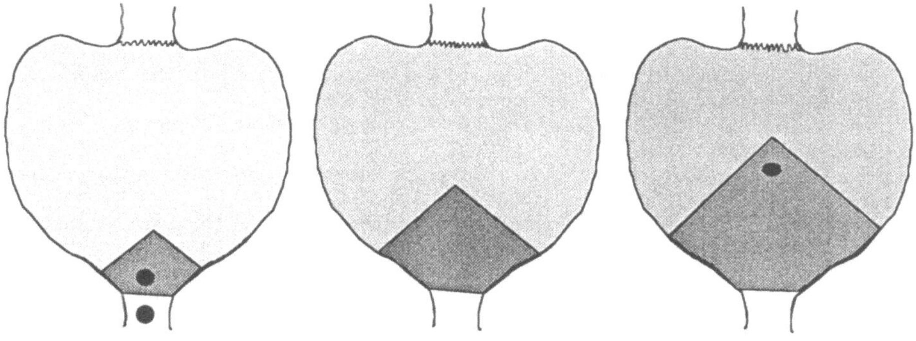

Fig. 11. Model of the hypothetical relevancy of the different areas of antral and body mucosa

that future investigations of gastric ulcer should separate the results of pyloric and other gastric ulcers.

At the present time we can only speculate regarding the possible reasons for these differences. The fact that the inflammatory reaction to HP in the body of the stomach is, overall, less pronounced than in the antrum can be explained by the neutralization of NH_3 by HCl in the body. The question as to whether the age of the patient and/or the duration of gastritis leads to a pylorocardial expansion, and thus to reduced acid secretion, less NH_3 neutralization, and more pronounced gastritis in the body, remains to be answered by longitudinal studies.

On the other hand, however, the question must also be reexamined as to whether, perhaps, all these differences might not be a reflection of an inherent variability in the size of the area of the body and antral mucosa [5, 6]: an antral HP infection in patients with a large area of body mucosa could stimulate an enormous increase in gastric acid output and thus lead to peptic ulcer in the pyloric or duodenal region (Fig. 11). A large area of antral mucosa could lead to an angular ulceration after HP infection, because in addition to the damage the infection does to the mucosa, high mechanical stress and poor blood supply in this region can be demonstrated. An intermediate division of the stomach mucosal surface would then lead to gastritis without ulcer. If this were the case, in addition to other ulcerogenic factors, the difference in antral and body mucosal area might possibly be a contributing factor to the solution of the problem as to why it is that, after HP infection, a patient can develop only gastritis, or a duodenal ulcer, or a pyloric ulcer, or a gastric ulcer. If this speculation were to be confirmed by longitudinal studies, it might prove possible to predict the development of an ulcer on the basis of the grade, activity, and topography of gastritis in the antrum and body.

References

1. Morris A, Nicholson G (1987) Ingestion of Campylobacter pyloridis causes gastritis and raised fasting gastric pH. Am J Gastroenterol 82:192–199
2. Mahoney MJ, Wyatt JI, Littlewood JM (1988) Campylobacter pylori gastritis. Arch Dis Child 63:654–655
3. Stolte MS, Eidt S, Ritter M, Bethke B (1988) Campylobacter pylori und Gastritis – Association oder Induktion? Pathologe 10:21–26
4. Börsch G (1987) Campylobacter pylori: new and renewed insights into gastritis associated ulcer disease (GAUD). Hepatogastroenterology 34:191–193
5. Oi M, Oshida K, Sugimura S (1959) The location of gastric ulcers. Gastroenterology 36:45–56
6. Oi M, Ito Y, Kumagi F, Yoshida K, Tanaka Y, Yoshikawa K, Miho O, Kilima M (1969) A possible dual control mechanism in the origin of peptic ulcer. Gastroenterology 57:280–293

Lymphocytic Gastritis

J. HAOT, A. JOURET, and P. MAINGUET

Definition and Diagnostic Criteria

Lymphocytic gastritis is a newly described clinicopathological entity first reported in 1985 and 1986 by Haot et al. It is characterized by a dense lymphocytic infiltration of surface and pit gastric epithelium often extending to the gastric glands where it is, however, less prominent [9, 10, 12].

The histological features of the disease are striking. At low magnification, the epithelium appears intensely basophilic and punctuated by numerous small dark dots; at higher magnification, these are small lymphocytes, with scanty cytoplasm and a dense, irregular, cleaved nucleus, most often closely apposed to the epithelial cell membrane and surrounded by a clear halo (Figs. 1, 2). The diagnostic hallmark of the disease is the presence of intraepithelial lymphocytes, the other histological features being less prominent. Although mucosal thickness is, in general, within the normal range, there is some focal edema and elongation of the glandular pits. A lymphoplasmacytic infiltrate, albeit slight in some observations, is present in the lamina propria. There is no correlation between the intensity of lamina propria inflammatory reaction and the epithelial lymphocytic infiltration. Erosion leads to a more complex picture; the inflammatory infiltration of the lamina propria is denser and often of a mixed type. Cryptitis and crypt abcesses [12] are found.

Statistical Data

Prevalence of Lymphocytic Gastritis

Two thousand archival cases diagnosed as chronic gastritis were reviewed for the presence of dense lymphocytic intraepithelial infiltrates to determine the prevalence of lymphocytic gastritis in a routine workload; approximately 4% were reclassified as lymphocytic gastritis, the rest being for the most part the usual types of chronic gastritis, poor in intraepithelial lymphocytes [10]. This prevalence has since been confirmed [5].

Service d'Anatomie Pathologique
Service de Gastroentérologie, Université Catholique de Louvain, Cliniques Universitaires St-Luc, Avenue Hippocrate 10, B-1200 Bruxelles, Belgique

Helicobacter pylori, Gastritis and Peptic Ulcer
Edited by P. Malfertheiner and H. Ditschuneit

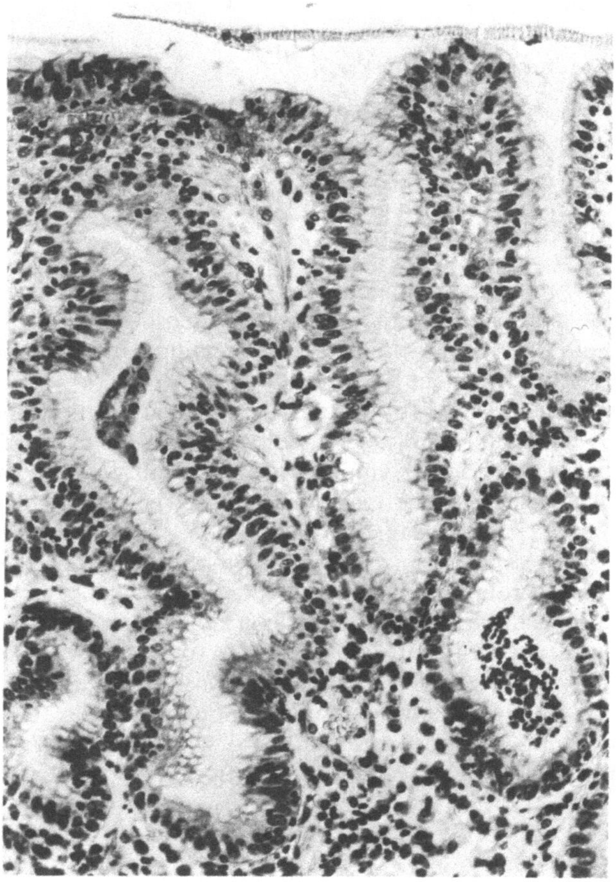

Fig. 1. Lymphocytic gastritis. Low magnification (×6). Surface and pit epithelium punctuated by numerous small *dark dots*

Age and Sex Incidence

Rarely seen in young children, lymphocytic gastritis appears mostly as a disease of late adulthood. In our series of 87 petients the average age was 54.9 ± 4.6 years which is comparable to the age ($54.2 + 3.7$ years) of patients with other types of chronic gastritis [10]. We have previously reported a males/females ratio of 0.8 [10]. Dixon et al. [5] also recorded a female preponderance (males/females: 1:1.8). However, in a larger series we obtained a ratio closer to 1/1 (unpublished data).

Lymphocyte Counts

Lymphocyte counts were performed on cases of lymphocytic gastritis, chronic active gastritis, and normal stomachs. The ratio of the number of lymphocytes to the number of surface epithelial cells was found to be seven times higher in

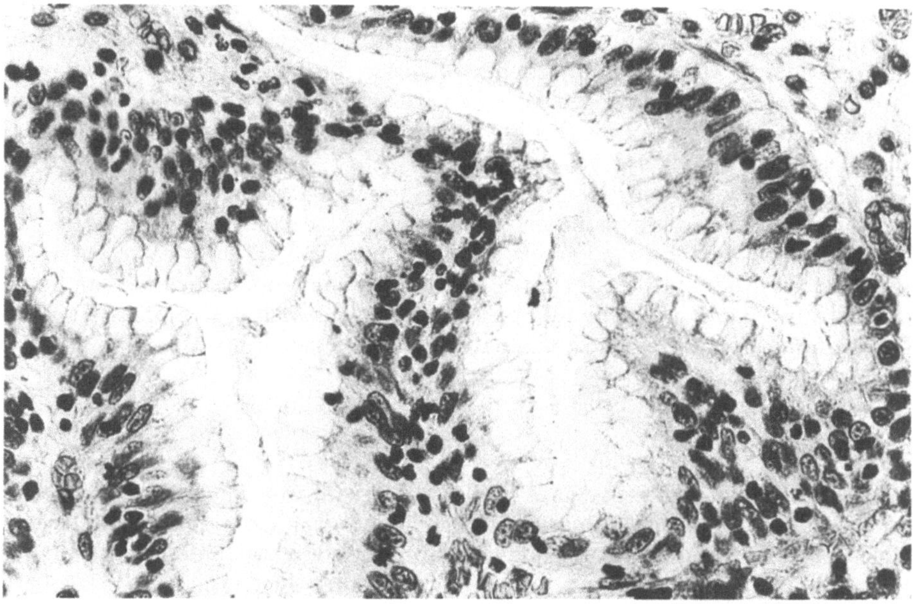

Fig. 2. Lymphocytic gastritis, higher magnification (× 40). Intraepithelial lymphocytes are surrounded by a clear halo; their nuclei are dark and irregular

lymphocytic gastritis than in chronic gastritis and 20 times higher than in normal mucosa [9].

More recently, we found an average of 57 lymphocytes per 100 epithelial cells in lymphocytic gastritis, while in chronic active gastritis the figures were 3.4 per 100 epithelial cells and in normal mucosa 2.5 per 100 epithelial cells. These results (Student's *t* test) were highly significant compared with both chronic gastritis and normal mucosa ($P < 0.001$). Even in this large series (54 cases of lymphocytic gastritis), there was no overlap in lymphocyte counts between lymphocytic gastritis and the other groups [14].

Dixon et al. [5] obtained similar results on 17 cases of lymphocytic gastritis with an average of 55.3 ± 19.5 lymphocytes per 100 epithelial cells while in chronic Helicobacter-positive gastritis the figures were 3.7 ± 1.3. Moreover, no significant differences were found when biopsies were repeated 6–22 weeks after a first endoscopy.

Anatomoclinical Correlations

Endoscopic Data

We observed at an early stage, that lymphocytic gastritis exhibited unusual endoscopic findings. A first retrospective study of 46 cases showed that the nature and topography of the endoscopic lesions were totally different from the usual chronic active gastritis. In chronic active gastritis, the inflammation was generally diffuse, giving a reddish tint to the mucosa, the vascular network of which appeared irregular and congestive; when erosions were present, they were usually flat with an irregular outline. Inflammation was maximal in the antrum while the fundus was relatively spared.

In lymphocytic gastritis, this type of appearance was uncommon. In the majority (67%), lesions predominated in the body and consisted of elevated nodules with an indurated base (aphthoid nodules) [11]. More recently, in a prospective study, we demonstrated the endoscopic appearance of lymphocytic gastritis to be a complex picture of thickened rugal folds often crossed by large serpiginous ulcerations and punctuated by numerous aphthoid nodules [12].

Clinical Data

Patients with chronic active gastritis presented with vague epigastric pains, those with lymphocytic gastritis often suffered, in addition, from anorexia and weight loss. Cases of rapid emaciation have been reported in which deterioration of the general condition was so prononced as to simulate gastric cancer [11].

Relationship with Varioliform Gastritis

Varioliform gastritis is an endoscopic condition first reported in 1947 by Moutier and Cornet [19] who described two cases; more recently, Lambert et al. [18] redefined the condition that they called diffuse varioliform gastritis as: "characterized by large folds in the fundus and erosive mucosal bulges disseminated in the fundus and antrum."

Lambert et al. [18] considered separately a second type of varioliform gastritis, limited to the antrum that received the name of antral varioliform gastritis. Descriptions clearly fitting those of varioliform gastritis have been reported by different authors under various names such as chronic erosive gastritis, octopus sucker gastritis, chronic verrucous gastritis [6, 8, 21].

It appeared clear from our earliest research that the endoscopic features of the majority of cases of lymphocytic gastritis were very similar to those of diffuse varioliform gastritis. Both retrospective and prospective studies were undertaken to compare these two conditions.

Retrospective Studies

From the files of the department of gastroenterology 192 cases were selected on the basis of a presumed diagnosis of varioliform or erosive gastritis. Two pathologists reviewed the biopsy material independtly; 92 cases were diagnosed as lymphocytic gastritis while 100 were classified as non-specific gastritis. A close correlation was found between the endoscopic diagnosis of varioliform gastritis and the histological diagnosis of lymphocytic gastritis (48 cases out of 58). The correlation improved when the endoscopic descriptions were reviewed. In contrast to "nonspecific gastritis," most of the cases of lymphocytic gastritis had a quite typical endoscopic pattern of enlarged folds and aphthoid nodules nearly always predominating in the fundic region [12].

In collaboration with Lyon Medical School, we had the opportunity of reviewing some of the cases of varioliform gastritis used by Lambert et al. [17, 18] and André et al. [1–3] in their studies on varioliform gastritis. Here, too, we found an excellent correlation between the histopathological diagnosis of lymphocytic gastritis and the clinical diagnosis of diffuse varioliform gastritis (26 out of 35 original cases were lymphocytic gastritis). The discrepancies were healed varioliform gastritis, Crohn's disease, and varioliform gastritis limited to the antrum [13].

Prospective Studies

Three endoscopists and two pathologists devised these studies to establish the correlation between lymphocytic gastritis and varioliform gastritis. They agreed on a clear definition of varioliform gastritis, as a complex pattern of enlarged rugal folds and of widespread (aphthoid) nodules most often disseminated along the rugae in a string-like fashion. According to the topography of the lesions, three subgroups were identified: diffuse when the whole stomach was involved and fundic when the lesions were restricted to the body or antral when restricted to the antrum. During a period of 3 years, the endoscopists diagnosed 66 cases of varioliform gastritis, 54 of which were lymphocytic gastritis. When the different subgroups of varioliform gastritis were considered separately, the correlation was even better. The 35 cases of diffuse varioliform gastritis and 18 out of 20 cases of fundic varioliform gastritis were histologically lymphocytic gastritis. Conversely, only 1 of 11 antral varioliform gastritis cases showed features of dense lymphocytic intraepithelial infiltration characteristic of lymphocytic gastritis.

A cross check was made by screening all the gastric biopsies performed by the three endoscopists during the same period (4840 biopsies) in search of lymphocytic gastritis. In addition to the 54 cases of varioliform gastritis, 13 cases were found bringing the total number of cases of lymphocytic gastritis in the series to 67. These results confirm, in accordance with the previous work of Lambert et al. [18], that there are two types of varioliform gastritis. The first predominates in the body or involves the whole stomach: this is the usual endoscopic expression of lymphocytic gastritis. The second is confined to the antrum and has no clear histological counterpart; it includes examples of chronic active Helicobacter (or Campylobacter) positive gastritis and reflux gastritis, cases of a poorly known condition of antral edematous and fibrotic nodules [7, 20], and even exceptional cases of lymphocytic gastritis. Although varioliform gastritis is the common endoscopic appearance of lymphocytic gastritis, our study demonstrates that a small proportion of lymphocytic gastritis (20%) has no clear endoscopic counterpart. This small group of "silent" lymphocytic gastrites is presently under study; unpublished data suggest that some of the cases correspond to "endoscopically" healed varioliform gastritis.

Relationship with Helicobacter Infection

Dixon et al. [5] found that 41% of 17 cases of lymphocytic gastritis were Campylobacter positive, as opposed to 90% of cases of chronic active gastritis. Our figures were lower, involving less than 20% of cases of lymphocytic gastritis and 65% of chronic gastritis [22]. It would appear that lymphocytic gastritis confers relative resistance to Helicobacter infection.

Immunological Data

Study of Mucosal Plasma Cells

André et al. [1–3] stated that varioliform gastritis was a form of delayed immunity expressed by a large increase of IgE plasma cells. Controversy has arisen about these results. Our own studies comparing groups of lymphocytic gastritis with varioliform features and chronic active gastritis could not reveal any differences as far as IgE plasma cells of the lamina propria were concerned [15].

Study of the Intraepithelial Lymphocytes

Dixon et al. [5], in a retrospective study on paraffin blocks have shown that in lymphocytic gastritis, the intraepithelial lymphocytes were T cells. Our studies on cryostat sections have confirmed these findings and demonstrated that most of these cells belong to the T8 "cytotoxic supressor" subtype with a rather small proportion of T4 "helper/inducer" subtype. In the lamina propria, the T lymphocytes were quite numerous with a slight predominance of T4 subtype. B cells were absent from the epithelium and were rare in the lamina propria [16].

Pathogenesis

Very little is known about the etiology of lymphocytic gastritis. The dense intraepithelial infiltrate of T lymphocytes and the relative proportion of T8 and T4 subtypes are comparable to observations in celiac disease suggesting a possible role of immunological factors. However, the progression of the two diseases is very different since lymphocytic gastritis resolves in about 50% of cases after period of 1–2 years [4], while (in adults at least celiac disease can only be cured by withdrawal of gluten. Studies are under way to identify the lymphocyte subtypes, possibly leading to a more precise interpretation of the pathogenesis of lymphocytic gastritis.

Conclusion

Lymphocytic gastritis is a new entity characterized by a dense infiltration of the gastric surface and pit epithelium by T lymphocytes. Its usual endoscopic appearance of enlarged folds and raised nodules is that of diffuse of fundic varioliform gastritis. There is no correlation between lymphocytic gastritis and antral varioliform gastritis which seems to constitute a different, heterogeneous condition with no precise histological counterpart. Lymphocytic gastritis can cure spontaneously in about 50% of cases after a period of 1–2 years. Its etiology is unknown. The histological similarity with celiac disease suggests a possible role of immunological factors.

References

1. André C, Moulinier B, Lambert R, Bugnon B (1976) Gastritis varioliformis, allergy, and disodium cromoglycate. Lancet: 964–965
2. André C, Vialard JL, Fargier MC (1979) IgE-mediated reactions in inflammation of the stomach. In: Pepys J, Edwards AM (eds) The most cell, its role in health and disease. Pitman, London, pp 642–646
3. André C, Gillon J, Moulinier B, Martin A, Fargier MC (1982) Randomised placebo-controlled double-blind trial of two dosages of sodium cromoglycate in treatment of varioliform gastritis: comparison with cimetidine. Gut 23:348–352
4. Delos M, Jouret-Mourin A, Wallez L, Willette M, Mainguet P, Haot J (1986) Evolution histologique d'une série de gastrites caractérisées par une infiltration lymphocytaire intraépitheliale. Acta Endosc 16:185–187
5. Dixon MF, Wyatt JI, Burke DA, Rathbonne BJ (1988) Lymphocytic gastritis-relationship to Campylobacter pylori infection. J Pathol 154:125–132
6. Elta GH, Fawaz KA, Dayal Y, McLean AM, Philipps E, Bloom SM, Paul RE, Kaplan MM (1983) Chronic erosive gastritis – recently recognized disorder. Dig Dis Sci 28:7–12
7. Franzin G, Manfrini C, Musola R, Rodella S, Fratton A (1984) Chronic erosions of the stomach – clinical, endoscopic and histological evaluation. Endoscopy 16:1–5
8. Green PHR, Fevre DI, Barrett PJ, Hunt JH, Gillespie PE, Nagy GS (1977) Chronic erosive (verrucous) gastritis. A study of 108 patients. Endoscopy 9:74–78
9. Haot J, Wallez L, Jouret-Mourin A, Hardy N (985) La gastrite "à lymphocytes." Une nouvelle entité? Acta Endosc 15:187–188
10. Haot J, Delos M, Wallez L, Hardy N, Lenzen B, Jouret-Mourin A (1986) Les lymphocytes intraépithéliaux en pathologie gastrique inflammatoire. Acta Endosc 16:61–65
11. Haot J, Jouret-Mourin A, Delos M, Wallez L, Melange M, de Galocsy C, Boemer F, Willette M, Mainguet P (1986) Etude anatomo-clinique d'une série de gastrites chroniques charactérisées par une infiltration lymphocytaire intraépithéliale. Acta Endosc 16:69–74
12. Haot J, Hamichi L, Wallez L, Mainguet P (1988) Lymphocytic gastritis: newly described entity. A retrospective endoscopic and histological study. Gut 29:1258–1254
13. Haot J, Berger F, André C, Moulinier B, Lambert R, Mainguet P (1989) Lymphocytic gastritis versus varioliform gastritis. A historical series revisited. J Pathol 158:19–22
14. Haot J, Jouret A, Willette M, Gossuin A, Mainguet P (1990) Lymphocytic gastritis: prospective study of its relationship with varioliform gastritis. Gut (in press)
15. Hardy N, Weynands B, Haot J (1990) Etude immunocytochimique des plasmocytes de la muqueuse digestive dans la gastrite varioliforme diffuse. Acta Endosc (in press)
16. Jouret A, Wallez L, Bellassai J, Mainguet P, Haot J (1989) La gastrite lymphocytaire: étude immunocytochimique des lymphocytes intra-épithéliaux et choriaux. Acta Endosc 19–24
17. Lambert R, Bugnon B, Moulinier B, André C (1977) Aspects endoscopiques de la gastrite varioliforme diffuse. Endosc Dig 2:81–86

18. Lambert R, André C, Moulinier B, Bugnon B (1978) Diffuse varioform gastritis. Digestion 17:159–167
19. Moutier F, Cornet A (1947) Les gastrites cachectisantes. Entretiens de Bichat: 177–178
20. Nesland AA, Berstad A, Serck-Hanssen A (1986) Histological findings in erosive prepyloric changes. Scand J Gastroenterol 21:239–245
21. Walk L (1975) Polyps caused by gastric erosions. Radiologe 15:354–355
22. Wallez L, Weynand B, Haot J (1987) Evaluation histologique de la présence d'organismes de type Campylobacter dans les gastrites lymphocytaires. Acta Endosc 17:283–288

Exocrine and Endocrine Epithelial Changes in Types A and B Chronic Gastritis *

E. SOLCIA, C. CAPELLA, R. FIOCCA, M. CORNAGGIA, G. RINDI, L. VILLANI, F. BOSI, and L. AMBROSIANI

Gastritis-associated epithelial changes are known to have a major role in the pathogenesis of gastric diseases such as peptic ulcer and exocrine or endocrine cell neoplasia [1, 26, 27, 31, 33]. Thus, a careful cytologic characterization of these epithelial changes at the light and electron microscopy level should prove helpful in understanding the cause and evolution of such diseases.

In the present investigation, atrophic, hyperplastic, or metaplastic lesions of exocrine and endocrine epithelia in type A, type B, or cancer-associated chronic gastritis are analysed using a panel of histochemical methods as well as electron microscopy to characterize their differentiation patterns.

Material and Methods

The following material was investigated: (a) Endoscopic biopsies or surgical specimens from 30 cases of type A chronic atrophic gastritis (CAG), 17 of which were with associated gastric argyrophil carcinoids and one with a pyloric G cell tumor [24, 33]. (b) Endoscopic biopsy specimens of antropyloric and corpus-fundus mucosa from 100 consecutive patients with type B chronic gastritis, mostly suffering from dyspepsia (66) or peptic ulcer (15 duodenal and 11 gastric). (c) Gastrectomy specimens of 68 patients suffering from early gastric cancer [32].

To paraffin sections of formaldehyde-fixed tissues the following methods were applied: (a) hematoxylin-eosin and Giemsa stain for typing of gastritis and search for *Helicobacter* (*Campylobacter*) *pylori* [10]; (b) alcian blue-periodic acid-Schiff (PAS), alcian blue high iron diamine (HID), and periodic acid-concanavalin A (PACONA) reactions for gastrointestinal mucins [9]; (c) Sevier-Munger's or Grimelius' silver for endocrine cells [14]; and (d) immunoperoxidase using antibodies against pepsinogen I and II, M 1 antigen of gastric surface cell mucin, M 3 antigen of intestinal goblet cells, CAR-5 intestinal antigen, chromogranin A, serotonin, gastrin, somatostatin, pancreatic polypeptide (PP), glucagon, entero-

* This work was supported in part by grants from the Italian National Research Council (Special Projects on Oncology and Biomedical Technologies; Gastroenterology Group) and the Health and Education Ministry.

[1] Department of Human Pathology, University of Pavia, 27100 Pavia, Italy; [2] Department of Pathology, IRCCS Policlinico S. Matteo, 27100 Pavia, Italy; [3] Department of Pathology, Multizonal Hospital, 21100 Varese, Italy

Helicobacter pylori, Gastritis and Peptic Ulcer
Edited by P. Malfertheiner and H. Ditschuneit

glucagon, cholecystokinin, motilin, secretin, gastric inhibitory polypeptide (GIP), and neurotensin [8, 12, 33].

Samples of gastric mucosa were fixed in formaldehyde-glutaraldehyde solution, postfixed in osmium tetroxide, and embedded in Epon. Ultramicrotomic sections were stained with uranyl and lead or with immunogold procedures for gastrin, pepsinogen I, or pepsinogen II, as previously reported [4].

Results

Type A Gastritis

In fundic-type mucosa, the dominant finding was oxyntic gland atrophy with disappearence of parietal and chief cells. However, hyperplastic changes of foveolar, mucous neck, and endocrine cells were also frequently found (Figs. 1–4). In addition, pyloric- or intestinal-type metaplasia of the gastric epithelium, simultaneously involving its endocrine and exocrine components, was observed.

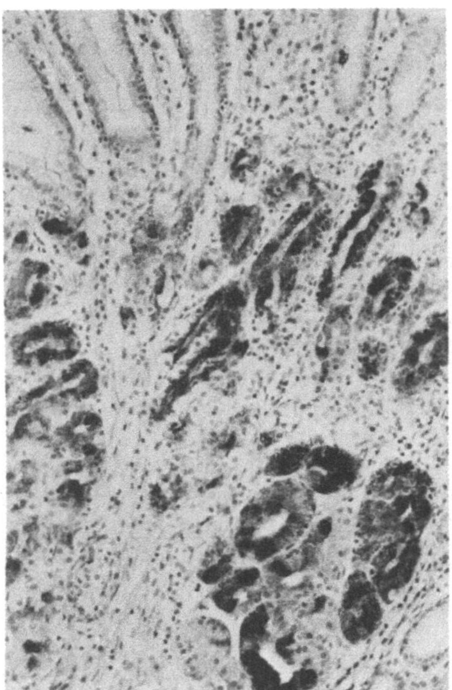

Fig. 1. Hyperplastic mucous neck cells stained using the PACONA technique for gastric mucin. Note nonreactivity of foveolar cell mucin in the *upper part* and lack of PACONA-negative chief and parietal cells in the *deeper part* of the mucosa; X 240

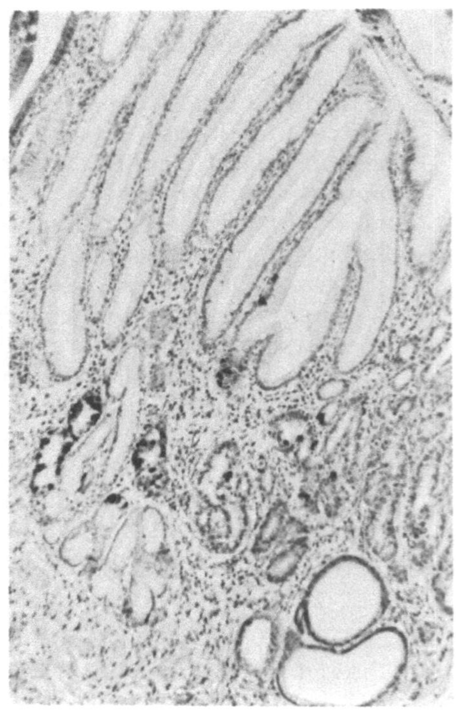

Fig. 2. Gastrin-immunoreactive cells in an area of pyloric-type metaplasia. Note foveolar hyperplasia in the *upper part* and mucous neck cell hyperplasia in the *right lower part* of the micrograph. Immunoperoxidasehematoxylin, X160

Both foveolar hyperplasia and intestinal metaplasia were easily recognized in conventional histologic preparations. Separation of mucous neck hyperplasia (so-called pseudopyloric metaplasia) from pyloric-type metaplasia was better achieved with the aid of histochemical markers (Figs. 1–3 and Table 1). The presence of mucous neck and argyrophil enterochromaffin-like (ECL) cells in the absence of gastrin cells was distinctive of the former lesion. Mucous neck cells were characterized by their simultaneous expression of pepsinogens I and II as well as of PACONA-reactive mucin [4, 9], while ECL cells were stained with

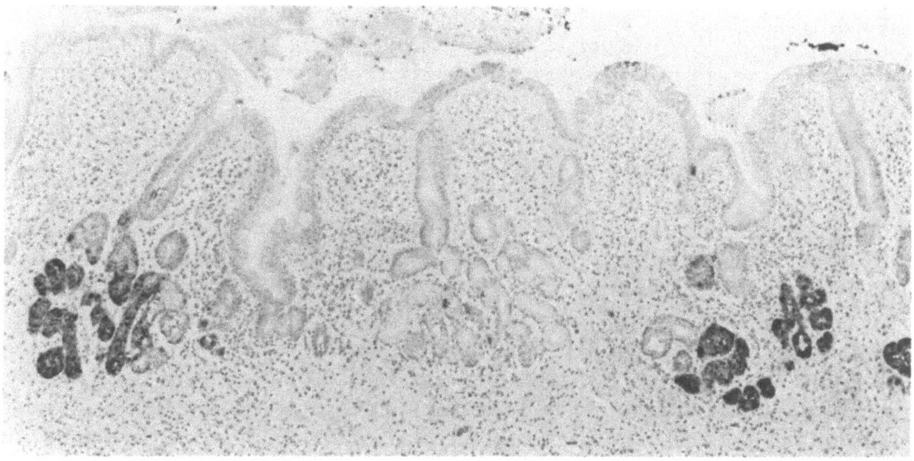

a

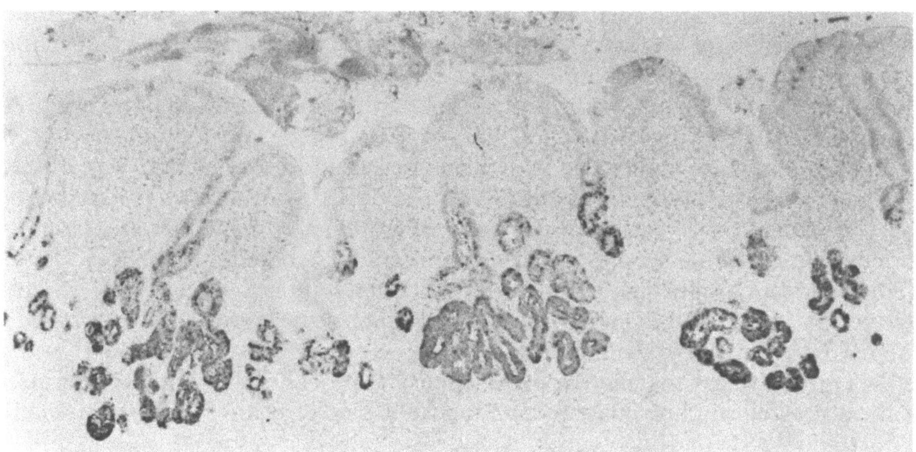

b

Fig. 3a, b. Pepsinogen I (**a**) and pepsinogen II (**b**) immunostaining of two consecutive sections of type A CAG. Note an area of pyloric-type metaplasia (*center* of the micrograph in **a**) lacking pepsinogen I while retaining pepsinogen II expression. Foci of mucous neck hyperplasia show both pepsinogen I and II, while the foveolar-superficial epithelium lacks both. Immunoperoxidase-hematoxylin; **a** = **b**, X90

Table 1. Gastric type markers in gastric metaplasias

Metaplasia	PGI	ECL cells	PGII	PACONA	Gastrin cells
Pyloric	NEG	NEG	POS	POS	POS
Pseudopyloric	POS	POS	POS	POS	NEG
Intestinal					
Complete	NEG	NEG	NEG	NEG	N/P
Incomplete	NEG	NEG	POS	POS	N/P
Normal mucosa					
Pyloric glands	NEG	NEG	POS	POS	POS
Fundic glands, neck	POS	POS	POS	POS	NEG
Fundic glands, body	POS	POS	POS	NEG	NEG
Intestinal crypts	NEG	NEG	NEG	NEG	N/P

NEG, negative; POS, positive; N/P, negative/positive

chromogranin A antibodies and Grimelius' or Sevier-Munger's silver, in the absence of reactivity with Masson's argentaffin reaction [30, 33].

The general morphology of the two histochemically characterized lesions was distinctive, pyloric-type metaplasia often showing more branched or coiled tubules formed of larger cells with clearer cytoplasm and more basally flattened nucleus. In less severely affected mucosa, transitional patterns between oxynticopeptic glands and pure mucous neck hyperplasia were found, either in the form of residual chief cells (large, basophilic, PACONA-negative, pepsinogen-positive cells) at the bottom of some glands or as surviving parietal cells scattered among the overwhelming population of mucous neck cells. More or less extensive mucous neck hyperplasia was found in all cases of type A atrophic gastritis. Morphologically defined pyloric-type metaplasia was observed in eight of the twelve cases investigated, mainly as multifocal lesions, with accompanying gastrin cells in four cases (Fig. 2).

In addition, a somewhat mixed pyloric/pseudopyloric lesion was observed, showing pyloric-type necks with G cells coupled to PGI-positive glands.

Intestinal metaplasia was mainly of complete, small intestine type, and composed of columnar enterocytes with straight, rooted microvilli, goblet cells, and intestinal-type endocrine cells (especially intestinal subtypes of enterochromaffin cells and cholecystokinin, motilin, neurotensin, and enteroglucagon cells) with or without Paneth cells.

Diffuse, linear, or micronodular hyperplasia of the argyrophil endocrine cells was a regular finding in the corpus and fundus mucosa affected by type A chronic atrophic gastritis (Fig. 4a).

In general, the nature of the endocrine cells was in keeping with that of their accompanying exocrine epithelium; however, the hyperplastic endocrine cells forming the micronodules disseminated in the lamina propria were mostly composed of gastric fundic-type cells, as suggested by the frequent detection of ECL cells at ultrastructural investigation. Minute hyperplastic micronodules were found in 11 of 12 type A CAG patients lacking endocrine tumors and in all the 15

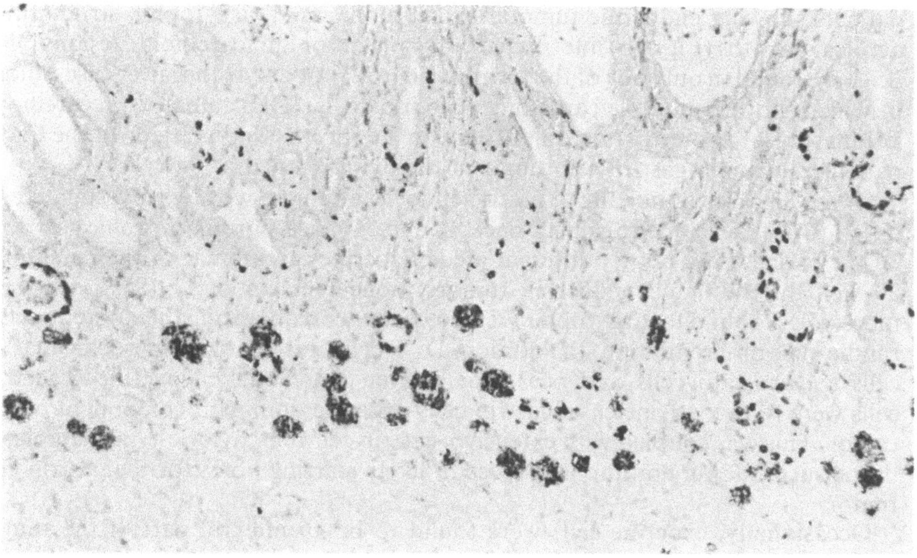

a

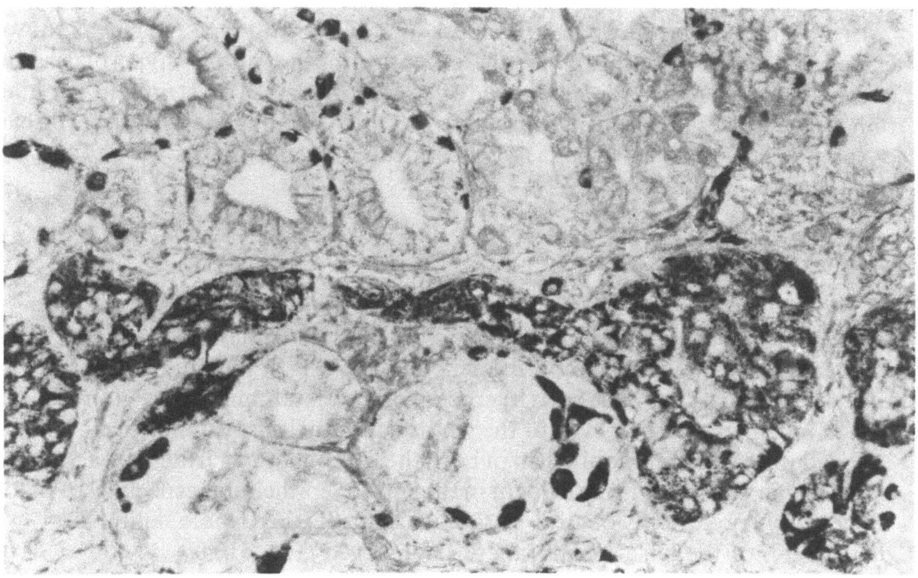

b

Fig. 4a, b. Micronodular and linear hyperplasia of argyrophil endocrine cells in type A CAG with extensive atrophy of oxyntic glands and sclerosis. **b** Enlarging micronodule infiltrating the lamina propria in between hyperplastic mucous neck glands: a putative "precarcinoid" lesion in a type A CAG with carcinoidosis. Grimelius' silver; **a** X 110; **b** X 300 patients showing endocrine tumors. In addition, small, argyrophil, structurally atypical growths (Fig. 4b) interpreted as dysplastic or "precarcinoid" lesions [33–35] were found in only one of the former and in all but one of the latter cases, most of which showed multiple carcinoids (carcinoidosis). Disseminated precarcinoid lesions (precarcinoidosis) were also found in the corpus-fundus region of the CAG patient showing a gastrin cell tumor in the pyloric mucosa.

Of the 33 cases of nonantral gastric endocrine tumors we have investigated so far, 21 (64%) developed against a background of type A chronic atrophic gastritis. Of these, one was a poorly differentiated endocrine carcinoma. Of the remaining 20 well differentiated endocrine tumors, none metastatic, 17 (85%) showed more or less abundant argyrophil ECL cells, thus corroborating their potential for fundic-type differentiation. In addition, D_1/P cells, gastric-type enterochromaffin cells, somatostatin cells, and, occasionally, even pancreatic polypeptide or gastrin cells were found. In one case, gastrin cells were the dominant component of the tumor. This was coupled with extensive, gastrin cell-rich, pyloric-type metaplasia in the surrounding nontumor mucosa, a likely starting point for the gastrin cell tumor.

Occasionally, exocrine cells were found to be an integral part of the tumor growth. In two tumors PAS-positive and M1 antigen-immunoreactive [12] foveolar-type cells with large secretory granules with a distinctive punctate substructure were found to be admixed with endocrine cells. Despite the extensive intestinal metaplasia observed in many cases of type A atrophic gastritis, none of the endocrine tumors observed in these patients showed a significant intestinal-type component.

In most cases, the antropyloric mucosa of patients with type A gastritis showed essentially normal findings, with the exception of gastrin cell hyperplasia. Cell counting in four such patients showed 192–259 (mean 217) gastrin cells per linear mm of pyloric mucosa. In four control cases, lacking peptic ulcer disease, dyspepsia, or hypergastrinemia, 41–93 (mean 68) gastrin cells were found.

Type B Gastritis

Of the 100 cases investigated, 87 showed *Helicobacter pylori* (HP) adhering to surface and foveolar epithelium (Fig. 5a). Two kinds of lesions were seen in this epithelium: (a) swelling of colonized cells with bulging of their luminal surface and formation of micropapillae and (b) erosion of the juxtaluminal part of the cell with loss of mucin granules or even of the whole supranuclear part of the cell. The first type of lesion was commonly seen in both the foveolar and surface epithelium, the erosive type was much more evident in the surface epithelium (especially of pyloric mucosa) where, occasionally, desquamation of the whole epithelium with denudation of the basal membrane was observed (Fig. 5b).

The relationship of these papillary-erosive lesions with HP colonization of the same biopsy specimen is shown in Table 2. The epithelial lesions were practically exclusively of HP-positive cases and specimens (and even of HP-positive areas inside specimens). Their severity and diffusion correlated with the amount of bacteria; however, 30 of the 87 cases displaying bacteria lacked epithelial lesions.

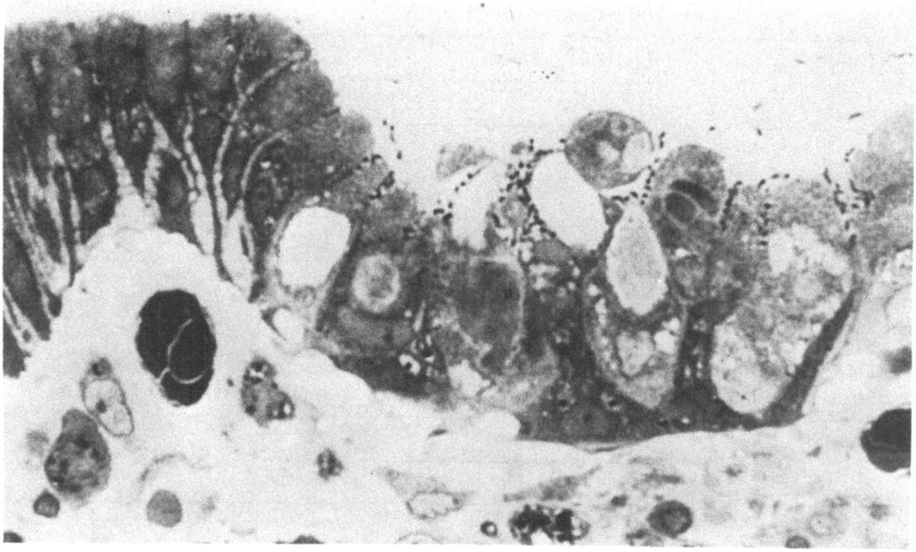

a

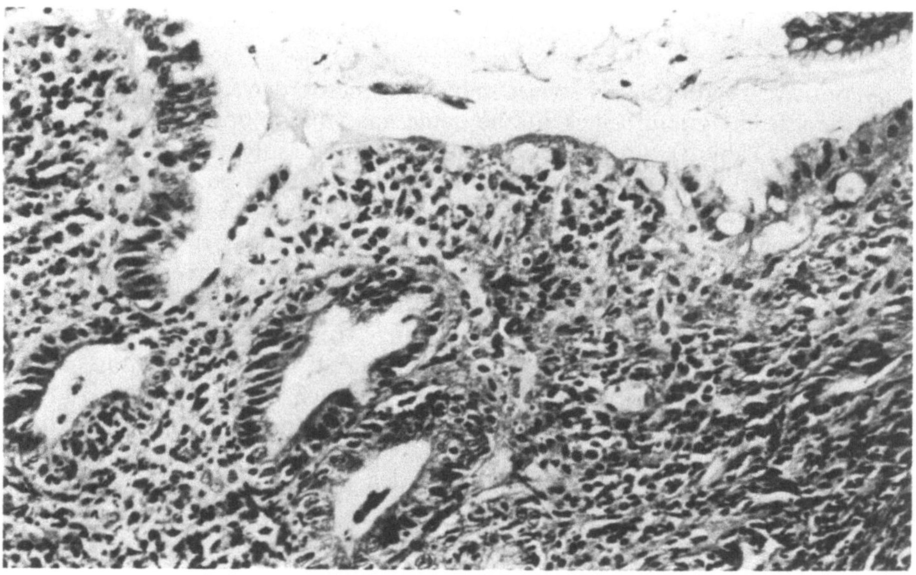

b

Fig. 5a, b. Type B chronic gastritis. **a** Semithin resin section of gastric superficial epithelium showing vacuoles as well as numerous *Helicobacter pylori* (HP) adhering to the luminal surface (*center* and *right*). **b** Complete erosion of an area of superficial epithelium (*center*) in an active gastritis. Hematoxylin-eosin; **a** X 1500; **b** X 280

Table 2. *Helicobacter pylori* (HP) and micropapillary/microerosive epithelial lesions in antral biopsy specimens from 100 patients with gastritis

HP	Epithelial lesions			Total
	–	+	++	
–	12	1	0	13
+	18	11	4	33
++	12	11	31	54
Total	42	23	35	100

Table 3. Relationship between epithelium-associated granulocytes and micropapillary/microerosive epithelial lesions in antral biopsy specimens from 100 patients with gastritis

Granulocytes	Epithelial lesions			Total
	–	+	++	
–	30	4	1	35
+	11	18	11	40
++	1	1	23	25
Total	42	23	35	100

A positive relationship between papillary-microerosive epithelial lesions and intraepithelial granulocytes of the same case and biopsy specimen was also observed (Table 3), although the topography of granulocytes did not necessarily overlap that of epithelial erosions and/or HP colonization.

Ultrastructural investigations showed early changes in the HP-colonized epithelia in the form of vacuolization of the juxtaluminal cytoplasm immediately underlying the luminal surface (Fig. 6). Larger vacuoles were easily detected with light microscopy of 0.5- to 1-µm thick resine sections (Fig. 5a); however, only rarely were such vacuoles seen in 5-µm paraffin sections. Vacuolized cells invariably showed HP adhering to their luminal surface, while in most cases there was no concomitant relationship with intraepithelial granulocytes and lymphocytes or severity of inflammatory changes in the underlying lamina propria.

Cellular swelling with endoluminal bulging as well as lesion of some juxtaluminal junctional complexes resulted in formation of intercellular intraepithelial pockets filled with bacteria. Rupture of the luminal cell membrane with loss of juxtaluminal mucin granules and cytoplasm was found in more severely affected cells of the surface epithelium.

Atrophic changes of the pyloric glands were usually moderate. Only in a minority of cases was intestinal metaplasia, especially of type I or II, a prominent finding. No consistent change was found in antropyloric gastrin cells, apart from their disappearance in areas of intestinal metaplasia, sometimes coupled with an

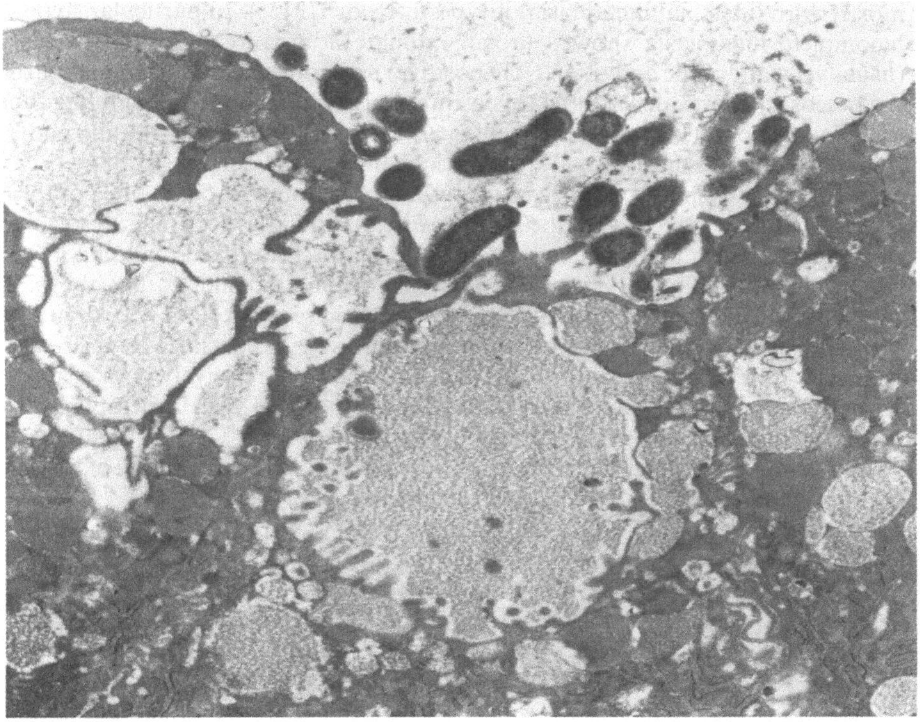

Fig. 6. Ultrastructure of vacuolar changes in the surface epithelium colonized with HP. Aldehyde-osmium fixation, uranyl-lead staining; X 12000

increase in the adjacent pyloric mucosa. Linear or micronodular hyperplasia of gastric-type enterochromaffin cells was observed, especially in areas of gland atrophy and sclerosis.

In most cases superficial gastritis was found in the corpus-fundus mucosa of stomachs with HP-positive type B gastritis. Atrophic or metaplastic changes were observed only in a few cases. Atrophic-subatrophic changes of the oxyntic glands in type B gastritis differed from those of type A gastritis for their lack of intraepithelial lymphocyte aggression and relative paucity of pyloric metaplasia and hyperplastic changes. Active gastritis with prominent intraepithelial granulocyte infiltration in the generative zone (neck) of the glands with resultant lack of glandular cell differentiation (rather than destruction of formed glands) seemed the key factor in such lesions.

Cancer-Associated Gastritis

In the juxtatumor mucosa of stomachs bearing early cancers (with special reference to the 36 cases of glandular type), chronic gastritis was found regularly, mostly with intestinal metaplasia of types I, II, or III [25]. In particular, type III incomplete metaplasia showed (a) an intimate admixture of mature foveolar (usually alcianophilic and often HID positive) and goblet cells in the upper part of the mucosa, (b) prevalence of relatively immature columnar cells resembling those of intestinal crypts and deep foveolae in the midgenerative stratum, and (c) frequently the presence of mature pyloric glands in the deeper stratum.

In addition, dysplastic changes were found in nontumor mucosa of 32 of the 68 cases investigated and in 29 of the 36 cases with glandular cancer. This dysplasia of flat mucosa was mainly composed of immature, atypical, columnar cells somewhat resembling (atypia apart) the cells forming the midstratum of type III metaplasia. The histologic resemblance between these two lesions is also outlined by the similarity of their histochemical patterns, with special reference to the reactivity of columnar cells for CAR-5 antigen and the relatively poor reactivity with other markers of gastrointestinal epithelia (Fig. 7). Type III metaplasia with marked expansion of its generative, poorly differentiated, CAR-5 immunoreactive stratum, usually coupled with mild cytologic and structural atypia of the same or upper and lower strata, was found in some cases.

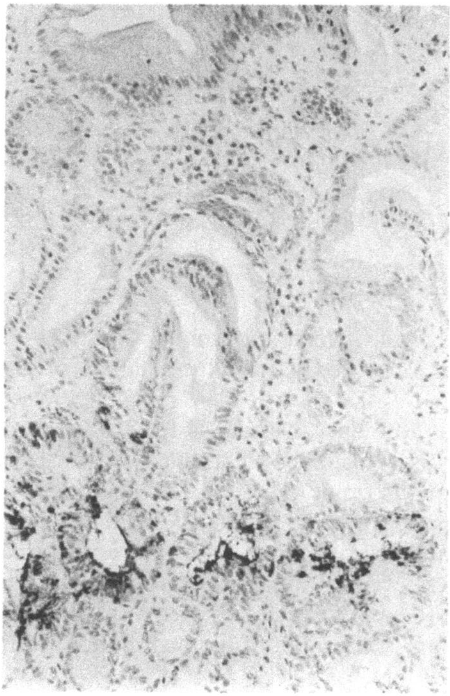

a

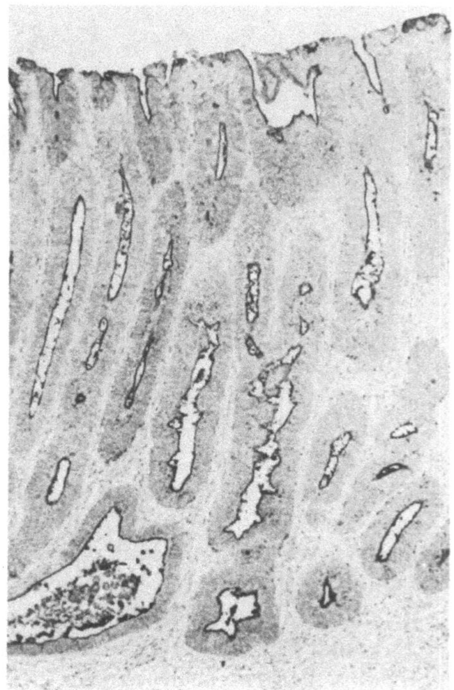

b

Fig. 7 a, b. Cancer-associated gastritis. **a** Type III intestinal metaplasia showing CAR-5 antigen expression by fairly immature columnar cells. Mature goblet cells scattered in the hyperplastic foveolae (*top* and *center*) as well as pyloric glands (*bottom*) lack the antigen. **b** Moderate dysplasia showing extensive, lumenal-type expression of CAR-5 antigen. Immunoperoxidase-hematoxylin; **a** X 240; **b** X 130

In general, dysplastic lesions, the generative zone of type III metaplasia as well as type II metaplasia, showed a rather minor endocrine component. Linear, palisade-forming hyperplasia of endocrine cells (especially intestinal-type enterochromaffin cells) was observed with some frequency in the deep part of type I metaplasia. No dysplasia limited to endocrine cells was found.

Discussion

As a rule normal, hyperplastic, and metaplastic epithelia occurring in chronic type A, type B, or cancer-associated gastritis show a parallel, reciprocally coherent differentiation and growth of their exocrine and endocrine components. However, in most cases endocrine and exocrine tumors arising from these epithelia have a separate histogenesis and very different behavior [31, 33]. Severe, long-standing, type A chronic atrophic gastritis is now known to be at risk for the development of argyrophil carcinoids restricted to the same body-fundus mucosa which is affected by the gastritis [1–3]. Despite a number of intestinal and pyloric-type metaplastic changes occurring in this disease, most such tumors (including 17 of our 20 cases) proved to be formed essentially of gastric fundic-type endocrine cells, including argyrophil, histamine-producing ECL cells which are normally exclusive of the oxyntic mucosa [16, 31]. No comparable characterization of tumor cells comprising gastric cancers, also known to arise with increased frequency in this disease [22, 29], has been performed so far. However, an extensive investigation of 316 gastric cancers arising in the general population [11] showed a striking predominance of pyloric gland differentiation (found in 47% of cases) over oxyntic gland differentiation (found in about 3% of cases).

Thus, a search for specific precursor lesions of endocrine tumors at the histologic level, as well as for specific inducing or promoting factors at the epidemiologic or clinical level, seems indicated. In the case of argyrophil body-fundus carcinoids possible precursor (precarcinoid) lesions have been recently described and differentiated from various types of hyperplastic changes [33, 34]. It seems interesting that we have found such precarcinoid lesions in nontumor mucosa of 14 out 15 patients bearing both type A CAG and argyrophil "ECL cell" carcinoids and in only one of 12 cases showing type A CAG without carcinoids. A role for the severe hypergastrinemia usually associated with type A CAG in promoting carcinoid development – through proliferation of endocrine cells which have undergone transformation during the long-standing atrophic, regenerative, and metaplastic process – seems likely considering the known trophic action of gastrin on gastric ECL cells [15].

Our characterization of true pyloric-type metaplasia with gastrin cells and its differentiation from pseudopyloric metaplasia (to be interpreted essentially as mucous-neck hyperplasia) may help in understanding the rare insurgence of gastrin-producing tumors in the corpus-fundus of patients with type A gastritis. An occasional gastrinoma arising from the pyloric mucosa of such patients, against a background of gastrin cell hyperplasia, has also been reported [24].

A general role for gastrin cells in the pathogenesis of HP-associated ulcer disease [20] seems unlikely from our purely morphologic investigation. However, a small group of patients (all of which proved to have extensive HP colonization of their pyloric mucosa) showing H_2-blocker resistant duodenal ulcer coupled with basal and/or stimulated hypergastrinemia [6], had gastrin cell counts intermediate between those of controls and of overt gastrin cell hyperplasia [35]. Thus, some role for bacterial ammonia in buffering intraglandular HCl, thereby causing gastrin cell stimulation and hyperfunction, cannot be ruled out, at least in a selected minority of patients.

A role for Helicobacter-induced damage (especially of the erosive type) of surface gastric epithelium in the pathogenesis of gastric peptic ulcer seems likely. Whether the surface lesion is due to direct cytotoxicity from bacterial toxins or from bacteria-evoked leukocyte infiltration of the epithelium, or to both mechanisms, remains to be clarified. Interestingly, very mild, apparently early lesions found ultrastructurally in superficial-foveolar epithelium are essentially of the vacuolar type and closely mimic the vacuolar changes induced in vitro on epithelial cell lines by the vacuolizing cytotoxin obtained from many HP strains in culture [7, 19]. On the other hand, a positive correlation between intraepithelial granulocyte infiltration and cytolitic-erosive epithelial damage has been also found, thus confirming the results of Raws et al. [23].

The possibility that some cytotoxin is directly involved in the pathogenesis peptic ulcer [7] is supported by the increased concentration of erosive lesions that we observed in the surface epithelium of the pyloric mucosa from gastric ulcer patients. A similar behavior of intraepithelial granulocyte infiltrates has been recently reported by Heilman and Niesch [17]. It appears likely that direct epithelial cytotoxicity is a key factor in the initiation of the lesion, with both direct and granulocyte-mediated cytotoxicity contributing to its progression up to complete erosion of the epithelial lining. The damaging action of lumenal acid and pepsin on the exposed lamina propria, as well as the impaired epithelial regeneration of nearby foveolar epithelium due to persistent HP infection and inflammation, will then promote the epithelial erosion developing into peptic ulcer.

The existence of HP strains devoid of significant pathogenicity with regard to the gastric epithelium (though inducing an inflammatory response in the lamina propria) seems likely, considering that no epithelial lesions were found in biopsy specimens of 34% of cases showing HP colonization. More investigations are needed to identify possible "ulcerogenic" and "nonulcerogenic" strains of HP.

In general, atrophy of glands and their replacement with hyperplastic-metaplastic epithelia was much more prominent in cancer-associated gastritis than in type B gastritis coupled with either peptic ulcer or dyspepsia. This finding is in keeping with the epidemiologic and clinicopathologic studies of Sipponen et al. [27, 28] showing a sharp increase of cancer risk in subjects with extensive intestinal metaplasia of the antrum. Our finding of an increased incidence of type III intestinal metaplasia in neoplasia-associated gastritis supports previous suggestions that this subtype of metaplasia is more significantly correlated with the development of cancer than is type I or type II metaplasia [25]. This conclusion is further supported by the histologic (frank atypia apart) and histochemical homologies we have observed between the generative zone of some type III metaplasia and conventional dysplastic lesions.

Although at present no conclusive evidence is available concerning the possible histogenetic links between type III metaplasia and frank dysplasia, a role of type III metaplasia – a histologically complex lesion involving various lines of differentiation – in cancer development would explain the heavily mixed cellular pattern found in up to 50% of gastric cancers [12].

The rather frequent occurrence of HP in neoplasia-associated gastritis [5] (personal unpublished observations) as well as recent epidemiologic studies [13] showing geographic overlapping of HP infection and gastric cancer risk, indicate the need for further clinicopathologic and epidemiologic studies to clarify the relationship between long-standing, HP-positive type B gastritis and neoplasia-associated gastritis. The capacity of HP to inactivate vitamin C and other antioxidants [18] known to be protective against the development of gastric cancer [21] may offer a pathogenetic basis linking HP to the development of cancer via the progressive derangement of regulatory factors modulating the renewal and differentiation of epithelial cells in the gastric mucosa.

References

1. Borch K, Renvall H, Liedberg G (1985) Gastric endocrine cell hyperplasia and carcinoid tumours in pernicious anemia. Gastroenterology 88:638–648
2. Capella C, Polak JM, Frigerio B, Solcia E (1980) Gastric carcinoids of argyrophil ECL cells. Ultrastruct Pathol 1:411–418
3. Carney JA, Go VLW, Fairbanks VF, Moore SB, Alport EC, Nora FE (1983) The syndrome of gastric argyrophil carcinoid tumors and nonantral gastric atrophy. Ann Intern Med 99:761–766
4. Cornaggia M, Riva C, Capella C, Solcia, E, Samloff IM (1987) Subcellular localization of pepsinogen II in stomach and duodenum by the immunogold technique. Gastroenterology 92:585–593
5. Correa P, Ruiz B (1989) Campylobacter pylori and gastric cancer. In: Rathbone BJ, Heatley RV (eds) Campylobacter pylori and gastroduodenal disease. Blackwell, Oxford.
6. Delle Fave G, Rindi G, Puoti M, Annibale B, Corleto V, Giordano E, Iannoni C, Solcia E (1988) Withdrawal of antisecretory therapy causes a regression of G and argyrophil cell hyperplasia. Gastroenterology 94:A94
7. Figura N, Guglielmetti P, Rossolini A, Barbieri A, Cusi G, Musmanno RA, Russi M, Quaranta S (1989) Cytotoxin production by Campylobacter pylori strains isolated from patients with peptic ulcers and from patients with chronic gastritis only. J Clin Microbiol 27:225–226
8. Fiocca R, Rindi G, Capella C, Grimelius L, Polak JM, Schwartz TW, Yanaihara N, Solcia E (1987) Glucagon, glicentin, proglucagon, PYY, PP and proPP-icosapeptide immunoreactivities of rectal carcinoid tumors and related non-tumor cells. Regul Pept 17:9–29
9. Fiocca R, Villani L, Tenti P, Solcia E, Cornaggia M, Capella C (1987) Characterization of four main cell types in gastric cancer: foveolar, mucopeptic, intestinal columnar and goblet cells. An histopathologic, histochemical and ultrastructural study of early and advanced tumours. Pathol Res Pract 182:308–325
10. Fiocca R, Villani L, Turpini F, Turpini R, Solcia E (1987) High incidence of Campylobacter-like organisms in endoscopic biopsies from patients with gastritis, with or without peptic ulcer. Digestion 38:234–244
11. Fiocca R, Cornaggia M, Villani L, Capella C, Solcia E, Samloff IM (1988) Expression of pepsinogen II in gastric cancer. Its relationship to local invasion and lymph node metastases. Cancer 61:956–962
12. Fiocca R, Villani L, Tenti P, Cornaggia M, Finzi G, Riva C, Capella C (1990) The foveolar cell component of gastric cancer. Hum Pathol 21:260–270

13. Forman D, Sitas F, Newell D, Yarnell J, Burr M, Elwood P, Buiatti E, Palli D, Lachlan G, Chen J (1989) Epidemiology of Campylobacter pylori. Klin Wochenschr [Suppl 18] 67:21
14. Grimelius L, Wilander E (1980) Silver stains in the study of endocrine cells of the gut and pancreas. Invest Cell Pathol 3:3–12
15. Håkanson E, Ekelund M, Sundler F (1984) Activation and proliferation of gastric endocrine cell. In: Falkmer S, Håkanson R, Sundler F (eds) Evolution and tumour pathology of the neuroendocrine system, Elsevier, Amsterdam, pp 371–398
16. Håkanson R, Bottcher G, Ekblad E, Panula P, Simonsson M, Dohlsten M (1986) Histamine in endocrine cells in the stomach. A survey of several species using a panel of histamine antibodies. Histochemistry 87:5–17
17. Heilman KL, Niesch I (1989) Different grade of activity of gastritis in duodenal and gastric ulcer. Klin Wochenschr [Suppl 18] 67:27–28
18. Johnson AW, Aleksandrowicz J, Tompkins DS, Rathbone BJ, Losowsky MS (1988) The effect of Campylobacter pylori on ascorbic acid and urea metabolism in vitro. In: Mégraud F, Lamouliatte H (eds) Gastroduodenal pathology and Campylobacter pylori. Elsevier, Amsterdam
19. Leunk RD, Johnson PT, David BC, Kraft WG, Morgan DR (1988) Cytotoxic activity in broth-culture filtrates of Campylobacter pylori. J Med Microbiol 26:93–99
20. Levi S, Beardshall K, Haddad G, Playford R, Chosh P, Calam J (1989) Campylobacter pylori and duodenal ulcers: the gastrin link. Lancet 1:1167–1168
21. Mirvish SS (1983) The etiology of gastric cancer. JNCI 71:629–647
22. Mosbech J, Vinebaek A (1950) Mortality from and risk of gastric carcinoma among patients with pernicious anemia. Br Med J 2:390–394
23. Rauws EAJ, Tytgat GNJ (1989) Campylobacter pylori. Academic Medical Centre, Amsterdam, pp 1–164
24. Russo A, Buffa R, Grasso G, Giannone G, Sanfilippo G, Sessa F, Solcia E (1980) Gastric gastrinoma and diffuse G cell hyperplasia associated with chronic atrophic gastritis. Digestion 20:416–419
25. Silva S, Filipe MI (1986) Intestinal metaplasia and its variants in the gastric mucosa of Portuguese subjects: a comparative analysis of biopsy and gastrectomy material. Hum Pathol 17:988–995
26. Sipponen P, Kekki M, Siurala M (1983) Atrophic chronic gastritis and intestinal metaplasia in gastric carcinoma. Comparison with a representative population sample. Cancer 52:1062–1068
27. Sipponen P, Seppala K, Aarynen M, Helske T, Kettunen P (1989) Chronic gastritis and gastroduodenal ulcer: a case control study on risk of coexisting duodenal or gastric ulcer in patients with gastritis. Gut 30:922–929
28. Sipponen P, Varis K, Fraki O, Seppala K, Jarvinen P, Korri UM, Siurala M (1990) Evidence that chronic gastritis precedes peptic ulcer. A follow-up of 454 patients with or without gastritis. Hepatogastroenterology (in press)
29. Siurala M, Lehtola M, Ihamaki T (1974) Atrophic gastritis and its sequelae. Scand J Gastroenterol 9:441–449
30. Solcia E, Capella C, Vassallo G, Buffa R (1975) Endocrine cells of the gastric mucosa. Int Rev Cytol 42:223–286
31. Solcia E, Capella C, Buffa R, Usellini L, Frigerio B, Fontana P (1979) Endocrine cells of the gastrointestinal tract and related tumors. Pathobiol Ann 9:163–203
32. Solcia E, Capella C, Fiocca R, Frigerio B, Riva C, Villani L (1984) Tipizzazione dell'early gastric cancer. In: Progressi in anatomia patologica. Istituto Anatomia Patologica, Perugia, pp 247–255
33. Solcia E, Capella C, Sessa F, Rindi G, Cornaggia M, Riva C (1986) Gastric carcinoids and related endocrine growths. Digestion [Suppl 1] 35:3–22
34. Solcia E, Bordi C, Creutzfeldt W, Dayal Y, Dayan AD, Falkmer S, Grimelius L, Havu N (1988) Histopathological classification of nonantral gastric endocrine growths in man. Digestion 41:185–200
35. Solcia E, Capella C, Buffa R, Rindi G, Fiocca R, Tenti P (1988) The endocrine system of the gut in health and disease. Ital J Gastroenterol 20:335–343

The Classification of Gastritis *

K. L. Heilmann and M. Stolte [1]

Until recently the histological diagnosis of gastritis was of little relevance for gastroenterologists or patients. Whether the inflammation was superficial or "atrophic" or with or without intestinal metaplasia had no real meaning. This attitude toward gastric histology changed dramatically when it became apparent that over 75% of all histological gastritis cases could be attributed to infection with *Helicobacter pylori*. Subsequent morphological studies stimulated by this discovery revealed well-recognizable differences between infectious gastritis and other forms of the gastritis process [3]. These sometimes striking morphological differences have shed new light on the pathogenesis of gastritis. Therefore, it was considered timely to propose a new classification of gastritis based on these new observations.

The gastroenterologic group of the German Society of Pathologists organized a workshop to classify and grade gastritis [2]. The results of this workshop are presented here. The participants were of the opinion that the descriptive diagnosis continues to be desirable and that – whenever possible or necessary – etiological factors might be "added."

Normal Findings

The normal gastric mucosa contains no lymphocytes, plasma cells, or granulocytes. This definition of "normal" gastric mucosa is necessary, for in numerous reports, mild infiltration of the lamina propria of the gastric mucosa with leukocytes and plasma cells is considered a "normal finding." This is the reason for many of the discrepancies in the findings reported in the literature on *H. pylori*.

Acute Gastritis

Acute gastritis, a very rare event, should be diagnosed whenever an inflammatory infiltrate made up only of neutrophilic granulocytes without round cells is present.

---

* Proposal of the German Society of Pathologists

Institute of Pathology, Städt. Krankenhaus Landshut, D-8300 Landshut, FRG

[1] Institute of Pathology, Klinikum Bayreuth, Prenschwitzer Straße 101, D-8580 Bayreuth, FRG

Helicobacter pylori, Gastritis and Peptic Ulcer
Edited by P. Malfertheiner and H. Ditschuneit

Depending upon the intensity of the infiltration by neutrophilic granulocytes, this "acute gastritis" can be further subdivided into "mild, moderate," or "severe."

Chronic Gastritis

Chronic gastritis is characterized by a pure round cell infiltrate with few eosinophils. The following grading of this chronic gastritis was proposed:

- Minimal chronic gastritis: scattered lymphocytes and plasma cells
- Mild chronic gastritis: sparse lymphoplasmacellular infiltrate
- Moderate chronic gastritis: moderately dense lymphoplasmacellular infiltrate
- Severe chronic gastritis: very dense lymphoplasmacellular infiltrate also invading the deeper glands bearing parts of the mucosa

A controversial point raised during discussion was whether the term superficial and "atrophic gastritis" should be abandoned. At present, there is no definitive evidence to show that the so-called superficial gastritis progressively develops into the atrophic form. Since, in the strict sense of the term, atrophic gastritis applies only to the autoimmune form in the fundus/body and the pathogenetic mechanisms differ in the antrum and corpus, the opinion prevailed that the two older terms "superficial gastritis" and "atrophic gastritis" should be abandoned in favor of "gastritis," purely and simply. The extent of a possible reduction of glands could be indicated by the addition of "partial" or "total" atrophy.

Chronic Active Gastritis

Chronic active gastritis is diagnosed when, in addition to the lymphoplasmacellular infiltrate, neutrophilic granulocytes are present. The density of the neutrophilic granulocytic infiltrate determines the degree of activity of the gastritis. Principally, grading of the chronic active gastritis is done as described above and is further qualified by indicating the degree of activity:

- Activity grade 1: scattered interstitial and intraepithelial granulocytes
- Activity grade 2: dense granulocytic infiltrate with numerous intraepithelial granulocytes in the region of the foveolae
- Activity grade 3: dense granulocytic infiltrate with development of "abscesses" in the foveolae

Additional Findings

Intestinal Metaplasia

If intestinal metaplasia is present, it should be differentiated into three different types, and the distribution and extent of the metaplasia should be indicated, as focal or diffuse:

- Type I: complete intestinal metaplasia (small intestinal type)
- Type II: incomplete intestinal metaplasia ("enterogastric type")
- Type III: incomplete intestinal metaplasia (colonic type)

Atrophy

Atrophy should be diagnosed if there is a reduction in the glandular mass. It should be graded into partial and total atrophy.

Foveolar Changes, Irregularities, or Hyperplasia

These are important morphological findings and should be considered in the histological diagnosis.

Intraepithelial Lymphocytes

Until know intraepithelial lymphocytes have not been a well-recognized histological marker. They should, however, be considered in the report and in the diagnosis when there are more than 20 intraepithelial lymphocytes/100 surface cells.

Erosions

The term "erosive gastritis" should be abandoned. It was recommended that, depending upon their extent, erosions should be mentioned as a qualifying addition to the diagnosis.

Helicobacter pylori

The pathologists should always indicate whether *H. pylori* is present or not. In the routine diagnostic evaluation, no special stain is necessary, for the alcoholic H & E stain and examination under high-power objectives ($\times 40$, $\times 63$) are sufficient.

Basal Lymphoid Aggregates and Lymphoid Follicles

Basal lymphoid aggregates and lymphoid follicles appear to develop as sequelae of a *H. pylori* associated gastritis. This additional finding should also be mentioned in the report and probably in the diagnosis.

Based on these morphological parameters it seems possible to propose a classification of gastritis considering the etiopathogenetic mechanisms (Table 1), as was done also by Wyatt and Dixon [3].

Type A Gastritis

The autoimmune gastritis is localized within the fundus and body of the stomach. In the end stage of the autoimmune process, the glands are atrophic and the

Table 1. Classification of gastritis

Type	Etiology/pathogenesis	Localization
A	Autoimmune	Corpus
B	Bacterial	Antrum/corpus
A/B	Autoimmune/bacterial	Antrum + corpus
C	Chemical injury	Anastomosis/antrum
Special forms of gastritis		
Lymphocytic gastritis		
Other forms of infectious gastritis		
Granulomatous gastritis		
"Crohn gastritis"		
Eosinophilic gastritis		
Other forms		

lamina propria shows a mild chronic inflammatory infiltrate. Frequently intestinal metaplasia is present. In the totally atrophic stage of autoimmune gastritis, basal neuroendocrine cell complexes are frequently found, together with an increase in glandular endocrine cells. In the "active" stage of autoimmune gastritis prior to the total destruction of the glands, this auto-aggressive disorder can be recognized by a periglandular lymphocytic infiltrate.

Type B Gastritis

Bacterial gastritis (type B) is induced almost exclusively by *H. pylori* organisms. Other infectious forms of gastritis are to be found in the group "special forms" (see below).

Type A/B Gastritis

Autoimmune gastritis (type A gastritis) usually leads to a complete atrophy of the glands and anacidity of the gastric juice only in advanced age. Some of these patients will presumably have acquired a type B gastritis at a younger age. This may explain how a mixed type A/B gastritis can develop from primary *H. pylori* gastritis and secondary autoimmune gastritis. This combination may also explain the reports in the earlier literature of a tendency of antral gastritis to heal with atrophy of the body mucosa.

Type C Gastritis

Gastritis of type C is probably for the most part induced by reflux of certain mucosa-damaging bile acids. This bile reflux gastritis is characterized histologically by edema, dilatation and hyperemia of the capillaries, a sparse inflammatory infiltrate comprising lymphocytes and plasma cells in the upper part of the lamina propria, and elongation and/or tortuosity of the foveolae. It is to be found not only in the region of the anastomosis of the operated stomach, but also in the antrum of the nonoperated stomach. In this bile reflux gastritis, a common endoscopic finding is a stripe of redness running from the pylorus in the cranial direction. Whether this bile reflux gastritis is an entity that can be unequivocally diagnosed histologically, and whether this form of gastritis can be differentiated from other chemically inducted inflammatory changes in the gastric mucosa is not yet clear.

Special Forms

Lymphocytic Gastritis

Lymphocytic gastritis [1] is a special type of gastritis that can occur within the framework of chronic or chronic active gastritis. This form of gastritis is characterized by the presence of intraepithelial T lymphocytes. The diagnosis lymphocytic gastritis should only be made when there is an involvement of at least 20% of the mucus producing surface epithelial cells. Lymphocytic gastritis probably exists in two forms; one is characterized by erosions and foveolar hyperplasia and corresponds endoscopically to the "varioliform" type of gastritis. The other form is associated with a normal endoscopic appearance of the mucosa. Lymphocytic gastritis is apparently a particular form of an immunological response and it is possible that a few patients with Helicobacter associated gastritis react in this manner.

Other Infectious Forms of Gastritis

In the search for *H. pylori*, it has been found that in rare cases (0.25%) other spiral bacteria that stain with silver nitrate may also be present in the stomach. All other infectious forms of gastritis are extremely rare and occur almost exclusively in immunodeficient patients, such as the virus induced forms of gastritis (in particular CMV) or parasitic gastritis.

Granulomatous Gastritis

In principle the diagnosis granulomatous gastritis must remain descriptive in most cases. Additional clinical investigations are required to establish whether the granulomas represent a partial substrate of a granulomatous systemic disease, or whether an idiopathic "granulomatous gastritis" is present.

Eosinophilic Gastritis

On the basis of forceps biopsy material, the extremely rare eosinophilic gastritis can be only suspected, since the eosinophilic granulocytes are, for the most part, located in the deep layers of the gastric mucosa. To clarify this diagnosis further clinical and biochemical investigations are necessary (e.g., eosinophilia, elevated serum levels of IgE, gastric wall thickening, gastric motility disorders, pyloric stenosis, other allergic underlying disorders, and gastrointestinal parasites).

Crohn-Associated Gastritis

The type of gastritis typical of Crohn's disease is – in analogy to the colon – a discontinuous inflammatory infiltration of the gastric mucosa. Patchy infiltrates located predominately around the pits and containing abundant neutrophilic granulocytes that often penetrate into the foveolae are found.

We think in accordance with Wyatt and Dixon [3] that this classification, based on the characteristic patterns of mucosal response to the different underlying mechanisms of injury, may enable the gastroenterologists to select patients for future clinical studies and probably for specific treatment.

References

1. Haot J, Hamichi L, Walez L, Mainguet P (1988) Lymphocytic gastritis: a newly described entity, a retrospective endoscopic and histological study. Gut 29:1258–1264
2. Heilmann KL, Stolte M, Borchard F, Heine M, Löning Th, Ottenjann R, Remmele W, Rühl G, Schaefer HE, Schlake W, Seib HJ, Stamm B, Steininger H, Wiebecke B (1988) Gastritis-Graduierung und Klassifikation. Pathologe 10:194–196
3. Wyatt JI, Dixon MF (1988) Chronic gastritis – a pathogenetic approach. J Pathol 154:113–124

Morphological Aspects of Gastritis – a Comment

A. B. Price and M. Stolte*

The Workshop covered a wide range of gastric inflammatory pathology. The reports published here represent but a selection, and this summary simply draws attention to particular aspects of interest or contention. The emphasis was on *Helicobacter pylori*, its associated gastritis, the natural history of gastritis, and any impact these have on the current classification of gastric inflammatory disease.

Pathologists are beholden to their clinical colleagues to provide endoscopic material and it is logical to start with Tytgat's paper, which reviews the spectrum of endoscopic gastritis. Whether the endoscopist is justified in using the term "gastritis," or whether this has to be solely the domain of the pathologist is more than just semantics. His paper draws attention to the poor correlation between endoscopic features and histology, this correlation improving with the severity of the changes seen. This highlights the problem of what constitutes "normality," to endoscopists and pathologists alike. Discordance is concentrated to the mild end of the spectrum. Indeed, many of the conflicting reports in the literature on whether *H. pylori* may or may not be seen on normal mucosa revolve around the observer variation associated with defining normality. Kirschner et al. point out that the normal gastric mucosa has an immune system in place. T-cell intraepithelial lymphocytes, up to 6 per 100 epithelial cells, can be found along with small numbers in the lamina propria. The small numbers of B-cells and plasma cells present in normal mucosa are restricted to the lamina propria. The problem of defining endoscopic normality also arises in the work of Czinn, in which children (5–18 years) were the subject of a study. Gastrointestinal mucosal nodularity associated with prominence of mucosal lymphoid follicles is probably within the normal range for a child. However, the nodularity seen in the antrums of the children in this study did not appear to be associated with follicular hyperplasia, and Czinn believes it to be a pathognomonic sign of pediatric *H. pylori* infection. Another problem area raised by Tytgat is whether biopsies for the assessment of gastritis should be random or targeted. To our knowledge this has never been addressed systematically and is one of considerable importance needing further research. The minimum assessment necessary to classify gastritis is two sets of two biopsies, preferably taken from the anterior and posterior aspects of the antrum and corpus respectively. The incisura and sites within 2 cm of the pylorus should be avoided for general purposes, though they must obviously be sampled if specific lesions are detected.

Dept. of Pathology, Northwick Park Hospital, Harrow Middlesex, U.K.
* Institute of Pathology, Klinikum Bayreuth, Prenschwitzer Straße 101, D-8580 Bayreuth, FRG

Helicobacter pylori, Gastritis and Peptic Ulcer
Edited by P. Malfertheiner and H. Ditschuneit

The contributions from Eidt and Stolte along with that from Sipponen look at behaviour of the gastric antrum and corpus in peptic ulcer disease but from different viewpoints. The apparent preference of *H. pylori* for antral mucosa emphasizes the need to document the two main gastric compartments, antrum and corpus, separately in order to study the differing role of damage at these sites on the natural history of peptic ulcer disease. Eidt and Stolte show that when either pyloric or duodenal ulcers are present antral mucosal inflammatory patterns are similar when colonization by *H. pylori*, activity, and severity of inflammation are compared. Changes in the corpus are less marked. When ulceration is present elsewhere in the stomach, then it is the mucosa of the corpus which is most affected, implying that pyloric and gastric ulcers warrant separate consideration. They speculate that it may be an inherent variability in the size of the antral and corporal gastric compartments, when inflamed, that determines at which site ulceration is found. This may have validity for it is known that the antral/corporal border moves upwards along the lesser curve with age [1]. However, very little is known about variations in size of the antral and corporal compartment in the normal population, Eidt and Stolte's data and hypothesis being a stimulus to gather such information.

There is unique experience in Finland with longitudinal studies on gastritis. Sipponen states a firm belief in the progressive nature of chronic gastritis over a long time course. This belief is anchored to the concept of developing atrophy in either the antral or corporal compartment causing altered gastric pathophysiology, namely, altered gastrin responses with antral disease and hypo- or achlorhydria with corporal disease. From these longitudinal studies the risk factors for the development of peptic ulcer disease are calculated. There is marked variation according to which gastric compartment is affected.

The asociation of *H. pylori* with gastritis and peptic ulcer disease raises the issue of the validity of the current classifications of gastritis. Tytgat presents, in brief, the main ones in use. The works of Eidt and Stolte when contrasted to that of Sipponen illustrate well the different approaches to the problem of morphology and classification and the need for uniformity. In the former's presentation attention is centered on inflammation and its activity while the latter centers on the role of atrophy. With an increasing number of papers appearing on gastritis and its significance, it is important that there can be cross-evaluation. A uniform classification and system of grading is therefore urgently required. At a minimum there must exist a working formulation allowing cross-reference from one classification system to another. The recent classification of the German pathologists [2] provides a standard approach to grading most of the important morphological parameters. It adheres to the established type A, type B, and type AB classification nomenclature as does that of Wyatt and Dixon [3]. Both introduce an all-important etiological limb to the system. There is a built-in rigidity to all of the current classifications which may no longer be warrented at the current moment of expanding and changing knowledge. Thus, in the system of Correa [4] and that proposed by Paull and Yardley [5], the appearance of intestinal metaplasia and atrophy place the gastritis in a category separate from that due to *H. pylori*. Sipponen et al. [6], on the other hand, use yet another system and believe in a step-wise progression of gastritis in which *H. pylori* and atrophy are an integral part. There is much to recommend a simple but flexible system using standard anatomical nomenclature rather than letters that will refer to the major changes in each anatomical compartment (antrum and/or corpus). Onto this an etiological qualification may be linked when known, e.g., *Helicobacter*-associated chronic gastritis of the antrum, autoimmune-type chronic gastritis of the corpus. Indeed there is a working party in progress that will report to the 1990 World Congress of Gastroenterology trying to devise a standard nomenclature or working formulation that will allow a comparison between different studies.

The meeting in Ulm was designed to give the participants multidisciplinary immersion in the affairs of *H. pylori*. Perhaps a word of caution is necessary so that histopathologists are not swept along and diagnose *Helicobacter*-associated gastritis at the merest sight of any organism on the gastric mucosa. Classically, *H. pylori* does have a seagull shape but other bacteria can be found on the gastric surface and perhaps more problematic, *H. pylori* is polymorphic [7]. Shape is therefore not an entirely reliable guide to its identification. The contribution of Loffeld et al. is relevent here, and a highly sensitive immunoperoxidase method is described using polyclonal antiserum. This is shown to be especially useful, over culture, when only small numbers of organisms are present. However, the specificity of polyclonal preparations remains a contentious issue. During the workshop Heilmann [8] drew attention to other organisms, in particular, *Gastrospirillum hominis*, as an additional cause of gastritis. It is worth remembering that changes in acidity will alter the gastric microenvironment and microbial ecology [9]. Strict criteria for the histological identification of *Helicobacter*-like organisms need to be maintained or other organisms, which may also be potential pathogens, may be overlooked.

Gastritis has etiologies and patterns other than that associated with *H. pylori*. In this respect pathologists need to recognize, among others, reflux gastritis [10] and lymphocytic gastritis, the latter described here by Haot et al. It seems likely that the changes found in enterogastric reflux can also be seen after other insults to the stomach such as certain chemicals and drugs [3]. The etiology of lymphocytic gastritis is unknown at the present time, its relation to *H. pylori* colonization also being unclear. Haot et al. suggest the condition might confer some colonization resistance to the gastric mucosa. Their paper also draws attention to the correlation between the histological picture and the anatomical distribution of the varioliform lesions seen endoscopically. Histological lymphocytic gastritis correlates best with fundic or diffuse involvement but not with endoscopic disease limited to the antrum. Varioliform lesions restricted to this area have a more variable microscopic picture.

This selection of papers on morphological aspects of gastritis serves to emphasize the stimulus given to inflammatory disease of the gastric mucosa by the discovery of *H. pylori*. As a result new data is emerging rapidly on all aspects of gastric morphology and pathophysiology. Besides the points discussed above, other interesting aspects of gastritis are addressed in the papers of Baczako and Solcia et al. The latter challenge microbiologists by suggesting there may be ulcerogenic and nonulcerogenic strains of *H. pylori*. The immunological approach of Kirchner et al. highlights differences between microbial and autoimmune patterns of gastritis that up to now have been difficult to appreciate on routine staining and light microscopy. Although the data is from a small sample only, it has useful diagnostic potential. Finally, it is a challenge to consider the impact on world health of the epidemiological data presented here and elsewhere in the Ulm Congress [11, 12], in particular the associations drawn between *H. pylori*, peptic ulcer disease, and gastric cancer.

References

1. Kimura K (1972) Chronological transition of the fundic-pyloric border determined by stepwise biopsy of the lesser and greater curvatures of the stomach. Gastroenterology 63:584–592
2. Stolte M, Heilmann KL (1989) Neue Klassifikation und Gradierung der Gastritis. Leber Magen Darm 5:220–226
3. Wyatt JI, Dixon MF (1988) Chronic gastritis – a pathogenetic approach. J Pathol 154:113–124
4. Correa P (1980) The epidemiology and pathogenesis of chronic gastritis: three aetiologic entities. Front Gastrointest Res 6:98–108
5. Paull C, Yardley JH (1989) Pathology of Campylobacter pylori-associated gastric and esophageal lesions. In Blaser MJ (ed) Campylobacter pylori in gastritis and peptic ulcer disease. Igaku-Shoin, New York, p 76
6. Sipponen P, Seppala K, Aarynen M et al. (1989) Chronic gastritis and gastroduodenal ulcer: a case control study on risk of coexisting duodenal or gastric ulcer in patients with gastritis. Gut 30:922–929
7. Bode G, Malfertheiner P, Lehnhardt G, Strohle A, Ditschuneit H (1989) Polymorphism of Campylobacter pylori due to aging and chemical induction. Klin Wochenschr 67 (Suppl XVIII):4–5
8. Heilmann KI, Borchard F, Nowottny V (1989). Gastritis caused by Gastrospirillum Hominis. Klin Wochenschr 67 (Suppl XVIII):27
9. Borriello SP, Reed PJ, Dolby JM, Barclay PE, Webster ADB (1985) Microbial and metabolic profile of achlorhydric stomach: comparison of pernicious anaemia and hypogammaglobulinaemia. J. Clin Pathol 38:946–953
10. Borchard F (1989) Reflux gastritis. Klin Wochenschr 67 (Suppl XVIII):6
11. Forman D, Sitas F, Stacey A et al. (1989) Epidemiology of Campylobacter pylori. Klin Wochenschr 67 (Suppl XVIII):21
12. Sitas F, Forman D, Chen J, Newell D, Stacey A (1989) Prevalence of infection with Campylobacter pylori in China: results from a large correlation study in 46 regions. Klin Wochenschr 67 (Suppl XVIII):64

Helicobacter pylori in Duodenal Ulcer Pathogenesis

Aetiological Factors in Duodenal Ulcer

J. J. Misiewicz

Epidemiology

Duodenal ulceration (DU) is a major disease in most countries of the world and is widely prevalent in the West. Unfortunately accurate epidemiological surveys are difficult or impossible to do and data relating to the incidence and prevalence of DU are at best approximate. This is because the diagnosis of DU can only be established with certainty by using expensive, or invasive investigations such as air-contrast radiology, or fibre-optic endoscopy. This precludes large scale investigation surveys. Unfortunately even the presence of typical symptoms (epigastric pain related to food and relieved by antacids) can not be relied upon to establish the diagnosis. This is because many patients with typical symptoms of DU will turn out on investigation not to have one. The same considerations apply to noninvasive techniques for the detection of *Helicobacter pylori*, because most individuals infected with this organism will not have an ulcer, or even symptoms of dyspepsia.

Information from necropsy surveys indicates that approximately 20% of men and 10% of women will have suffered from a peptic ulcer at some time [1, 2]. These data, however, are heavily confounded by such considerations as the frequency with which autopsies are done, the selected populations subjected to this procedure, or the diligence and skill of the pathologist. Mortality rates give information on the fraction of the ulcer population subject to lethal complications, but this is heavily age related and must represent an unknown proportion of the whole [3]. Mortality rates may also be affected by the patterns of prescribing drugs in the population studied. For example, frequent prescription of NSAIDs may increase mortality from DU in the elderly [4, 5]. It is generally believed that approximately 10% of people in the West suffer from DU; the incidence in the emergent countries is less certain, but appears to be rising in the urbanised population.

Thus environmental factors may affect the prevalence, incidence and also perhaps the clinical attributes of DU. Assignment of relative importance to the various influences thought to be active in this area is difficult. Gastric ulcer (GU) was probably the predominant form of ulceration in the upper gut at the turn of the century, but this was followed by an increasing rise in frequency of DU. Whether this may be in some way related to an increased prevalence of *Helicobacter pylori* is difficult to say. One hypothesis postulates that the rapid rise

Department of Gastroenterology, Central Middlesex Hospital, London NW10 7NS. UK

Helicobacter pylori, Gastritis and Peptic Ulcer
Edited by P. Malfertheiner and H. Ditschuneit

in the incidence of DU at the start of the present century was a cohort phenomenon triggered by deprivation and stress secondary to the economic recession present at that time [6]. Bad housing conditions and overcrowding could have led to the spread of infection with *Helicobacter* at that period in history.

More recently the trends detected in the epidemiology of DU have been somewhat conflicting. The frequency of GU seems to be declining in the Western countries, while surveys conducted in Denmark and in the USA suggest that incidence of DU is not changing [7, 8]. Data from the United Kingdom indicate that the incidence of perforated DU has decreased in the younger, but increased in the older population; the increased perforation rate seems to affect women more than men.

Environmental changes can be expressed, for example as differences in perforation rates, even within relatively small areas and to affect DU and GU in different ways (Table 1). As all those areas have uniform standards of provision of health care, and as most patients with perforation are liable to be admitted to hospital and accurately diagnosed at laparotomy, these figures are likely to be reliable [9].

Table 1. Regional admission rates per 10000 population with perforated peptic ulcer in the U.K.

	Duodenal	Gastric
South		
East Anglia	0.07	0.08
Wessex	0.08	0.08
Oxford	0.13	0.04
South-West	0.14	0.02
North		
Leeds	0.22	0.07
Liverpool	0.23	0.03
Manchester	0.23	0.10
Newcastle	0.31	0.10
Scotland	0.45	0.07

Diet

It is very difficult to show convincingly any role for diet in the pathogenesis of DU. Nevertheless, diet is very important for the patient with DU, and many of them are emphatic that certain foods, such as high fat or fried meals provoke their ulcer pain. This may be related to stimulation of gastric secretion or changes in the rate of gastric emptying, but experimental confirmation of this common observation has been elusive. Prospective studies have correlated the high incidence of gastric cancer in Japan with low intake of vegetables and with smoking [10]. High salt intake and high incidence of GU coexist in Japan and a correlation between high dietary salt content and high mortality from GU (but not DU) has been put forward for western areas [11]. There was said to be a difference in the incidence of DU between northern and southern India, possibly related to the chewable diet in the north and a mushy vegetarian diet in the south, but this difference in the incidence of DU is not based on firm epidemiological evidence. Another hypothesis postulates that the reason for the apparently declining rate of serious complications of DU is the increased dietary consumption of polyunsaturated fatty acids, which act as precursors of prostaglandins, which in turn may have a role in cytoprotection and in inhibition of acid secretion [12]. If diet is important in the pathogenesis of DU, it is clear that we do not know how it operates.

Genes and Gender

DU is commoner in men than in women, although the incidence in women may be inreasing. It does increase in women after the menopause, suggesting that sex hormones, perhaps oestrogens, have a protective role. Oestrogens have been used in the medical treatment of DU in men in one trial, but the side effects were unacceptable.

Patients with a family history of DU are more likely to develop the disease than others, and the ulcer may be more severe. There are other genetic markers of the ulcer diathesis, such as the nonsecretion of blood group polysaccharides in the saliva [13].

Attack and Defence

The many factors that eventually result in the development of a chronic DU are often thought of as attack and defence mechanisms, the imbalance favouring the "attack" eventually resulting in the appearance of an ulcer crater. Attack factors have been more extensively studied than defence factors.

Secretion of Acid

Patients with DU as a group secrete more acid than controls, while those with GU have normal or low acid outputs. Various mechanisms have been proposed to account for the high acid secretion in DU. Patients with DU tend to have a high parietal cell mass and their parietal cells may be more sensitive to gastrin (or pentagastrin). The gastrin negative-feedback switch-off mechanism may be inefficient, leading to inappropriate hypergastrinaemia. As gastrin is a trophic hormone for the parietal cells, it has been proposed that this is the way in which the large parietal cell mass in DU comes about.

There are some very interesting recent findings to link this hypothesis with the near universal colonisation of the antrum and of gastric metaplasia islands in the duodenal cap with *Helicobacter pylori* in DU disease. It has been shown that asymptomatic normal young males with *Helicobacter pylori* have a high integrated 24-h gastrin response to meals [14], and this was followed by the observation of normogastrinaemia in *Helicobacter pylori* negative DU patients; *Helicobacter pylori* positive patients had the hypergastrinaemia and a higher peak acid output [15]. The mechanism through which *Helicobacter pylori* operates to trigger the inappropriate secretion of gastrin is not clear. It may be the consequence of antral mucosal inflammation or perhaps due to alkalinisation of the microenvironment of the G cells in the antrum consequent on the production of ammonia from urea by the *Helicobacter pylori* urease. This would disrupt the normal operation of the negative feedback loop controlling the release of gastrin from the antral G cells. This hypothesis is attractive because it accounts for the observed experimental changes under one factor present almost uniformly in DU, namely colonisation by *Helicobacter pylori.*

There are other abnormalities of gastric acid secretion in DU, namely high nocturnal acid output, high acid secretion in response to cephalic stimulation or to stimulation by sham feeding or the ingestion of food. A high vagal drive has been postulated, but has so far eluded unequivocal experimental proof. Indeed, it has been difficult to show abnormalities of gastric acid secretion in all the patients included in the various published studies (Table 2) [16]. This may be due to experimental errors, such as transpyloric losses of gastric juice, that are inherent in the techniques of collection of gastric acid, and which are not corrected for in all the studies. Such errors may be especially important at low rates of acid output, for example when nocturnal secretion is measured. Measurements of acid output require unpleasant intubation techniques and are therefore only possible to conduct for relatively short periods of time. Measurements of intragastric acidity are not subject to those constraints to the same extent, but they provide data on pH, not on the volume of acid secreted. In one interesting study meal-stimulated and nocturnal acid secretion was measured over a 24-h period in a small series of patients with DU, who had a significantly higher acid output than controls [17]. It is not clear however, whether these data can be extrapolated to all patients with DU disease.

It is clear that acid is a key factor in the pathogenesis of ulcer, but its ranking in the order of importance in a multifactorial situation is not clear, especially as the equation may differ from patient to patient. At one end of the spectrum, in the presence of gross acid hypersecretion driven by aberrant gastrin release by a gastrinoma, acid is clearly the main agent responsible for the severe, intractable, and often ectopic, ulceration. At the other end is the individual with acid secretion

Table 2. Percentage of subjects with duodenal ulcer (DU) showing an abnormal acid secretory response in various studies. [Modified from 16]

Variable of acid secretion	Studies (*n*)	Range of subjects with abnormal response (%)
Maximal acid output following histamine or pentagastrin stimulation > normal	17	16–56
Basal acid output > normal	7	10–38
Cephalic acid output > normal	4	7–55
Low D_{50} to pentagastrin	6	0–45
Meal-stimulated acid output > normal	8	10–100

in the normal range, who has a DU and in whom other, perhaps multiple, factors come into prominence.

Acid secretion is certainly important in preventing the healing of DUs, as has been abundantly shown by numerous trials reporting significantly better rates of ulcer healing during treatment with agents that neutralise secreted acid (antacids) or that prevent its secretion (anticholinergics, H_2 histamine receptor blockers, H^+K^+ATPase inhibitors). There is also a correlation between healing rates of DU and the degree of acid inhibition [18, 19]. Inhibition of acid by maintenance treatment with H_2 receptor antagonists significantly lowers the relapse rates of DU, thus emphasising the central role of acid in maintaining the chronicity of DU [20].

Pepsin

Pepsin has been less intensively studied than acid; in general its secretion moves in parallel with that of HCl. The role of pepsin in the causation of DU is believed to be based on its capacity to disrupt the mucus-bicarbonate barrier, although it must be admitted that data pertaining to this have been collected mainly through studies of the gastric, rather than the duodenal, mucus [21]. Precursors of pepsin, pepsinogens, can be measured in the serum by radioimmunoassay, and thus lend themselves to epidemiological studies. Serum pepsinogens (Pgs) can be divided into PgI, PgII and a slow-migrating protease [22]. About 50% of DU patients have high serum concentration of PgI, which is higher in more severe forms of the disease, and in those with high acid output [23]. It is inherited as an autosomally dominant trait and can be used as a marker in some families with a strong history of DU [24]. These observations emphasise the role of constitutional factors in ulcer diathesis.

Quantitative measurement of pepsin secreted into the gastric juice is not easy. Pure pepsin I – which is associated with DU – degrades gastric mucus glycoprotein faster and remains active at a higher pH than pepsin III, the main enzyme found in human gastric juice. Mucolytic activity of gastric juice from patients with DU resembles that of pepsin I, pepsin III-like activity is exhibited by juice from healthy controls [25].

Motility

Patients with DU are said to have faster gastric emptying, but overlap with other groups is considerable. Faster emptying might increase the acid load entering the duodenum, thus contributing to duodenal mucosal injury.

Smoking and Alcohol

It is not clear whether smoking plays a part in initiating the process that leads to duodenal ulceration, but it is established that smoking affects adversely the relapse rate of ulcers healed on medical treatment. The mechanism of the accelerated relapse rate associated with smoking is unknown [26].

Alcohol so far has no clearly defined role in the pathogenesis of DU.

Drugs

Drugs can cuse local lesions in the gastrointestinal mucosa if tablets lodge locally on the mucosal surface and then dissolve, forming a local hyperosmolar environment; this process has no part to play in the aetiology of DU, although it may lead to clinically dangerous situations.

Aspirin and nonsteroidal anti-inflammatory drugs (NSAIDS) increase gastrointestinal blood loss and this has been shown repeatedly in various animal models and in man. It is not certain how these observations relate to the development of the chronic DU diathesis. The vast majority of patients taking NSAIDS for arthropathies do so without serious unwanted effects; a minority, (predominantly the elderly), develop potentially serious complications such as haemorrhage or perforation. It is not clear whether these are complications of pre-existing ulcers, or whether new ulcer formation takes place. Even if it did, the place of drugs in the aetiology of DU must be necessarily very restricted, although the unwanted effects of NSAIDS are clinically very important [4, 5].

Psychological Factors

Most pre-existing diseases may get worse during periods of personal stress – or the patient is more conscious of them. Acute emotions can affect gastric mucosal blood flow, as documented by early studies on one or two subjects with chronic gastric fistula. Two patients have documented decreased gastric acid secretion after a major life event had been resolved. It is not possible to say how important or how often psychological factors enter into the causation of DU.

Protective Factors

Mucus and Bicarbonate

The insoluble layer of mucus gel that coats the gastric and duodenal epithelium is thought to be very important in preserving the integrity of the epithelial cells. Intact mucus is impermeable to larger molecules, such as pepsin, which itself can solubilise mucus by cleaving the nonglycosylated component of the mucus molecule. Breakdown of mucus by enzymic action, or by drugs, or by bile salts leads to the loss of its physicochemical properties and to the shedding of mucus into the gastric lumen, where it has no protective action.

Although impermeable to large molecules, mucus is permeable to hydrogen ions. It has been suggested that bicarbonate secreted by the mucosa is held in an unstirred layer by the mucus; direct measurements of pH gradients with microelectrodes have shown it to rise sharply as the epithelial surface is approached by the electrode tip. This has been termed "the mucus/bicarbonate barrier", which is thought to protect the epithelium from the action of acid. It also forms the narrow ecosphere where *Helicobacter pylori* has its habitat – a doubtful benefit to humankind [27–30].

Bicarbonate is also secreted into the duodenal lumen and patients with DU secrete less of it in response to luminal perfusion with acid, than controls [31, 32]. Another source of bicarbonate is pancreatic juice. It is not known how abnormally low duodenal bicarbonate fits into the defence vs attack equation, but it constitutes yet another abnormality that is present in DU disease.

Prostaglandins

The importance of prostaglandins (PGs) in the aetiology of DU is uncertain. Measurement of PGs in biopsy specimens from patients with DU has yielded discordant results. Treatment with NSAIDS would tend to inhibit the synthesis of prostanoids and may be important in the mechanisms of mucosal injury produced by those drugs. Cytoprotective mechanisms may operate to prevent acute mucosal injury and to mediate mucosal restitution, but they may not be relevant to chronic recurrent DU [33].

Conclusions

DU is a multifactorial disease, with many influences interacting in each affected individual, eventually to produce the ulcer crater in the duodenum. *Helicobacter pylori* can be detected in most patients with DU, but it is also present in many who do not have an ulcer or indeed in symptomatic individuals. The very high prevalence of the organism in emergent countries [34–36], for example, has not produced a pandemic of duodenal ulcer disease. It seems therefore that, in addition to the presence of *Helicobacter*, other factors must be present for the disease to manifest itself.

References

1. Levi IS, de la Fuente AA (1963) A post mortem study of gastric and duodenal peptic lesions. Gut 4:349–359
2. Watkinson G (1960) The incidence of chronic peptic ulcer found at necropsy. Gut 1:14–31
3. Coggon D, Lambert P, Langman MJS (1981) Twenty years of hospital admissions for peptic ulcer in England and Wales. Lancet 1:1302–1304
4. Collier DSJ, Pain JA (1985) Non steroidal anti-inflammatory drugs and peptic ulcer perforation. Gut 26:359–363
5. Walt R, Katschinski B, Logan R, Ashley J, Langman M (1986) Rising frequency of ulcer perforation in elderly patients in the United Kingdom. Lancet 1:489–492
6. Susser S, Stein Z (1962) Civilisation and peptic ulcer. Lancet 1:115–118
7. Bonnevie O (1980) Peptic ulcer in Denmark. Scand J Gastroenterol [Suppl 63] 15:163–174
8. Kurata JH, Honda GD, Frankl H (1982) Hospitalisation and mortality rates for peptic ulcers: a comparison of large health maintenance organisation and United States data. Gastroenterology 83:1008–1016

9. Brown RC, Langman MJS, Lambert PM (1976) Hospital admissions for peptic ulcer during 1968–72. Br Med J 1:35–37
10. Hirayama T (1981) Methods and results of gastric cancer screening. In: Fielding JW, et al. (eds) Gastric cancer. Pergamon, Oxford, pp 77–84
11. Sonnenberg AS (1986) Dietary salt and gastric ulcer. Gut 27:1138–1142
12. Hollander D, Tarnawski A (1986) Dietary essential fatty acids and the decline in peptic ulcer disease: a hypothesis. Gut 27:239–242
13. Clarke CA, Edwards JW, Haddock DRW, Howel Evans AW, McConnell PB, Sheppard PM (1956) AB0 blood groups and secretor character in duodenal ulcer. Br Med J 2:725–731
14. Smith JTL, Pounder RE, Evans DJ, Graham DY, Evans DG (1989) Inappropriate 24 hour hypergastrinaemia in symptomatic Campylobacter pylori infection. Gut 30:A732
15. Levi S, Beardshall K, Haddad G, Playford R, Ghosh P, Calam J (1989) Campylobacter pylori and duodenal ulcer: the gastrin link. Lancet 1:1167–1168
16. Lam SK (1984) Pathogenesis and pathophysiology of duodenal ulcer. Clin Gastroenterol 13:447–472
17. Feldman M, Richardson CT (1986) Total 24-hour gastric acid secretion in patients with duodenal ulcer: comparison with normal subjects and effects of cimetidine and parietal cell vagotomy. Gastroenterology 90:540–544
18. McIsaac R, McCanless I, Summers K, Wood JR (1987) Ranitidine and cimetidine in healing of duodenal ulcer: meta analysis of comparative clinical trials. Aliment Pharmacol Ther 1:369–381
19. Jones DB, Howden CW, Burget DW, et al. (1987) Acid suppression in duodenal ulcer: a meta-analysis to define optimal dosing with antisecretory drugs. Gut 28:1120–1127
20. Gough KR, Bardhan KD, Crowe JP, et al. (1984) Ranitidine and cimetidine in prevention of duodenal ulcer relapse. Lancet 2:659–662
21. Venables CW (1986) Mucus, pepsin and peptic ulcer. Gut 27:233–238
22. Taggart TR, Samloff IM (1987) Immunochemical, electrophoretic and genetic heterogeneity of pepsinogen 1. Gastroenterology 92:143–150
23. Samloff IM, Liebman WL, Donitch NM (1975) Serum group 1 pepsinogen by radioimmunoassay in control subjects and patients with peptic ulcer. Gastroenterology 69:83–90
24. Rotter JL, Sones JQ, Samloff IM, et al. (1979) Duodenal ulcer disease associated with elevated serum pepsinogen 2: an inherited autosomal dominant disorder. N Engl J Med 300:63–73
25. Pearson JP, Ward R, Allen A, Robert NB, Taylor WH (1986) Mucus degradation by pepsin: comparison of mucolytic activity of human pepsin 1 and pepsin 2. Gut 27:243–249
26. Sontag S, Graham DY, Belsito A, et al. (1984) Cimetidine, cigarette smoking and recurrence of duodenal ulcer. N Engl J Med 311:689–695
27. Hollander F (1954) Two two-component mucus barriers. Arch Intern Med 93:107–120
28. Heatley NG (1959) Mucosubstance as a barrier to diffusion. Gastroenterology 37:313–323
29. Allen A, Garner A (1980) Gastric mucus and bicarbonate secretion and their possible role in mucosal protection. Gut 21:249–262
30. Rees WDW, Turnberg LA (1982) Mechanism of gastric mucosal protection: a role for the "mucus-bicarbonate" barrier. Clin Sci 62:343–348
31. Isenberg JI, Hogan DL, Koss MA, Selling JA (1986) Human duodenal mucosal bicarbonate secretion: evidence for basal secretion and stimulation by hydrochloric acid and a synthetic prostaglandin E1 analogue. Gastroenterology 91:370–379
32. Isenberg JI, Selling JA, Hogan DL, Koss MA (1987) Impaired proximal duodenal mucosal bicarbonate secretion in patients with duodenal ulcer. N Engl J Med 316:374–378
33. Hawkey CJ, Walt RP (1986) Prostaglandins for peptic ulcer: a promise unfulfilled. Lancet 2:1084–1086
34. Graham DY, Opekun AR, Sekeley S, Evans DG, Evans DJ (1989) High prevalence of Campylobacter pylori (Cp) infection in poor and rich Peruvian children determined by ^{13}C urea breath test (^{13}C-UBT). Gastroenterology 96:A260
35. Dehesa M, Dooley CP, Cohen H, Fitzgibbons P, Perez-Perez GI, Blaser MJ (1989) High prevalence of Campylobacter pylori (Cp) in an asymptomatic hispanic population. Gastroenterology 96:A115
36. Megraud F, Brassens-Rabbe MP, Denis F, Touhami M, Hoa DQ (1989) Campylobacter pylori infection – evidence for a worldwide distribution and high prevalence in developing countries. Gastroenterology 96:A338

The Histogenesis of Gastric Metaplasia in Chronic Non-Specific Duodenitis

K. C. Liu [1] and N. A. Wright [2]

Introduction

It has long been appreciated that the epithelial cells which clothe the small intestinal villi originate in the crypts and migrate upwards. They acquire the ability to secrete and absorb, and are finally cast off into the intestinal lumen. The ability to label proliferating crypt cells with relatively stable markers, such as tritiated thymidine [1] and bromodeoxyuridine [2] shows unequivocally that the overall direction of movement is upwards towards the villus tip.

However, there has been no general agreement about the specific pathways taken by these migrating epithelial cells on the villus, nor the pathway taken by other cell on the villus, nor the pathway taken by cells leaving the crypts before they enter the villus epithelium proper; three-dimensional studies reveal that, in man, as many as six crypts supply each villus [3], and there has been considerable discussion of the possibility raised by Potten et al. [4] that the emergent cells from a single crypt might be directed to more than one villus. In man, of course, several crypts fuse to form a vestibule which then communicates with the intervillous space [5] (see Fig. 1).

On these questions, opinion has been sharply divided: Smith and Peacock [6], after studies on neonatal pigs, considered that villus cells migrated as isolated, confined cohorts, while Carr, Hamlet and Watt [7], who observed spiral ridges circumventing human villi in scanning electron micrographs, proposed that these ridges themselves were the migration pathways and that cells spiralled around the villus. Others have conceived a much less ordered arrangement [8, 9]; considerable cell mixing as cells left the crypts was envisaged by Partridge and Simpson [8] with the formation of "traffic jams" at the villus bases as cells of different migration velocity intermingled. The work of Bjerknes and Cheng [9] also indicated much mixing of cells once these were on the villus: in the crypts, some 80% of newly formed mucous cells appear in pairs, while on the villi the distribution of mucous cells is random, as cells migrate and become mixed with the migrating progeny of other crypts.

The appropriate experimental model to solve this riddle would be where the progeny of a single crypt could be selectively labelled and followed as the cells migrate onto the villus and interrelate with cells leaving adjacent crypts. The

[1] Department of Histopathology, Royal Postgraduate Medical School, Du Cane Road, London W12 OHS, U.K.; [2] ICRF/RCS Histopathology Unit, 35–43 Lincoln's Inn Fields, London WC2A 3PN, U.K.

Helicobacter pylori, Gastritis and Peptic Ulcer
Edited by P. Malfertheiner and H. Ditschuneit

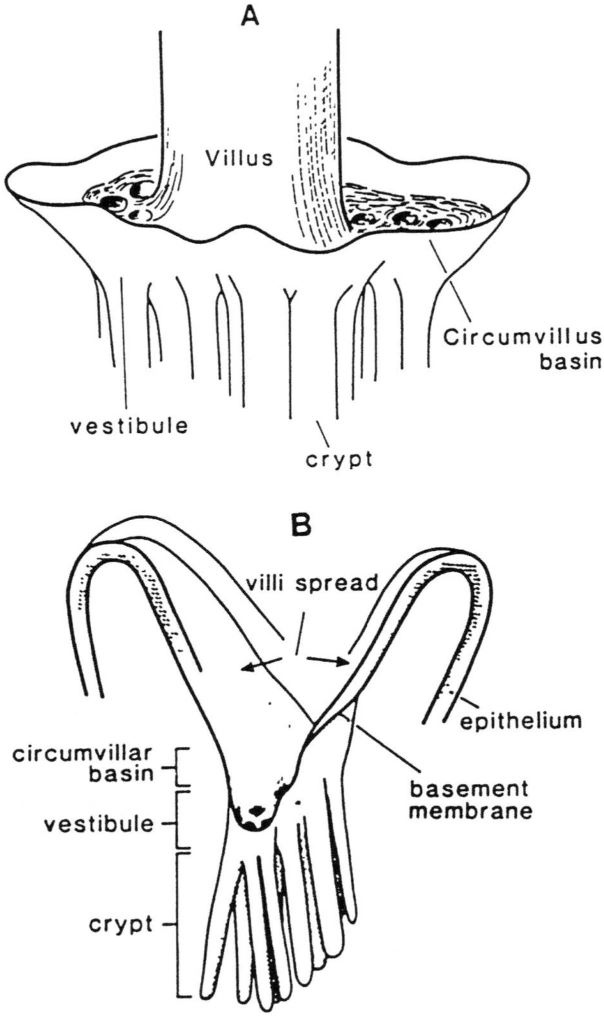

Fig. 1 a, b. Three-dimensional representations of the crypt showing **a** villus relationship in the human small intestine. Several crypts fuse to form the vestibules. **b** The vestibules coalesce to form the circumvillus or circumvillar basin, and higher still the basins intercommunicate to become a continuous intervillous space (Redrawn from Cocco et al. [5])

migration pathway would then be revealed. For the mouse, the puzzle was apparently solved using mouse embryo aggregation chimaeras [10–12] and a carbohydrate polymorphism that allowed the intestinal epithelium of the different strains of inbred mice which were used to make the chimaeras to be discerned by the binding of the lectin *Dolichos biflorus* agglutinin [11, 12]. In these chimaeras, the crypts are clonal and consist of cells from a single parental strain; mixed crypts are not seen, and hence the progeny of each crypt can be readily discerned by the presence or absence of lectin binding. Mapping serial sections, with three-dimensional computer reconstruction, Wilson, Ponder and Wright [13, 14] showed that epithelial cells migrate from the crypts in staight lines and as tightly bound cohorts onto and through the villus epithelium; moreover, a single crypt can indeed feed cells to more than one villus.

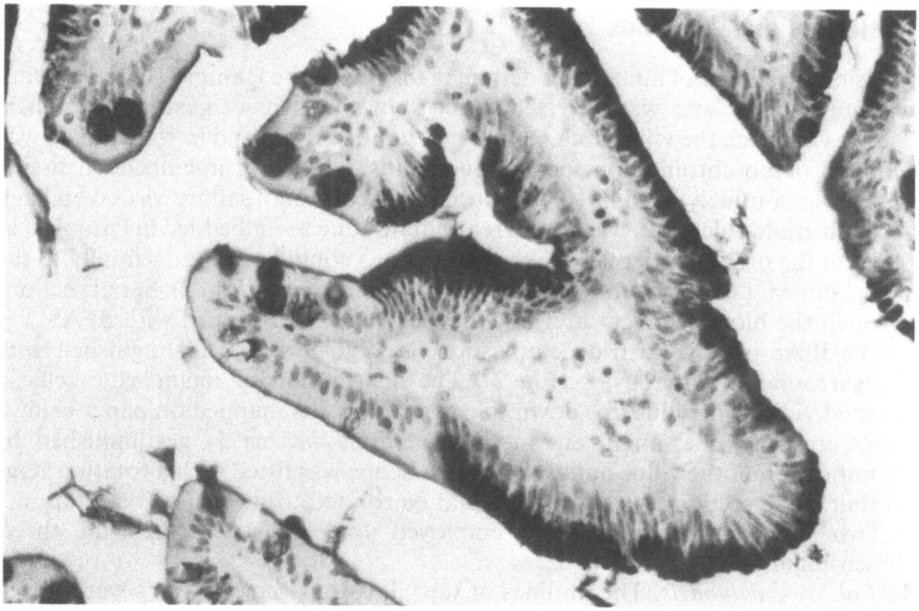

Fig. 2. A section of a duodenal villus showing the intense dPAS-positive staining of the metaplastic cells. The metaplastic cells can be readily distinguished from the enterocytes and goblet cells

However, the situation in the human small bowel remains obscure. It seems intrinsically unlikely that the possibility of a similar chimaeric model would exist in the human situation. We noted that gastric metaplasia, a well-recognized histological appearance in the human duodenum [15], is often seen focally on villi cut in tangential section, and shows remarkably sharp demarcation from the villus enterocytes and goblet cells when stained with diastase periodic acid–Schiff (dPAS) (Fig. 2). In fact it is not dissimilar in pattern to villi in chimaeric animals [14]. Because these metaplastic cells stand out so sharply from the normal background of villus cells, it is possible to trace their migration pathway. Such observations should also reveal the origin of these metaplastic cells, for which there is no general agreement; whether from the differentiated villus epithelial cells [16, 17], or from cells in the crypt area.

In this paper, we follow the pathway of migration of gastric metaplastia cells on human duodenal villi, and show (i) that the straight migration pathway is also seen in humans, (ii) that migration occurs in relatively tight cohorts, but there is some mixing at the margins of the migration pathway, (iii) that the progeny from one vestibule or crypt group can supply cells to more than one villus, and (iv) that gastric metaplasia takes its origin from cells in the basal crypt area or in Brunner's glands ducts.

Materials and Methods

Excision specimens of human duodenum in the files of the Hammersmith Hospital were reviewed. Cases were selected which showed extensive gastric metaplastic changes in which the villus architecture was not disturbed and in which there was minimal or no chronic non-specific duodenitis. The paraffin-embedded tissue, which was routinely fixed in 10% neutral buffered formol saline, was taken back through graded alcohols, water, and reorientated and re-embedded in Paraplast at 90 °C to the original orientation, so that sections could be cut tangentially to the bowel lumen, i.e., transverse sections of villi could be cut (Fig. 2). Serial sections through the block were cut at a thickness of 4 μm, and stained with dPAS.

The dPAS-positive gastric metaplastic cells could be readily distinguished from the surrounding enterocytes (Fig. 2). The distribution of metaplastic cells in selected villi, from villus tips down to crypt bases, was mapped on paper using a Leitz drawing tube; alignment in these sections was easily accomplished by identification of the villus outlay. The microscope was fitted with a rotating stage so that if the section was askew, it could be rotated to its correct orientation.

Two different methods were employed to reconstruct the villi three-dimensionally:

1. *Polystyrene models*. The outlines of the villus transverse sections were traced onto polystyrene tiles and cut out with a hot wire. The areas occupied by the metaplastic cells were shaded red. The shapes were aligned as indicated by the serial section analysis and glued together.
2. *Computer-aided reconstruction*. The three-dimensional reconstruction package produced by Contextvision (Sweden) was used.

Results

Effectively all the conclusions can be derived considering the two villi shown in section in Fig. 3, and modelled in Fig. 4. In Fig. 3a, the tips of the two adjacent villi are shown, each showing metaplastic cells; note that the villus tips, with their metaplastic cells are some distance from each other. Figure 3b shows a section about halfway down the two villi. Figures 3c and 3d show levels progressively closer to the vestibular-villous junction. It can be readily appreciated that the two areas of metaplasia, while still on different villi, became closer together until finally they met face to face (Fig. 3d) and fused to form a single vestibule (Fig. 3e).

Thus this single vestibule is supplying metaplastic cells to two villi at the level of the vestibule, where cells emerge onto the surface. The migrating cells evidently divide into two streams, one destined for each villus.

The migration pathway on the villi can be best appreciated by inspecting the three-dimensional reconstruction (Fig. 4). Here the model consists of one whole villus and part of a second villus. With the model opened out to display the facing surfaces of the two adjacent villi, the two trains of gastric metaplastic cells can be seen going upwards from the same vestibule. Migration is seen to be in straight lines, and these migration lines are relatively tight with little mixing of metaplastic cells with enterocytes once the cells are on the villus surface proper, so that the

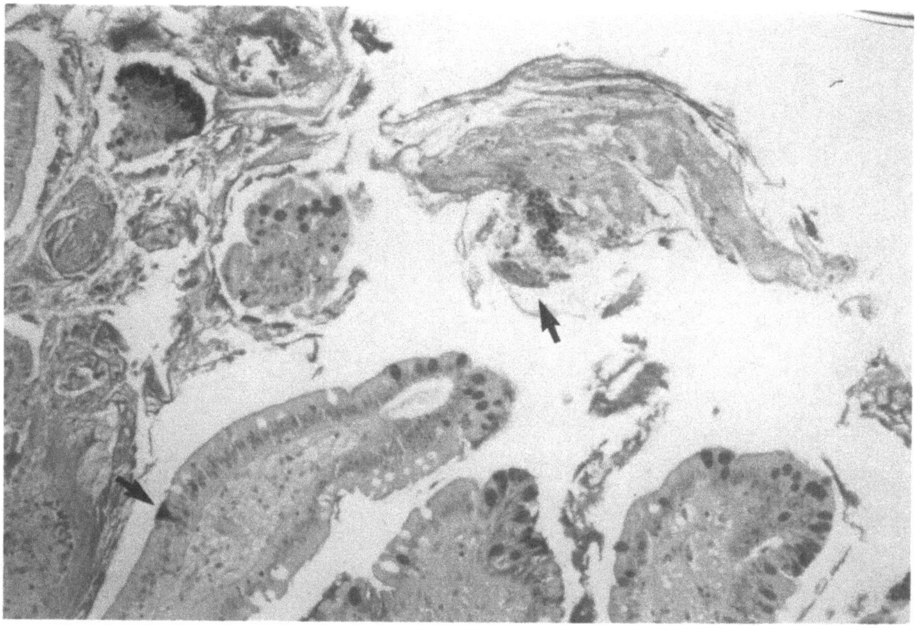

a

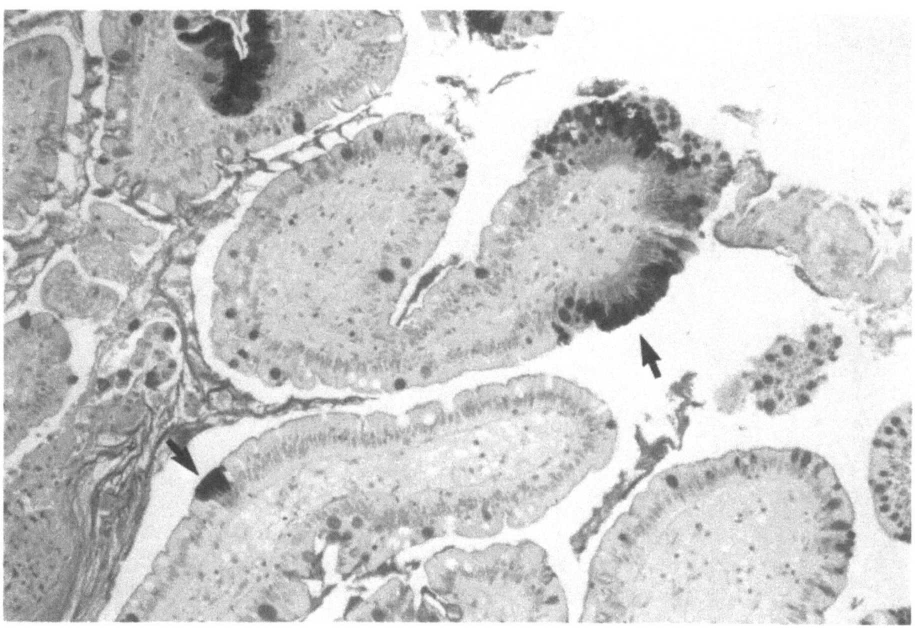

b

Fig. 3a–f. A selection of levels through the villi modelled and discussed in the text: **a** about midvillus level, **b** at the vestibular-villus junction, **c** within the vestibule. Note that the areas of metaplasia on the two villi (*arrowed*) **d** move towards each other **e** as the vestibular-villous junction is reached, and **f** join to form the same vestibule

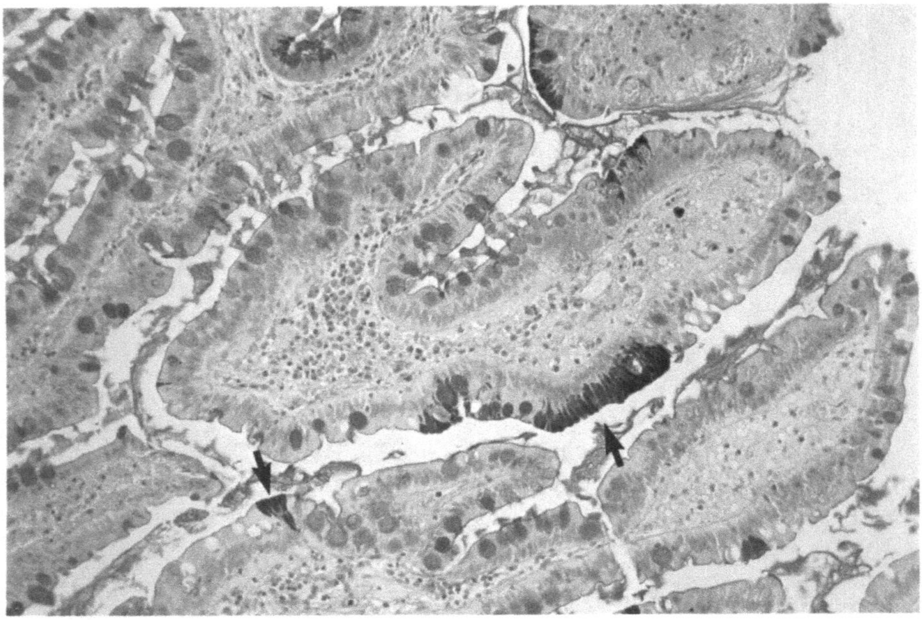

Fig. 3c

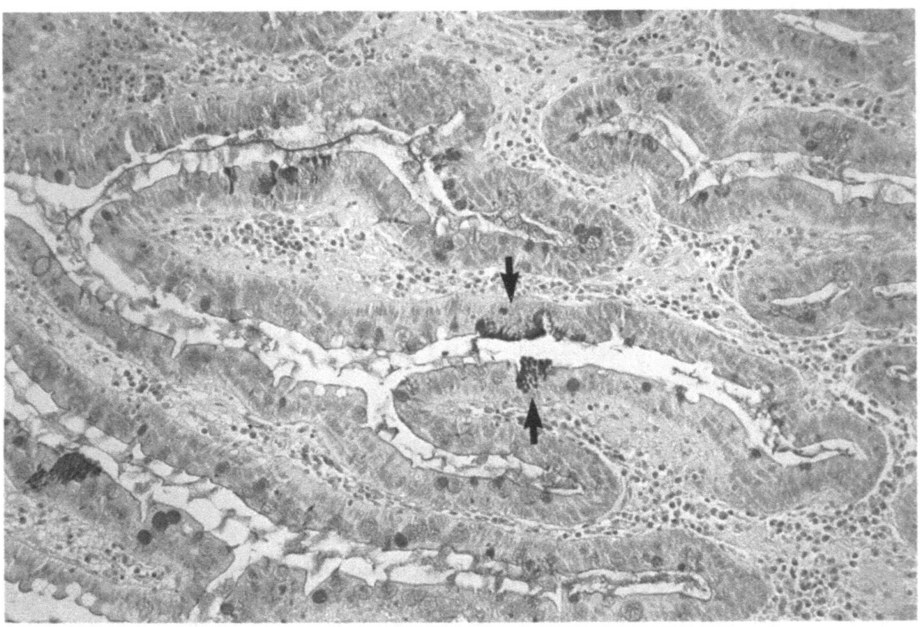

Fig. 3d

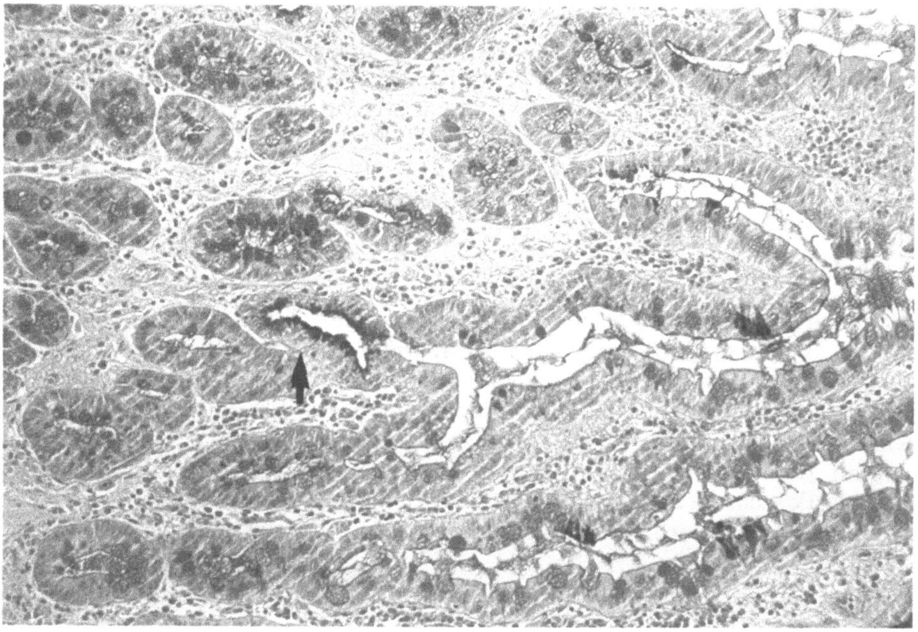

Fig. 3e

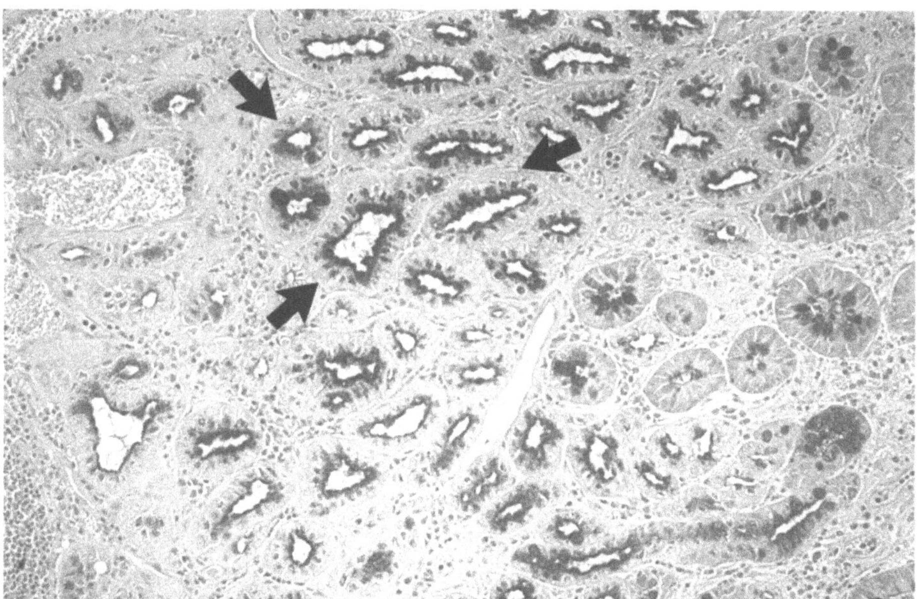

Fig. 3f

migration pathway stands out distinctly. Around the villus base, however, there is a small degree of cell mixing with some single metaplastic cells isolated from the main migration streams; indeed in Figs. 4a and 4c, there are some few metaplastic cells around the villus base completely isolated and not apparently associated with any main metaplastic cell stream. In some places, the migration pathway of metaplastic cells appears to be squeezed by the migrating enterocytes at each side of the stream. At the mid-villus level, some metaplastic cells leave the stream and move onto the migration pathway of the neighbouring enterocytes (Fig. 4c). Similarly there is a focal tendency for enterocytes to invade the migration pathway of the metaplastic cells (Figs. 4a–c).

Another face of the complete villus (Fig. 4a) shows a different pattern: two cohorts of migrating metaplastic cells are seen moving upwards and sandwiching a column of enterocytes; note that all three migration pathways are tight and relatively straight. Eventually, the enterocyte migration pathway becomes blocked off by metaplastic cells, while the extreme left-hand column of metaplastic cells in Fig. 4a then comes to an abrupt end, with cells possibly rejoining the main column. At the very tip of the complete villus, a large area is occupied by metaplastic cells at the junction of two migrating metaplastic streams. The computer-aided reconstruction showed basically the same arrangement.

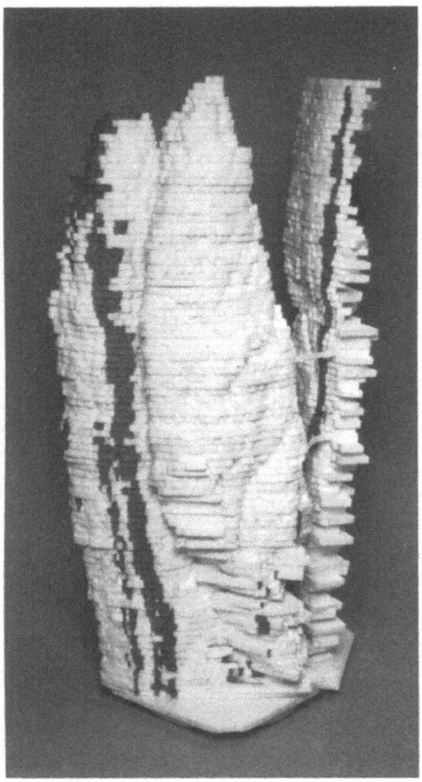

Fig. 4a–c. Serial views of the polystyrene villus model

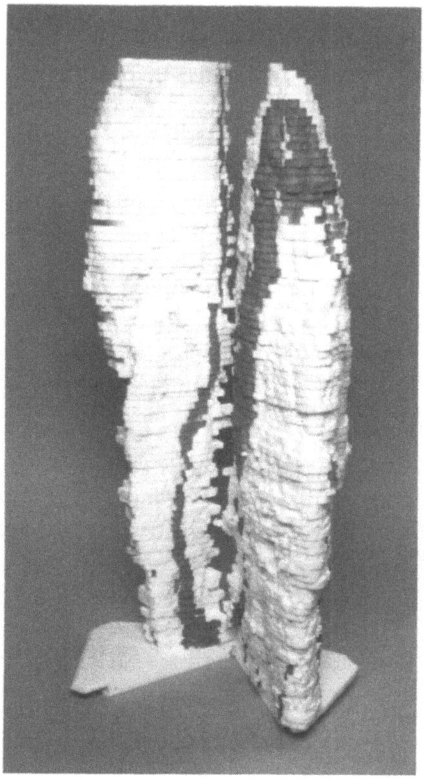

Fig. 4b

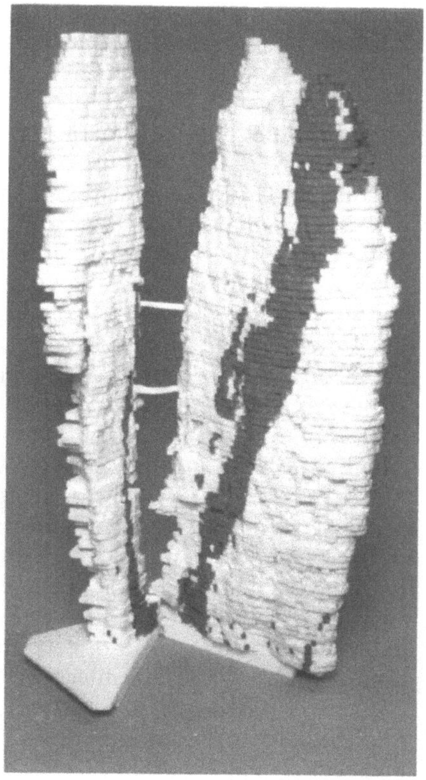

Fig. 4c

When the serial sections were continued below the level of the vestibule, the vestibule was seen to divide into several crypt-like tubules still lined with gastric metaplastic cells (Fig. 3f). Further down, where normal duodenal crypts end in zones of Paneth's cells, these tubules, still lined with cells having the morphology of metaplastic cells, are seen to penetrate the muscular mucosa (Fig. 3f).

Discussion

These observations allow several conclusions to be drawn about migrating cells in the human small intestine. Firstly, that migrating cells from the crypts and vestibules largely remain associated with each other, and that as they move onto the villus, they migrate in relatively tight cohorts. While there is some intermingling of cells on the lower villus, there is certainly no extensive mixing, as suggested by Smith and Peacock [6] and Partridge and Simpson [8]. Thus, as cells migrate from the vestibule towards the villus base, which can be some distance [14], they remain relatively closely associated with each other. This is particularly interesting since studies in both chimaeric mice [14] and mice which are heterozygous for glucose-6-phosphate dehydrogenase [18] show a similar arrangement. It might be expected that cells meeting the migrating progeny of other crypts, particularly at the rapid migration velocity recorded for the small intestine [19], would freely intermingle and any migration pathway would rapidly become effaced. Despite some intermingling in human duodenal villi (Fig. 3), the migration pathway remains very distinct and, of course, there must be some good reason for this: Wilson et al. [14] speculated that the spot desmosomes [20], which act like rivets to hold the epithelial cells together at button-like points of contact, are so strong as to make a very rigid and organised cell sheet, and thus allow little or no mixing. Perhaps in the human case, some of these contacts are less strong: hence minor mixing at the base. However, note the very thin pathway of the right-hand villus in Fig. 4a. Often this is only one or two cells in thickness, but it still maintains its uninterrupted integrity as a defined migration pathway. On the other hand, even on the villus, some cells become isolated as groups or as single cells, so there is indeed some scope for lateral migration.

It is also revealing that migration is in relatively straight lines. There is no evidence for the spiral pathway suggested by Carr et al., [7], neither have we found any evidence of interrupted migration streams or short migration streams that are such a feature of the migration pathway in the chimaeric mouse [11, 12]. It is true that minor columns appear to peter out (Fig. 4a), but here cells may be moving slightly laterally on the villi to rejoin another stream, and there is no need to involve periodic cessation of crypt cell production, as has been suggested in the mouse model, to explain interrupted columns.

However, we might ask what has happened to those cells at the base of the villus not associated with any migration pathway, and indeed the nature of their fate: have they merely become isolated from major metaplastic migration pathways by more pronounced lateral migration and are migrating in the enterocyte escalator instead? Quite possibly, but we should note the proposal of Partridge and Simpson [8] that traffic jams of cells exist at villus bases, in which some cells, unable to get onto the main migration pathway, were held up for protracted periods, perhaps indefinitely.

We have shown that a single vestibule (Fig. 1) can supply cells to more than one villus, as is also the case in the mouse [13, 14], and in the case described in detail here, the migrating progeny of this vestibule divide apparently unevenly, with most cells moving onto the right hand villus, as seen in Fig. 4c. The factors which control this division are of course presently unknown.

The question must be put as to whether the demonstrated migration pattern of these metaplastic cells is also representative of migrating enterocytes. Of course, it may be that metaplastic cells have special desmosomal contacts which link them inexorably together, and that villus enterocytes migrate in quite a different manner. We concede that this is possible but unlikely; the enterocytes in Fig. 4a also migrate in a straight line (albeit bounded by metaplastic cells). Moreover, the pattern shown here is remarkably similar to that described in the mouse [14, 12]. However, we have confined our definitive conclusions to metaplastic cells.

A further question must be about the nature of the motive force for migrating cells on the villi. This is an old problem for which there is still no satisfactory answer. Perhaps at this time we can merely say that so-called mitotic pressure is unlikely to be the cause, since migration continues in the presence of total mitotic inhibition [21], and it is likely that migration is an active process on the part of the villus epithelial cell. In this respect, a largely neglected field is the geometry of cell migration in this villus sheet. Individual cells are probably hexagonal in shape, and the introduction of a specific, defined and recognizable cell type, such as gastric metaplastic epithelium, into an epithelial sheet, is a tissue geometry problem that has received considerable theoretic treatment [21] but little experimental attention in the mammalian case. Opening out the villus epithelium into a notional sheet, and modelling the movement of intruding metaplastic cells using the approaches suggested by Bjerknes [22] is one possible approach.

Finally, our observations cast some light on the fascinating question of the nature of metaplasia in gastrointestinal epithelial cells. There are two common types of epithelial metaplasia – intestinal metaplasia as seen often in the stomach, and the gastric metaplasia seen here. A similar cell lineage which occurs in the stomach, small intestine and colon is often called pyloric or pseudopyloric metaplasia. In the small intestine, pyloric metaplasia is frequently associated with ulceration, and in the duodenum, with chronic active duodenitis and duodenal ulcer disease [16]. However, it is also well described in non-inflamed mucosa. The relatively large area of metaplastic cells on the villus tip is consistent with the observation, frequently made on routine histological material, that gastric metaplasia is most often seen on, or confined to, the villus tips [16, 17] and consequently takes its origin from differentiated villus enterocytes. However, any longitudinal section which includes the large metaplastic area on the tip of the villus (as shown in Fig. 4) and avoids a migration pathway (as most longitudinal sections through this villus would) gives the impression that gastric metaplasia is confined to the tip, which is of course patently not the case. On the other hand, most investigators would ascribe this metaplasia to an induced change in the direction of stem cell differentiation in duodenal crypts, towards a cell lineage which resembles, or in fact is, the mucous cells on the surface of the stomach foveolars. In fact, Slack [23] has suggested that metaplasia is a clonal phenomenon due to activation and transcription of homeotic genes in stem cells. It could be, of course, that duodenal crypt stem cells undergo a metaplastic change and repopulate not only their associated crypts and villi with gastric metaplastic cells, but also the underlying Brunner's glands ducts, which of course enter and are in direct continuity with the crypt bases. We consider this to be unlikely because

a) the tubules, lined with metaplastic epithelium, which communicate with Brunner's glands ducts and are continuous with the metaplastic cells on the villi, do not contain mitotic figures, unlike the nearby crypts (Fig. 3f),
b) the metaplastic cells show distinct light microscopic and electron microscopic resemblance to Brunner's glands ducts epithelium [24, 25], and
c) Rhodes [25] in the histamine-stimulated cat, showed that gastric metaplasia was in direct continuity with Brunner's glands ducts.

Our observations indicate, thus, that it is distinctly possible that so-called gastric metaplasia is nothing more than an overgrowth of Brunner's glands duct epithelium onto the villi. It is well known that Brunner's glands secrete bicarbonate, mucus and epidermal growth factor/urogastrone (EGF/URO). EGF/URO acts to inhibit gastric acid secretion and is also cytoprotective.

However, our three-dimensional studies [26] on the histogenesis of pyloric metaplasia in the small intestine in Crohn's disease show that the tubules take their origin from the crypt base, ramify in the lamina propria to form a small gland, and then reach the surface by growing up the core of an adjacent villus, opening into the surface via a pore. Metaplastic cells migrate through this pore onto the villus surface, clothing the surface where they resemble exactly the gastric metaplastic cells of the duodenum. The appearances in Fig. 3f would also be consistent with this mode of histogenesis.

From a teleological viewpoint, it is possible that gastric metaplasia may represent an attempt by a duodenal mucosa, assaulted by excess gastric acid to redress the balance. A major examination of the phenotype of these metaplastic cells and their resemblance to Brunner's glands duct epithelium on the one hand, and the gastric foveolar-surface mucous cell, on the other, seems warranted, as does an exploration of the relationship between the extent of gastric metaplasia and duodenal EGF/URO content. Whatever its nature, it is clear that following the pathways of migration of this easily defined epithelium is a considerable tool for investigating migration patterns in human intestinal epithelial sheets.

Acknowledgements. The authors wish to thank Mrs. Yvonne Price who prepared the serial sections.

References

1. Quastler H, Sherman FG (1959) Cell population kinetics in intestinal epithelium in the mouse. Exp Cell Res 17:420–438
2. Wynford-Thomas D, Williams ED (1986) Use of bromodeoxyuridine for cell kinetic studies in intact animals. Cell Tissue Kinet 19:179–182
3. Wright NA, Alison MR (1984) The biology of epithelial cell populations. Vol 2. Clarendon, Oxford
4. Potten CS, Chwalinski S, Khokhar MT (1982) Spatial interrelationships in surface epithelia; their significance in proliferation control. In: Pitts JD, Finbow ME (eds) Functional integration of cells in animal tissues, pp 286–231 Cambridge University Press, Cambridge
5. Cocco AE, Dohrmann MJ, Hendrix TR (1986) Reconstruction of normal jejunal biopsies: three-dimensional histology. Gastroenterology 51:24–31
6. Smith JMW, Peacock MA (1980) Anomalous replacement of foetal enterocytes in the neonatal pig. Proc R Soc B 206:411–420
7. Carr KE, Hamlet P, Watt E (1981) Scanning electron microscopy, autolysis and irradiation as techniques in studying small intestinal morphology. J. Microsc 123:161–173
8. Partridge BJ, Simpson LO (1980) Duodenal epithelial cell migration and loss in W2B mice. Micron 11:63–75
9. Bjerknes M, Cheng H (1985) Mucous cells and cell migration in the mouse duodenal epithelium. Anat Rec 212:59–73
10. Mintz B (1971) Allophenic mice of multi-embryo origin. In: Daniel J Jr (ed) Methods in mammalian embryology. Freeman, San Francisco, pp 186–202
11. Ponder BAJ, Wilkinson M (1983) Organ related differences in binding of Dolichos biflorus agglutinin to vascular endothelium. Dev Biol 96:535–547
12. Schmidt GH, Wilkinson MM, Ponder BAJ (1985) Cell migration pathway in the intestinal epithelium: an in situ marker system using mouse aggregation chimaeras. Cell 40:425–429

13. Wilson G, Ponder B, Wright NA (1984) How epithelial cells move in sheets; villus migration patterns on the small intestine. J Anat 138:583
14. Wilson JJG, Ponder BAJ, Wright NA (1985) Use of a mouse chimaeric model to study migration patterns in the small intestinal epithelium. Cell Tissue Kinet 18:333–343
15. Liber AG (1951) Abberant, pyloric glands in regional ileitis. AMA Arch Pathol 51:205–212
16. Whitehead R (1986) Mucosal biopsy of the gastrointestinal tract. 3rd Edn. Saunders, Philadelphia, PA, USA
17. Wyatt J, Rathbone BJ, Heatley RV, Dixon MF (1987) Campylobacter pyloridis and acid-induced gastric metaplasia in the pathogenesis of duodenitis. J Pathol 151:73A
18. Thomas GA, Williams D, Williams ED (1988) The demonstration of tissue clonality by X-linked enzyme histochemistry. J Pathol 155:101–108
19. Lipkin M, Sherlock B (1963) Cell population kinetics in the gastrointestinal tract of man. II. Cell renewal in stomach, ileum, colon and rectum. Gastroenterology 45:721–729
20. Alberts B, Bray D, Lewis J, Raff M, Robert K, Watson JD (1983) Molecular biology of the cell. Garland, New York, p 673
21. Potten CS, Moore JV, Chwalinski S (1983) Cytotoxic effects as exemplified by small intestine. In: Potten CS, Hendry JH (eds) Cytotoxic results to tissue. Churchill Livingstone, Edingburgh, pp 231–279
22. Bjerknes M, Allen MJ, Totafurno J (1989) Cell flows under conditions of changing tissue geometry. J Math Biol (in press)
23. Slack J (1985) Homeotic transformations in man: implications for the mechanisms of embryonic development and for the organisation of epithelia. J Theor Biol 144:463–490
24. James AH (1964) Gastric epithelium in the duodenum. Gut 5:285–294
25. Patrick WJA, Denham D, Forrest APM (1974) Mucous change in the human duodenum: a light and electron microscopical study and correlation with disease and acid secretion. Gut 15:767–776
26. Wright NA, Pike C, Elia G (1990) Mucosal ulceration induces a novel, epidermal growth factor-secreting cell lineage in gastrointestinal stem cells. Nature Vol. 343, p 82–85

Gastric Metaplasia in Duodenal Mucosa – Key Factor for *H. pylori* Colonization and Duodenal Ulcer Pathogenesis?

P. Malfertheiner, A. Stanescu, G. Bode, K. Baczako, Ph. Kühl, and H. Ditschuneit

Surface mucus cells (SMCs) of the gastric epithelium are the target cells for *H. pylori* colonization [1]. In contrast epithelial cells with absorptive function lining the duodenal and intestinal mucosa were never found to be colonized by *H. pylori*. The specific interrelationship (i.e. adhesion) between *H. pylori* and its gastric target cell is responsible for the unique location of *H. pylori* in the stomach (antrum and body area) with the only exception of its ectopic localization on metaplastic gastric SMCs in different parts of the gastrointestinal tract. Outside of the stomach *H. pylori* has been detected on gastric type epithelium in the duodenum [2–4], in the esophagus [5], Meckel's diverticula [6] and the rectum [7].

The postulated contributing role of *H. pylori* in the pathogenesis of duodenal ulcer and its specific affinity to gastric SMCs have focused the interest on gastric metaplasia in the duodenum [8, 9]. The following questions concerning gastric metaplasia in the duodenum will be discussed in this chapter:

1. Is gastric metaplasia a well-defined entity?
2. Is gastric metaplasia in the duodenum a variation from normal or is it abnormal?
3. What mechanisms induce the formation of gastric metaplasia in the duodenum?
4. What is the pathological significance of gastric metaplasia in the duodenum?

Gastric Metaplasia: Definition and Characterization

Gastric type mucosa, microscopically recognized by positive staining with the periodic acid-Schiff reagent (PAS) may be detected as small foci in different regions of the gastrointestinal tract. Gastric metaplasia is most frequently observed in duodenal mucosa with a patchy distribution and preferentially in close proximity to the pyloroduodenal junction [10, 11]. More exceptionally gastric metaplasia is detected in other parts of the gastrointestinal tract including the esophagus, rectum, and gallbladder [5–7], and the same locations have been reported for gastric heterotopias [12, 13]. In conditions other than in association with intestinal diseases a definite answer is not available whether gastric metaplasia is a congenital or acquired condition. The association of gastric metaplasia with pathological conditions will be considered separately.

Universität Ulm, Abteilung Innere Medizin II, Robert-Koch-Str. 8, 7900 Ulm, FRG

Helicobacter pylori, Gastritis and Peptic Ulcer
Edited by P. Malfertheiner and H. Ditschuneit

A clear distinction has to be made between gastric metaplasia that is related to intestinal mucosa containing exclusively SMCs from gastric heterotopia, i.e. tissue containing SMCs as well as parietal and chief cells. Various recent studies have dealt with the issue of heterotopic gastric mucosa, less frequently observed in duodenal mucosa than gastric metaplasia [9, 14, 15]. Gastric heterotopia appears as well-structured gastric tissue and is thought to be congenital in origin [17].

Gastric metaplasia if by definition restricted to gastric SMCs may be distinguished on a cellular level in complete metaplastic cells and transitional metaplastic cells. Complete gastric metaplastic cells appear as prismatic epithelial cells with stunted microvilli and a high density of secretory granules in the apical cellular region. The transitional type cell combines the aspects of SMCs and absorptive type cells (enterocytes) with more pronounced microvilli and few secretory granules (Figs. 1, 2).

Based on histological criteria gastric metaplasia is considered complete if villi are totally affected or incomplete when only parts of the duodenal villi are lined with gastric type SMCs (Figs. 3, 4). Using light microscopy techniques early changes of the normal duodenal mucosa and the beginning transformation into gastric metaplasia are observable with alkaline phosphatase staining. Alkaline phosphatase may get lost before other structural changes become apparent. The distinction between fully developed gastric metaplasia and gastric heterotopia is easily made using conventional histology (Fig. 5).

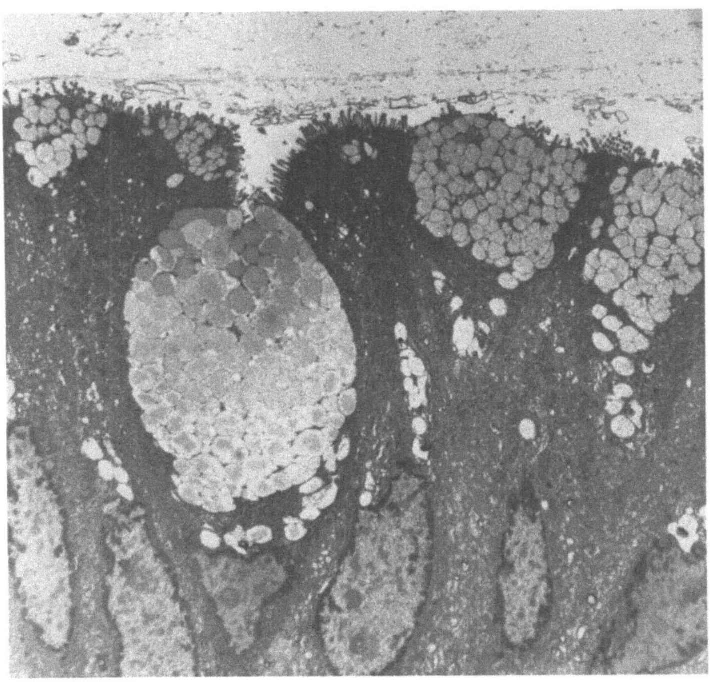

Fig. 1. Complete metaplastic surface mucus cells (SMCs) of the gastric type, next to incomplete SMC and a goblet cell (EM, × 3800)

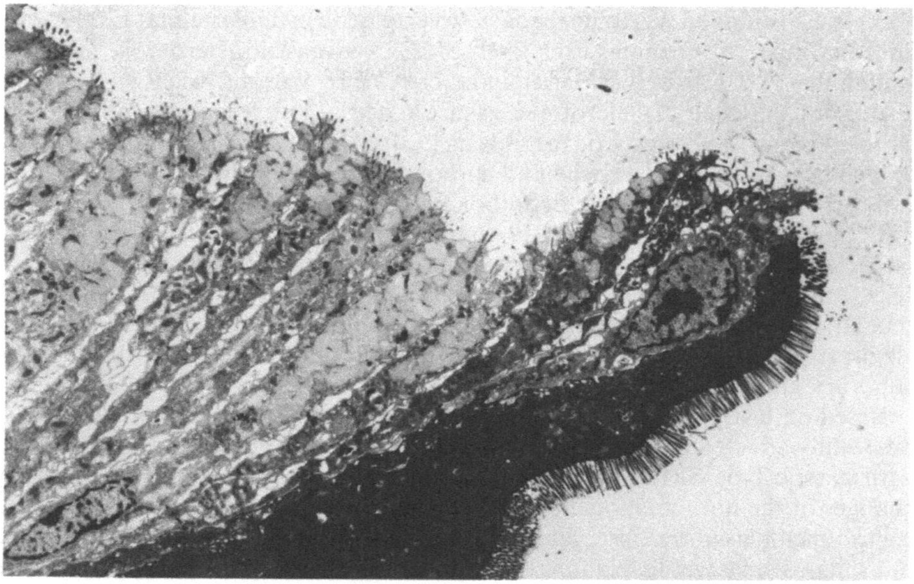

Fig. 2. Gastric metaplasia in continuity with duodenal type mucosa (enterocytes; EM, ×3800)

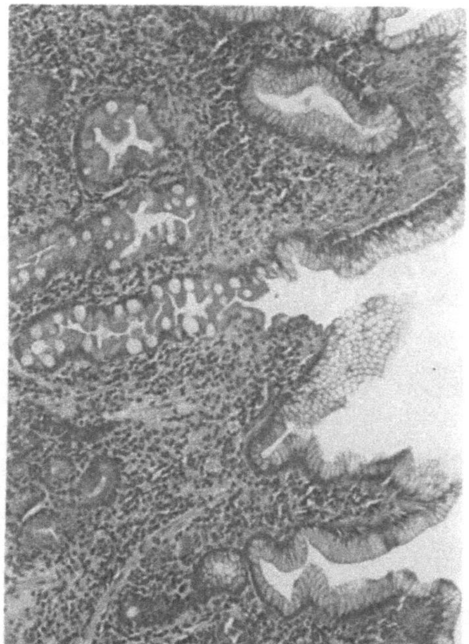

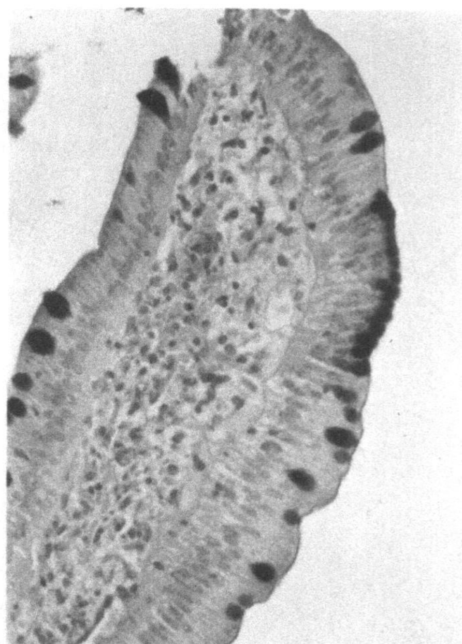

Fig. 3. Complete gastric metaplasia on duodenal villi (Hematoxylin-Eosin, ×20)

Fig. 4. Incomplete gastric metaplasia on a duodenal villus (PAS, ×20)

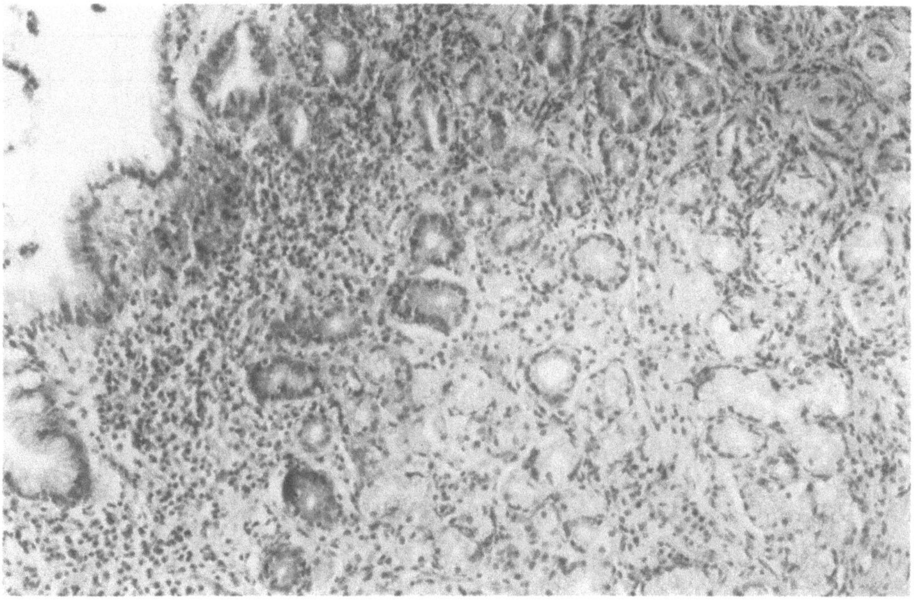

Fig. 5. Gastric heterotopia in duodenal mucosa (Hematoxylin-Eosin, ×20)

Is Gastric Metaplasia in the Duodenum a Normal Variation or Is It Abnormal?

A definite answer to this question is currently not available because of conflicting data reported in the literature. These inconclusive data are influenced by patient selection bias and by sampling of duodenal biopsy specimens mostly obtained from patients with macroscopically abnormal duodenal mucosa. Moreover, an insufficient number of biopsy specimens were taken by most investigators and these may not be sufficient for correctly interpreting a condition that is characterized by its patchy nature. In a carefully performed prospective study Kreuning et al. [10] detected gastric metaplasia in 64% of healthy controls in one out of six biopsy specimens taken from different sites of the proximal duodenum. In the same study, however, only a smaller group of patients was found to have gastric metaplasia in more than one and only few in more than two of the duodenal biopsy specimens [10]. This study confirms the patchy nature of gastric metaplasia in the duodenum and explains discrepancies that may arise due to different sampling sites. To further address this question Lawson [11] contributed by studying gastroduodenal mucosal strips taken from resected specimens during Whipple's procedure. He found that the antroduodenal junction until 1 cm beyond the pylorus was partly lined with gastric SMCs but no further SMCs could be detected beyond this restricted area. Different authors have reported a high prevalence of gastric metaplasia associated with duodenitis and duodenal ulcer

Table 1. Prevalence of gastric metaplasia in the duodenum in health and disease

			GM detected (%)
Controls		Ref.	
Kreuning et al.	1978	[10]	64
Lawson	1988	[11]	0
Wyatt et al.	1987	[18]	16
Duodenitis			
Shousha et al.	1983	[19]	72
Wyatt et al.	1987	[18]	85
Duodenal ulcer			Ulcer border (%)
James	1964	[17]	90
Patrick et al.	1974	[20]	78
Gregory et al.	1982	[21]	100
Steer	1985	[22]	58
Malfertheiner et al.	1985	[23]	100
Marshall et al.	1988	[4]	92
Carrick et al.	1989	[16]	91

GM = gastric metaplasia

but infrequently in healthy controls (Table 1). To contribute to this debated issue we studied biopsy specimens obtained from the mid- and distal portion of the duodenal bulb in 129 patients with normal macroscopic findings of the duodenal mucosa and found gastric metaplasia to be present in 18% of these patients. Contrary to the low prevalence of SMCs in the duodenum of individuals with an intact mucosa, we found gastric metaplasia to be constantly present at the edge of duodenal ulcers [23, 24]. The actual available data show that gastric metaplasia prevails in patients with duodenal ulcer and duodenitis.

The detection of gastric metaplasia in healthy people depends mainly on the duodenal site from which the biopsy specimens are taken. Critical for the pathophysiological role of gastric metaplasia in duodenal ulcer pathogenesis may be the extent of gastric metaplasia in the duodenal bulb. Extended areas of SMCs in the duodenum may offer a greater surface and consequently higher risk for infection with *H. pylori* than small foci of SMCs. James found 16% of the duodenal villi affected by gastric metaplasia in his study on ten patients with duodenal ulcer. More data on the extent of gastric metaplasia in patients with duodenal ulcer are eagerly awaited.

What Mechanisms Induce the Formation of Gastric Metaplasia in the Duodenum?

Gastric metaplasia may be considered an nonspecific response to mucosal injury of any kind, since it is observed in association with duodenal ulcers as well as with intestinal ulcerations during the course of inflammatory bowel diseases [9, 25, 26]. The major aggression to the duodenal mucosa appears to be mediated by the duodenal acid load. Experiments on cats, rats, and pigs have shown that gastric metaplasia can be induced by an increased exposure of the duodenal mucosa to acid [27–29]. Some evidence for the role of gastric acid hypersecretion for the induction of gastric metaplasia is found also in humans. Few patients with Zollinger-Ellison syndrome have been reported to have a marked presence of gastric metaplasia in the duodenum [30, 31]. Patrick et al. [20] and Wyatt et al. [18] suggested from their studies that gastric metaplasia of the duodenal bulb is closely correlated with gastric acid output. In favor of the acid theory concerning the formation of gastric metaplasia in the duodenum are observations that report a reversion of gastric metaplasia following highly selective vagotomy [32]. Short-term treatment with H2-receptor antagonists was not effective in reversing gastric metaplasia [33]. Although acid hypersecretion is known to occur in one- to two-thirds of patients with duodenal ulcer, disease gastric metaplasia in the duodenum is present in 90%–100% of these patients. This points to the importance of hyperacidity in the duodenal bulb as net effect, which may only partly result from acid hypersecretion, but partly from a disturbed bicarbonate secretion of the duodenal mucosa or from defective clearing mechanisms (i.e. motor disorders). Few other factors, such as sex (prevalence of males) and old age, are discussed as factors involved in the genesis of gastric metaplasia, but only few epidemiological data exist to support them [32].

In a prospective study on 129 patients presenting with dyspepsia and macroscopically normal duodenal mucosa, we found *H. pylori* in antral mucosa and the histological activity of antral gastritis not to be related to the occurrence of gastric metaplasia in the duodenum (Fig. 6). This observation makes it rather unlikely that *H. pylori* infection itself leads to the formation of gastric metaplasia in the duodenum.

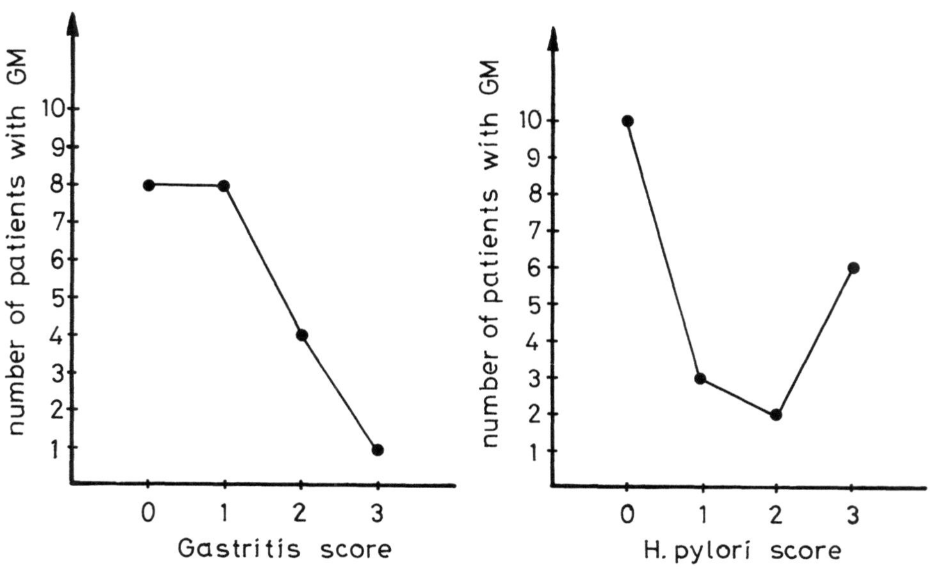

Fig. 6. Relationship between occurrence of gastric metaplasia in duodenum and inflammatory activity and *H. pylori* colonization in antral mucosa in patients with endoscopical normal duodenal mucosa

What Is the Pathological Significance of Gastric Metaplasia in the Duodenum?

According to the broad concept of gastric metaplasia as a nonspecific response to any kind of harmful agent in any part of the entire intestine, gastric metaplasia may be considered as an "emergency" tissue build-up for protection from mucosal disease. More specifically the formation of gastric metaplasia in the duodenum is intended to conteract the increased acid load. In fact SMCs of the gastric epithelium appear most effective in the protection against acid corrosion by producing mucus and by participating in the transport and entrapment of bicarbonate within the mucus [34, 35]. The increased defensive quality of the metaplastic transformed duodenal mucosa becomes obsolete when these metaplastic SMCs become infected by *H. pylori.*

There are considerable data to show that *H. pylori* attaches to duodenal mucosa only at the site of gastric metaplasia (Fig. 7) [4, 18, 24]. It has also been shown that gastric metaplasia in the duodenum becomes infected only in the presence of infected antral mucosa [36]. This suggests that antral infection spreads to the duodenum when the condition of gastric metaplasia preexists. It is rather unlikely that antral colonization with *H. pylori* induces the transformation of the duodenal epithelium, as we did not find a significant correlation between antral infection with *H. pylori* and the occurrence of gastric metaplasia in the duodenum. When gastric metaplasia becomes infected duodenal inflammation results [18, 37] and the inflamed mucosa may be less resistent to aggressive factors. The mucus secreting function of SMCs infected by *H. pylori* appears altered with an increase in sialic acid and a decrease in fucose [38].

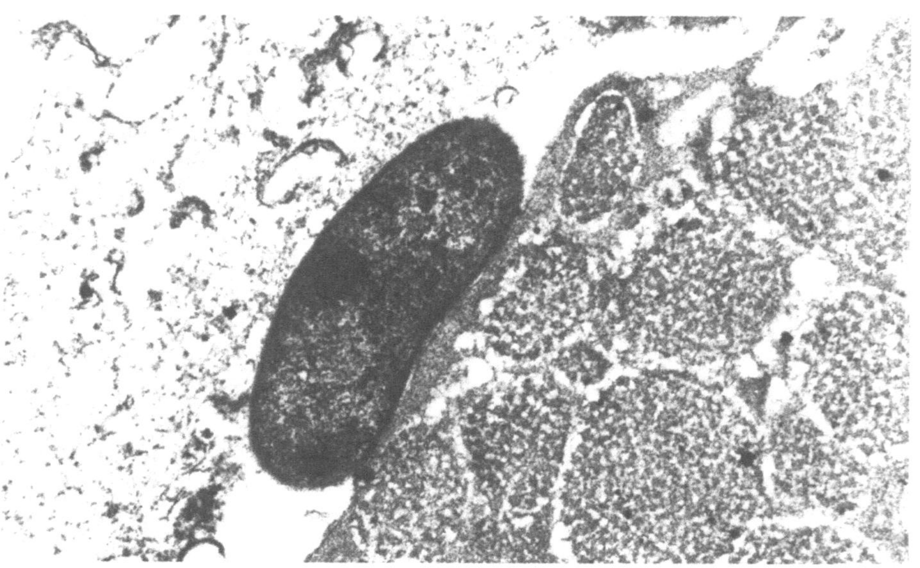

Fig. 7. Adhesion of *H. pylori* on metaplastic gastric SMC in duodenal mucosa

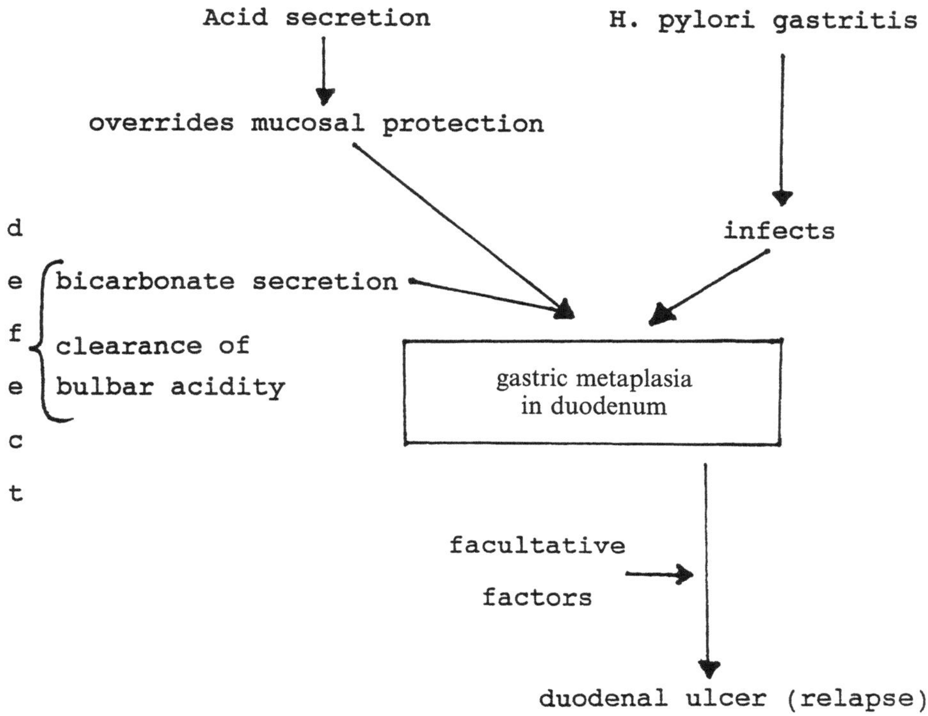

Fig. 8. Flowgram of the hypothesis for the role of gastric metaplasia as a factor contributing to the formation of duodenal ulcers

The hypothesis for the role of gastric metaplasia as a factor contributing to duodenal ulcer formation is summarized in the flowgram (Fig. 8). The persistence of gastric metaplasia following treatment with conventional acid-inhibitory drugs may predispose to frequent duodenal ulcer relapses, whereas the reversibility of gastric metaplasia following proximal gastral vagotomy may explain the low relapse rate of duodenal ulcers following the surgical procedure. More data are needed for confirmation whether gastric metaplaia is an obligatory condition for *H. pylori* infection of the duodenum and whether that is an essential cofactor in duodenal ulcer formation. Future successful therapeutic strategies may try either to reverse the metaplastic tissue or to eradicate *H. pylori* infection, or both.

References

1. Bode G, Malfertheiner P, Ditschuneit H (1988) Pathogenetic implications of ultrastructural findings in Campylobacter pylori related gastroduodenal disease. Scand J Gastroenterol [Suppl 142] 23:25–39
2. Wyatt JI (1989) Relationship of C. pylori to duodenal disease. (in press)

3. Malfertheiner P, Bode G, Vanek E, Stanescu A, Lutz E, Blessing J, Ditschuneit H (1987) Campylobacter pylori – besteht ein Zusammenhang mit der peptischen Ulcuskrankheit? Dtsch Med Wschr 112:493–497
4. Marshall BJ, Goodwin CS, Warren JR, Murray R, Blincow ED, Blackbourn SJ, Phillips M, Waters TE, Sanderson CR (1988) Prospective double-blind trial of duodenal ulcer relapse after eradication of Campylobacter pylori. Lancet I:1437–1442
5. Hazell SL, Carrick J, Lee A (1988) Campylobacter pylori can infect the esophagus when gastric tissue is present (abstract). Gastroenterology 94:178
6. DeCothi GA, Newbold KM, O'Connor HJ (1989) Campylobacter-like organisms and heterotopic gastric mucosa in Meckel's diverticula. J Clin Pathol 42:132–134
7. Pambianco DJ, Dye KR, Marshall BJ et al. (1988) Gastritis in the rectum: campylobacter-like organisms in heterotopic inflamed gastric mucosa. Gastroenterology 94:A 340
8. Goodwin CS (1988) Duodenal ulcer, Campylobacter pylori, and the "leaking roof" concept. Lancet 1467–1469
9. Graham DY (1989) Campylobacter pylori and peptic ulcer disease. Gastroenterology 96:615–625
10. Kreuning J, Bosman FT, Kuiper G, Wall AM, Lindeman L (1978) Gastric and duodenal mucosa in "healthy" individuals. J Clin Pathol 31:69–77
11. Lawson HH (1988) Definition of gastroduodenal junction in healthy subjects. J Clin Pathol 41:393–396
12. Wolff M (1971) Heterotopic gastric epithelium in the rectum. A report of three new cases with a review of 87 cases of gastric heterotopia in the alimentary canal. Am J Clin Pathol 55:604–616
13. Willis RA (1968) Some unusual heterotopias. Br Med J 3:267–272
14. Lessells AM, Martin DF (1982) Heterotopic gastric mucosa in the duodenum. J Clin Pathol 35:591–595
15. Ikeda K, Sannohe Y, Murayama H et al. (1982) Heterotopic gastric mucosa in the duodenum – reaction to congo red under fiberscopic observation. Endoscopy 14:168
16. Carrick J, Lee A, Hazell S, Ralston M, Daskalopoulos G (1989) Campylobacter pylori, duodenal ulcer, and gastric metaplasia: possible role of functional heterotopic tissue in ulcerogenesis. Gut 30:790–797
17. James AG (1964) Gastric epithelium in the duodenum. Gut 5:285–294
18. Wyatt JI, Rathbone BJ, Dixon MF, Heatley RV (1987) Campylobacter pyloridis and acid induced gastric metaplasia in the pathogenesis of duodenitis. J Clin Pathol 40:841–848
19. Shousha S, Parkins RA, Bull TB (1983) Chronic duodenitis with gastric metaplasia: electron microscopic study including comparison with normal. Histopathology 7:873
20. Patrick WJA, Denham D, Forrest PM (1974) Mucous changes in the human duodenum: a light and electron microscopic study and correlation with disease and gastric acid secretion. Gut 15:767–776
21. Gregory MA, Moshal MG, Spitaels JM (1982) Changes in the morphology of villar epithelial cells adjacent to duodenal ulcers during the process of healing. Scand J Gastroenterol 17:441–448
22. Steer HW (1984) Surface morphology of the gastroduodenal mucosa in duodenal ulceration. Gut 25:1203–1210
23. Malfertheiner P, Bode G, Mader U, Baczako K, Stanescu A, Ditschuneit H (1985) Histologische und ultrastrukturelle Befunde in der Abheilungsphase des Ulcus duodeni. Klin Wochenschr 63:1061–1070
24. Malfertheiner P, Bode G, Stanescu A, Ditschuneit H (1989) Gastric metaplasia and Campylobacter pylori in duodenal ulcer disease: an ultrastructural analysis. Gastroenterol Clin Biol 13:71 B–74 B
25. Antonius JI, Gump FE, Lattes R, Lepore M (1960) A study of certain microscopic features in regional enteritis, and their possible prognostic significance. Gastroenterology 38:889–905
26. Nawaz K, Graham DY, Fechner RE, Eiband JM (1974) Gastric heterotopia in the ileum with ulceration and chronic bleeding. Gastroenterology 66:113–117
27. Rhodes J (1964) Experimental production of gastric epithelium in the duodenum. Gut 5:454–458

28. Gaskin RJ, Gad A, Barros AAJ, Jeffe SN, Baron JM (1975) Natural history and morphology of secretagogue-induced ulcers in rats. Gastroenterology 69:903
29. Florey HW, Jennings MA, Jennings DA, O'Connor RC (1939) The reactions of the intestine of the pig to gastric juice. J Pathol Bacteriol 49:105–123
30. James AH (1962) Gastric epithelium in the duodenum of a patient with gastric hyperacidity. II World Congress of Gastroenterology, Munich, 11:540–543
31. Parrish JA, Rawlings DC (1965) Intestinal mucosa in the Zollinger-Ellison syndrome. Gut 6:286–289
32. Wyatt JI, Rathbone BJ (1989) Gastric metaplasia in the duodenum and Campylobacter pylori. Gastroenterol Clin Biol 13:78 B–82 B
33. Malfertheiner P, Bode G, Stanescu A, Ditschuneit H (1989) Gastric metaplasia in the duodenum – a persistent condition following ulcer healing. Gastroduodenal Disease and Campylobacter pylori. Charlottesville, Virginia 21 (Abstract)
34. Isenberg JI, Selling JA, Hogan DL, Koss MA (1987) Impaired duodenal mucosal bicarbonate secretion in duodenal ulcer patients. N Engl J Med 316:374–379
35. Flemström G, Turnberg LA (1984) Gastroduodenal defence mechanisms. Clin Gastroenterol 13:327–354
36. Burette A, De Prez C, Glupzynski Y, Nyst JF, Deltenre M (1989) Prevalence of gastric metaplasia and Campylobacter pylori infection in the duodenum and relationship to infection. In: Megraud F, Lamouliatte H (eds) Gastroduodenal pathology and Campylobacter pylori. Elsevier, New York, pp 327–331
37. Johnston BJ, Reed PI, Ali MH (1988) Prevalence of Campylobacter pylori in duodenal and gastric mucosa – relationship to inflammation. Scand J Gastroenterol [Suppl 142] 23:69–75
38. Bode G, Malfertheiner P, Kullmann C, Ditschuneit H (1989) Gastritis and duodenal ulcer: lectin-binding characteristics of Campylobacter pylori infected and noninfected surface mucous cells. In: Megraud F, Lamouliatte H (eds) Gastroduodenal pathology and Campylobacter pylori. Elsevier, New York, pp 353–359

The Relationship Between Gastric Metaplasia and Inflammation in the Duodenum

J. I. Wyatt

Gastric metaplasia (GM) is a response of intestinal mucosa to injury, and occurs at various sites along the gastrointestinal tract in response to mucosal ulceration. Examples of this have already been discussed (see p. 279, 292).

In this contribution, the evidence that the mucosal damage occurring in the context of inflammation in the duodenum may itself be a factor in leading to further development of GM is reviewed. While this would enhance healing at other sites, perhaps by the production of growth factors, in the patient with duodenitis and *H. pylori* the expansion of GM in the duodenum would increase the area available for colonisation by *H. pylori*.

In animals, GM develops at the site of healing duodenal mucosa, whether the damage has been caused by physical [1] or chemical [2] injury, or by excess acid [3–5]. In cats treated with subcutaneous histamine to stimulate acid secretion, those with more severe mucosal damage, resulting in ulceration, showed more extensive GM than those treated in the same way which did not develop ulceration [5]. Tatsuta et al., studying chemical (NaOH) injury in rats [2], found that GM occurred in the context of healing of mucosal damage in the presence of acid.

GM occurs in the duodenum in otherwise normal subjects. We have found it in 22% of patients without duodenitis if one biopsy is examined [6]. Sampling error is clearly important in such a study, and Kreuning et al., examining multiple biopsies, stated the frequency of GM in asymptommatic subjects to be 64%, infrequently associated with inflammation [7]. However, in the absence of inflammation it is unusual to find more than isolated foci of GM, typically at the tips of occasional villi.

A number of authors have described a quantitative relationship between the presence of inflammation and the extent of GM [8, 9]. Shousha et al. [8], in a study of 120 patients with non-ulcer dyspepsia, stated that the extent of GM was directly related to the severity of inflammation. The detailed morphometric study by Jenkins et al. [9] examined the correlation between different histological parameters in the duodenal mucosa. Highly significant correlations were found between GM and neutrophil infiltration, and to a lesser extent with villous blunting. No association was found, however, between chronic inflammatory cell infiltration and the presence of GM. Some of our own experience relating the extent of GM to duodenal inflammation is summarised in Fig. 1. There is a

Pathology Department, St. James's University Hospital, Leeds LS9 7TF, UK

Helicobacter pylori, Gastritis and Peptic Ulcer
Edited by P. Malfertheiner and H. Ditschuneit

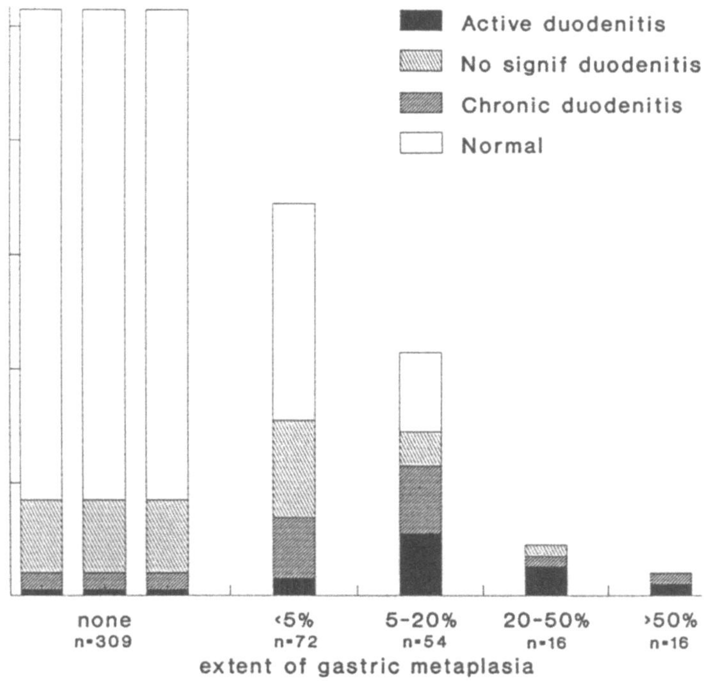

Fig. 1. Relation between the likelihood of duodenal inflammation and the extent of GM in duodenal biopsies from 467 patients.

significant association between the degree of GM and the presence of active duodenitis, although the similar trend for chronic inflammation did not reach significance.

We have found GM in 31% of single duodenal biopsies, and shown that the frequency of finding GM is similar in *H. pylori* positive and *H. pylori* negative subjects (h). However, the extent of GM in the *H. pylori* positive subjects was significantly greater than in the *H. pylori* negative ones. This difference is largely accounted for by the subgroup of *H. pylori* positive persons with duodenal inflammation (Fig. 2), which is consistent with the inflammation in this group being a stimulus to metaplastic change.

The most consistant site for finding GM in the duodenum is at the margin of a duodenal ulcer [10]. This suggests that ulceration develops within (or at the edge of) an area of GM, and that ulceration occurs when there is a breakdown in the integrity of the mucosa in an area of metaplasia. GM forms during healing of duodenal ulceration [11, 12] and in addition GM tends to persist at the site of the ulcer after healing [13], which may account for the tendency for recurrent ulcers to form at the site of previous ulceration.

The above observations may be explained by a positive feedback mechanism whereby *H. pylori* infection of metaplastic foci furthers the duodenal mucosal

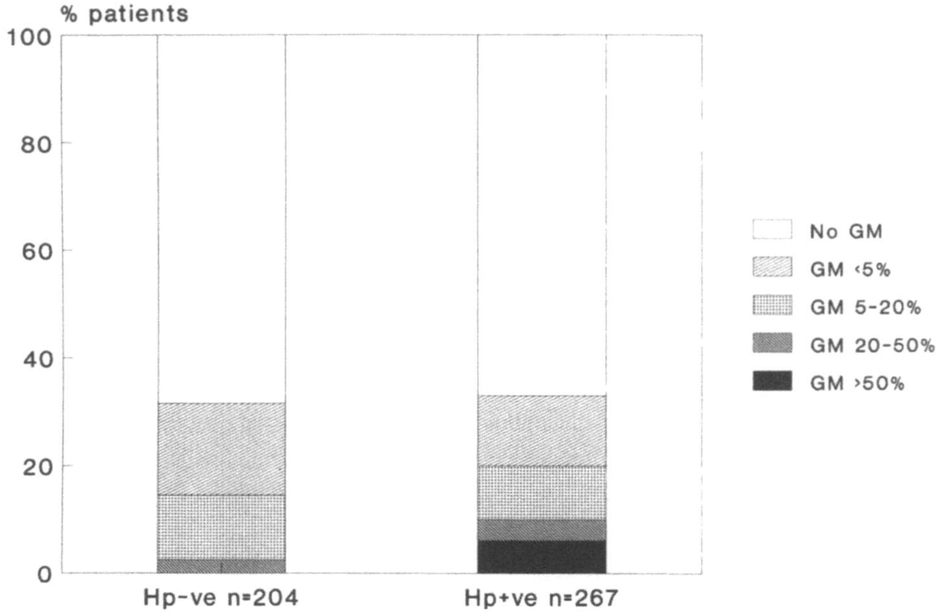

Fig. 2. The extent of GM in the duodenum in patients according to the absence or presence of *H. pylori* associated gastritis ($p = 0.002$ for *H. pylori* negative vs. *H. pylori* positive-patients)

damage, which extends the area of metaplasia and thus increases susceptibility to wider *H. pylori* colonisation, and increasing severity of duodenitis. This effect would tend to weight the balance of aggressive and protective factors at the mucosal surface towards developing ulceration.

If this mechanism is operating in vivo, then eradication of *H. pylori* would result in healing of the duodenitis followed by a reduction in the extent of GM. Follow-up studies of biopsies from the duodenum will be important to determine whether eradication of *H. pylori* is accompanied by healing of duodenitis and resolution of GM. Animal models suggest that gastric metaplasia is reversible and resolves after healing of the mucosa [1, 2].

References

1. Florey HW, Harding HE (1935) The healing of artificial defects of the duodenal mucosa. J Pathol 40:211–218
2. Tatsuta M, Iishi H, Yamamura H, Yamamoto R, Taniguchi H (1989) Enhancement by tetragastrin of experimental induction of gastric epithelium in the duodenum. Gut 30:311–315
3. Florey HW, Jennings MA, Jennings DA et al. (1939) The reactions of the intestine of the pig to gastric juice. J Path Bact 49:105–123
4. Gaskin RJ, Gad A, Barros AAJ et al. (1975) Natural history and morphology of secretagogue-induced ulcers in rats. Gastroenterology 69:903–910

5. Rhodes J (1964) Experimental production of gastric epithelium in the duodenum. Gut 5:454–458
6. Wyatt JI, Rathbone BJ, Dixon MF, Heatley RV (1987) Campylobacter pyloridis and acid induced gastric metaplasia in the pathogenesis of duodenitis. J Clin Pathol 40:841–848
7. Kreuning J, Bosman FT, Kuiper G et al. (1978) Gastric and duodenal mucosa in "healthy" individuals. An endoscopic and histopathological study of 50 volunteers. J Clin Pathol 31:69–77
8. Shousha S, Spiller RC, Parkins RA (1983) The endoscopically abnormal duodenum in patients with dyspepsia: biopsy findings in 60 cases. Histopathology 7:23–34
9. Jenkins D, Goodall A, Gillett FR et al. (1985) Defining duodenitis: quantitative histological study of mucosal responses and their correlations. J Clin Pathol 38:1119–1126
10. Marshall BJ, Warren JR, Blincow ED et al. (1988) Prospective double-blind trial of duodenal ulcer relapse after eradication of Campylobacter pylori. Lancet II:1437–1441
11. Zoli C, Bonvicini F, Ercoli C, Gasbarrini G, Laschi R (1984) Ultrastructural aspects of duodenal mucosa repair during treatment of peptic disease. Acta Physiol Hungarica 64:385–392
12. Malfertheiner P, Bode G, Mader K (1985) Histologische und ultrastrukturelle Befunde in der Abheilungsphase des Ulcus duodeni. Klin Wochenschr 63:1061–1070
13. Fullman H, Van Deventer G, Schneidman D et al. (1985) "Healed" duodenal ulcers are histologically ill. Gastroenterology 88:1390A

Gastroduodenal Motility in Relation to Peptic Ulcer Pathogenesis and *Helicobacter pylori* Infection

O. PIERAMICO and P. MALFERTHEINER

Multiple factors and in varying degrees are involved in the etiology of peptic ulcer disease. Gastric acid secretion, mucosal defense mechanisms, gut hormones (i.e., hypergastrinemia), and motor disorders of the upper gastrointestinal tract have been associated with the pathogenesis of peptic ulcer. The relevance of motility disorders, however, as well as their cause-effect relationship in the pathogenesis of ulcers are still unclear.

Gastric and duodenal ulcers are considered to be two separate clinical entities because of the differing relative influence of etiologic factors and pathogenic mechanisms. For this reason interdigestive and postprandial motility will be considered separately in the case of gastric and duodenal ulcers.

Gastric Ulcer

Increased duodenogastric reflux [1], fasting antral hypomotility [2], and a diminished frequency of phase III activity in the stomach [1] are motor disorders reported in the fasting state in association with gastric ulcer (GU; Table 1). The clinical significance of the altered interdigestive motor patterns is unclear. Phase III is recognized as exercising a "housekeeper" function in the GI tract [3, 4] and its frequent absence in patients with GU may be responsible for the prolonged contact of acid secretions and biliary reflux constituents with the gastric mucosa. The finding that normal interdigestive motility is restored in some patients after ulcer healing could support the hypothesis of a motor disorder secondary to the ulcer [2] but no studies are available to clarify this crucial point.

Gastric emptying of solid and liquid meals seems to be different in GU. While emptying of liquids is not modified, emptying of solids is delayed [5]. Since antral motility is important in the fragmentation and successive emptying of solid

Table 1. Gastrointestinal motility in gastric ulcer disease

• Fasting antral hypomotility	Garrett et al. [2]
• Less recurrence of phase III in the stomach	Miranda et al. [1]
• Increased duodenogastric reflux	Miranda et al. [1]
• Delayed gastric emtying of solids	Miller et al. [5]

Dept. of Internal Medicine and Gastroenterology, University of Ulm, Robert-Koch-Str. 8, D-7900 Ulm, FRG

Helicobacter pylori, Gastritis and Peptic Ulcer
Edited by P. Malfertheiner and H. Ditschuneit

particles [6], a delayed emptying of solids correlates well with the reported antral postprandial hypomotility [22]. In summary, GU is associated with prolonged contact with the gastric mucosa of either secretions in the fasting state or digested materials in the postprandial state.

Duodenal Ulcer

Interdigestive Motility

The occurrence of motor and secretory cycles are not modified during the interdigestive state in duodenal ulcer (DU) patients. Only in patients with gastric acid hypersecretion the duration of the migrating myoelectrical complex (MMC) is prolonged because of a prolonged phase II [7, 8] (Table 2). The cause for the prolonged phase II seems to be the modified secretory status in these patients since i.v. bolus of H2 blocker restores a normal MMC duration [7]. An impaired orad transport of bicarbonate in the duodenal bulb could result from the low frequency of retrograde motor events in the duodenum and may explain the insufficient increase of duodenal pH during each phase III activity [9]. Furthermore, reduced duodenal motility during fasting leads to an impaired clearance of HCl from the bulbus [10].

Table 2. Gastrointestinal motility in duodenal ulcer disease

Interdigestive motility	
• Prolonged MMC duration	Itoh et al. [8]
• Prolonged MMC duration in DU with gastric acid hypersecretion	Bortolotti [7]
• Low frequency of retrograde motor events in duodenum	Borgstrom et al. [9]
• Low fasting antral motility	Johnson [10]
Postprandial motility	
• Increased duodenal acid load	Malagelada et al. [16] Lam et al. [17]
• Shorter lag phase for emptying of solids and faster emptying of liquids	Maddern et al. [11]
• No difference between DU and controls during and after healing	Holt et al. [24]
• Faster emptying of liquids in healed DU	Williams et al. [13]
• Normal gastric emptying in healed DU	Moore et al. [12]

Postprandial Motility

The results of different studies reporting gastric emptying in DU are conflicting [11–14]. This is a consequence of the different techniques employed and also the different compositions of test meals used in these studies.

It has been a commonly held belief that gastric emptying is shorter in DU and that it provokes premature contact of the digested substances with the duodenal mucosa. However, studies supporting this view have often ignored the dilutional effect produced by gastric secretion and have not measured the postprandial delivery of acid into the duodenum [15, 16].

Malagelada et al. [16], using a perfusion technique with a nonadsorbable marker, found an abnormally prolonged secretory response to food with concomitantly high rates of acid delivery into the duodenum in patients with DU. Since total gastric emptying of *food* was not impaired, it is unlikely that DU is the result of a primary gastric motor disorder but it was speculated to be a failure of the duodenal receptors regulating secretion and emptying of *acid* via a feedback mechanism. To assess this hypothesis, Lam et al. [17] examined the sensitivity of duodenal pH receptors by giving liquid meals at different pH and measuring concurrent gastric emptying using a dilution method. Both the duodenal acid load and gastric emptying were found to be increased in DU compared with controls. This phenomenon was independent from the pH of test meals.

The discrimination between emptying of solids and liquids has become possible by the introduction of scintigraphic methods. Using different markers (radioisotopes) specific for liquids and solids, Maddern et al. [11] have quantitated emptying in DU. The total emptying of solids was not found to be changed in DU but altered in its dynamic pattern. DU patients showed a premature beginning of emptying as demonstrated by a shorter lag phase. The exponential emptying of liquids is not quantitatively different from controls but faster within the first 30 min. From these results we learn that the gastric emptying pattern is particularly modified during the first postprandial hour, which supports the hypothesis of a disturbed sensitivity of duodenal receptors.

Although the nature of motor disorders observed with DU seems to be defined, the role of motility disorders in the pathogenesis of DU is still not clear. Are the motor disturbances a primary disorder concomitant with the formation of DU or are they a consequence of the preexisting morphological lesions?

Attempts to answer this question have been undertaken by recording motility during active ulceration and after healing, but results have been contradictory [13, 14]. A more rapid emptying of liquids only is reported in healed DU, indicating a primary motor disorder or a long-lasting motor disorder as consequence of the ulceration.

Helicobacter pylori Infection and Gastroduodenal Motility

Little is known about the influence of *Helicobacter pylori* on motility of the upper gastrointestinal tract. The discovery of a bacterial etiology of chronic type B gastritis is recent [18]. The association of DU with chronic *H. pylori* positive antrum gastritis in more than 90% of patients [19], has led to the hypothesis that *H. pylori* may be involved in the pathogenesis of DU but the relationships among *H. pylori* infection, peptic ulcer disease, and gastroduodenal motility remain far from being understood. To evaluate the relationship between antral inflammation and motility we studied 16 patients with non-ulcer dyspepsia (NUD) and chronic antrum gastritis, divided into two groups according to their *H. pylori* status. In accordance with other studies [21, 22], we found no change in the duration of the MMC and of the individual interdigestive phases in these patients. After food ingestion, 8 of the 16 NUD patients showed a decreased postprandial motor index in the antrum. Despite a more severe grade of inflammatory activity of the antral mucosa in *H. pylori* positive patients, the decrease of the antral postprandial motility was similar between *H. pylori* positive and *H. pylori* negative patients with chronic gastrini (Fig. 1).

The motor abnormalities described in chronic gastritis are not restricted to the inflamed areas (i.r. stomach) but could affect the entire small intestine [23] (Table 3).

The association between chronic gastritis and impaired motility could support a primary role of motility in peptic ulcer disease (Fig. 2). A deranged gastrointes-

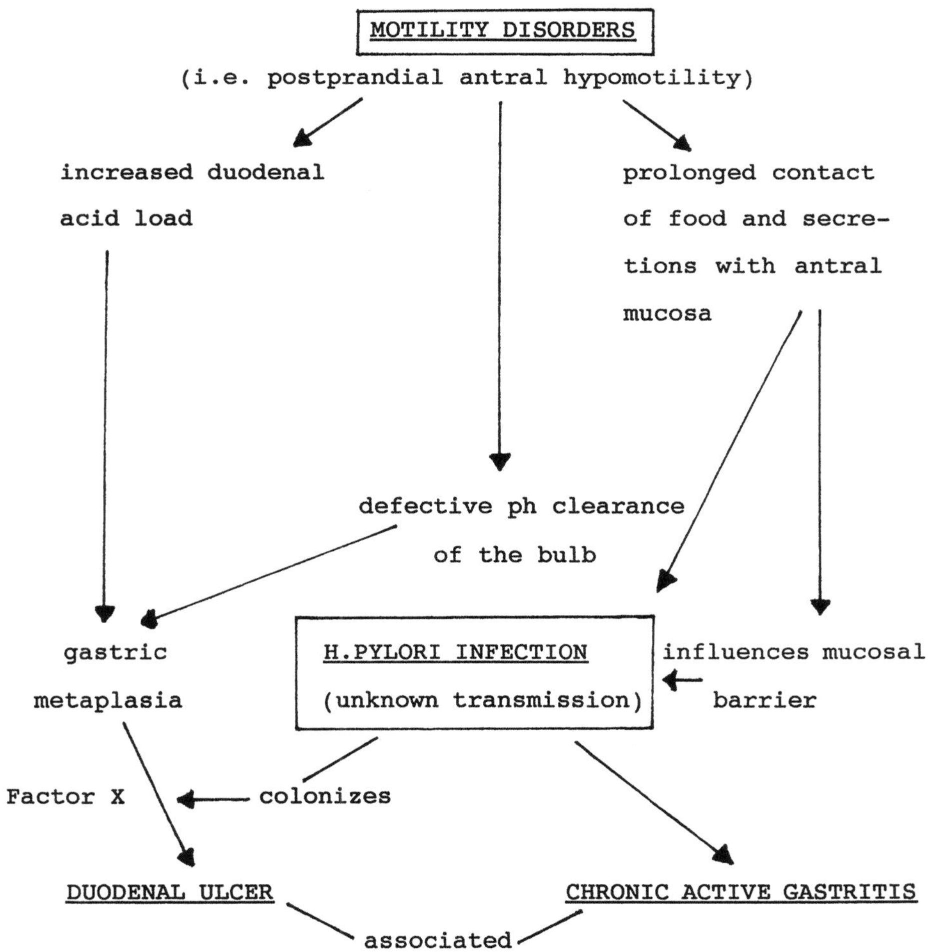

Fig. 1. Hypothesis for Pathogenesis of duodenal ulcer

Table 3. Gastrointestinal motility in NUD and chronic gastritis

• Normal recurrence and duration of the MMC	Moore et al. [22] Malagelada and Stanghellini [20] Pieramico et al. [21]
• Postprandial antral hypomotility in NUD	Malagelada and Stanghellini [20]
• Postprandial antral hypomotility in chronic gastritis	Moore et al. [22]
• Postprandial antral hypomotility in chronic type B gastritis; no difference between Cp+ and Cp−	Pieramico et al. [21]
• Delayed gastric emptying in NUD; no difference between Cp+ and Cp−	Wegener et al. [25]
• Normal gastric emptying in type B chronic gastritis Cp+	Prakash et al. [26]
• Delayed orocecal transit time in chronic type B gastritis; no difference between Cp+ and Cp−	Wilberg et al. [23]

tinal motility (i.e., antral postprandial hypomotility), by inducing a prolonged contact of food and secretions with antral mucosa, could contribute to the chronic inflamation of the antral mucosa and predispose it to *H. pylori* colonization. On the other hand, gastrointestin motor disorders could favor the induction of gastric metaplasia in the duodenal mucosa through an increased duodenal acid load and a decreased acid clearance of the bulb. This hypothesis, however, is still in search of further corroborating data.

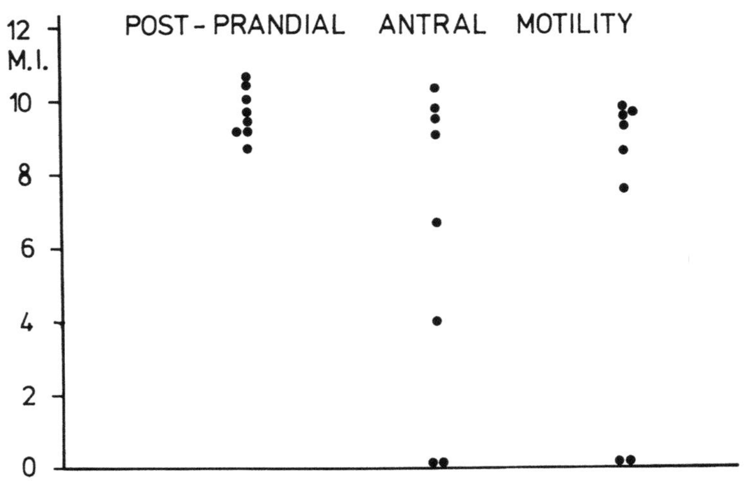

Fig. 2. Postprandial antral motility in sixteen patients with NUD and chronic gastritis and in eight control subjects.

References

1. Miranda M, Defilippi C, Valenzuela JE (1985) Abnormalities of interdigestive motility complex and increased duodenogastric reflux in gastric ulcer patients. Dig Dis Sci 30:16–21
2. Garret JM, Summerskill WHJ, Code CF (1966) Antral motility in patients with gastric ulcer. Am J Dig Dis 11:780–789
3. Sarna S (1985) Cyclic motor activity; migrating motor complex. Gastroenterology 89:894–913
4. Code CF, Marlett JA (1975) The interdigestive myoelectric complex of the stomach and small bowel of dogs. J Physiol (Lond) 246:289–309
5. Miller CJ, Malagelada JR, Longstreth GF, Go VLW (1980) Dysfunctions of the stomach with gastric ulceration. Dig Dis Sci 25:857–864
6. Minami H, McCallum RW (1984) The physiology and pathophysiology of gastric emptying in humans. Gastroenterology 86:1592–1610
7. Bortolotti M (1989) Interdigestive gastroduodenal motility in duodenal ulcer: role of gastric acid hypersecretion. Am J Gastroenterol 84:17–21
8. Itoh Z, Sekiguchi T (1983) Interdigestive motor activity in health and disease. Scand J Gastroenterol [Suppl] 121–134, Vol. 18, Suppl. 87
9. Borgstrom S, Arborelius M (1978) Duodenal motility pattern in duodenal ulcer disease. Scand J Gastroenterol 13:349–352
10. Johnson AG (1979) Peptic ulcer and the pylorus. Lancet 1:710–712
11. Maddern GJ, Horowitz M, Hetzel DJ, Jamieson GG (1985) Altered solid and liquid gastric emptying in patients with duodenal ulcer disease. Gut 26:689–693
12. Moore JG, McIntyre B, Alazraki N (1985) Gastric emptying of combined liquid-solid meals in healed duodenal ulcer. Dig Dis Sci 30:1134–1138
13. Williams NS, Elashoff J, Meyer JH (1986) Gastric emptying of liquids in normal subjects and patients with healed duodenal ulcer disease. Dig Dis Sci 31:943–952
14. Griffith EH, Owen GM, Campbell H (1968) Gastric emptying in health and in gastroduodenal disease. Gastroenterology 54:1–7
15. Cobb JS, Bank S, Marks IN (1971) Gastric emptying after vagotomy and pyloroplasty. Am J Dig Dis 16:207–215
16. Malagelada JR, Longstreth GF, Deering TB, Summerskill WHJ, Go VLW (1977) Gastric secretion and emptying after ordinary meals in duodenal ulcer. Gastroenterology 73:989–994
17. Lam SK, Istenberg JI, Grossman MI, Lane WH, Hogan DL (1982) Rapid gastric emptying in duodenal ulcer patients. Dig Dis Sci 27:598–604
18. Warren JR, Marshall BJ (1983) Unidentified curved bacilli on gastric epithelium in active chronic gastritis. Lancet 1:1273–1275
19. Blaser MJ (1985) Gastritis Campylobacter-like organisms, gastritis and peptic ulcer disease. Gastroenterology 93:371–383
20. Malagelada JR, Stanghellini V (1985) Manometric diagnosis of the motor disturbances of the upper gastrointestinal tract. Gastroenterology 88:1225–1231
21. Pieramico O, Wilberg S, Glasbrenner B, Malfertheiner P (1989) Antroduodenal motility in Campylobacter pylori associated chronic type B gastritis. Klin Wochenschr [Suppl 18] 67:54
22. Moore SC, Malagelada JR, Shorter RG, Zinsmeister AR (1986) Interrelationship among gastric mucosal morphology, secretion, and motility in peptic ulcer disease. Dig Dis Sci 31:673–684
23. Wilberg S, Pieramico O, Malfertheiner P (1989) Role of antral inflammation on orocecal transit time in patients with non-ulcer dyspepsia. Klin Wochenschr [Suppl 18] 67:72
24. Holt S, Heading RC, Taylor TV, Forrest JA, Tothill P (1986) Is gastric emptying abnormal in duodenal ulcer? Dig Dis Sci 31:685–692
25. Wegener M, Borsch G, Schaffstein J, Schulz-Flache C, Mai U, Leverkus F (1988) Are dyspeptic symptoms in patients with Campylobacter pylori-associated Type B gastritis linked to delayed gastric emptying? Am J Gastroenterol 83:737–740
26. Prakash C, Marshall BJ, Plankey MW, Guerrant R, McCallum RW (1987) Gastric emptying of solids in patients with Campylobacter pylori gastritis. Am J Gastroenterol 32:935

Relationship Between *Helicobacter pylori* and Gastroduodenal Physiology: Acid Secretion and Pepsinogens

K. J. HENGELS

It is now accepted that *Helicobacter pylori* (*H. pylori*) causes type B gastritis. Apart from its putative role as a cofactor in ulcerogenesis, *H. pylori* infection seems to be a predictive factor for ulcer recurrence [1]. Although *H. pylori* is known to alter gastric and duodenal morphology and function, the pathogenetic mechanisms behind the observed disturbances are only partly elucidated.

Two of the key factors in ulcerogenesis that might be affected by *H. pylori* infection are gastric acid and pepsin secretion.

Relationship Between Gastric Acid and Pepsin Secretion

The secretion of gastric acid and pepsin is determined by genetic, endogenous and environmental factors such as parietal and chief cell mass, stimulation by neurocrine, paracrine, and endocrine mechanisms, dietary habits, and pharmacological agents, etc. In a wide variety of situations parietal and chief cell functions are closely linked. On the other hand, significant disturbances in mucosal integrity will usually affect both systems. For instance atrophic gastritis will reduce both parietal and chief cells to nearly the same extent [2].

Serum Pepsinogens

With the introduction of the serum pepsinogen determination test [3, 4], new insights into dynamic changes in the gastroduodenal mucosa became available [5].

Pepsinogens are the main digestive proenzymes secreted by the stomach. Immunologically, two groups of pepsinogens can be distinguished [6]; pepsinogen A (Pg A), which is only secreted by cells of the gastric body, and pepsinogen C (PG C), which is additionally produced by the cardiac and pyloric glands in the stomach and by Brunner's glands in the proximal duodenum. Upon secretion pepsinogens are acid-converted in the stomach to active pepsins. A minute portion of the secreted pepsinogens enters the circulation where it can be detected by sensitive methods such as radioimmunoassays. Serum pepsinogens are cleared by the kidney. Individual pepsinogen serum levels remain constant. A significant

Department of Gastroenterology, Heinrich-Heine University, Moorenstr. 5, 4000 Düsseldorf, FRG

Helicobacter pylori, Gastritis and Peptic Ulcer
Edited by P. Malfertheiner and H. Ditschuneit

positive correlation exists between peak acid output and the basal serum Pg A level [7]. The secretion of pepsinogens can be stimulated by gastrin [8].

In patients with superficial gastritis serum Pg A and predominantly Pg C levels are elevated indicating an increased release of pepsinogens into the circulation. Due to the loss of chief cells, serum Pg A concentrations in patients with atrophic gastritis are low. Since the normal mucosa of the gastric corpus is replaced by Pg C-producing metaplasia, serum Pg C levels are maintained. Accordingly, there is a progressive drop in the Pg A/Pg C ratio in the course of gastritis [5].

The integrity of the gastric mucosa and thereby the steady state of serum pepsinogens is profoundly disturbed when noxious agents such as nonsteroidal anti-inflammatory drugs (NSAID) act on the gastric mucosa [9].

Thus, the factors determining serum pepsinogen levels are (a) peptic cell mass, (b) the degree of stimulation of the peptic cells, (c) the permeability of the internal and external cell membranes of the peptic cells, and (d) the renal excretion and metabolism of pepsinogens.

Parietal and Chief Cell Function During Acute Infection with *H. pylori*

Epidemic gastritis is an acute disease that has been observed in patients and in volunteers participating in studies involving gastric intubation [10, 11]. Recently it has been suggested that the disease is often an iatrogenic *H. pylori* infection [12], while Marshall et al. have shown that *H. pylori* causes the acute upper gastrointestinal symptoms observed in epidemic gastritis [13]. Histologically, acute *H. pylori*-positive antral gastritis was observed which extended to the body of the stomach within 1 week. The antral phase of the infection was accompanied by a transient but considerable increase in basal acid output and pepsin secretion, and when the infection reached the gastric corpus, achlorhydria and hyposecretion of pepsin were observed [12]. Since subsequent histology revealed that *H. pylori* and *H. pylori* antibody titers had increased [14], the outbreak of epidemic gastritis with hypochlorhydria that occurred in 1979 in healthy volunteers participating in secretory studies [11] may retrospectively be attributed to *H. pylori* infection [15]. Serum Pg A levels were markedly elevated during hypochlorhydria and returned to normal when gastritis and acid secretion improved [11]. Fasting serum gastrin concentration and the integrated gastrin response to a meal were similar during hypochlorhydria and after recovery.

The mechanisms by which these functional disturbances were induced remain unclear. Parietal cell mass and gastric permeability remained normal [11]. Obstruction of the gastric glands by mucus, granulocytes, and sloughed epithelium have been suggested as causes of reduced acid secretion and hyperpepsinogenemia, as well [16]. However, this theory must still be supported by further evidence. Previously Cave and Vargas [17] demonstrated that *H. pylori* possesses a powerful, noncytotoxic, protein-containing inhibitor of acid secretion.

Since hyperpepsinogenemia can be associated with various normochlorhydric conditions such as gastritis [5] and NSAID-induced gastropathy [9], etc., it should be ascertained whether factors other than mechanical ones can explain the elevated serum Pg A levels in epidemic gastritis. Gastrin does not seem to be involved [11]. A toxin (as yet hypothetical) similar to the one described by Cave and Vargas, which affects chief cell function directly would be a perfect candidate to explain the hyperpepsinogenemia observed in epidemic gastritis.

Whether acute *H. pylori* infection occasionally develops into atrophic gastritis with permanent hypochlorhydria is unknown. Sustained hypochlorhydria with clinical improvement has been observed in patients with the Zollinger-Ellison syndrome [17, 18]. Usually, however, when gastritis improves gastric acid secretion recovers to normal levels after a mean of 4 months [11].

Parietal and Chief Cell Function During Chronic Infection with *H. pylori*

Patients with duodenal ulcer (DU) disease are almost invariably *H. pylori* positive [19]. They tend to have inappropriate gastrin release and higher mean basal and stimulated acid secretion than control subjects [20]. The pathogenetic role of *H. pylori* in this context remains to be defined.

Recently Levi et al. [21] proposed that the ammonia produced by *H. pylori* from urea raises antral pH, thereby increasing basal as well as postprandial gastrin release and stimulating gastric acid secretion. This assumption was supported by McColl and coworkers [22] who demonstrated that eradication of *H. pylori* in patients with DU disease resulted in lower gastrin levels and reduced acid responses to meals. In 25 out of 30 children who were *H. pylori* negative after treatment, serum gastrin levels fell [23]. Unfortunately the level of gastric acid secretion was not determined. The hypothesis that *H. pylori* causes hypersecretion has not remained unchallenged. Several groups failed to find consistent relations between *H. pylori* status, acid secretion, and serum gastrin concentration [15, 24, 25], leaving the question open to further investigations and debate.

Serum pepsinogen levels have been found to be increased in adults with gastritis and peptic ulcer disease [5, 26]. Children with *H. pylori*-associated gastritis had higher serum Pg A concentrations than children without *H. pylori*-associated gastritis [27]. We recently observed a marked fall in Pg C serum concentrations in all *H. pylori*-positive patients with active gastritis who become *H. pylori*-negative with normalized gastric mucosal histology after therapy (K. J. Hengels et al., unpublished data). In contrast, serum Pg C levels remained unchanged in patients with persistent *H. pylori*-positive, active gastritis. These data suggest that serum pepsinogen levels reflect short-term changes in the gastric mucosa associated with *H. pylori* status.

Not only the effect of *H. pylori* on gastric secretion, but also the influence of luminal acidity on *H. pylori* colonization has aroused considerable interest. It has been suggested that *H. pylori* is acid-sensitive [28], but other studies have demonstrated that the organism is uncommon in achlorhydria with and without pernicious anemia [29–31]. A positive correlation between the prevalence of *H. pylori* and acid secretion has been found, indicating that hypochlorhydria creates an unfavorable milieu for the growth of *H. pylori* [29]. However, other factors such as changes in mucus production may also play an important role for *H. pylori* colonization of the stomach in patients with gastritis and various degrees of atrophy. The assumption that *H. pylori* depends on acid secretion is further supported by the observation that *H. pylori* is cleared from the antrum of patients with DU disease after the administration of omeprazole [32].

Conclusions

Acute *H. pylori* infection is followed by hypochlorhydria and hyperpepsinogenemia. Chronic *H. pylori* infection is associated with normochlorhydria and hyperpepsinogenemia. The pathogenetic mechanisms that cause these functional disturbances and the role of *H. pylori* hereby are unknown to a great extent.

References

1. Borody T, Cole P, Noionan S, et al. (1988) Long-term Campylobacter pylori recurrence post-eradication. Gastroenterology 94:43
2. Hengels KJ, Heinrich G, Strohmeyer G (1986) Relationship between peptic cell mass, mucosal pepsinogen concentration, and pepsinogen secretion in normal gastric mucosa and gastritis. Gastroenterology 90:1457
3. Samloff IM, Liebman WM (1974) Radioimmunoassay of group I pepsinogens in serum. Gastroenterology 66:494–502
4. Samloff IM (1982) Pepsinogen I and II. Purification from gastric mucosa and radioimmunoassay in serum. Gsstroenterology 82:26–33
5. Samloff IM, Varis K, Ihamäki T, et al. (1982) Relationship among serum pepsinogen I, and II, and gastric mucosal histology. Gastroenterology 83:204–209
6. Samloff IM, Taggart RT, Hengels KJ (1985) Molecular variants of human aspartic proteinases. In: Kostka V (ed) Aspartic proteinases and their inhibitors. de Gruyter, Berlin, pp 79–85
7. Samloff IM, Secrist DM, Passaro E (1975) A study of the relationship between serum group I pepsinogen levels and gastric acid secretion. Gastroenterology 69:1196–1200
8. Hirschowitz BI (1967) Secretion of pepsinogen. In: Code CF (ed) Handbook of Physiology. Alimentary Canal II. Williams and Wilkins, Baltimore, pp 889
9. Hengels KJ, Halfkann M, Strohmeyer G (1988) Detection and control of spontaneous and drug-induced gastroduodenal lesions by determination of serum pepsinogens. Gastroenterology 94:183
10. Gledhill H, Leischester RJ, Lightfoot N, et al. (1985) Epidemic hypochlorhydria. Br Med J 290:1383–1388
11. Ramsey EJ, Carey KV, Peterson WL, et al. (1979) Epidemic gastritis with hypochlorhydria. Gastroenterology 76:1449–1457
12. Graham DY, Alpert LC, Smith JL, et al. (1988) Iatrogenic Campylobacter pylori infection is a cause of epidemic achlorhydria. Am J Gastroenterol 83:974–980
13. Marshall BJ, Armstrong JA, McGechie DB et al. (1985) Attempt to fulfill Koch's postulates for pyloric campylobacter. Med J Aust 142:436–439
14. Peterson W, Lee E, Skoglund M (1987) The role of campylobacter pyloridis in epidemic gastritis with hypochlorhydria. Gastroenterology 92:1575
15. Brady CE, Hadfield TL, Hyatt JR, et al. (1988) Acid secretion and serum gastrin levels in individuals with campylobacter pylori. Gastroenterology 94:923–927
16. Spiro HM, Schwartz RD (1958) Superficial gastritis. A cause of temporary achlorhydria and hyperpepsinogenemia. N Eng J Med 259:682–684
17. Cave DR, Vargas M (1989) Effect of a campylobacter pylori protein on acid secretion by parietal cells. Lancet II:187–189

18. Wiersinga WM, Tytgat GN (1977) Clinical recovery owing to target parietal cell failure in a patient with Zollinger-Ellison syndrome. Gastroenterology 73:1413–1417
19. Goodwin CS, Armstrong JA, Marshall BJ (1986) Campylobacter pyloridis, gastritis, and peptic ulceration. J Clin Pathol 39:353–365
20. Walsh JH, Richardson CT, Fordtran JS (1975) pH dependence of acid secretion and gastrin release in normal and ulcer subjects. J Clin Invest 55:462–468
21. Levi S, Beardshall K, Haddad G et al. (1989) Campylobacter pylori and duodenal ulcers: the gastrin link. Lancet I:1167–1168
22. McColl KEL, Fullarton GM, El Nujumi AM et al. (1989) Lowered gastrin and gastric acidity after eradication of campylobacter pylori in duodenal ulcer. Lancet II:499–500
23. Odera G, Vaira G, Holton J et al. (1989) Amoxycillin plus tinidazole for campylobacter pylori gastritis in children: assessment by serum IgG antibody, pepsinogen I, and gastrin levels. Lancet II:690–692
24. Staub P, Jost R, Eberle Ch et al. (1989) Campylobacter pylori Besiedlung des Antrums: Einfluß auf Gastrin, Somatostatin, pankreatisches Polypeptid und Neurotensin. Schweiz Med Wochenschr 119:765–767
25. Wagner S, Freise J, Bucholz S et al. (1989) Campylobacter pylori and gastric acidity. Am J Gastroenterol 84:201–202
26. Samloff IM, Liebman WM, Panitch NM (1975) Serum group I pepsinogens by radioimmunoassay in control subjects and patients with peptic ulcer. Gastroenterology 69:83–90
27. Odera G, Vaira D, Holton J (1989) Serum pepsinogen I and IgG antibody to campylobacter pylori in non-specific abdominal pain in childhood. Gut 30:912–916
28. Langenberg ML, Tytgat GNJ, Shipper MEI, et al. (1984) Campylobacter-like organisms in the stomach of patients and healthy individuals. Lancet I:1348
29. Siurala M, Sipponen P, Kekki M (1988) Campylobacter pylori in a sample of Finnish population: relations to morphology and function of the gastric mucosa. Gut 29:909–915
30. Flejou JF, Bahame P, Smith AC (1989) Pernicious anemia and campylobacter like organisms; is the gastric antrum resistant to colonisation? Gut 30:60–64
31. Charasz N, Roucayrol AM, Chaplain C et al. (1989) Campylobacter pylori, pernicious and non pernicious anemia, achlorhydric atrophic gastritis and atrophic gastritis without achlorhydria. In: Megraud F, Lamouliatte H (eds) Gastroduodenal pathology and campylobacter pylori. Excerpta Medica, Amsterdam, pp 529–531
32. Mainguet P, Delmee M, Debougmie JC (1989) Omeprazole, campylobacter pylori and duodenal ulcer. Lancet II:389–390

Helicobacter pylori and Duodenal Ulcer – The Gastrin Link

J. Calam, S. Levi, K. Beardshall, R. Playford, P. Ghosh, I. Swift, W. Foulkes, and G. Haddad

Introduction

Our work represents an attempt to understand the relationship between the "traditional" view that increased acid secretion is an important cause of duodenal ulcers (DUs) and the new evidence that DU is caused by *Helicobacter pylori* (HP).

Gastric Acid Secretion in Duodenal Ulcer Disease

Mean rates of gastric acid secretion are higher in patients with DU disease than in normal individuals [1–4]. Gastric secretion is important in duodenal ulcerogenesis: acid-pepsin is both corrosive and strongly proteolytic. DUs hardly ever occur in individuals with gastric secretion rates below the normal range [5], and ulcers heal rapidly if gastric secretion is suppressed [6]. Duodenal-gastric metaplasia, which allows *Helicobacter pylori* to colonise the duodenum, appear to be caused by gastric hypersecretion [7].

Patients with DU disease have, on average, more parietal cells than normals, i.e. a greater parietal cell mass [4]. This is reflected by an increased mean peak acid output (PAO), measured during maximal stimulation [1, 4]. In addition there is evidence that patients with DU disease secrete a greater proportion of their PAO after meals [2–4], suggesting that the parietal cells are stimulated more strongly, or for longer, after eating.

Familial Hyperpepsinogenaemia I

Samloff and his collaborators used serum group I pepsinogens as a measure of the secretory cell mass to study the origin of this abnormality in DU disease. About 60% of patients with DU disease had pepsinogen I levels above the normal range [8]. Interestingly, about one-half their first degree relatives also had elevated serum pepsinogen I and these relatives seemed to be predisposed to DU disease [9]. At the time it appeared most likely that an increased secretory cell mass was an inherited trait which predisposes to DU disease.

Departments of Medicine and Surgery, Royal Postgraduate Medical School, Hammersmith Hospital, Du Cane Road, London W12 ONN

Helicobacter pylori, Gastritis and Peptic Ulcer
Edited by P. Malfertheiner and H. Ditschuneit

Research into the greater stimulation of parietal cells in DU disease has focussed on gastrin.

Gastrin and Duodenal Ulcer Disease

Gastrin is a peptide hormone which is released from the gastric antrum, and to a lesser extent from the proximal duodenum, by food [4]. Gastrin is the most potent known stimulant of gastric acid secretion and appears to be responsible for most of the increase in acid secretion which occurs after eating [10]. In addition gastrin has a trophic effect on the parietal cell mass [11].

Mean postprandial plasma gastrin concentrations have been found to be increased in patients with DU disease [12]. This is remarkable because a low intragastric pH normally inhibits gastrin release [13]. Thus, in DU patients who tend to secrete excessive acid, even normal gastrin levels could be regarded as inappropriately high. This suggested a defect in the inhibition of gastrin release by a low intragastric pH in patients with DU disease. Walsh et al. showed by intragastric titration that gastrin release and acid secretion is indeed inhibited less by a low intragastric pH in DU patients than in normal individuals [13].

While investigating the cause of inappropriate gastrin release in DU disease, Taylor et al. studied an apparent subgroup of patients who had marked postprandial hypergastrinaemia together with gastric hypersecretion [14]. It was found that about one-half of the first degree relatives of these patients also have postprandial hypergastrinaemia with hyperpepsinogenaemia I. These results suggested that the defect which leads to excess gastrin release in DU disease is familial. Moreover this familial defect might, through the trophic effect of gastrin [11], be one cause of the familial increase in secretory cell mass revealed by studies of pepsinogen I [4, 8, 9].

In recent years the main thrust of work on the aetiology of DU disease has turned from physiology to microbiology.

Evidence that *Helicobacter pylori* (HP) Causes Duodenal Ulcer Disease

Recent epidemiological studies have provided strong evidence that HP plays a major role in the aetiology of DU disease: First, the prevalence of HP is much greater in DU patients than in controls [15]. Second, the duration of remission of DU disease after different treatments is proportional to the percent eradication of HP that they produce. Third, ulcers tend not to recur in a particular patient until recolonization has taken place [16–18]. These findings have raised the question of how HP might cause DU disease.

The Role of Gastric Metaplasia Within the Duodenum

HP only colonises gastric epithelium and is most abundant in the gastric antrum [15]. How therefore does HP cause ulcers in the duodenum? The most popular explanation is based on the presence of metaplastic gastric epithelium in the duodenum of most patients with DU disease. These patches of gastric metaplasia are believed to result from gastric hypersecretion and may be the sites of previous ulcers [7]. They become colonised with HP which is believed to produce ulcers by a local cytotoxic effect diminishing resistance to lumenal aggressive factors, including acid [19].

We considered the possibility that HP in the gastric antrum might also contribute to ulcerogenesis by increasing gastrin release.

Our Hypothesis: "the Gastrin Link"

We postulated that HP itself, rather than one or more inherited traits, might be responsible for hypergastrinaemia and acid hypersecretion in DU disease [20]. This idea was based on consideration of the likely effect of HP on the pH within the antral mucus layer. HP produces a powerful urease which splits urea to produce one carbon dioxide molecule and two ammonia molecules, with a net production of alkali [15]. This reaction forms the basis of the biopsy urease test [21]. Gastrin is released from specialised G cells located in the midregion of the antral gastric pits [14]. HP organisms exist, attached to epithelial cells, on the surface and within the pits of the gastric antrum [22]. Therefore it seemed possible that the production of alkali by HP might cause the impaired inhibition of gastrin release by intragastric acid in DU disease [13] (Fig. 1).

If this were the case then the majority of DU patients who have a positive urease test would be expected to have higher plasma gastrin concentrations than the minority in whom the urease test was negative. It was necessary to study the patient's acid secretion rate to determine that any increase in gastrin release was indeed inappropriate, particularly as HP is capable of diminishing gastric acid secretion [15, 23, 24].

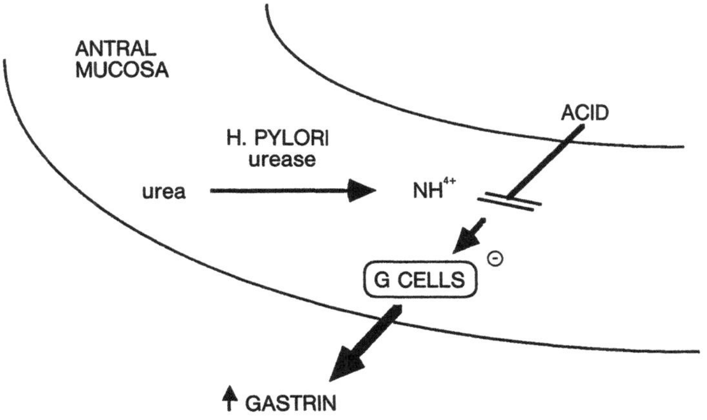

Fig. 1. The hypothesis that prompted the work: local production of alkaline ammonia by *Helicobacter pylori* urease may prevent inhibition of gastrin release by intragastric acid

Comparison of DU Patients with Positive and Negative Urease Tests

Our initial results were published in *The Lancet* [20]. The study has now been extended and a total of 51 patients with DU disease have been studied [25]. Seven were urease negative (−ve) and 44 had a positive (+ve) biopsy urease test [21]. Basal plasma gastrin concentrations were significantly higher in urease +ve compared with in urease −ve patients (13.1 (S.E.M. 2.0) compared with 6.3 (2.6) pmol/l, $P < 0.05$). The integrated plasma gastrin response to a standard meal was also significantly greater in the urease +ve patients than in those in whom the test was −ve [20, 25] (Fig. 2). Furthermore the PAO, which reflects parietal cell mass, was significantly higher in the +ve than in the −ve patients (45 (3) compared with 30 (4) mmol/h, Fig. 3).

These results were consistent with our hypothesis that HP causes the hypergastrinaemia of DU disease. In addition they supported the idea that HP induced hypergastrinaemia is responsible for the increased mean parietal cell mass in DU disease. However at this stage we also considered an alternative explanation for our results: that the gastrin driven hypersecretion might cause HP colonization, rather than vice versa. We therefore repeated studies of acid and gastrin after treatment of DU patients with a +ve urease test with a regime known to suppress the growth of HP [25]. Patients recieved tri-potassium di-citrato bismuthate (DeNol, Gist Brocades) 120 mg q.d.s. for 4 weeks, plus metronidazole 400 mg t.d.s. for the last 2 weeks of treatment with DeNol. Repeat studies were performed within 1 week of completing this treatment [25].

The Effect of Suppressing the Growth of *Helicobacter pylori*

Of ten patients studied, one remained urease +ve (but the test took 20 h, compared with 2 h before treatment, to change colour). Duodenal ulcers healed in

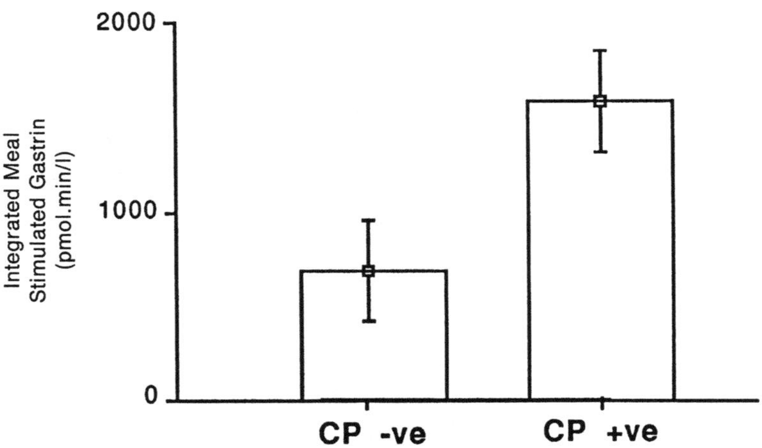

Fig. 2. Integrated meal stimulated plasma gastrin responses in patients with duodenal ulcers with (CP+) and without (CP−) antral *Helicobacter pylori* ($n = 44$ and $n = 7$ respectively, $P < 0.05$)

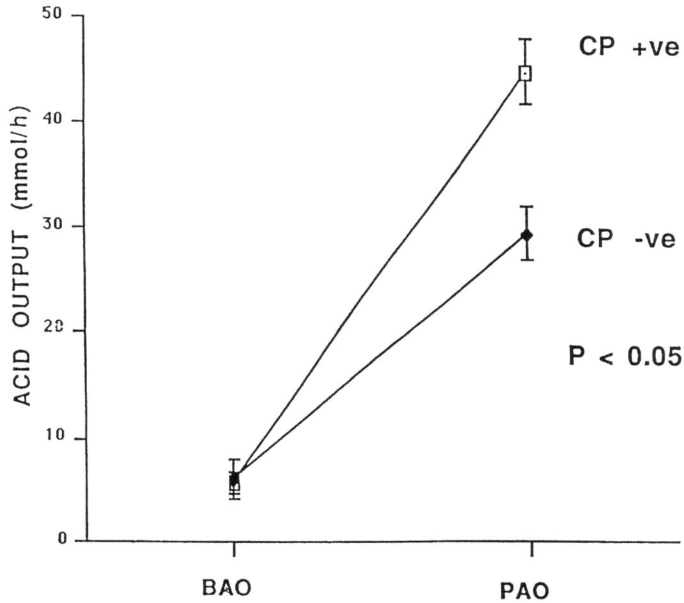

Fig. 3. Basal and peak acid output in patients with duodenal ulcers with (CP+) and without (CP−) antral *Helicobacter pylori* ($n = 44$ and $n = 7$ respectively)

all but one patient. Treatment led to a significant fall in the integrated gastrin response, but PAO did not change significantly (Fig. 4). A slight fall in basal gastrin levels from 7.5 (1.8) to 6.7 (1.9) pmol/l after treatment was not statistically significant. This result supports the view that HP is responsible for the hypergastrinaemia of DU disease. We are currently extending the study to include control groups treated with ulcer-healing regimes which do not affect HP. The lack of fall in PAO does not exclude the possibility that HP associated hypergastrinaemia eventually increases the parietal cell mass because it is likely that the study did not allow sufficient time for the trophic effect of gastrin to be reversed. Patients will be restudied 6 months after treatment.

Discussion

Our results support the hypothesis that HP causes the increases in mean postprandial plasma gastrin concentration and acid secretion rates that occur in patients with DU disease [20, 25]. It will be interesting to determine the relationship between HP colonisation and elevated serum group I pepsinogens [9] and hypergastrinaemia within families [14]. Many other questions need to be addressed.

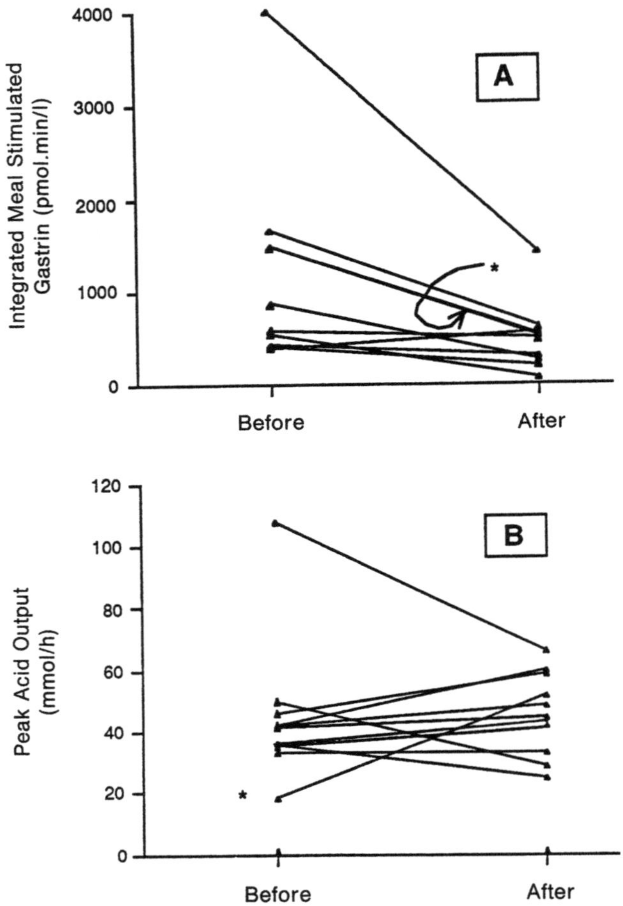

Fig. 4 A, B. Changes in integrated postpradial gastrin response (**A**) and peak acid output (**B**) in patients before and after treatment with colloidal bismuth subcitrate (4 weeks) and metronidazole (2 weeks); * indicates the patient who remained weakly positive for antral urease after treatment. The change in gastrin response was significant ($P < 0.005$) but the change in peak acid output was not

How Does *Helicobacter pylori* Increase Gastrin Release?

We have postulated that it does so by altering the pH within the antral microenvironment. This hypothesis is attractive because it offers a specific explanation for the defective inhibition of gastrin release by a low intragastric pH that Walsh et al. originally described in patients with DU disease [13]. In addition prolonged alkalinization of the antrum is already known to produce major increases in gastrin release, including greater postprandial release, as seen in patients with pernicious anaemia [26].

Wyatt et al. have suggested that HP might increase gastrin release by producing antral inflammation. They found elevated basal plasma gastrin levels in association with antral inflammation, whether HP was present or not [27]. Acid was not measured so that it remains possible that the four HP – ve patients with antritis in that study had increased plasma gastrin levels because of diminished acid secretion. However there is already evidence that immunological activation releases gastrin: exposure of immunised animals to immunogen causes gastrin release [28] and the cytokines gamma interferon and interleukin 2 stimulate release of gastrin from the isolated canine antrum [29]. Cooper et al. found that patients with hypergastrinemic DU disease have an antrum which is abnormally sensitive to weak stimulants of gastrin release, at constant pH [30]. This finding is consistent with synergistic stimulation of gastrin release by meals and some other intra-epithelial factor.

At present it seems possible that HP increases gastrin release by more than one mechanism.

How Does DeNol Heal Duodenal Ulcers?

Our results, and those of McColl et al. [31], indicate that eradication of HP diminishes gastrin release. McColl was able to show than the intragastric pH after a standard meal was significantly more alkaline after eradication but there was no obvious change in pH throughout a 24-h period. Ulcer healing on DeNol is as rapid as on ranitidine which has a much greater effect on intragastric pH [31, 32]. Therfore it seems likely that DeNol heals ulcers by other mechanisms. DeNol may produce rapid healing by its antimicrobial effect on HP organisms within the duodenum [33], allowing repair of "the leaking roof" [19]. DeNol combines with ulcer slough which may protect the ulcer crater itself [34]. DeNol also stimulates gastric prostaglandin synthesis producing cytoprotection [35], although it is not clear whether this also occurs in the duodenum.

What Causes Duodenal Ulcers in Patients with a Negative Urease Test

In our study 7 out of 51 (14%) patients had a negative urease test. We used McNulty's modified method which was found by McNulty et al. to have a specificity of 99.8% and has a sensitivity of 90%, compared with the accumulated result of Gram stain, histology and culture [21]. However when organisms presented in patients with a negative urease test they were scanty. At present it is not clear whether the scanty organisms, if present, contribute to ulcerogenesis, or whether ulcers in such individuals are due to one of the other abnormalities that are present in patients with DU disease, such as rapid gastric emptying of acid into the duodenum [4], or diminished duodenal prostaglandin synthesis [36], perhaps due to cigarette smoking [37].

Why Does HP Apparently Cause No Disease in Some Inidviduals and Different Diseases in Others?

First infection with HP is known to produce symptomatic gastritis [38, 23] with temporarily diminished acid secretion [24], but it is not clear whether these events always occur on first exposure to HP. Also the proportion of patients who become chronically colonised with HP after initial exposure is not known.

Once the individual has become colonised he is very likely to have histological type B gastritis, but to have no symptoms. Type B gastritis affects the gastric antrum, but the not the body and fundus which secrete acid. The prevalence of HP colonisation in type B gastritis is about 70% [39]. At this stage, according to the work of Smith et al., the individual will have normal gastric acid secretion but elevated plasma gastrin concentrations [18].

What proportion of individuals progress from HP associated type B gastritis to duodenal ulcer disease is not known. An obsessional autopsy study in Leeds showed active or healed DUs in about 13% of individuals (males 17%, females 8%) aged 40–49 years [40]. About 33% of an American population were colonised with HP in this age [41].

The sequence of events between type B gastritis and DU disease remains open to speculation. Possibly the hypergastrinaemia of HP associated type B gastritis [18] increases acid secretion through its trophic [11] and stimulatory [10] effects, producing duodenal gastric metaplasia [7] which becomes colonised with HP, leading to ulcers [19].

Whether or not this sequence of events takes place may depend partly on the presence or absence of type A gastritis. Type A gastritis is associated with anti-parietal cell antibodies, affects the body and fundus of the stomach and diminishes acid secretion [42]. The prevalence of type A gastritis increases with age [43]. Thus DU disease may be initiated if HP colonisation occurs before type A gastritis develops and prevents hypersecretion. This is consistent with the typical presence of HP and type B gastritis and the absence of type A gastritis in patients with DU disease [44]. However colonisation with HP in the presence of type A Gastritis may predispose to other diseases: chronic gastric ulcer and carcinoma of the stomach are both associated with the presence of type A gastritis together with type B gastritis [45]. Recent evidence suggests that HP may predispose to both of these conditions [46, 47].

The most severe form of type A gastritis leads to complete absence of gastric acid secretion. This state appears to protect the stomach from colonisation with HP, perhaps because acid is needed to neutralise the ammonia produced by urease, or because other organisms which colonise the anacid stomach compete with HP [48].

What is quite clear is that the isolation of HP by Marshall in 1982 [49] has led to a revolution in our understanding of several common diseases which is both fascinating scientifically, and offers the prospect of greatly improved prevention and treatment.

References

1. Baron JH (1978) Clinical tests of gastric secretion. MacMillan, London
2. Lam SK, Ong GB (1980) Identification of two subgroups of familial early-onset duodenal ulcers. Ann Intern Med 93:540–546
3. Malagelada JR, Longstreth GF, Deering TB, Summerskill WH, Go VL (1977) Gastric secretion and emptying after ordinary meals in duodenal ulcer. Gastroenterology 73:989–993

4. Soll AH (1989) Duodenal ulcer and drug therapy. In: Sleisenger MH, Fordtran JS (eds) Gastrointestinal disease, 4th ed. Saunders, Philadelphia, pp 814–879
5. Baron JH (1982) Current views on pathogenesis of peptic ulcer. Scand J Gastroenterol [Suppl 80] 17:1–10
6. Clissold SP, Campoli-Richards DM (1986) Focus on omeprazole. Drugs 32:15–47
7. Wyatt JI, Rathbone BJ, Dixon MF et al. (1987) Campylobacter pylori and acid-induced gastric metaplasia in the pathogenesis of duodenitis. J Clin Pathol 40:841–848
8. Samloff MI, Leibman WM, Panitch NM (1975) Serum group I pepsinogens by radioimmunoassay in control subjects and patients with peptic ulcer. Gastroenterology 69:83–90
9. Rotter JI, Sones JQ, Samloff IM, Richardson CT, Gursky JM, Walsh JH, Rimoin DL (1979) Duodenal ulcer disease associated with elevated serum pepsinogen I; an inherited autosomal dominant disorder. N Engl J Med 300:327–377
10. Maxwell V, Eysselein VE, Kleibeuker J, Reedy T, Walsh JH (1984) Glucose perfusion intragastric titration. Dig Dis Sci 29:321–326
11. Johnson LR (1976) The trophic action of gastrointestinal hormones. Gastroenterology 70:278–282
12. Taylor IL, Dockray GJ, Calam J, Walker RJ (1979) Big and little gastrin responses to food in normal and ulcer subjects. Gut 20:957–962
13. Walsh JH, Richardson CT, and Fordtran JS (1975) pH dependence of acid secretion and gastrin release in normal and ulcer subjects. J Clin Invest 55:462–468
14. Taylor IL, Calam J, Rotter J, Vaillant C, Samloff I, Cook A, Simkin E, Dockray G (1981) Family studies of hypergastrinaemic, hyperpepsinogenemic I duodenal ulcer disease. Ann Intern Med 95:421–425
15. Dooley CP, Cohen H (1988) The clinical significance of Campylobacter pylori. Ann Intern Med 108:70–79
16. Coghlan JG, Gilligan D, Humphries H et al. Campylobacter pylori and recurrence of duodenal ulcers: a 12-month follow-up study. The Lancet II:1109–1111
17. Marshall BJ, Goodwin CS, Warren JR et al. (1988) Prospective double-blind trial of duodenal ulcer relapse after eradication of Campylobacter pylori. The Lancet II:1467–1469
18. Smith JT, Pounder RE, Evans DJ, Graham DY, Evans DG (1989) Inappropriate 24 h hypergastrinaemia in asymptomatic C. pylori infection. Gut 30:A 732–733
19. Goodwin CS (1988) Duodenal ulcer, Camplyobacter pylori, and the "leaking roof" concept. Lancet II:1467–1469
20. Levi S, Beardshall K, Playford R, Ghosh P, Haddad G, Calam J (1989) Campylobacter pylori and duodenal ulcers: the gastrin link. Lancet I:1167–1168
21. McNulty CA et al. (1989) Detection of Campylobacter pylori by the biopsy urease test: an assessment in 1445 patients. Gut 30:1058–1063
22. Price AB, Levi J, Dolby JM, Dunscombe PL, Smith A, Clark J, Stephenson ML (1985) Campylobacter pyloridis in peptic ulcer disease: microbiology, pathology, and scanning electron microscopy. Gut 26:1183–1188
23. Morris A, Nicholson G (1987) Ingestion of Campylobacter pyloridis causes gastritis and raised fasting gastric pH. Am J Gastroenterol 82:192–199
24. Graham DY, Alpert LC, Lacey Smith J, Loshimura HH (1988) Iatrogenic Campylobacter pylori infection is a cause of epidemic hypochlorhydria. Am J Gastroenterol 83:971–980
25. Levi S, Beardshall K, Swift I, Foulks W, Playford R, Ghosh P, Calam J (1989) Antral Campylobacter pylori, hypergastrinaemia, and duodenal ulcers: effect of eradicating the organism. Br Med J 299:1504–1505
26. Walsh JH (1989) Gastrointestinal peptide hormones. In: Sleisenger MH, Fordtran JH (eds) Gastrointestinal disease, 4th ed. Saunders, Philadelphia, pp 78–107
27. Wyatt JI, Rathbone BJ, Green DM, Primrose J (1989) Raised fasting serum gastrin in chronic gastritis is independent of Campylobacter pylori status and duodenal ulceration. Gut 30:A 1483
28. Teichmann RK, Andress HJ, Leibich H, Seifert J, Brendel W (1984) Possible role of Ia-positive cells in the antrum in gastrin secretion. Eur Surg Res 16:64–65
29. Reinhard K, Teichmann E, Pratschke J, Grab J, Hammer C, Brendel W (1986) Gastrin release by interleukin-2 and gamma-interferon in vitro. Can J Physiol Pharmacol 64:62

30. Cooper RG, Dockray GJ, Calam J, Walker RJ (1985) Acid and gastrin responses during intragastric titration in normal subjects and duodenal ulcer patients with G-cell hyperfunction. Gut 26:232–236
31. McColl KE, Fullarton GM, ElNujumi AM, Macdonald AM, Brown IL (1989) Lowered gastrin and gastric acidity after eradication of Campylobacter pylori in duodenal ulcer. Lancet II:499–500
32. Walt R, Male PJ, Rawlings J, Hunt RH, Milton-Thompson GJ, Miziewicz JJ (1981) Comparison of the effects of ranitidine, cimetidine and placebo on the 24 h intragastric acidity and nocturnal acid secretion in patients with duodenal ulcer. Gut 22:49–54
33. Goodwin CS, Blake P, Blinkcow E (1986) The minimum inhibitory and bactericidal concentrations of antibiotics and anti-ulcer agents, against Campylobacter pyloridis. J Antimicrob Chemother 17:309–314
34. Hall DW (1989) Review of the modes of action of colloidal bismuth subcitrate. Scand J Gastroenterol [Suppl 157] 24:3–6
35. Hall DW, Van Den Hoven WE (1987) Gastric mucosa protective effects of colloidal bismuth subcitrate. Int J Tissue React 9:427–432
36. Ahlquist DA, Dozois RR, Zinmeister AR, Malagelada JR (1983) Duodenal prostaglandin synthesis and acid load in health and duodenal ulcer disease. Gastroenterology 85:522–528
37. Quimb QF, Bonnice CA, Burnstein SH, Eastwood GL (1986) Active smoking depresses prostaglandin synthesis in human gastric mucosa. Ann Intern 104:616–619
38. Marshall JB, Armstrong JA, McGechie DB, Glancy RJ (1985) Attempt to fulfil Koch's postulates for pyloric campylobacter. Med J Aust 142:436–439
39. Steininger H, Schneider U, Bartz K, Simmler B (1989) Campylobacter pylori and gastritis – Besiedelungsdichte und Gradder Entzündung. Semiquantitative und morphometrische Untersuchung. Leber Magen Darm 19:70–78
40. Langman MJ (1987) Aetiology of peptic ulcer. In: Misiewicz JJ (ed) Diseases of the gut and pancreas, 1st edn. Blackwell, Oxford, pp 268–281
41. Graham DY, Klein PD, Opekun AR, Boutton TW (1988) Effect of age on the frequency of active Campylobacter pylori infection diagnosed by the [^{13}C] urea breath test in normal subjects and patients with peptic ulcer. J Infect Dis 157:777–780
42. Strickland RG, Mackay I (1973) A reappraisal of the nature and significance of chronic atrophic gastritis. Am J Dig Dis 18:426–440
43. Varis K, Ihamaki T, Harkonen M, Samloff IM, Siurala M (1979) Gastric morphology, function and immunology in first degree relatives of probands with pernicious anaemia. Scand J Gastroenterol 14:129–139
44. Kekki M, Sipponen P, Siurala M (1984) Progression of antral and body gastritis in active and healed duodenal ulcer and duodenitis. Scand J Gastroenterol 19:382–388
45. Glass GB, Pitchumoni CS (1975) Atrophic gastritis. Hum Pathol 6:219–250
46. O'Connor HJ, Oixon MF, Wyatt JI, Axon AT, Dewar EP, Johnston D (1987) Campylobacter pylori and peptic ulcer disease. Lancet II:633–634
47. Forman D, Sitas F, Stacey A, Newell D, Yarnell J, Burr M, Elwood P, Buiatti E, Pali D, Lachlan G, Chen J. Epidemiology of Campylobacter pylori (oral presentation friday am Ulm meeting)
48. Flejou J-F, Smith AC, Stockbrugger RW, Rode J, Price AP (1989) Pernicious anaemia and campylobacter like organisms; is the gastric antrum resistant to colonization? Gut 30:60–64
49. Warren JR, Marshall B (1983) Unidentified curved bacilli on gastric epithelium in active chronic gastritis. Lancet I:1273–1275

Early Relapse Rate after Healing of *Helicobacter pylori* Positive duodenal Ulcers. Munich Duodenal Ulcer Trial*

G. A. Mannes [1], E. Bayerdörffer [2], and the German Ulcer Study Group

Introduction

During the past 10 years it has been noted in several clinical trials that patients with duodenal ulcers (DU) treated with bismuth compounds experienced lower relapse rates than DU patients treated with H_2-receptor antagonists [1]. With the redetection of *Helicobacter (Campylobacter) pylori* (HP) in 1982 and its subsequent culture from gastroduodenal biopsy material it has emerged that the main action of bismuth salts may be bactericidal. Bismuth may lead to clearance of HP and subsequent healing of gastritis. The colonization of gastric metaplasia in the duodenal bulb by HP provides the missing link between antral gastritis in many patients and the development of DU in only some of them.

Recent clinical trials suggest that eradication of HP from the gastroduodenal mucosa may prevent relapse of DU and thus may heal DU disease [2–6]. However, many questions are left unanswered by these preliminary studies, possibly due to the small numbers of patients.

We wanted to test these findings from the first clinical studies by using a larger series of DU patients to thereby try to get further insight into the conditions that determine DU relapse.

[1] Medical Department, Krankenhaus der Barmherzigen Brüder
[2] Medical Department II, Klinikum Großhadern, University of Munich
[3] Gastroenterological Practitioners
[4] Medical Department Innenstadt, University of Munich
[5] Gastroenterological Endoscopy, Department of Surgery, University of Munich
[6] 2. Medical Department, Krankenhaus Bogenhausen
[7] Medical Department, Krankenhaus Fürstenfeldbruck
[8] Medical Department II, University of Ulm
[9] Department of Microbiology, University of Munich
[10] Department of Pathology, Klinikum Bayreuth
[11] Department of Microbiology, University of Hamburg
[12] Institute for Biomathematics and Epidemiology, University of Munich

* W. Höchter [3], J. Weingart [3], W. Heldwein [4], S. Müller-Lissner [5], O. Kuntzen [6], W. Bornschein [3], J. Wilkening [7], H. Oertel [2], P. Malfertheiner [8], Ch. Blendinger [1], G. Ruckdeschel [9], P. Pfaller [9], M. Stolte [10], H. von Wulffen [11], and W. Köpcke [12].
The study was sponsored by the companies Röhm Pharma, Grünenthal, Cascan, Heumann, and Becton & Dickinson

Helicobacter pylori, Gastritis and Peptic Ulcer
Edited by P. Malfertheiner and H. Ditschuneit

Methods

Patient Selection

Patients aged 18–75 years having an active DU with a diameter of at least 5 mm were selected; those with gastric or pyloric ulcers, those who took H_2-receptor antagonists, bismuth compounds, antibiotics, or any other antiulcer medication (daily) during the last 4 weeks before endoscopy, and those who took corticosteroids, NSARs, or regularly more than 100 mg acetylsalicylic acid or any other antiinflammatory medication regularly were not eligible for the study.

Further exclusion criteria were ulcer surgery including PSV (except where there was a history of ulcer perforation), pregnancy, renal insufficiency (creatinine > 2.5 mg/dl), and any presenting contraindications for biopsy (Table 1).

Table 1. Inclusion and exclusion criteria

Inclusion
- All patients between 18 and 75 years who were HP positive in the HP culture and/or WS stain

Exclusion
- Patients with bismuth, antibiotic, and H_2-receptor antagonist treatment during the past 4 weeks
- Patients with renal insufficiency (creatinine > 2.5 mg/dl)
- Patients with signs of ulcer bleeding during the past week
- Patients with resected stomachs
- Patients with known allergy against penicillin
- Pregnant patients
- Patients taking NSARs, more than 100 mg ASS daily, steroids, or other antiulcer drugs

Endoscopy and Biopsy

At each endoscopic examination one antral biopsy specimen was taken for the rapid urease test and two antral biopsy specimens for the HP culture. Four additional biopsy specimens, two from the antrum and two from the body, were taken for histologic examination. Since most of the ulcer patients were outpatients they were randomly assigned to the treatment groups immediately after endoscopy. Patients who proved to be HP negative in both HP culture and histology were withdrawn from the study and treated conventionally. Each endoscopic examination was accompanied by laboratory tests for hemoglobin, creatinine, serum amylase, GGT, SGPT, and WBC. Further serum samples were collected to determine bismuth levels and HP antibodies.

Microbiology and Histology

Two biopsy specimens for microbiological investigation were kept and transported at environmental temperature in a transport medium (Port-A-Cul, Becton & Dickinson, Heidelberg, FRG) to the Department for Microbiology within four hours. The specimens were processed immediately and inoculated onto two different blood agar media (blood agar base 2, Oxoid, Wesel, FRG, with 10% human erythrocytes). One medium was supplemented by 3.64 mg amphotericin B only, the other additionally with 10 mg vancomycin, 5 mg trimethoprim, and 0.31 mg polymyxin B. Plates were incubated for 5–7 days at 37° in an atmosphere as produced by the gas pack Anaerocult C (E. Merck, Darmstadt, FRG). HP was identified by the usual morphological and biochemical criteria, especially by urease, catalase and oxidase activity.

All strains are kept frozen in liquid nitrogen in pepton yeast extract broth with 5% dimethylsulfoxide for later studies.

Two biopsies from the antrum and two from the body were placed in neutral buffered formalin. All mucosal specimens were stained by H&E to grade gastritis and by Warthin-Starry (WS) stain to grade the mucosal colonization by HP. The microbiologist and the pathologist were blinded to the therapy by coding the samples with random numbers.

Randomization and Therapy

Three treatment groups (Table 2) were compared with respect to the 6-week healing rates, eradication of HP 4 weeks after cessation of therapy, and DU relapse during the follow-up period. Randomization was carried out by a central study secretariat. Each center had its own randomization list. Before starting therapy the patients history was recorded. Complaints were recorded every day during the first 14 days of treatment, at the 6- and 10-week control, during the follow-up when a symptomatic relapse occurred, and after 1 year.

Table 2. Therapy groups

1. Bismuth subsalicylate
3 × 600 mg as chewable tablets,[a] 2 h after meals, for 6 weeks
2. Bismuth subsalicylate + amoxicillin
3 × 600 mg as chewable tablets, 2 h after meals, for 6 weeks 2 × 1000 mg as dissolvable tablets[b] with meals for the first 14 days
3. Ranitidine
1 × 300 mg[c] at bedtime

[a] Jatrox, Röhm Pharma, Darmstadt, FRG
[b] Amoxypen, Grünenthal, Stolberg Rhld., FRG
[c] Sostril, Cascan, Wiesbaden, FRG

Antacids (Trigastril tablets, Heumann Pharma, Nürnberg, FRG) were given on request in all therapy groups

Patient Monitoring

Patients had to grade their complaints as low, medium, and intense, and record them in a diary for the first 14 days. All patients were interviewed by phone with regard to how they were managing their therapy and possible side effects. Information about length and intensity of pain was recorded before therapy, at the 6- and 10-week control, when a symptomatic relapse occurred, and after 1 year. The antacid consumption was also recorded during the entire therapy period.

Results

One hundred four patients were randomized till the end of June 1989, 35 each for the ranitidine and the bismuth subsalicylate (BSS) treatment, and 34 for BSS + amoxicillin. Six patients, two in each therapy group, did not fulfill the inclusion criteria because they were HP negative in both the culture and histology.

The baseline data of patients with respect to sex, age, length of ulcer history, average number of previous ulcers, and smoking and drinking habits are shown in Table 3.

Table 3. Baseline comparison of patients

	BSS	BSS + amoxicillin	Ranitidine
Mean age (years)	46.2	46.3	50.2
Male/female	24/11	21/13	25/10
Ulcer history >1 year	79%	80%	73%
Mean number of previous ulcers	3.1	2.7	2.8
Smokers (>10/day)	41%	48%	55%
Alcohol (>10 × /month)	47%	45%	39%

Effect of Treatment on DU Healing and HP Status

Of 29 ulcers 24 (83%) were healed in the BSS group at the 6-week control, as were 23/25 (92%) of the ulcers in the BSS + amoxicillin group, and 19/31 (61%) of the ulcers in the ranitidine group. HP was cleared in 41% of the patients treated with BSS, in 68% of the BSS + amoxicillin group, and in none (0%) of the ranitidine group. The eradication rate as judged 4 weeks after cessation of therapy was 13% in the BSS-treated group and 48% in the BSS + amoxicillin group (Table 4).

Effect of Treatment on DU Relapse

Patients whose DU were healed at the 6-week endoscopy were followed for an average period of 6 months. The cumulative relapse rates are shown in Fig. 1. After BSS treatment the cumulative relapse rate was 44% within 6 months, after BSS + amoxicillin treatment it was 9% and after ranitidine therapy 68%.

Table 4. Healing rates and HP status

Therapy	6 weeks	10 weeks
BSS ($n = 29$) healed:	24/29(83%)	22/24(92%)
HP negative	12/29(41%)	3/24(13%)
HP positive	17/29(59%)	21/24(87%)
BSS + amoxicillin ($n = 24$) healed:	23/25(92%)	22/23(96%)
HP negative	17/25(68%)	11/23(48%)
HP positive	8/25(32%)	12/23(52%)
Ranitidine ($n = 31$) healed:	19/31(61%)	11/22(50%)
HP negative	0/31(0%)	0/22(0%)
HP positive	31/31(100%)	22/22(100%)

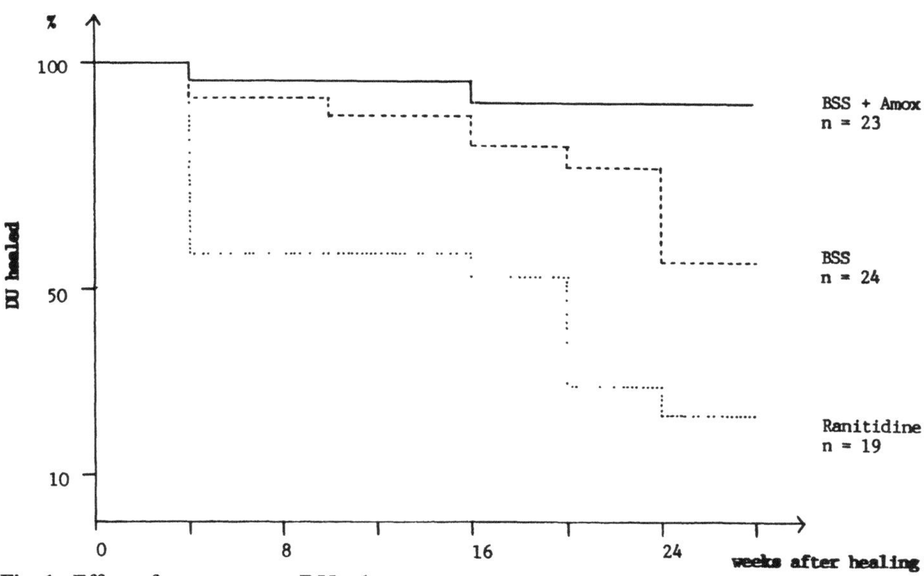

Fig. 1. Effect of treatment on DU relapse

Effect of HP Status on DU Healing and DU Relapse

Patients whose HP was cleared at the 6-week endoscopy had healed DU in 93% of cases, whereas those who were still HP positive had a healing rate of 69% (Table 5).

Patients with persistent HP infection after ulcer healing were further subdivided according to the different preceding therapies. The subsequent relapse rates during follow-up were 36% after BSS treatment, 8% after BSS + amoxicillin, and 45% after ranitidine treatment (Table 6).

Table 5. Effect of HP status on DU healing

	Healing at 6 weeks
HP positive patients	42/69 (69%)
HP negative patients	27/29 (93%)

Table 6. Effect of treatment and HP status on DU relapse

HP status	Treatment	Relapses (*n*)	(%)	Median follow-up (months)
HP negative	BSS + amox/BSS	15/none	0	6.0
HP positive	BSS + amox	12/ 1	8	5.9
	BSS	21/ 8	36	6.0
	Ranitidine	19/10	45	6.2

Effect of Treatment on Pain Relief and Side Effects

Two of 33 patients in the ranitidine group had to be withdrawn because they additionally took bismuth salts. In the bismuth group two patients did not keep to the study protocol and two discontinued the therapy because of side effects. The combination therapy of bismuth + amoxicillin was not carried out as prescribed by four patients, and three were excluded from the study because of side effects (diarrhea which was not associated with clostridium difficile (2) and mycotic stomatitis (1); Table 7). All patients with side effects induced by amoxicillin continued to take the bismuth tablets for 6 weeks.

Relief of pain occurred faster in patients in the ranitidine group than in those treated with bismuth. However, the pain score during the follow-up period was lower after BSS or combined BSS + amoxicillin therapy as compared with the ranitidine treatment (Table 8).

Table 7. Side effects

Therapy	Total		Excluded	
	n	%	*n*	%
BSS	12	36	2	6
Increased pain (8), meteorism and constipation (6), weakness (3), nausea (3), sleep disturbance (2), metal taste (2), eruption of vesicles on the tongue (1)				
BSS + amoxicillin	14	44	5	16
Diarrhea (9), stomatitis (4), vomiting (1), constipation (1), allergy against amoxicillin (1), burning foot sole (1)				
Ranitidine	3	9	0	0
Unrest (1), sleep disturbance (1), feeling of tiredness (1)				

Table 8. Effect of treatment on pain score

Therapy	Mean duration for relief of acute symtoms (days)	Pain score at week 10
BSS	8.1	4.5
BSS + amoxicillin	9.3	3.5
Ranitidine	5.3	5.0

Discussion

In 1987 the first duodenal ulcer (DU) trial was published by Coghlan et al. [2]. They reported that eradication or clearance of HP by colloidal bismuth subcitrate (CBS) resulted in a lower relapse rate of 27% compared with 79% after cimetidine treatment during a 1-year follow-up period. The association between HP status after treatment and DU relapse rate was masked in this study since true eradication of HP had not been checked 4 weeks after cessation of therapy [7]. The study of Marshall et al. [3] which checked HP eradication 14 days after the end of treatment found a relapse rate of 25% in patients in whom HP had been eradicated compared with 92% after cimetidine treatment. In two of four patients recolonization with HP was detected very early in this study suggesting, that recrudescence had occurred rather than reinfection. Patients declared as eradicated in this study did not relapse so far. The study of Rauws et al. [4] confirmed the association between the eradication of HP and the absence of DU relapses during follow-up.

In this study the healing rate of DU in patients with persistent HP infection was 61% compared with 92% in those patients with cleared HP infection. These data are very similar to those of Marshall et al. [3]. Eradication as checked 4 weeks after cessation of treatment was 13% with BSS monotherapy, 48% after combined BSS + amoxicillin treatment, and 0% after ranitidine – as expected. The eradication rate with BSS monotherapy was similar to that reported by Börsch et al. [5]. In principle, the eradication rates as checked 4 weeks after therapy are in the same range with BSS as those achieved with CBS [4]. In this study, combined BSS + amoxicillin treatment eradicated HP to the same extent (48%) as with the combination CBS + amoxicillin [3, 4]. The cumulative relapse rates during 6 months of follow-up so far were dependent on the preceding therapy and were 9% in the BSS + amoxicillin group, 44% in the BSS group, and 68% in the ranitidine group (Fig. 1). These relapse rates are very similar to those reported by Marshall et al. [3]. Thirteen patients of the BSS + amoxicillin group and two of the BSS group in whom HP was eradicated did not relapse, clearly supporting the pathogenic role of HP in DU disease. The clear identification of eradication may be improved by the seven biopsies which were performed to detect HP even if patchy colonization was present [8]. Similar results concerning the relation ship between eradication of HP and DU relapse were obtained by Börsch et al. [5] and Borody and Carrick [6]. Recrudescence or reinfection [4] did not occur.

DU relapse rates in patients with *persistent* HP infection also depended on the preceding therapy and were 8% after BSS + amoxicillin, 36% after BSS, and 45% after ranitidine (Table 6). This finding agrees with earlier bismuth therapy studies [1] and suggests that suppression of the HP infection may also decrease DU relapse rates.

The compliance of patients was better in the ranitidine group than in the group treated with BSS or BSS + amoxicillin. Pain relief was faster with ranitidine compared with the other groups but was not accompanied by an increased antacid consumption. However, the symptom scores at the 10-week control were lower after BSS or BSS + amoxicillin therapy compared with ranitidine (Table 8).

We conclude that DU do not reoccur after eradication of HP which confirms the dominant role of HP in DU disease. Even patients with persistent HP infection showed lower relapse rates after BBS or BSS + amoxicillin therapy during the 6-month follow-up.

References

1. Lane MR, Sum PL (1988) Recurrence of duodenal ulcer after medical treatment. Lancet I:1147–1149
2. Coghlan JG, Gilligan D, Humphries H, McKenna D, Dooley C, Sweeney E, Keane C, O'Morain CO (1987) Campylobacter pylori and recurrence of duodenal ulcers – a 12-month follow-up study. Lancet I:1109–1111
3. Marshall BJ, Goodwin CS, Warren JR, Murray R, Blincow ED, Blackbourn SJ, Phillips M, Waters TE, Sanderson CR (1988) Prospective double-blind trial of duodenal ulcer relapse after eradication of Campylobacter pylori. Lancet I:1437–1441
4. Rauws E, Langenberg W, Houthoff HJ, Zanen HC, Tytgat GN (1988) Campylobacter pyloridis-associated chronic active antral gastritis. Gastroenterology 94:33–40
5. Börsch G, Wegener M, Reitemeyer E, Mai U, Opferkuch W (1989) Clinical results and medium term follow-up studies after eradication of Campylobacter pylori. In: Mégraud F, Lamouliatte H (eds) Gastroduodenal pathology and Campylobacter pylori. Elsevier, Amsterdam, p 599
6. Borody TJ, Carrick J (1989) Long term Campylobacter pylori recurrence following apparent eradication. In: Mégraud F, Lamouliatte H (eds) Gastroduodenal pathology and Campylobacter pylori. Elsevier, Amsterdam, p 591
7. Bayerdörffer E, Ottenjann R (1988) The role of antibiotics in Campylobacter pylori associated peptic ulcer disease. Scand J Gastroenterol [Suppl 142] 23:93–100
8. Bayerdörffer E, Oertel H, Lehn N, Kasper G, Mannes GA, Sauerbruch T, Stolte M (1989) Topographic association between active gastritis and Campylobacter pylori colonisation. J Clin Pathol 42:834–839

The Role of Surgery in Patients with *Helicobacter pylori* Infection

M. Büchler [1], P. Malfertheiner [2], H. Friess [1], F. Flock [1], A. Stanescu [2], K. Baczako [3], and H. G. Beger [1]

Since the detection of *Helicobacter pylori* by Warren and Marshall in 1983 [1], this bacterium has significantly changed our opinion on the pathogenesis of gastritis and peptic ulcer disease. Recent investigations have shown, that chronic type B gastritis is almost exclusively induced by *H. pylori* infection [2]. The role of *H. pylori* infection with regard to the development of ulcer disease has still not been completely determined. In duodenal ulcer disease, *H. pylori* infection of the antral mucosa has been demonstrated in 70%–100% of patients [3], and the bacteria have also been found in the duodenal mucosa of up to 83% of patients [4]. Generally *H. pylori* infection within the duodenum is linked with gastric metaplasia in the duodenal mucosa. In the majority of patients, eradication of *H. pylori* infection leads to a long-lasting healing of duodenal ulcer disease [5–7].

To date, surgical procedures play a minor role in the selective treatment of duodenal ulcer disease, although selective proximal vagotomy has been shown to provide the lowest recurrency rates in comparison with all other single treatment protocols. The rate of ulcer recurrency after selective proximal vagotomy was between 6% and 12% 5 years after surgery, when larger patient populations were followed [8–10]. Very few data exist on the effect of surgical treatment on *H. pylori* infection. There is one contribution in the literature, the only study with a prospective design, showing a reduction in *H. pylori* colonization following highly selective vagotomy [11].

The aim of our investigation was to study prospectively the influence of selective proximal vagotomy on *H. pylori* infection, gastric metaplasia, the degree of inflammation, and glycoprotein synthesis of the gastric mucosa.

Patients

Eight patients suffering from chronic duodenal ulcer disease were recruited in a prospective design. There were seven men and one women with a median age of 34 years (range 21–58 years). Exclusion criteria were previous treatment with colloidal bismuth and/or antibiotics. All patients were scheduled for selective proximal vagotomy with ($n = 2$) and without ($n = 6$) pyloroplasty. After informed consent, the patients underwent three upper gastrointestinal (GI) endoscopy

[1] Department of General Surgery, University Hospital, Steinhövelstr. 9, 7900 Ulm
[2] Department of Gastroenterology, Robert-Koch-Str. 8, 7900 Ulm
[3] Institute for Pathology, Heidenheimerstrasse, Ulm, FRG

Helicobacter pylori, Gastritis and Peptic Ulcer
Edited by P. Malfertheiner and H. Ditschuneit

investigations with multiple biopsies: the first one preoperatively, the second 2 weeks postoperatively, and the third 3 months after selective proximal vagotomy. To date, six patients have been followed up for 3 months.

Methods

In summary, 11 biopsies were taken during each endoscopic examination (one corpus, eight antrum, two duodenum). We took four of these biopsies for histology (one corpus, one antrum, two duodenum), a further one from the antrum to carry out an *H. pylori* rapid urease test, two from the antrum for culture studies, and the remainder to analyze glycoprotein synthesis in the gastric antrum.

Histology. Biopsies were fixed in formalin (4%) and processed for routine paraffin embedding. To evaluate and estimate the degree of mucosal inflammation, hematoxylin-eosin stain was employed. The degree of *H. pylori* colonization was analyzed following silver staining.

Culture. H. pylori was cultured for 3–5 days (37 °C) on Wilkens-Chalgren blood agar with Skirrow supplement. The earliest time to consider a culture to be negative was after 10 days.

H. pylori Rapid Urease Test. We used a purchasable test system (CUT, Temmler Pharma Marburg, FRG) which is based on the cleavage of urea by urease produced by *H. pylori*.

Glycoprotein synthesis. Glycoprotein synthesis of gastric surface mucus cells was determined indirectly according to the method used by Baczako and coworkers [12]. The test system is based on the incorporation of radioactive glucosamine (D-[6-^{3}H]glucosamine HCl, Amersham, Braunschweig, FRG). To express glycoprotein synthesis, we calculated glucosamine incorporation (area under the curve) during 2–4-h, and 6-h incubation times.

Definition of Scores. The degree of *H. pylori* colonization (antrum) was scored 0–8. The lowest score, 0, means no *H. pylori* was detectable either by histology, by the rapid urease test, or by culture. The highest score, 8, means massive HP colonization, proven in all three test systems.

The degree of mucosal inflammation was scored 0–3, considering the extent of polymorph nuclear leukocytes (PML) and plasma cellular (PZ) infiltration, according to Whitehead [13], where 0 means no PML and 3 means massive infiltration with PML and PZ.

Results

Preoperatively, all patients were infected with *H. pylori*. The degree of *H. pylori* colonization was 6.75 ± 1.5 ($\bar{x} \pm SD$) (Table 1). Two weeks after selective proximal

Table 1. Incidence and degree of *H. pylori* colonization in gastric antrum

	Prevalence *n* Patients	(%)	Degree (score: 0–8)
preoperatively	8/8	100	6.7 ± 1.5
2 weeks postoperatively	5/8	62	2.5 ± 3.0
3 months postoperatively	2/6	33	2.3 ± 3.7

vagotomy, the rate of *H. pylori* infection was 62%. The degree of *H. pylori* colonization decreased to 2.5 ± 3.0 (Table 1). Three months postoperatively, only two of six patients (33%) were still *H. pylori* positive. The degree of colonization was again decreased (2.3 ± 3.7).

Gastric metaplasia. Preoperatively, five out of eight patients showed gastric metaplasia in their duodenal mucosa. Two weeks after selective proximal vagotomy, the prevalence of gastric metaplasia was 25% (two of eight patients). In only one (17%) out of six patients included in the late follow-up (3 months) were we still able to detect gastric metaplasia.

Degree of Mucosal Inflammation. Preoperatively, the mean score of mucosal inflammation was 1.38 (± 0.52). Two weeks postoperatively, this score decreased to 0.63 ± 0.74. Three months after selective proximal vagotomy, the mean degree was 0.67 ± 0.52.

Glycoprotein Synthesis. Preoperatively, the total integrated glycoprotein synthesis in vitro over a 360-min period was 23.3 ± 11.1 cpm/µg protein. Two weeks postoperatively glycoprotein synthesis was slightly increased to 27.8 ± 13.2 cpm/µg protein. Three months postoperatively, there was an increase in mucus production (31.4 ± 9.7 cpm/µg protein) (Table 2).

Table 2. Glycoprotein synthesis

	Glycoprotein synthesis $\bar{x}$ integrated cpm/µg ± SD
preoperatively	23.3 ± 11.1
2 weeks postoperatively	27.8 ± 13.2
3 months postoperatively	31.4 ± 9.7

Final Considerations

Although our data are based on a small patient population, selective proximal vagotomy seems to influence *H. pylori* infection and the consequences of *H. pylori* colonization such as mucosal inflammation and the presence of gastric metaplasia in the duodenum. Our data confirm the results of the only prospective study which has been carried out so far [11]. The authors of that study showed a significant reduction in *H. pylori* colonization 3–12 months after selective proximal vagotomy. By contrast, there are two retrospective studies, reported 4 years and 1.5–4 years after selective proximal vagotomy in which the authors demonstrated a higher incidence of *H. pylori* infection than in nonoperated patients [14, 15].

Our data demonstrate that selective proximal vagotomy influences the state of *H. pylori* infection by reducing the incidence of infection, the degree of mucosal inflammation, and the presence of gastric metaplasia in the duodenum.

These features correlate with an increase in gastric mucus production as shown by the results of glycoprotein synthesis in gastric epithelial cells. The discrepancy between our data and the retrospective data might be due to a secondary or reinfection occurring a long time after selective proximal vagotomy. It certainly seems worthwhile to follow up our patients for several years in order to observe these possible long-term effects.

On the other hand, our preliminary data describe a new and evidently important effect of selective proximal vagotomy. This procedure may induce ulcer healing not only by reducing acid secretion capacity but also by influencing the state of *H. pylori infection.*

References

1. Warren JR, Marshall BJ (1983) Unidentified curved bacilli on gastric epithelium in chronic gastritis. Lancet I:1273–1275
2. McNulty CAM (1989) Pathogenicity of Campylobacter pylori – A causative factor in gastritis? Scand J Gastroenterol 24 [Suppl 160]:184–187
3. Wyatt JI (1989) The role of Campylobacter pylori in the pathogenesis of peptic ulcer disease. Scand J Gastroenterol 24 [Suppl 157]:7–11
4. Carrick J, Lee A, Hazell S, Ralston M, Daskalopoulos G (1989) Campylobacter pylori, doudenal ulcer, and gastric metaplasia: possible role of functional heterotopic tissue in ulcerogenesis. Gut 30:790–797
5. Marshall BJ, Goodwin CS, Warren JR, Murray R, Blincow ED, Blackbourn SJ, Phillips M, Waters TE, Sanderson CR (1988) Prospective double-blind trial of duodenal ulcer relapse after eradication of Campylobacter pylori. Lancet II:1437–1442
6. Coghlan JG, Gilligan D, Humphries H, McKenna D, Dooley C, Sweeney E, Keane C, O'Morain C (1987) Campylobacter pylori and recurrence of duodenal ulcer-a 12 month follow-up study. Lancet II:1109–1111
7. Borody T, Noonan S, Cole P, Hyland L, Morgan A, Lenne J, Moore-Jones D, Brandl S (1989) Duodenal ulcer recurrence in patients remaining C. pylori-negative long term post-eradication. Gastroenterology 96:A52
8. Holle GE, Frey KW, Thieme C, Hoole FK (1988) Recurrence of peptic ulcer after selective proximal vagotomy and pyloroplasty in relation to changes in clinical signs and symptoms between 1969 and 1983. Surg Gynecol Obstet 167:271–281
9. Byrne DJ, Brock BM, Morgan AG, McAdam WAF (1988) Highly selective vagotomy: a 14-year experience. Br. J Surg 75:869–872
10. Primose JN, Axon ATR, Johnston D (1988) High selective vagotomy and duodenal ulcers that fail to respond to H2 receptor antagonists. Br Med J 296:1031–1035
11. Steer HW, Hawtin P, Newell DG (1988) The effect of surgical treatment of chronic duodenal ulceration on Campylobacter pylori colonization of the stomach. Workshop on Gastroduodenal pathology and Campylobacter pylori. Bordeaux, October (Abstract P153)
12. Baczako K, Melzner I, Malfertheiner P, Pietrzyk C, Brass B (1986) In vitro glycoprotein synthesis and secretion by gastric mucosal tissue of patients with gastritis. Am J Gastroenterol 81:346–353

13. Whitehead R, Truelove SC, Gear MWL (1972) The histological diagnosis of chronic gastritis in fibreoptic gastroskopy biopsy specimens. J Clin Pathol 25:1–11
14. O'Connor HJ, Dixon MF, Wyatt JI, Axon ATR, Ward DC, Dewar EP, Johnston D (1986) Effect of duodenal ulcer surgery and enterogastric reflux on Campylobacter pylori. Lancet II:1178–1181
15. Hölscher AH, Petkaneschkov G, Bollschweiler E (1989) Effekt der proximal-gastrischen Vagotomie auf die Campylobacter Besiedelung des Magens. Z Gastroenterol 27:A563

Is *Helicobacter pylori* Negative Duodenal Ulcer a Separate Disease?

N. J. Talley

Introduction

The etiology of chronic duodenal ulcer, defined as a break in the duodenal mucosa that extends through the muscularis mucosa, is not established although acid is clearly an essential factor [2]. However, the realization that *H. pylori* antral gastritis is very strongly associated with chronic duodenal ulceration has led to new theories regarding the pathogenesis of this disease.

H. pylori Gastritis and Chronic Duodenal Ulcer

In our experience, nearly all patients with a chronic duodenal ulcer crater have *H. pylori* gastritis detected, and others have consistently reported similar findings [9, 13, 20]. While a cause-and-effect relationship is not yet accepted, several lines of evidence strongly support the contention that *H. pylori* is of etiological importance in at least relapsing chronic duodenal ulceration [3, 14].

A possible mechanism by which gastric *H. pylori* infection could induce ulceration in the duodenum has been put forth. It has been postulated that *H. pylori* infects areas of gastric metaplasia found in the duodenum [24] resulting in inflammation of the mucosa and perhaps degradation of the mucus bicarbonate layer [22]. This could allow noxious luminal contents such as acid an opportunity to diffuse closer to the epithelium and thus eventually lead to ulceration. However, while this theory is attractive, it does not appear to explain why large areas of infected antral mucosa do not slough off, or why duodenal ulceration is focal and intermittent.

In an attempt to account for the apparent discrepancies, it has been hypothesized that as *H. pylori* significantly damages the mucus bicarbonate barrier, the bacteria, which are very sensitive to acid, migrate away from the most compromised areas of the gastric mucosa to areas where the mucus layer is still protective [17]. The injured epithelial foci left behind may then be able to regenerate the mucus layer and heal. However, if the bacteria cannot migrate far, as in a very tiny island of gastric metaplasia in the duodenum, or if other factors do not allow the healing process to occur, then focal ulceration develops. This ulceration eliminates the infected area of gastric metaplasia. Evidence to support this hypothesis has been presented elsewhere [17].

Department of Gastroenterology, Mayo Clinic, Rochester, Minnesota 55905 USA

Helicobacter pylori, Gastritis and Peptic Ulcer
Edited by P. Malfertheiner and H. Ditschuneit

H. pylori-Negative Chronic Duodenal Ulcer

Only a minority of patients with chronic duodenal ulcer disease do not have *H. pylori* detected by routine diagnostic methods. Assuming there is truly a duodenal ulcer crater identified, there are five possible explanations for the absence of organisms in the stomach of such patients.

Misclassification of *H. pylori* status

H. pylori gastritis is known to be patchy and therefore sampling error may result in investigators missing the presence of the bacteria. The gold standards for the diagnosis of *H. pylori* infection have traditionally been histology and culture. Histology is reported to have a sensitivity and specificity ranging between 85% and 100%, while culture generally has a lower sensitivity but is 100% specific [6, 18]. However, the gold standard to which these methods should be compared has not been defined; many authors have inappropriately compared histology or culture with combined histology and culture results. Whatever method is used, it is inevitable that a few cases of duodenal ulcer disease will be misclassified; this may account for a substantial portion of *H. pylori*-negative ordinary duodenal ulcer disease reported in the literature. As more sensitive and specific tests become available for the diagnosis of *H. pylori*, which can reliably detect the presence of low numbers of organisms, it will be possible to determine if this is indeed the case.

Extreme Gastric Acid Hypersecretion

Another possible explanation for *H. pylori*-negative duodenal ulcer disease is the presence of the Zollinger-Ellison syndrome. Approximately 90%–95% of patients with a gastrinoma develop ulceration at some time during their disease course [16]. Our group has retrospectively reviewed resected gastric specimens in 18 patients with the Zollinger-Ellison syndrome, of whom 17 had duodenal ulcer disease [8]. The results were compared with 18 controls who had an antrectomy for ordinary chronic duodenal ulcer disease. All specimens were stained with hematoxylin and eosin and Giemsa and were reviewed by an independent, experienced pathologist who was unaware of the diagnosis. Gastric acid secretory studies were available in 17 of the 18 Zollinger-Ellison syndrome patients. It was found that while 89% of the chronic duodenal ulcer patients had *H. pylori* gastritis, only 44% of the Zollinger-Ellison patients had *H. pylori* identified. Peak acid output was significantly reduced in those Zollinger-Ellison syndrome patients who had *H. pylori* infection. Similarly, Koop et al. [10] reported that 6 of 17 patients with the Zollinger-Ellison syndrome had antral *H. pylori* on biopsy, while Saeed et al. [21] found 6 of 20 patients with a gastrinoma had serological evidence of infection. The cumulative data suggest that the prevalence of *H. pylori* in the Zollinger-Ellison syndrome with duodenal ulcer is much less than in ordinary chronic duodenal ulcer disease. Whether this also applies in other rare conditions associated with excessive gastric acid secretion and ulceration, such as systemic mastocytosis, is unknown [1].

Nonsteroidal Antiinflammatory Drug (NSAID) Induced Ulceration

While some have suggested that NSAIDs explain *H. pylori*-negative duodenal ulceration [13], the role of NSAIDs in duodenal ulcer disease is controversial [11, 12]. Acute duodenal damage can be induced by NSAIDs and NSAIDs may increase the risk of duodenal ulcer perforation [4]. On the other hand, Duggan et al. [7] reported that the relative risk of chronic duodenal ulcer in NSAID users was increased only 10% above those not exposed and this increase was not statistically significant. Others have suggested that NSAIDs may protect patients with *H. pylori* from developing duodenal ulcers although the data are inconclusive [15]. It appears likely that, on their own, NSAIDs are unable to explain *H. pylori* negative chronic duodenal ulcer disease.

Not Peptic Ulceration

A number of uncommon diseases can cause ulceration in the duodenum that can be misdiagnosed as ordinary duodenal ulcer disease. Crohn's disease, tuberculosis, lymphoma or carcinoma of the duodenum, carcinoma of the pancreas eroding into the duodenum, or cytomegalovirus infection in patients with AIDS can all cause intractable duodenal ulceration [19].

H. pylori Negative Ordinary Duodenal Ulcer Disease – a True Subgroup?

It has been suggested that *H. pylori*-negative duodenal ulcer disease may form a distinct subgroup. Sobala et al. [23] reported that 5 of 56 patients with duodenal ulcer did not have *H. pylori* infection; these patients smoked less than infected patients. Debongnie et al. [5] found 32 of 183 patients with duodenal ulcer disease were *H. pylori* negative, and these patients were older than infected patients. However, other explanations for the absence of *H. pylori* were not carefully addressed in these studies, and thus the existance of such a subgroup remains to be verified.

Conclusion

There is evidence that *H. pylori* is one of the most important risk factors for the presence of chronic duodenal ulcer disease. Despite this, chronic duodenal ulceration is likely to be a multifactorial disease as most persons with *H. pylori* gastritis never develop an ulcer. The absence of identified *H. pylori* in a patient with chronic duodenal ulceration is most likely to be the result of the bacteria having been missed. Rarely, the Zollinger-Ellison syndrome may explain

H. pylori-negative duodenal ulceration. Whether there is a very small group of chronic duodenal ulcer patients without the Zollinger-Ellison syndrome or other disease who truly have *H. pylori*-negative ulcer disease is not established. Perhaps such cases are rare in referral centers where a high rate of relapsing duodenal ulcer disease is seen; it is conceivable that *H. pylori*-negative duodenal ulcer could be more prevalent in the community or in those who do not have relapsing ulcer disease. Identification of such a group could yield important clues about the mechanisms by which all duodenal ulcers develop.

References

1. Ammann RW, Vetter D, Deyhle P, Tschen H, Sulser H, Schmid M (1976) Gastrointestinal involvement in systemic mastocytosis. Gut 17:107–112
2. Blair AJ, Feldman M, Barnett C, Walsh JH, Richardson CT (1987) Detailed comparison of basal and food-stimulated gastric acid secretion rates and serum gastrin concentrations in duodenal ulcer patients and normal subjects. J Clin Invest 79:582–587
3. Coghlan JG, Gilligan D, Humphries H, McKenna D, Dooley C, Sweeney E, Keane C, O'Morain C (1987) Campylobacter pylori and recurrence of duodenal ulcers – a 12-month follow-up study. Lancet 2:1109–1111
4. Collier DS, Pain JA (1985) Non-steroidal anti-inflammatory drugs and peptic ulcer perforation. Gut 26:359–363
5. Debongnie JC, Legros G, Beyaert C (1989) C. pylori negative duodenal ulcer: an age related phenomenon? Klin Wochenschr [Suppl 18] 67:15
6. Deltenre M, Glupczynski Y, DePrez C, et al. (1989) The reliability of urease tests, histology and culture in the diagnosis of Campylobacter pylori infection. Scand J Gastroenterol [Suppl 160] 24:19–24
7. Duggan JM, Dobson AJ, Johnson H, et al. (1986) Peptic ulcer and non-steroidal anti-inflammatory agents. Gut 27:929–933
8. Fich A, Talley NJ, Shorter RG, Phillips SF (1990) The Zollinger-Ellison syndrome: Relation to Helicobacter pylori – associated chronic gastritis and gastric acid secretion. Dig Dis Sci (in press)
9. Graham DY (1989) Campylobacter pylori and peptic ulcer disease. Gastroenterology 96:615–625
10. Koop H, Stumpf M, Eissele R, Lamberts R, Creutzfeldt W, Arnold R (1989) Antral Campylobacter pylori in different disease states of gastric acid secretion. Gastroenterology 96:A267
11. Kurata JH (1989) Ulcer epidemiology: an overview and proposed research framework. Gastroenterology 96:569–580
12. Langman MJS (1989) Epidemiologic evidence on the association between peptic ulceration and anti-inflammatory drug use. Gastroenterology 96:640–646
13. Marshall BJ, McGechie DB, Rogers PA, Glancy RJ (1985) Pyloric Campylobacter infection and gastroduodenal disease. Med J Aust 142:439–444
14. Marshall BJ, Goodwin CS, Warren JR, et al. (1988) Prospective double-blind trial of duodenal ulcer relapse after eradication of Campylobacter pylori. Lancet 2:1437–1442
15. Martin DF, Montgomery E, Dobek AS, Patrissi GA, Peura DA (1989) Campylobacter pylori, NSAIDs, and smoking: risk factors for peptic ulcer disease. Am J Gastroenterol 84:1268–1272
16. McGuigan JE (1989) The Zollinger-Ellison syndrome. In: Sleisenger MH, Fordtran JS (eds) Gastrointestinal disease, 4th edn. Saunders, Philadelphia, pp 909–924
17. Ormand JE, Talley NJ (1989) Campylobacter pylori, mucus and peptic ulceration: a dynamic interaction. J Clin Gastroenterol 11:492–495
18. Ormand JE, Talley NJ (1990) Helicobacter pylori: controversies and an approach to management. Mayo Clin Proc 65:414–426
19. Pounder RE (1987) What is an intractable duodenal ulcer and how should it be managed? Aliment Pharmacol Ther 1:439S–446S

20. Rauws EAJ, Langenberg W, Houthoff HJ, et al. (1988) Campylobacter pyloridis-associated chronic active antral gastritis. A prospective study of its prevalence and the effects of antibacterial and anti-ulcer treatment. Gastroenterology 94:33–40
21. Saeed ZA, Evans DJ, Evans DG, Maton PN, Graham DY, Jensen RT (1989) Campylobacter pylori and the Zollinger-Ellison syndrome. Gastroenterology 96:A433
22. Slomiany BL, Bilski J, Sarosiek J, et al. (1987) Campylobacter pyloridis degrades mucin and undermines gastric mucosal integrity. Biochem Biophys Res Commun 144:307–314
23. Sobala GM, Dixon MF, Axon ATR (1989) Campylobacter pylori negative peptic ulcers. Klin Wochenschr [Suppl 18] 67:64
24. Wyatt JI, Rathbone BJ, Dixon MF, et al. (1987) Campylobacter pyloridis and acid induced gastric metaplasia in the pathogenesis of duodenitis. J Clin Pathol 40:841–848

Helicobacter pylori in the Pathogenesis of Peptic Ulcer – Evidence in Favour

A. T. R. Axon

Although the concept of contagious illness was well recognised in mediaeval times, the medical establishment has been slow to accept infectious aetiologies for diseases. The Hungarian obstetrician, Semmelweis, working in Vienna in the 1850s showed that he could reduce the mortality rate of purpural sepsis in his wards from 18 % to 1 % if doctors and medical students were obliged to wash their hands before examining patients, yet he was disbelieved, hounded from Vienna, and eventually committed to a mental institution.

Marshall and Warren's [1] hypothesis that *Helicobacter pylori* (HP) plays a pivotal role in the pathogenesis of duodenal ulcer (DU) is still disbelieved by many, but the evidence in its support is now very compelling.

Epidemiological data [2] from developed countries suggest that around 40 % of the adult population is colonised with HP compared with 95 % of those who have duodenal ulcer. Conversely, duodenal ulcer is virtually unknown in the aboriginal community of Australia where the prevalence of colonisation is below 5 %. If an individual is infected with HP the risk of developing peptic ulcer is 20 times that of an uninfected person. The high prevalence of HP in duodenal ulcer can be accounted for by only one of three explanations, either (a) duodenal ulcer causes HP infection, (b) HP causes duodenal ulcer; or (c) both duodenal ulcer and HP arise from a third common factor.

As HP persits after duodenal ulcer healing and is present in "normal" individuals it cannot possibly be secondary to ulceration, so either explanation (b) or (c) is dorrect.

Duodenal ulcer is almost invariably associated with antral gastritis. The risk of developing peptic ulcer is 20–30 times higher in individuals with antral gastritis than in normals [3]. Antral gastritis is also known to be associated with HP. Could this be the common factor responsible for both HP and duodenal ulcer? This theory is not tenable because there is irrefutable evidence that HP is the underlying cause of the chronic gastritis, not vice versa [4]. HP associated gastritis (type B gastritis) is pathologically distinct. The organism is almost invariably associated with it, but not with other types of gastritis and this implies it is a primary not a secondary invader. Its presence is associated with a cellular inflammatory reaction by the host and organisms are found within polymorphonuclear leukocytes; local and systemic antibody production is stimulated and antibody has been detected coating the organism in situ. Clearance of the organism by pharmacalogical

Gastroenterology Unit, The General Infirmary, Leeds, LS1 3EX, U.K.

Helicobacter pylori, Gastritis and Peptic Ulcer
Edited by P. Malfertheiner and H. Ditschuneit

Table 1. Helicobacter in duodenal biopsy specimens [from 5]

HP+/total duodenal ulcer	HP+/total duodenitis	HP+/total normal	Reference
45/53	36/39	0/58	Johnson 1986
8/11	–	–	Steer 1984
8/13	11/21	0/256	Wyatt 1987
7/24	–	–	Bode 1987
14/14	5/16	–	Caselli 1988
Total 82/115 (71%)	52/76 (68%)	0/314 (0%)	

means leads to histological resolution of the gastritis. Volunteers taking the organism by mouth and others accidentally infected have developed a self-limited gastritic illness associated with histological gastritis which progresses to typical type B chronic gastritis. A similar histological picture can be induced by feeling the organsism to experimental animals. These data confirm that HP is a pathogen and is the cause, not the effect, of this commonly seen gastritis.

The association between antral gastritis and peptic ulcer has been recognised for many years, but the presence of duodenitis is an even stronger predictor for duodenal ulcer, being much less common in the general population than gastritis, and HP appears to be responsible for duodenitis as well as type B gastritis. Table 1 [5] shows the prevalence of duodenal infection with HP in duodenal ulcer and duodenitis. In the series of five papers where the organism was sought, duodenal infection was found in 82 of 150 cases of duodenal ulcer (71%) and 52 of 76 cases of duodenitis (68%). This compared with a prevalence of 0 of 314 (0%) in normal individuals. Only a minority of patients whose stomachs are colonised with HP develop duodenitis because HP is able to infect only gastric epithelial cells. These are not found in the duodenum except where gastric heterotopia exists or, more commonly, when gastric metaplasia has occurred. In a group of 269 dyspeptic patients studied by Wyatt et al. [6], 34 had active duodenitis and of these 31 (91%) had gastric metaplasia in the duodenum. This compares with only 12 of the 235 (5%) patients who had no inflammation. From these data it is clear that for duodenitis to arise, gastric metaplasia must be present within the duodenum, but inflammation will still not occur unless the patient's stomach is also colonised by HP; naturally all colonised patients have antral type B gastritis too. Under these circumstances the organism normally resident in the stomach infects the metaplastic tissue in the duodenum to give rise to duodenitis. This association is clearly shown in a Venn diagram (Fig. 1) and accounts for the two observations: first, that duodenal ulcer is nearly always associated with HP infection (and therefore antral gastritis) and, second, that only a small proportion of those with gastric HP develop duodenitis and duodenal ulceration.

The mechanism whereby duodenal HP infection leads to duodenal ulcer is unknown. The organism produces cytotoxins and proteases which may modify mucin. The organism adheres to the underlying cells and appears to cause degenerative changes within them. The immune reaction stimulated may lead to

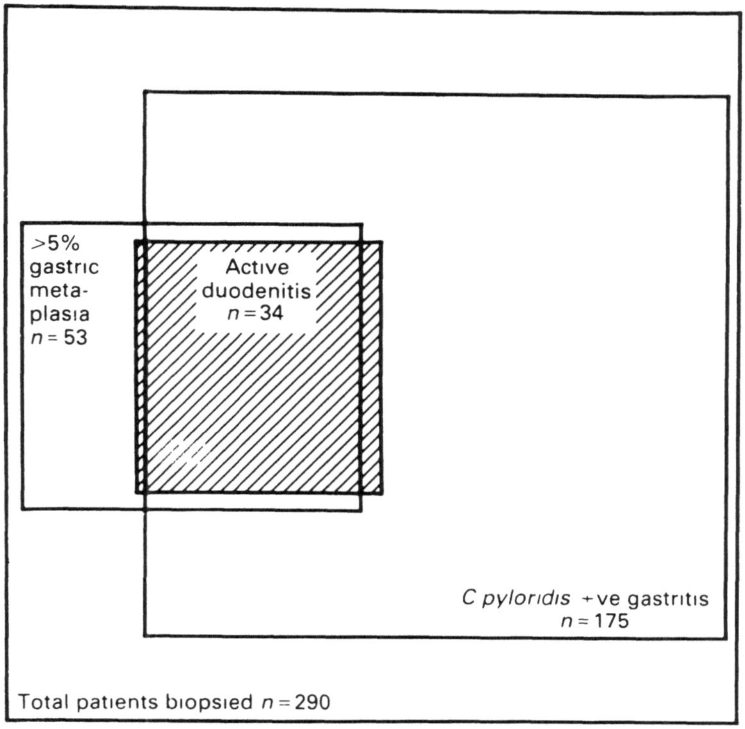

Fig. 1. Active duodenitis, gastric metaplasia in the duodenum and HP-associated gastritis [6]

autogenous tissue damage. Clearly there are a number of ways in which the organism might undermine the mucosal protection mechanisms [7]. In addition recent work has shown that gastric antral infection with HP leads to an increase of plasma gastrin [8] and a higher postprandial acid output. This may be responsible for the generally higher acid secretion seen in populations with duodenal ulcer which in turn may increase the likelihood of ulceration either by a direct effect or through the induction of gastric metaplasia within the duodenum, thus allowing duodenitis to develop.

The most persuasive evidence in favour of the role of HP in duodenal ulceration, however, comes from studies where the organism has been eliminated and patients have been followed for ulcer relapse. A number of earlier trials had shown that duodenal ulcer patients healed with H2-receptor antagonists relapsed more quickly than those given colloidal bismuth subcitrate. Bismuth is known to clear HP temporarily in around 40% of individuals and to eliminate it in 10%–20%. This antibacterial action might account for the reduced relapse rate. More recent trials have been undertaken in patients whose ulcers have been healed by a variety of regimes and who have then been followed up both for HP status and recurrence. Three studies (Table 2) have assessed HP status 1 month after initial healing. When HP was present 62 of 84 (74%) relapsed, compared with only 10 of

Table 2. Eradication of HP (1 month after treatment) and ulcer relapse at 1 year or 6 months [from 9]

Author	HP+	HP−
Coghlan	19/24	4/15
Marshall	37/44	6/26
Rauws [a]	6/16	0/15
Total	62/84 (74%)	10/56 (18%)

[a] Rauws' folow-up was for 6 months; the others 1 year

Table 3. Eradication of HP (assessed at end point) and ulcer relapse [from 9]

Author	HP+	HP−
Coghlan	22/29	1/10
Marshall	38/47	5/23
Smith	25/36	0/8
Lambert [a]	25/33	0/12
Borody	2/3	0/28
Rauws [a]	6/16	0/15
Total	118/164 (72%)	6/96 (6%)

[a] Lambert and Rauws follow-up was for 6 months; the others 1 year

56 (18%) when HP had apparently been eradicated. Amalgamated data from six studies (Table 3) show that in those where HP was eradicated the relapse rate was 6 out of 96 (6%) compared with 118 of 164 (72%) where HP persisted. The 6% recurrence rate is what would be predicted on the basis of epidemiological studies which suggest that a small proportion of ulcers are not due to infection with HP.

It might be argued that the lower relapse rate in those without Helicobacter is merely the effect of the use of bismuth salts. A recent study [9] however, shows this not to be the case. A total of 36 individuals with resistant ulcer were randomised to either colloidal bismuth subcitrate or colloidal bismuth subcitrate plus antibiotics. Of the 31 ulcers which healed, 15 remained HP negative and 16 HP positive. At 6 months, none of the first group had relapsed whereas 37% of the HP-positive group had developed recurrent ulcers. This was in spite of the fact that both groups received an equivalent course of colloidal bismuth subcitrate.

The evidence in favour of HP as being the prime aetiological factor in over 90% of duodenal ulcers is very persuasive and further carefully controlled clinical trials are needed to confirm the hypothesis.

Joseph Lister, impressed by the work of the great French chemist Louis Pasteur, hypothesised that wound infection and gangrene might be caused by microorganisms and that aseptic surgery would reduce these complications. His findings were published in *Lancet* in 1867 [10]. However, this work which was to revolutionise the practice of surgery was disbelieved in some parts of the world. Nine years after publication, Samuel Gross, the virtually unchallenged leader of American surgery, wrote "little if any faith is placed by any enlightened or experienced surgeon on this side of the Atlantic in the so called carbolic acid treatment of Professor Lister" [11].

References

1. Marshall BJ, Warren JR (1984) Unidentified curved bacilli in the stomach of patients with gastritis and peptic ulceration. Lancet I:1311–1314
2. Dwyer B, Kaldor J, Tee W, Raios K (1989) The prevalence of Campylobacter pylori in human populations. In: Rathbone BJ, Heatley RV (eds) Campylobacter pylori and gastro-duodenal disease. Blackwell, Oxford, pp 190–196
3. Sipponen P, Seppala K, Aarynen M, Helske T, Kettunen P (1989) Chronic gastritis and gastroduodenal ulcer: a case control study on risk of coexisting duodenal or gastric ulcer in patients with gastritis. Gut 30:922–929
4. Axon ATR (1988) Campylobacter pylori. In: Pounder RE (ed) Recent advances in gastroenterology, no 7. Livingstone, Edinburgh, pp 225–244
5. Wyatt JI (1989) Campylobacter pylori, duodenitis and duodenal ulceration. In: Rathbone BJ, Heatley RV (eds) Campylobacter pylori and gastrodudenal disease. Blackwell, Oxford, pp 117–124
6. Wyatt JI, Rathbone BJ, Dixon MF, et al. (1987) Campylobacter pyloridis and acid-induced gastric metaplasia in the pathogenesis of duodenitis. J Clin Pathol 40:841–848
7. Rathbone BJ, Heatley RV (1989) Possible pathogenic mechanisms in Campylobacter pylori infection. In: Rathbone BJ, Heatley RV (eds) Campylobacter pylori and gastrodudenal disease. Blackwell, Oxford, pp 203–208
8. Levi S, Beardshall K, Haddad G, Playford R, Gosh P, Calam J (1989) Campylobacter pylori and duodenal ulcers: the gastrin link. Lancet I:1167–1168
9. Rauws EAJ, Tytgat GNJ (eds) Campylobacter pylori. Department of Gastroenterology and Hepatology, Academic Medical Centre, Amsterdam
10. Lister J (1867) On the antiseptic principle in the practice of surgery. Lancet Sept 21:353–356
11. Lyons AS, Petrucelli RJ Medicine – an illustrated history. Abrodel, New York

Helicobacter pylori as a pathogenetic Factor in Peptic Ulcer – the Argument Against

W. L. PETERSON

Helicobacter (H.) pylori burst upon the scene (as *Campylobacter pyloridis*, later *pylori*) in 1983 and has already become the most rapidly moving area in the field of peptic ulcer disease. Most investigators now accept that *H. pylori* is the cause of chronic (type B) gastritis, even though a satisfactory animal model is not yet available. Where controversy remains, however, is whether *H. pylori* gastritis can produce gastrointestinal symptoms (i.e., nonulcer dyspepsia) or whether *H. pylori* gastritis is an important prerequisite to the development of peptic ulcers. If it were, then the approach to ulcer therapy would be drastically altered. Indeed, it is my perception that many investigators and clinicians, especially outside the United States, have accepted this role for *H. pylori* and are already treating ulcer patients with antimicrobial therapy. It is my contention that the available evidence to support the role of *H. pylori* as a pathogenetic factor in peptic ulcer is inadequate to justify the use of antimicrobials.

Proponents of the importance of *H. pylori* in peptic ulcer point to several observations to support their case: a) most (probably all) patients with duodenal ulcer or gastric ulcer not associated with NSAID use can be found to have *H. pylori* gastritis; b) duodenal ulcers are believed to occur in areas of gastric metaplasia, often infected with *H. pylori*; c) eradication of *H. pylori* leads to prolonged remission from duodenal ulcer recurrence. Each of these points can be, in my opinion, effectively argued.

Most Patients with Peptic Ulcer have *H. pylori* Gastritis

There is no question that this is the case. However, a large proportion of healthy subjects with no history of ulcer and no endoscopic evidence of ulcer also are infected with *H. pylori*. Indeed, close to 50% or more of healthy humans over 60 years of age can be found to have *H. pylori* [1, 2]. Why do only a small proportion of subjects with *H. pylori* gastritis develop ulcers? The answer is clearly not related to excess acid secretion. Although at least some acid is necessary for the development of a peptic ulcer, only about one-third of duodenal ulcer patients have acid secretory values above the upper limits of normal and many patients with gastric ulcer have low levels of acid secretion [3]. Either there are factors other than *H. pylori* and acid or *H. pylori* gastritis is an unrelated phenomenon.

Professor of Medicine, University of Texas Southwestern, Medical School, Dallas, TX

Helicobacter pylori, Gastritis and Peptic Ulcer
Edited by P. Malfertheiner and H. Ditschuneit

Duodenal Ulcers Are Believed to Occur in Areas of Gastric Metaplasia, Often Infected With *H. pylori*

Since it is already accepted that patients with duodenal ulcer are infected with *H. pylori*, it should come as no surprise that where there are areas of metaplastic gastric epithelium in the duodenal bulb, such mucosa may harbor *H. pylori*. However, in at least one study where duodenal ulcers were biopsied, only 50% of such tissue was infected [4]. Furthermore, data have been presented which suggest that gastric metaplasia in the duodenal bulb may be a *consequence* of ulceration and not a prerequisite. Therefore, the presence of gastric metaplasia, even if infected, may represent an epiphenomenon rather than a pathogenetic factor.

Eradication of *H. pylori* Is Associated with Prolonged Remission from Duodenal Ulceration

There are several reports of studies in which patients with duodenal ulcer were treated with a regimen which not only healed the ulcer but eradicated *H. pylori* [4–8]. Such reports suggest that these patients, when followed for up to 12 months, have a lower incidence of ulcer recurrence than patients with healed ulcers in whom *H. pylori* was not eradicated. Unfortunately, only three of these reports represent randomized trials [4–6], and only two have been published in complete form [4, 5]. Results of these two trials are shown in Table 1. These results are, admittedly, exciting and provocative if taken at face value and, if confirmed, offer the type of evidence necessary to support antimicrobial therapy of duodenal ulcer. However, caution is needed in interpretation of these data for the following reasons: a) These studies are poorly blinded, a well-recognized flaw in controlled trials which can lead to bias. b) Only 20 patients treated with a bismuth regimen, but in whom *H. pylori* were still present, have been followed. Bismuth clearly has effects on gastroduodenal tissue apart from whatever effects it has on *H. pylori* [9, 10]. Indeed, bismuth alone eradicates *H. pylori* only about 25% of the time, and yet ulcers heal 70% of the time. Thus, it is possible that the long-term benefit of bismuth is related not to eradication of *H. pylori* but to its other properties. We

Table 1. Twelve-month recurrence of duodenal ulcer as a function of *H. pylori* status after initial healing with cimetidine alone, cimetidine ± tinidazole, bismuth alone, or bismuth ± tinidazole

Reference	Initial therapy	Recurrent ulcer	
		Hp+	Hp−
[5]	Cimetidine	11/14	1/4
	Bismuth [a]	8/10	3/11
[4]	Cimetidine ± tinidazole	30/34	0/1
	Bismuth [a] ± tinidazole	7/10	6/25
		56/68 (82%)	10/41 (24%)

[a] DeNol

need follow-up of larger numbers of patients whose ulcers were healed with bismuth but in whom *H. pylori* was not eradicated. c) Ulcers still recur in almost one-fourth of patients in whom *H. pylori* is eradicated.

What Is Needed

In today's scientific and medical community, one cannot justify the use of a therapeutic agent simply because it has not been shown to be ineffective. Rather, the burden is on those who would use such therapy to provide scientifically valid data to support their point. Anecdotes, unblinded, and unrandomized trials with small sample sizes are simply unacceptable. The literature is replete with examples (e.g., gastric freezing) of therapy believed effective until proper, randomized controlled trials showed no difference from placebo or sham therapy. What is needed are well-designed, randomized studies with large numbers of patients where stringent attempts are made to blind all involved. If such studies confirm that patients treated with a particular regimen and in whom *H. pylori* is eradicated have significantly lower recurrence than patients treated with the same regimen and in whom *H. pylori* is not eradicated, all but the most skeptical will be convinced as to the importance of *H. pylori* in ulcer disease.

What to Do Until More Data Are Available

There may be a tendency to go ahead and employ powerful antimicrobial therapy in patients with peptic ulcer while waiting for more data to come forth. In my opinion, this is to be condemned except in the situation of randomized trials or very special circumstances. Not only is the widespread use of "unproven" therapy scientifically unappealing, patients may develop antibiotic-induced colitis and other microbes may acquire resistance to the antibiotics employed.

Perspective

Even if *H. pylori* is shown by the studies described above to be as important a factor in the formation of peptic ulcer as in gastric acid, there still must be other important factors. For example, sucralfate has no effect on gastric acidity or *H. pylori* and yet some data suggest it may reduce the incidence of ulcer recurrence. As another example, *H. pylori* gastritis tends to be rather diffuse, and yet gastric ulcers are focal. For that matter, why do patients with duodenal ulcer who have plenty of *H. pylori* infected antral epithelium not develop gastric ulcers? Thus, the resolution one way or the other of the *H. pylori* chapter will in no way close the book on peptic ulcer disease. The chapter itself, however, is certainly an exciting one.

References

1. Dooley CP, Cohen H, Fitzgibbons PL, et al. (1989) Prevalence of Helicobacter pylori infection and histologic gastritis in asymptomatic persons. N Engl J Med 321:1562–1566
2. Graham DY, Klein PD, Opekun AR, Boutton TW (1988) Effect of age on the frequency of active Campylobacter pylori infection diagnosed by the [^{13}C] urea breath test in normal subjects and patients with peptic ulcer disease. J Inf Dis 157:777–780
3. Soll AH (1989) Duodenal ulcer and drug therapy. In: Sleisenger MH, Fordtran JS (eds) Gastrointestinal disease, 4th edn. Saunders Philadelphia, pp 824–825
4. Marshall BJ, Goodwin CS, Warren JR, et al. (1988) Prospective double-blind trial of duodenal ulcer relapse after eradication of Campylobacter pylori. Lancet II 1437–1442
5. Coghlan JG, Humphries H, Dooley C, et al. (1987) Campylobacter pylori and recurrence of duodenal ulcers – a 12-month follow-up study. Lancet II:1109–1111
6. Smith AC, Price A, Borriello P, Levi AJ (1988) A comparison of ranitidine and tripotassium dicitrato-bismuth (TDB) in relapse rate of duodenal ulcer. The role of Campylobacter pylori. Gastroenterology 94:A431
7. Lambert JR, Borromeo M, Korman MG, Hansky J, Eaves ER (1987) Effect of colloidal bismuth (De-Nol) and healing relapse of duodenal ulcer – role of Campylobacter pyloridis. Gastroenterology 92:1489
8. Borody TJ, Cole P, Noonan S, et al. (1988) Long term Campylobacter pylori recurrence post-eradication (abstract). Gastroenterology 94:A43
9. Konturek SJ, Dembinski A, Warzecha Z, Bielanski W, Brzozowski T, Drozdowicz D (1988) Epidermal growth factor (EGF) in the gastroprotective and ulcer healing actions of colloidal bismuth subcitrate (De-Nol) in rats. Gut 29:894–902
10. Shorock CJ, Crampton JR, Gibbons LC, Reed WDW (1989) Effect of bismuth subcitrate on amphibian gastroduodenal bicarbonate secretion. Gut 30:917–921

Helicobacter pylori in Duodenal Ulcer Pathogenesis – a Comment

J. WYATT

Duodenal ulcer is well known to be a multifactorial disease, and a wide collection of aetiological agents and pathogenetic mechanisms have been claimed to be related to its development. The arguments concerning the interaction of *H. pylori* and other factors in ulcer pathogenesis are introduced by an overview of the various aetiological agents currently known (see p. 271).

The object of the papers of this part is to determine the relation of *H. pylori* to these factors. Starting from the morphological evidence that duodenal inflammation develops when *H. pylori* infects gastric metaplasia (GM) in the duodenum (Fig. 1), the hypothesis is exponded to encompass other contributory factors in duodenal ulceration. Finally, by treating duodenal ulcer by acid suppression or by eradicating *H. pylori*, the relationship of acid and the bacteria to the development of duodenal pathology can be tested.

The nature of the epithelial cells of gastric metaplasia is discussed in detail (see p. 279, 292). The cells derive from the mucosal crypts and migrate up the villi to cover the tip of the villus. Similar "metaplasia" occurs elsewhere in the gastrointestinal tract, and may there represent a mechanism for delivery of epidermal growth

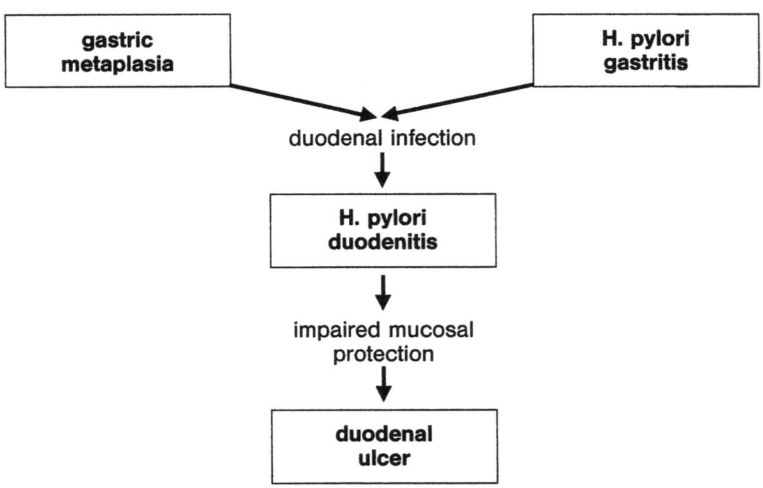

Fig. 1. Hypothesis: the interaction of *H. pylori* and gastric metaplasia in the duodenum

Pathology Department, St. James's University Hospital, Leeds LS97TF, UK

Helicobacter pylori, Gastritis and Peptic Ulcer
Edited by P. Malfertheiner and H. Ditschuneit

factor (EGF) in sites of ulceration. EGF is normally produced in the duodenum by Brunner's glands, and further work is needed to determine the relation of the extent of GM to duodenal EGF content.

Although GM in the duodenum is like gastric foveolar epithelium in most respects, there are some subtle differences detectable by, for example, lectin labelling of carbohydrate moieties. However, from the point of view of specificity of *H. pylori*, the localisation of the bacterium on areas of GM in the duodenum parallels its specificity for gastric epithelium in the stomach. Whatever substances subserve the receptor/ligand function of the interaction, the specificity is the same at both sites.

The development of GM as a response to mucosal injury may be related to acid injury in the duodenum, and also in theory to other causes of mucosal damage. There is as yet no evidence relating it to nonsteroidal antiinflammatory drug (NSAID) consumption or smoking. It is acquired and commoner in men than women, which may be a reflection of greater acid secretion in males. A further factor in extending GM may be the mucosal damage consequent on duodenal inflammation itself, such that once duodenitis developes the resulting damage increases the extent of GM, leading to a vicious circle of increasing mucosal damage.

There are a number of physiological differences in patients with duodenal ulcer compared to normals. These could have a role in duodenal ulcer pathogenesis, and they may be related either as cause or effect to the presence of antral inflammation and *H. pylori*. The questions of motility, pepsinogen secretion and acid secretion in relation to gastrin are considered in detail.

A number of studies of disturbed gastroduodenal motility are reported see p. 306. While abnormal motility could be demonstrated, its pathogenetic significance in duodenal ulcer disease remains to be determined. Furthermore, no differences have so far been demonstrated between *H. pylori* positive and negative subjects.

Abnormalities of pepsin secretion are well recognised in patients with duodenal ulcer disease, both in the type and rate of secretion (see p. 312). Serum pepsinogen measurements reflect the pepsin secretion of the gastric fundic mucosa, and increased serum pepsinogen is seen transiently in acute gastric damage, reflecting its release into the blood stream on breakdown of chief cells, and chronically in patients with hypersecretion of pepsin. Typically this is present in duodenal ulcer patients in parallel with their increased parietal cell mass. The effect on serum pepsinogen of eradication of *H. pylori* would determine whether hyperpepsinogenaemia is a manifestation of a patient's constitutional predisposition to duodenal ulcer disease, or is a result of *H. pylori* gastritis.

The ratio of pepsinogen I and II provides a marker for the extent and degree of atrophy of gastritis. An increased proportion of pepsinogen II (produced from antral type mucosa and the duodenum) occurs when atrophic gastritis developes, as seen in populations with high risk of gastric carcinoma; this profile is rare in duodenal ulcer.

Disturbances in dynamics of gastrin secretion and gastric acidity are well recognised in duodenal ulcer and have recently been linked to the presence of *H. pylori*. Inappropriate fasting hypergastrinaemia in duodenal ulcer patients appears to be dependent on the presence of *H. pylori* and is corrected if *H. pylori* is eradicated. It is not yet clear whether this "gastrin link" between *H. pylori* and acid secretion in duodenal ulcer disease is due to the urease activity of *H. pylori* leading to an increase in pH at the antral mucosal surface, or whether it is indirectly linked to mucosal inflammation (see p. 317).

The role of *H. pylori* in duodenal ulcer disease and its interaction with other pathogenetic factors can be tested by the effects of eradication of the bacteria on gastroduodenal pathophysiology. Treatment trials reported to date confirm the central role of *H. pylori* in ulcer relapse (see p. 327). The effect of eradication on GM, pepsinogen and motility remains to be determined.

Similarly, the role of acid secretion in the development of GM and inflammation can be tested by the effect of medical or surgical measures in reducing acid secretion and the reduction in GM shown to occur following highly selective vagotomy (see p. 335); however, there is also an unexpected decrease in the frequencey of *H. pylori* following the operation, which remains to be explained. It is probable that, besides the reducing acid secretion, selective vagotomy alters the gastric milieu on other respects to one less favourable for *H.pylori* colonisation. Finally, if *H. pylori* is aetiologically important in most duodenal ulcers, then ulcers in *H. pylori* negative patients should show different, distinct aetiological factors. This is the case, with these rare ulcers being associated with NSAID use or Zollinger-Ellison Syndrome (ZES) in most cases (see p. 340). Interestingly, ZES patients who were also *H. pylori* positive presented with ulceration at lower acid

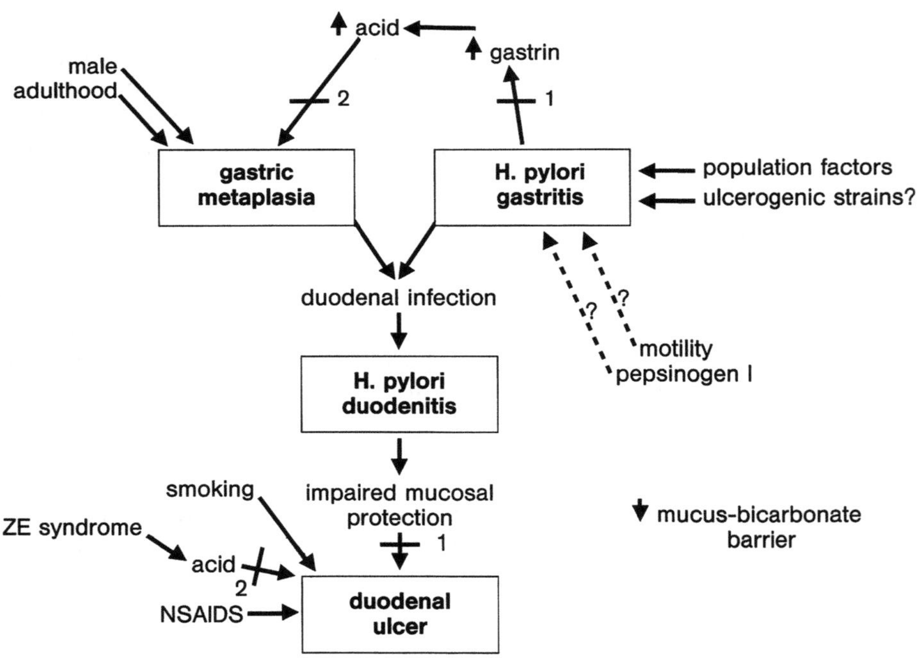

Fig. 2. Possible expansion of hypothesis in Fig. 1 to include other factors in duodenal ulcer pathogenesis. *1*, Effect of eradication of *H. pylori* on pathogenetic pathway. *2*, Effect of surgically treating hyperchlorhydria on pathogenetic pathway

output, suggesting that in this group the presence of *H. pylori* lowers the threshold of acid hypersecretion necessary for ulceration to develop.

A scheme for including all proposed factors in duodenal ulcer is shown in Fig. 2. This is a testable model, since the effect on the interactions of removing *H. pylori* would separate those aspects of duodenal ulcer pathogenesis which represent a constitutional predisposition to the disease, those which require the presence of *H. pylori* to be manifested, and those which are direct effects of the presence of the organisms.

Nonulcer Dyspepsia and Motility

Is *Helicobacter pylori* a Cause of Nonulcer Dyspepsia?

N. J. TALLEY

Introduction

The term "dyspepsia" means different things to different physicians. While no internationally accepted definition exits, one group of investigators defined dyspepsia to be upper abdominal or retrosternal pain, discomfort, heartburn, nausea, vomiting, or other symptom considered to be referrable to the proximal alimentary tract [8]. Although such a definition is probably too broad to be clinically useful, other groups have approached the problem in this way [2, 10, 22]. The majority of patients who present to physicians with chronic or recurrent dyspepsia, however it is defined, do not have a peptic ulcer or definite evidence of organic disease when investigated; such patients have been generally labelled as suffering from nonulcer dyspepsia (NUD) [48].

The etiology of nonulcer dyspepsia is not established. However, there has recently been enormous interest in the relationship between *Helicobacter (Campylobacter) pylori* gastritis and nonulcer dyspepsia. Indeed, the role of gastritis in the genesis of dyspeptic symptoms has been argued about for decades [45, 48], and the argument has been resurrected following the realization that a bacterial infection is very strongly associated with and very likely causes chronic inflammation in the stomach [26, 28]. *H. pylori* is found in approximately 50% of patients with nonulcer dyspepsia although the exact figure has varied from study to study reflecting the different populations evaluated [1, 5, 9, 15, 24, 26, 32, 35, 43] (Table 1). This has led to the hypothesis that *H. pylori* is the cause of symptoms in such patients.

The Concept of Cause

A cause can be defined as an event, condition, or characteristic that plays an essential role in producing the occurrence of a disease or disorder [38]. However, it can be argued that the cause of any disease or effect must consist of a constellation of components that act together. For example, turning on a light switch causes the light to go on, indicating a cause and effect relationship; yet, there are unseen component causes that act in concert to produce the effect (such as the presence of electrical current and a functioning light bulb), and the absence of any of these will

Assistant Professor of Medicine, Mayo Medical School, Rochester, MN 55905 USA

Helicobacter pylori, Gastritis and Peptic Ulcer
Edited by P. Malfertheiner and H. Ditschuneit

Table 1. Symptoms associated with *Helicobacter pylori*-positive nonulcer dyspepsia

Author	Nonulcer dyspepsa (*n*)	*H. pylori* positive (%)	Individual symptoms(s) associated with *H. pylori*[a]
Marshall and Warren [26]	65	58	"Burping"
Rokkas et al. [35]	55	45	Postprandial belching
Rathbone et al. [32]	193	54	Regurgitation (not esophagitis)
Andersen et al. [1]	33	39	Nil
Loffeld et al. [24]	109	56	Nil
Börsch et al. [5]	69	52	Flatulence, a *negative* predictor
Sobala et al. [43]	186	41	Nil
Collins et al. [9]	18	50	Nil
Guerre et al. [15]	96	40	Nil

[a] Significant differences between nonulcer dyspepsia patients with and without *H. pylori*

prevent the effect. A sufficient cause can be considered to be a set of minimal conditions and events that inevitably produce disease; all component causes must be present before disease develops [37]. The same concept can be applied to patients with any chronic disorder including nonulcer dyspepsia. In each case there is almost certainly a constellation of component causes that together lead to symptoms, and it is conceivable that *H. pylori* is one of these components (Fig. 1).

To help judge if a factor such as *H. pylori* is causal, Hill [17] described nine criteria; the more criteria that are present, the more reasonable is the case for causality. The relationship between these criteria and *H. pylori* in chronic nonspecific gastritis and nonulcer dyspepsia is presented in Table 2 [44]. Before a cause and effect relationship between *H. pylori* and nonulcer dyspepsia can be accepted, at the very least it needs to be shown that *H. pylori* is truly associated

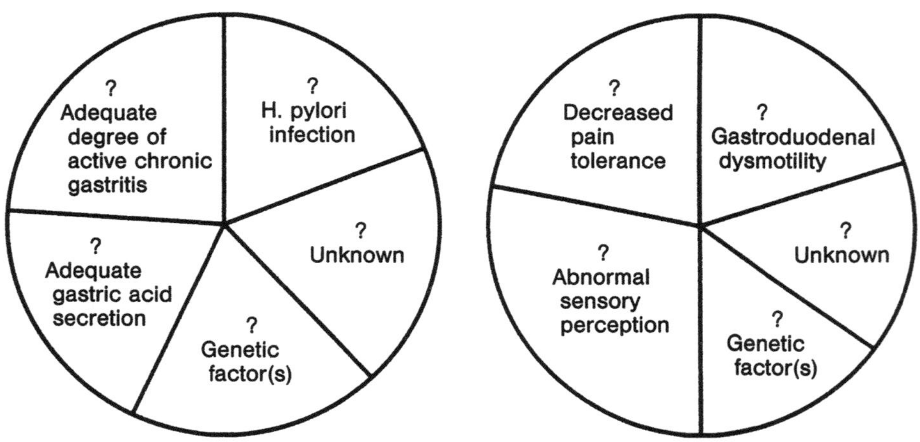

Fig. 1. Conceptual schematization of two groups of sufficient causes for nonulcer dyspepsia. If any component cause is absent, disease is absent. All component causes must be present for disease to occur. *Question mark* indicates these are hypothetical causes

Table 2. Epidemiological criteria that suggest an association is causal [From 44]

Criteria	*H. pylori* and chronic (nonspecific) gastritis	*H. pylori* and nonulcer dyspepsia
1. Strength of association (a large relative risk)	++++	±
2. Consistency (finding a consistent association in different countries at different times and between different research teams)	++++	+
3. Specificity (an almost one-to-one relationship between the association and disease)	+++	0
4. Appropriate time relationship (association precedes disease onset)	++	?
5. Biological gradiant (dose-response curve)	0	?
6. Biological plausibility	+++	?
7. Coherence of the evidence (data should not seriously conflict with known facts relating to disease's natural history and biology)	++++	±
8. Experiment (disease is abolished or diminished by removing the association)	+++	±

+ to ++++, evidence of a positive association varying from mild to marked: ±, equivocal association; 0, no association; ?, no data available

with a subgroup of patients with nonulcer dyspepsia, that infection precedes the onset of symptoms, and that symptoms are abolished by the eradication of the bacteria [46].

Evidence for an Association Between *H. pylori* and Nonulcer Dyspepsia

Prevalence of *H. pylori* in Nonulcer Dyspepsia

While *H. pylori* gastritis is found in 50% of patients with nonulcer dyspepsia, there is also evidence that *H. pylori* is highly prevalent in otherwise apparently asymptomatic persons [11] although symptom measurement has been suboptimal. Moreover, the seroprevalence of *H. pylori* has been clearly shown to increase with age [6, 11, 19, 29] (Fig. 2). Several studies from Europe where biopsy specimens were obtained from representative samples of the general population have also shown that chronic nonspecific gastritis is highly prevalent and increases with age

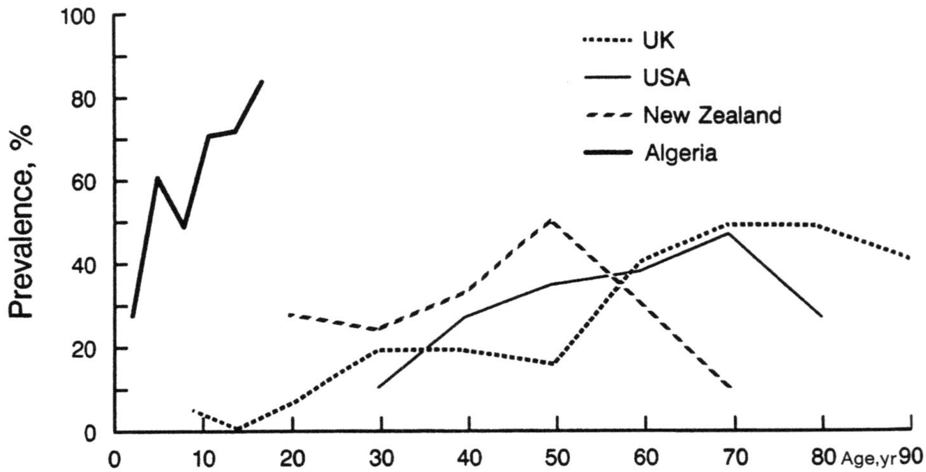

Fig. 2. *Helicobacter pylori* prevalence according to age in three industrialized nations (USA, UK, New Zealand) and one nonindustrialized nation (Algeria). [From 30]

[7, 18, 41, 51]. In industralized nations, both superficial antral and body gastritis may be found in more than 50% of the population after their fourth decade. Rauws and Tytgat [33] have similarly reported that the prevalence of *H. pylori* gastritis in nonulcer dyspepsia is age related in Caucasians but not in other ethnic groups, although the population evaluated was highly selected.

Despite the high prevalence in the community, it has been suggested that *H. pylori* occurs more frequently in patients with nonulcer dyspepsia than in volunteers [31, 34]. Shallcross et al. [39] reported that infection was significantly more common in subjects with dyspepsia younger than 30 and between 40 and 49 years of age when compared with blood donors in the same city; however, this difference failed to reach significance in the other age groups. What seems to have been forgotten is that blood donors and volunteers are not necessarily representative of the normal population. Indeed, in a study from Colombia, *H. pylori* was more common in volunteers that in patients with nonulcer dyspepsia [16]. No appropriate case-control studies have yet been reported that take into account age, ethnic background, and selection of subjects.

Association of *H. pylori* with a Specific Symptom Complex in Nonulcer Dyspepsia

While it has been suggested that *H. pylori* is associated with specific symptoms in patients with nonulcer dyspepsia [26, 32, 35], most studies indicate that symptoms are no different between those infected and those who are not [1, 5, 9, 15, 24, 43] (Table 1). An association between antacid use and *H. pylori* serology has also been reported [42]. Unfortunately, the quality of most of these studies is questionable, as usually no comprehensive validated measure of symptoms was used.

Importantly, studies of the association of *H. pylori* with nonulcer dyspepsia symptoms have usually failed to consider that there may be other causes for the dyspepsia in these patients, such as the irritable bowel syndrome [49]. In a study of 100 consecutive patients with a clinical diagnosis of the irritable bowel syndrome who underwent endoscopy and gastric biopsy, 62% had histological evidence of chronic gastritis; of these patients, only one-third had upper intestinal symptoms, while one-fifth with dyspepsia had no gastritis [12]. Indeed, Ihamäki et al. [18] found that while subjects from the community with antral gastritis did have significantly more upper abdominal symptoms than those without gastritis, if patients with other diseases such as peptic ulcer or gallstones were excluded, the association was no longer significant.

Experimental Evidence for a Causal Relationship Between *H. pylori* and Nonulcer Dyspepsia

Despite the equivocal evidence for an association between *H. pylori* and dyspepsia, the hypothesis that *H. pylori* is a cause of symptoms has gained attention. Uncontrolled trials have suggested that symptoms may improve after eradication of *H. pylori* [3, 33], but at best such uncontrolled data are preliminary and may be quite misleading. Perhaps the strongest support for *H. pylori* being causal or not in nonulcer dyspepsia will come from rigorously conducted therapeutic trials.

A small number of randomized controlled trials have attempted to determine whether symptoms can be reduced by treating *H. pylori* infection with bismuth. The first study was reported by McNulty et al. [27]. They compared bismuth subsalicylate and placebo in patients with *H. pylori* gastritis who were attending for endoscopy; the study was investigator blind, and symptoms were assessed before and immediately after 3 weeks of treatment. Eleven of 12 symptomatic patients whose infection was suppressed improved, compared with 21 of 32 symptomatic patients whose infection persisted. This difference was not statistically significant, presumably because the numbers studied were too small (a type II error). Borody et al. [4] in a 4-week double-blind trial compared bismuth subcitrate with placebo in 43 *H. pylori* infected patients. Diary cards were used to score symptoms, and infection was reassessed 2 weeks after stopping therapy. While significant symptomatic improvement was reported with bismuth, no data on the relationship to *H. pylori* status was given.

Rokkas et al. [36] compared bismuth subcitrate and placebo for 8 weeks in a double-blind study of 52 patients. This was the first published trial using bismuth that enrolled both *H. pylori*-positive and -negative patients; 83% of *H. pylori*-positive patients had their infection suppressed by the bismuth. They found that 92% of infected patients significantly improved on bismuth compared with only 25% on placebo; 60% of uninfected patients also improved on bismuth compared with 25% on placebo. It is important to note that the placebo response was remarkably lower in this study compared with the responses reported by other investigators. Lambert et al. [23] evaluated bismuth subcitrate and placebo in 48 nonulcer dyspepsia patients with *H. pylori* and 30 without infection. They reported that symptoms improved significantly only in the *H. pylori* infected patients given bismuth but the absolute improvement was small. In contrast, Loffeld et al. [25] found that bismuth subcitrate and placebo resulted in very similar symptom responses in patients with *H. pylori* gastritis. In that study 50 patients were randomized; while a significant reduction in the gastritis score occurred on bismuth but not placebo, 66% on bismuth and 71% on placebo reported symptom improvement. Finally, Kang et al. [20] randomized *H. pylori*-positive and -negative patients to receive either bismuth subcitrate or placebo; 51 completed the trial. A statistically significant benefit in *H. pylori* infected patients was found, but overall bismuth was not superior to placebo in reducing symptoms.

Single antibiotics have also been evaluated in a few randomized controlled trials but the results have been uniformly disappointing. McNulty et al. [27] found erythromycin was no better than placebo in symptomatic *H. pylori*-positive patients, but very few had suppression of their infection by this antibiotic. A trial of furazolidone, nitrofurantoin, and placebo in 69 *H. pylori* infected patients also demonstrated no significant difference in symptom response between the three treatment groups [13]. A high rate of adverse effects was encountered with nitrofurantoin, and only 58% of the patients on this treatment had suppression of their infection. Furazolidone suppressed *H. pylori* in 86% of cases, but only small numbers were randomized to this treatment arm. Glupczynski et al. [14] evaluated amoxycillin in 45 *H. pylori* infected patients; while 91% had suppression of *H. pylori* on therapy, symptomatic improvement was not different from those who were receiving placebo.

The experimental evidence therefore suggests that bismuth but not antibiotics alone may improve symptoms in patients who have *H. pylori* infection. However, the placebo response could explain much of the benefit observed. Moreover, all the studies have suffered from potentially serious biases which has made interpretation of the results difficult [47]. The adequacy of blinding with bismuth, for example (which results in black stools and teeth), was not carefully addressed by any of the trials. The symptom ratings used in each of the studies were also not validated; this led Veldhuyzen et al. [50] to conclude that no treatment is of established benefit in the relief of symptoms in *H. pylori*-associated nonulcer dyspepsia. Indeed, individual symptom improvement has been highly variable in all of the trials and no long-term follow-up results have been published. Bismuth also has other actions on the gastric mucosa in addition to suppressing *H. pylori*; it forms a physical barrier to acid and pepsin, increases alkali secretion by an independent mechanism, and possibly enhances prostaglandin formation [21, 40].

Unfortunately, none of the controlled trials have evaluated the effect of eradication of *H. pylori* on symptoms. It would be expected that if *H. pylori* is a component cause then its removal would abolish symptoms when gastritis eventually heals, but this remains to be established.

Conclusions

While there is convincing evidence that *Helicobacter pylori* is a cause of chronic nonspecific gastritis, the relationship between gastritis and dyspepsia is unclear. Although there is some very preliminary evidence that supports the hypothesis that *H. pylori* is of etiological importance in nonulcer dyspepsia, the quality of the data are questionable and the results conflicting. If *H. pylori* does play a role, the organism by itself is very unlikely to be sufficient to cause symptoms. This is because dyspepsia can occur in the absence of infection, and infection can be present in the absence of symptoms. However, *H. pylori* could comprise one of the causal components in some patients. In the hypothetical models presented in Fig. 1, the presence of all the component causes is necessary to induce symptoms. These models would explain why infection can be asymptomatic or why dyspepsia can occur in the absence of infection. If the models are correct, then in the vast majority of cases *H. pylori* will be a benign infection, as the other components necessary to induce clinical illness will be missing. Rigorously conducted randomized controlled trials and longitudinal studies are now needed to evaluate the hypothesis that *H. pylori* causes dyspepsia before treatment recommendations can be made. The search for other component causes is also important because we may then be able to predict who is likely to benefit from treatment.

References

1. Andersen LP, Elsborg L, Justesen T (1988) Campylobacter pylori in peptic ulcer disease. III. Symptoms and paraclinical and epidemiologic findings. Scand J Gastroenterol 23:347–350
2. Barbara L, Camilleri M, Corinaldesi R, et al. (1989) Definition and investigation of dyspepsia. Consensus of an international ad hoc working party. Dig Dis Sci 34:1272–1276
3. Borody TJ, Carrick J, Hazel ISL (1987) Symptoms improve after the eradication of gastric Campylobacter pyloridis (Letter). Med J Aust 46:450–451
4. Borody T, Hennessy W, Carrick J, Daskalopoulas G, Hazell S (1987) Double-blind trials of De-Nol in non-ulcer dyspepsia associated with Campylobacter pyloridis gastritis (Abstr.). Gastroenterology 92:1324
5. Börsch G, Schmidt G, Wegener M, et al. (1988) Campylobacter pylori: prospective analysis of clinical and histological factors associated with colonization of the upper gastrointestinal tract. Eur J Clin Invest 18:133–138
6. Mégraud F, Brassens-Rabbe MP, Denis F, Belbouri A, Hoa DQ (1989) Seroepidemology of Campylobacter pylori infection in various populations. J Clin Microbiol 27:1870–1873
7. Cheli R, Simon L, Aste H, et al. (1980) Astrophic gastritis and intestinal metaplasia in asymptomatic Hungarian and Italian populations. Endoscopy 12:105–108
8. Colin-Jones DG (1988) Management of dyspepsia: report of a working party. Lancet I:576–579
9. Collins JS, Knill-Jones RP, Sloan JM, et al. (1989) Comparison of symptoms between non-ulcer dyspepsia patients positive and negative for C. pylori using a single bias computer system for history taking. Klin Wochenschr [Suppl 18] 67:11
10. Crean GP, Card WI, Beattie AD, et al. (1982) "Ulcer-like dyspepsia". Scand J Gastroenterol [Suppl 79] 17:9–15
11. Dooley CP, Cohen H, Fitzgibbons P, Bauer M, Appleman MD, Perez-Perez GJ, Blaser MJ (1989) Prevalence of Helicobacter pylori and histologic gastritis in asymptomatic persons. N Engl J Med 321:1562–1566

12. Fielding JF, Doyle GD (1982) The prevalence and significance of gastritis in patients with lower intestinal irritable bowel (irritable colon) syndrome. J Clin Gastroenterol 4:507–510
13. Gastrointestinal Physiology Working Group (1988) Nitrofurans in the treatment of gastritis associated with Campylobacter pylori. Gastroenterology 95:1178–1184
14. Glupczynski Y, Burette A, Labbe M, Deprez C, de Reuck M, Deltenre M (1988) Campylobacter pylori associated gastritis: a double-blind placebo-controlled trial with amoxycillin. Am J Gastroenterol 83:365–372
15. Guerre J, Berthe Y, Chaussade S, et al. (1989) Has Campylobacter pylori gastritis a specific clinical symptomatology? Klin Wochenschr [Suppl 18] 67:25
16. Gutiérrez D, Sierra F, Gómez MC, Camargo H (1988) Campylobacter pylori in chronic environmental gastritis and duodenal ulcer patients. Gastroenterology 94:A163
17. Hill AB (1965) The environment and disease: association or causation? Proc R Soc Med 58:295–300
18. Ihamäki T, Varis K, Siurala M (1979) Morphological, functional and immunological state of the gastric mucosa in gastric carcinoma families. Comparison with a computer-matched family sample. Scand J Gastroenterol 14:801–812
19. Jones DM, Eldrige J, Fox AJ, Sethi P, Whorwell PJ (1986) Antibody to the gastric compylobacter-like organism ("Campylobacter pyloris") – clinical correlations and distribution in the normal population. J Med Microbiol 22:57–62
20. Kang JY, Tay HH, Wee A, Guan R, Math MV, Yap I (1989) Effect of colloidal bismuth subcitrate on symptoms and gastric histology in non-ulcer dyspepsia – a double-blind placebo controlled study. Gut 29:A1440
21. Konturek SJ, Radecki T, Piastucki I, Brzozowski T, Drozdowicz D (1987) Gastrocytoprotection by colloidal bismuth subcitrate (De-Nol) and sucralfate. Role of endogenous prostaglandins. Gut 28:201–205
22. Lagarde S, Spiro HM (1984) Non-ulcer dyspepsia. Clin Gastroenterol 13:437–446
23. Lambert JR, Dunn K, Borromeo M, Korman MG, Hansky J (1989) Campylobacter pylori – a role in non-ulcer dyspepsia? Scand J Gastroenterol [Suppl 160] 24:7–13
24. Loffeld RJLF, Potters HVPJ, Arends JN, Stobberingh E, Flendrig JA, von Spreeuwel JP (1988) Campylobacter associated gastritis in patients with non-ulcer dyspepsia. J Clin Pathol 41:85–88
25. Loffeld RJLF, Potters HVJP, Stobberingh E, Flendrig JA, van Spreeuwel JP, Arends JW (1989) Campylobacter associated gastritis in patients with non-ulcer dyspepsia: a double-blind placebo-controlled trial with colloidal bismuth subcitrate. Gut 30:1206–1212
26. Marshall BJ, Warren JR (1984) Unidentified curved bacilli in the stomach of patients with gastritis and peptic ulceration. Lancet I:1311–1315
27. McNulty CAM, Gearty JC, Crump B, et al. (1986) Campylobacter pyloridis and associated gastritis: investigator blind placebo-controlled trial of bismuth salicylate and erythromycin ethyl-succinate. Br Med J 293:645–649
28. Morris A, Nicholson G (1987) Ingestion of Campylobacter pyloridis causing gastritis and raised fasting pH. Am J Gastroenterol 82:192–199
29. Morris A, Nicholson G, Lloyd G, Haines D, Rogers A, Taylor D (1986) Seroepidemiology of Campylobacter pyloridis. N Z Med J 99:657–659
30. Ormand JE, Talley NJ (1990) Helicobacter pylori: controversies and an approach to management. Mayo Clin Proc 65:414–426
31. Pettross CW, Appleman MD, Cohen H, Valenzuela JE, Chandrasoma P, Laine LA (1988) Prevalence of Campylobacter pylori and association with antral mucosal histology in subjects with and without upper gastrointestinal symptoms. Dig Dis Sci 33:649–653
32. Rathbone BJ, Wyatt J, Heatley RV (1988) Symptomatology in C. pylori-positive and negative non-ulcer dyspepsia. Gut 29:A1473
33. Rauws EAJ, Tytgat GNJ (1989) Campylobacter pylori. den Ouden, Amsterdam
34. Rauws EAJ, Langenberg W, Houthoff HJ, Zanen HC, Tytgat GNJ (1988) Campylobacter pyloridis-associated chronic active antral gastritis. Gastroenterology 94:33–40
35. Rokkas T, Pursey C, Uzoechina E, et al. (1987) Campylobacter pylori and non-ulcer dyspepsia. Am J Gastroenterol 82:1149–1152

36. Rokkas T, Pursey C, Uzeochina E, et al. (1988) Non-ulcer dyspepsia and short-term De-Nol therapy: a placebo controlled trial with particular reference to the role of Campylobacter pylori. Gut 29:1386–1391
37. Rothman KJ (1976) Causes. Am J Epidemiol 104:587–592
38. Rothman KJ (1986) Modern epidemiology. Little, Brown, Boston, pp 115–129
39. Shallcross TM, Rathbone BJ, Heatley RV (1989) Campylobacter pylori and non-ulcer dyspepsia. In: Rathbone BJ, Heathley RV (eds) Campylobacter pylori and gastroduodenal disease. Blackwell, Oxford, pp 155–166
40. Shorrock CJ, Crampton JR, Gibbons LC, Rees WDW (1989) Effect of bismuth subcitrate on amphibian secretion. Gut 30:917–921
41. Siurala M, Isokoski M, Varis K, Kekki M (1968) Prevalence of gastritis in a rural population. Bioptic study of subjects selected at random. Scand J Gastroenterol 3:211–223
42. Skoglund ML, Whalen JW (1988) Correlation of C. pylori and upper GI symptoms. Gastroenterology 94:A430
43. Sobala GM, Dixon MF, Axon ATR (1989) C. pylori is not associated with a distinct dyspeptic syndrome. Gut 30:A733
44. Talley NJ (1988) Campylobacter pylori-associated gastritis. Is antibiotic therapy now justified? J Clin Gastroenterol 10:10–12
45. Talley NJ (1989) Chronic (non-specific) gastritis. Dig Dis 7:61–75
46. Talley NJ (1990) Non-ulcer dyspepsia: epidemiology, natural history, and association with Campylobacter pylori. In: Marshall BJ, et al. (eds) Campylobacter pylori. Blackwell, Cambridge (in press)
47. Talley NJ (1990) Optimal study design for therapeutic trials in Campylobacter pylori associated non-ulcer dyspepsia. In: Menge H, et al. (eds) Campylobacter pylori. Springer, Berlin Heidelberg New York (in press)
48. Talley NJ, Phillips SF (1988) Non-ulcer dyspepsia: potential causes and pathophysiology. Ann Intern Med 108:865–879
49. Talley NJ, Piper DW (1985) The association between non-ulcer dyspepsia and other gastrointestinal disorders. Scand J Gastroenterol 20:896–900
50. Veldhuyzen SJO, Tytgat KMAJ, Jalali S, Goodacre RL, Hunt RH (1989) Can gastritis symptoms be evaluated in clinical trials? An overview of treatment of gastritis, non-ulcer dyspepsia and Campylobacter-associated gastritis. J Clin Gastroenterol 11:496–501
51. Villako K, Tamm A, Savisaar E, Ruttas M (1976) Prevalence of antral and fundic gastritis in a randomly selected group of an Estonian rural population. Scand J Gastroenterol 11:817–822

Gastritis and Altered Motility; the Ability of a Mucosal Inflammatory Reaction to Alter Enteric Nerve and Smooth Muscle in the Gut

S. M. COLLINS

The Association Between Gastritis and Altered Gastric Motility

Dyspeptic symptoms such as pain, nausea, and vomiting, that accompany gastritis, and which are sometimes refractory to gastric acid suppression, may arise from disturbances in gastric motility. Indeed, motility disturbances have been described in several studies on patients with active gastric ulceration [6, 11]. In addition, Liebermann-Meffert and Allgöwer [9] demonstrated hypertrophy of muscle and structural alterations in enteric nerves in specimens resected from patients with gastric ulcer; Hiatt and Katz [7] showed increased numbers of mast cells in the muscle layers below gastric ulcer craters.

Motility changes have also been documented in patients with healed gastric ulcers, but in whom gastritis persists; a study by Moore et al. [12] showed an inverse correlation between the degree of gastritis and postprandial motor activity recorded manometrically from the antrum. These findings suggest that the presence of inflammation in the gastric mucosa is accompanied by alterations in gastric motility. Testoni et al. [19] found altered interdigestive motor activity in patients with bile reflux gastritis, but did not consider the motility changes to be secondary to gastritis since they were also observed in patients with bile reflux without gastritis. It should be noted that this study examined only interdigestive motility and it is possible that patients with gastritis exhibit motor changes on provocation (e.g., after food).

Altered Motility in Other Mucosal Inflammatory Conditions

Abnormal motor activity is a well-recognized feature of peptic esophagitis, and there is experimental evidence to suggest that motor disturbances may be secondary to the inflammation [2]. Alterations in motility have also been described in association with peptic ulcer disease [6, 11], untreated celiac disease [17], and inflammatory bowel disease [8, 16]. In the latter, motor changes in the colon are reversible as the inflammatory activity decreases, suggesting that inflammation alters motility [8]. However, the mechanisms whereby inflammatory cells induce motility changes are unknown. Since motility is determined by the contractile status of smooth muscle, under the influence of enteric nerves and gut hormones, it

Associate Professor of Medicine, Director, Intestinal Diseases Research Unit McMaster University Medical Centre, Hamilton, Ontario, Canada L8N 3Z5

Helicobacter pylori, Gastritis and Peptic Ulcer
Edited by P. Malfertheiner and H. Ditschuneit

is likely that inflammation alters motility by modulating the function and/or structure of these tissues.

To better understand the ability of inflammatory reactions to alter the motor apparatus of the gut, it has been necessary to use animal models of gut inflammation.

The Choice of Animal Model for Study

Previous studies, performed largely in this laboratory have shown that infection of animals with enteric parasites (nematodes) is associated with alterations in gastrointestinal transit [3], myoelectrical activity [15], and the propulsion of lumenal fluid [1]. Since these observations were made during the enteric phase of the primary infection, which is associated with an acute mucosal inflammatory reaction in the jejunum, these infections serve as useful models to study the impact of inflammatory processes on the motor apparatus of the gut: the smooth muscle and enteric nerves. The studies by Alizadeh at al. [1] showing altered propulsive activity in extrinsically denervated gut segments from rats infected by *Trichinella spiralis* suggest that the locus of effect of the inflammatory process on the motor apparatus lies within the gut wall.

The Effect of Inflammation on Smooth Muscle Contraction

Trichinella spiralis infection in the rat produces an inflammatory infiltrate that is largely restricted to the mucosa and submucosa of the proximal small intestine, and consists of a mixture of acute inflammatory cells [13]. Inflammation reaches a maximum within the first week of infection.

In longitudinal muscle from the jejunum of infected rats, there is a time-dependent change in contractility. There is a decrease in the degree of stretch necessary for the generation of maximum tension, as well as an increase in the maximum activity generated [20]. The increased contractile responsiveness is pharmacologically nonspecific and occurs when muscle is stimulated by the muscarinic agonist carbachol or by 5-hydroxytryptamine. These observations, together with the results of receptor-binding studies, indicate that the changes in contractility are receptor independent. Ligand binding of ^{3}H-*N*-methyl-scopolamine to muscle cell membranes actually revealed a decrease in the total number of muscarinic receptors in tissues from the inflamed intestine, and no significant alteration in affinity for either antagonists or agonists [14].

Further study of muscle cell membranes from infected rats revealed a substantial (>80%) decrease in the activity of the enzyme p-nitrophenyl-phosphatase (pNPP), a convenient biochemical marker for the ATP-dependent Na^+-K^+ pump. Suppression of the Na^+ pump was demonstrated by a decrease in the ouabain-sensitive uptake of ^{86}-rubidium by muscle from infected rats compared with controls. In many excitable tissues the Na^+ pump is electrogenic and contributes to the polarity across the plasma membrane. Suppression of the pump serves to partially depolarize the membrane, bringing it closer to its threshold for excitation. Thus, the biochemical evidence of pump suppression would be consistent with increased contractile responsiveness observed in muscle from the inflamed intestine if the pump is indeed electrogenic. Evidence in support of this was obtained by either stimulating or inhibiting the pump and observing predictable changes in tension generation by the muscle [14]. The attenuation of tension generation following pump inhibition in muscle from infected rats was taken as evidence of preexisting pump suppression. Taken in conjunction with the biochemical data, these findings indicate that Na^+ pump suppression contributes to the increased contractility of muscle from the inflamed intestine.

It is likely that other mechanisms contribute to the increased contractility and these are presently under study in our laboratory. In addition, it should be emphasized that inflammation is also associated with trophic changes in muscle in this model, as well as in man [9]. Morphometric analysis revealed both hypertrophy and hyperplasia of jejunal longitudinal muscle from the jejunum of Trichinella-infected rats [24]. Since the above-described changes in contractility were expressed per unit cross-sectional area of muscle, and are therefore independent of trophic changes, the increased thickness of the muscle would serve to amplify the force generated in the inflamed gut wall.

The Effect of Inflammation on Enteric Nerve Function

Studies in man have shown structural damage to enteric nerves in association with peptic ulcer disease [9] and inflammatory bowel disease [5]. Recent studies have examined the impact of mucosal inflammation on the function of enteric nerves in the gut. Using myenteric plexus/longitudinal muscle preparations from which the overlying mucosa and circular muscle had been dissected, measurements were made of acetylcholine (ACh) and noradrenaline release. ACh release was measured after preincubating tissues with ^{3}H-choline, washing, and stimulating with KCl, veratridine or electrical field stimulation. In control preparations, each stimulant caused a significant increase in ACh release. However, in tissues from rats infected with *T. spiralis* 6 days previously, ACh release was suppressed by 80%. The suppression of ACh release lasted over 23 days, and returned to normal only after more than 40 days postinfection [4]. Similar results have recently been obtained with ^{3}H-noradrenaline release from this preparation. On the 6th day postinfection, there was a $>70\%$ reduction in the evoked release of the sympathetic neurotransmitter from the inflamed intestine. In rats pretreated with betamethasone $3\,mg \cdot kg^{-1}$ s.c., the suppression of ^{3}H-NA release was attenuated, suggesting that it was a consequence of the host inflammatory response rather than a direct effect of the parasite [18].

Underlying Immunologic Mechanisms

Initial experiments examined the extent to which the above-described changes in muscle function require the local presence of the worm. Segments of jejunum were excluded from the gut prior to infection by *T. spiralis*. After the 6th day postinfection, contractility of muscle from the excluded segments was examined. Muscle from excluded segments of infected rats generated more tension than did that from the excluded loop of control rats [10]. These findings suggest that a systemic mechanism contributes to the inflammation-induced changes in smooth muscle function. The logical source of this mechanism is the immune system.

To investigate the role of lymphocytes in the development of smooth muscle changes, similar experiments were performed in athymic (nude rnu/rnu) rats and their thymus bearing littermates. In the thymus bearing (rnu/+) rats, Trichinella infection was accompanied by the same increase in muscle responsiveness as had been observed in Sprague Dawley rats on the 6th day postinfection. However, such changes were absent from athymic rats on the 6th day postinfection, suggesting thymic dependence of the muscle changes. This was confirmed by reconstituting lymphocyte capacity following the i.v. administration of splenic lymphocytes from a thymus bearing rat. Reconstitution of lymphocyte capacity was accompanied by the appearance of muscle hyperresponsiveness on the 6th day postinfection with *T. spiralis* [23]. These findings suggest that T lymphocytes play a role in the development of changes in smooth muscle contractility in the inflamed gut.

Mast cells may also interact with the motor apparatus of the gut. Mast cells have been observed in gastric muscle in association with peptic ulcer disease [7] and inflammatory bowel disease [5]. In the nematode model, mastocytosis in gut muscle is prominent and has served as a model to evaluate mast cell interactions. In this model, degranulation of connective tissue mast cells contracts muscle via the release of 5-hydroxytryptamine acting directly on the muscle [23]. In human tissue, mast cells contract muscle via the release of several mediators that include platelet activating factor [23]. It is not known, however, whether mast cell activation is part of the immune response to infection by Helicobacter.

Summary and Conclusion

There is ample clinical data from a variety of sources that demonstrate altered gastrointestinal motility in association with mucosal inflammatory diseases. While data on the existence of motility changes occurring in conjunction with Helicobacter infection are forthcoming (see contributions to this volume by Pieramico and by Marzio), it is reasonable to anticipate that the gastritis associated with this infection will produce motor changes. The foregoing has provided a description of recent work regarding the ability of a superficial inflammatory reaction to alter the function of smooth muscle and enteric nerves and will hopefully serve as a basis from which to evaluate and exploit motor changes in Helicobacter-associated gastritis.

References

1. Alizadeh H, Weems WA, Castro GA (1986) Immunologically induced alterations in the intrinsic state of guinea pig jejunum (Abstr). Gastroenterology 91:A1042

2. Biancani P, Barwick K, Selling J, McCallum R (1984) Effects of acute experimental "esophagitis" on mechanical properties of the lower esophageal sphincter. Gastroenterology 87(1):8–16
3. Castro GA, Badial-Aceves F, Smith JW, Dudrick SJ, Weisbrodt NW (1976) Altered small bowel propulsion associated with parasitism. Gastroenterology 71:620–625
4. Collins SM, Blennerhassett P, Vermillion DL, Blennerhassett M (1990) Altered myenteric plexus function in Trichinella-infected rats. Am J Physiol (in press)
5. Dvorak AM, Monahan RA, Osage JE, Dickerson GR (1980) Crohn's disease: transmission electron microscopic studies. II. Immunologic inflammatory response. Alterations in mast cells, basophils, eosinophils and the microvasculature. Hum Pathol 11(6):606–619
6. Garrett JM, Summerskill WJH, Code CF (1966) Antral motility in patients with gastric ulcer. Am J Dig Dis 11:780–789
7. Hiatt RB, Katz J (1962) Mast cells in inflammatory conditions of the gastrointestinal tract. Am J Gastroenterol 37:541–545
8. Kern F, Almy TP, Abbott FK, Bogdonoff MD (1952) The motility of the distal colon in nonspecific ulcerative colitis. Gastroenterology 19:492–503
9. Lieberman-Meffert D, Allgöwer M (1977) The morphology of antrum-pylorus in gastric ulcer disease. Prog Surg 15:109–139
10. Marzio L, Blennerhassett P, Chiverton S, Vermillion D, Collins SM (1986) Systemic factors are responsible for altered smooth muscle function in the inflamed gut. Gastroenterology 94(5):A286
11. Miller LJ, Malagelada JR, Longstreth GF, Go VL (1980) Dysfunction of the stomach with gastric ulceration. Dig Dis Sci 25:857–864
12. Moore SC, Malagelada JR, Shorter RG, Zinsmeister AR (1986) Interrelationships among gastric mucosal morphology, secretion, and motility in peptic ulcer disease. Dig Dis Sci 31(7):673–684
13. Moqbel R (1986) Helminth-induced intestinal inflammation. Trans R Soc Trop Med 80:719–723
14. Muller MJ, Huizinga JD, Collins SM (1990) Altered smooth muscle contraction and sodium pump activity in the inflamed rat intestine. Am J Physiol (in press)
15. Palmer JM, Weisbrodt NW, Castro GA (1984) Trichinella spiralis: intestinal myoelectrical activity during enteric infection in the rat. Exp Parasitol 57:132–141
16. Snape WJ, Matarazzo SA, Cohen S (1980) Abnormal gastrocolic response in patients with ulcerative colitis. Gut 21:392–396
17. Spiller RC, Lee YC, Edge C, et al. (1987) Delayed mouth cecum transit time of a lactulose labelled test meal in patients with steatorrhea caused by partially controlled celiac disease. Gut 28:1275–1282
18. Swain M, Blennerhassett P, Collins SM (1989) Intestinal inflammation alters sympathetic nerve function in the rat myenteric plexus. Gastroenterology 96:A498
19. Testoni PA, Fanti L, Bagnolo F, Passaretti S, Guslandi M, Masci E, Tittobello A (1989) Manometric evaluation of the interdigestive antroduodenal motility in subjects with fasting bile reflux, with and without antral gastritis. Gut 30(4):443–448
20. Vermillion DL, Collins SM (1988) Increased responsiveness of jejunal longitudinal muscle in Trichinella infected rats. Am J Physiol 254:G124–G129
21. Vermillion DL, Collins SM (1989) Mast cells contract human intestinal muscle via the release of P.A.F. and leukotriene D_4 in patients with or without Crohn's disease. Gastroenterology 96:A527
22. Vermillion DL, Ernst PB, Scicchitano R, Collins SM (1988) Antigen-induced contraction of jejunal smooth muscle in the sensitized rat. Am J Physiol 255 18:G701–G708
23. Vermillion DL, Ernst PB, Collins SM (1989) Altered smooth muscle function in the inflamed gut is T lymphocyte dependent. Gastroenterology 96:A528
24. Vignjevic P, Blennerhassett MG, Blennerhassett P, Vermillion DL, Collins SM (1989) Hypertrophy and hyperplasia of smooth muscle in the inflamed intestine. Gastroenterology 96:A530

Influence of Chronic Antral and Duodenal Inflammation on Motor Disorders

L. MARZIO and D. PIERAMICO

Motor disorders involving electrical control activity (ECA), fasting, and fed gastroduodenal motility and alteration in gastric emptying have been identified in patients with dyspeptic symptoms [2], chronic type B gastritis [3], and duodenogastric reflux [4]. The most frequent abnormality of antroduodenal motility during fasting is a longer than normal duration of migrating myoelectrical complexes (MMCs) mainly due to a prolongation of phase II and shortening of phase I activity [1]. This abnormality should increase the possibility of generating a gastroduodenal bile reflux which was earlier considered to be the main pathophysiologic process involved in the genesis of chronic gastritis. The demonstration that MMC prolongation and duodenogastric reflux is present also in patients without gastritis [1] may suggest the hypothesis that gastritis is more a consequence than cause of the motor abnormality, such as in case with esophagitis and of gastroesophageal reflux.

During the fed state patients with antral gastritis and type I gastric ulcer present a hypomotile antrum while the duodenum and jejunum are apparently normal. The degree of gastritis is inversely related to the postprandial motility index, while during fasting no difference is noted [3]. Antral hypomotility is present also with gastritis alone without ulcer disease, suggesting that gastric mucosal inflammation and antral hypomotility rather predispose to ulceration than simply accompany it [3].

Antral hypomotility during the postprandial period in patients with antritis should result in a delay in gastric emptying with abnormally prolonged stasis of food and acid in the stomach with increased likelihood of the formation of gastritis and ulcer. Conversely gastric emptying seems to be accelerated in duodenal ulcer patients who show an abnormally prolonged secretory response with excessive gastric emptying of unbuffered acid, especially in the late postprandial period [5]. Other motor abnormalities reported in duodenal ulcer patients include reduced frequency of mixing type duodenal motor waves and increased contractility in the distal duodenum [6]. A recent study on gastric emptying in duodenal ulcer patients however failed to detect any abnormality in comparison with the control population [7].

It may be summarized that gastric ulcer, duodenal ulcer, type B gastritis, and duodenitis are associated with abnormalities of motility during fasting an fed state. The presence of similar abnormalities in patients with dyspeptic symptoms

---

Institute of Medical Physiopathology, G. D'Annunzio University, 66100 Chieti, Italy

Helicobacter pylori, Gastritis and Peptic Ulcer
Edited by P. Malfertheiner and H. Ditschuneit

without any evidence of inflammation or ulcer suggests that motor abnormalities may precede the inflammatory process. In patients with nonulcer dyspepsia, Rees et al. [8] found low amplitude antral waves with a marked delayed gastric emptying after a solid meal. Malagelada and Stangellini [2] in a study of 104 patients with nonulcer dyspepsia found 27% with normal motor pattern, 41% with motor abnormality localized in the antrum, and the remaining had motor abnormalities in the antrum and small intestine. These include a complete absence of MMCs, nonpropagating or retrograde MMC, and marked hypomotility during fasting and the fed state. However, the symptoms did not predict the presence, severity, anatomic distribution of any manometric abnormality. The MMC cycle also appeared to be abnormally long or without the appearance of phase III activity during the period of observation.

The electrical counterpart of these motility alterations, that is, antral hypocontractility and delayed gastric emptying is the so-called antral tachigastria or disrhythmia. These abnormalities in the ECA have been shown in patients with severe abdominal pain, nausea, and vomiting [9], and the syndrome is characterized by increased frequency of ECA to up to 8–11 cycles per min in the stomach (tachigastria) with periodic absence of electrical depolarization (arrhythmia). In these patients electrical response activity is completely absent, therefore no contraction appears. Antral tachigastria has been reproduced experimentally in humans with the induction of motion sickness through labyrinthine stimulation [10].

With regard to *Helicobacter pylori*, type B gastritis, and motility both a close relationship between *H. pylori* gastritis and duodenitis [11] and antroduodenal motility abnormalities has been found. Therefore, it has been suggested that the altered motility abnormalities encountered in *H. pylori*-positive antral gastritis may be due to the infection itself. Studies from our laboratory confirm that in patients with *H. pylori*-positive gastritis there is a delayed gastric emptying of a solid meal in comparison with controls [12]. In another study, however, Wegener et al. [13] found that gastric emptying was delayed similarly in patients with nonulcer dispepsia both with and without *H. pylori*. These results suggest that factors other than *H. pylori* are involved in the genesis gastritis and related motility abnormalities. Studies evaluating motility alterations before and after eradication of *H. pylori* and healing of antral gastritis will help to elucidate the problem.

References

1. Testoni PA, Fanti L, Bagnolo F, Passaretti, Guslandi M, Masci E, Tittobello A (1989) Manometric evaluation of the interdigestive antroduodenal motility in subjects with fasting bile reflux, with and without antral gastritis. Gut 30:443–448
2. Malagelada J-R, Stangellini V (1985) Manometric evaluation of functional upper gut symptoms. Gastroenterology 88:1223–1231
3. Moore SC, Malagelada J-R, Shorter RG, Zinsmeister AR (1986) Interrelationships among gastric mucosal morphology, secretion, and motility in peptic ulcer disease. Dig Dis Sci 31:673–684
4. Miranda M, Defilippi C, Valenzuela JE (1985) Abnormalities of interdigestive motility complex and increased duodenogastric reflux in gastric ulcer patients. Dig Dis Sci 30:16–21

5. Malagelada J-R, Longstreth GR, Deering TB, Summerkill WHJ, Go VLW (1977) Gastric secretion and emptying after ordinary meals in duodenal ulcer. Gastroenterology 73:989–994
6. Monto GL, Ashworth WD, Malecki M, Thompson AB, Englert E (1976) Duodenal contraction wave pattern in patients with and without ulcer. Am J Gastroenterol 65:52–56
7. Holt S, Heading RC, Taylor TV, Forrest JA, Tothill P (1986) Is gastric emptying abnormal in duodenal ulcer? Dig Dis Sci 31:685–692
8. Rees WDW, Miller LJ, Malagelada J-R (1980) Dyspepsia, antral motor disfunction, and gastric stasis of solids. Gastroenterology 78:360–365
9. You CH, Lee KY, Chey WY, Menguy R (1980) Electrogastrographic study of patients with unexplained nausea, bloating and vomiting. Gastroenterology 79:311–314
10. Stern RM, Koch KL, Stewart WR, Lindblad IM (1987) Spectral analysis of tachygastria recorded during motion sickness. Gatsroenterology 92:92–97
11. Coelho LGV, Das SS, Payne A, Karim QN, Baron JH, Walker MM (1989) Campylobacter pylori in the esophagus, antrum, and duodenum: a histological and microbiological study. Dig Dis Sci 43:445–448
12. Marzio L, DiFelice F, Laico M, Celiberti V, DiTommaso C, DiIorio P, Seccia G, Cuccurullo F (1989) Gastric and gallbladder emptying in patients affected by idiopathic dyspepsia and positive. Klin Wschr 67 (suppl. XVIII):44
13. Wegener M, Borch G, Schaffstein, Schulz-Flake C, Mai U, Lewerkus F (1983) Are dyspeptic symptoms in patients with Campylobacter pylori-associated type B gastritis linked to delayed gastric emptying? Am J Gastroenterol 83:737–740

Therapeutic Approaches

Why Should We Treat *Helicobacter pylori* Infection?

M. DELTENRE[1], E. DEKOSTER[1], J. F. NYST[1], and Y. GLUPCZYNSKI[2]

Some years ago, the treatment of infection by *Helicobacter pylori* (HP) used to be an ethically embarrassing problem because no one could be sure whether the organism was a commensal or a true pathogen for gastroduodenal mucosa.

Nevertheless, most of what we have learned about the true responsibility of HP in various upper GI tract diseases has resulted from therapeutic trials. In 1989, we know that HP is a key factor in the development of chronic active gastritis and the maintenance of duodenal ulcer disease. The role of HP in patients with gastric ulcer relapses remains to be definitively established, even if Rauws [36] has reported a 100% incidence rate in patients with gastric ulcer (with alcoholics, NSAID users, and patients with "ischemic" ulcer excluded). In cases of "idiopathic dyspepsia," commonly named "non-ulcer dyspepsia" (NUD), the precise role of HP-related gastritis remains to be determined – because it must be underlined that most HP carriers with gastritis are symptom free [1].

As far as epidemic gastritis is concerned, most physicians encounter some cases of acute dyspepsia in previously asymptomatic patients in whom endoscopy with biopsies revealed gastritis with polynuclear infiltration and HP. But other organisms are able to give such clinical signs and the precise proportion of HP-related acute gastritis among acute dyspepsias is unknown, as is the mode of contamination. The invasive phase may be characterized by an acute achlorhydric gastritis, but so far there is no controlled trial showing the possible advantages of treatment at that stage.

We discuss here the advantages of HP suppression or, when possible, eradication in chronic gastritis, duodenal ulcer disease, and NUD.

Chronic Type B Gastritis

The responsibility of HP in the development of type B gastritis is almost certain. The organism is cultured, found at the site of the disease, and experimental infection has been successfully performed in animals. Histological features allow differentiation of this type of chronic gastritis from bile reflux gastritis [33] or autoimmune atrophic gastritis of Biermer's disease.

The clearance of HP improves the gastritis, but active gastritis reappears with HP reactivation [15, 29, 37]. This point was confirmed in a placebo-controlled

[1] Gastro Enterology Dpt (Prof. M. Deltenre), [2] Microbiology Dpt (Prof. E. Yourassovsky) University Hospital Brugmann, Free University Brussels (ULB-VUB)

Helicobacter pylori, Gastritis and Peptic Ulcer
Edited by P. Malfertheiner and H. Ditschuneit

study with amoxycillin [19]. When HP eradication is achieved, gastritis may progressively heal [36].

Karnes et al. [23] suggested that atrophy, a well-known precancerous condition, could be an end-stage of HP-related gastritis: among individuals over 60 years old with atrophic gastric mucosa, antibodies may be still present when the organism is no longer found in the antrum. This could be the first reason for which we would recommend eradication of HP.

Duodenal Ulcer

The second reason justifying a treatment is the clear relationship between HP and duodenal ulcer (DU) disease. HP is found in 100% of DU patients in whom Crohn's disease, NSAID intake, or carcinoma have been ruled out.

Among etiologic factors of DU, stress, smoking, NSAID intake, and genetics seem to be accessory (or poorly known) factors. Mucosal defense, optimal vascularization, and the amount and components of mucus are obviously important, but in clinical studies prostaglandin analogs were shown to be poorly effective and frequent side effects have been described. Probably because of available therapeutic means, acid and HP are currently considered to be prominent factors in the etiology of DU.

Acid may be suppressed by gastrectomy (with a high risk of dysplasia developing or even cancer after the 13th year) or long-term anti-H2 or Omeprazole treatment, but the risk of developing malignancy, although probably very low, does exist with long-term acid suppression. Finally, there is a 10%–15% failure rate of selective vagotomy after 5 years: these percentages are in the range of relapse rates after eradicating HP, the other key factor in duodenal ulcer disease.

The suppression of the organism might speed up the healing of the acute phase of DU. Bayerdoerffer et al. [3] using Ofloxacine, and Graham et al. [21], with a tritherapy associated with ranitidine, found that HP clearance accelerates healing of the acute phase of DU. The point may be of minor importance because Omeprazole alone is able to heal duodenal ulceration in 2 or 3 weeks; however, no data are currently available about the speed of DU healing with a combination of Omeprazole and antimicrobials.

Tremendously more important is the dramatic reduction of the relapse rate of DU after HP eradication. Colloidal bismuth subcitrate (CBS) is known to give a longer remission period than anti-H2. According to Chiverton and Hunt [12] this is not related to a rebound induced by anti-H2 treatment, since the mean relapse rate is equivalent both for placebo and anti-H2.

The importance of eradicating HP has been demonstrated by several independent studies [13, 30, 39] and the minimal relapse rate of 6% was observed by Borody et al. [7], only 3 out of 53 eradicated patients receiving oral tritherapy who returned for follow-up had relapsed in a mean period of 16.3 months. In the same studies, the relapse rate of still positive patients was 72%–81%. Eberhardt et al. [16] reported a significant reduction of the relapse rate of gastric ulcer after healing with bismuth subsalicylate compared with cimetidine.

"Idiopathic Dyspepsia" or NUD

Dr. Axon 1989, personal communication interestingly pointed out that among NUD patients, the prevalence of HP is perhaps 10% to 20% higher than in normal individuals and varies with age, race, and social condition. There is no specific symptomatic profile of HP+ NUD and a majority of HP carriers with gastritis are asymptomatic. Therefore, is there any benefit in suppressing or eradicating HP in NUD patients and if yes, which patients might benfit from this treatment? This question is especially important in children.

HP is known to be closely related to gastritis, but we overlook to what extent gastritis produces symptoms. Moreover, since 7% of patients with HP had no proven gastritis in our series [14], HP might, at least theoretically, itself produce symptoms.

In fact, the question of utmost importance is: what do we mean by NUD? One possible definition is a chronic (more than 4 weeks?) syndrome with upper abdominal complaints without ulcer, biliary, or pancreatic disease. But what about tiny mucosal erosions? If it seems reasonable to exclude from NUD, alcoholics, NSAID users, irritable bowel syndrome, gastroesophageal reflux, and operated patients as well as pregnant women, how should we consider patients using sedatives (or any other drug), diabetics with motility disorders, or simply "social" drinkers?

About the benefit of HP clearance in NUD, several studies comprising more than 200 patients [6, 24, 28, 38] revealed at least partial a benefit while the results of McNulty et al. [32], Kang et al. [22], Deltenre et al. [15], Bamford et al. [2] and Borsch and Opferkuch [8] in a total of 272 patients did not show any advantage. The discrepancies in the results are due to different scoring techniques and the difficulty in measuring "improvement" and "worsening." Moreover, because of double-blind randomization, very few HP-positive patients have been treated with active drugs in clinical trials. Among them, around one-third became asymptomatic, perhaps because of HP suppression but possibly because of other beneficial action of the drugs used. Finally, NUD is a long-term disease, and other diseases may supervene during its evolution. Studying NUD requires long-term follow-up and compliance of patients may be a problem.

Treatments and Side Effects

The discussion of various therapeutic combinations to attempt to eradicate HP is beyond the goal of the present paper. Nevertheless, while discussing the benefits of a treatment, possible complications and side effects must be kept in mind and balanced with potential advantages.

Monotherapies are usually safe but useless: anti-H2, sucralfate, prostaglandin analogs, metoclopramide, acetomiphene, propranolol, and all antimicrobial drugs tested in monotherapy have proved ineffective in terms of eradication. Side effects have been frequently reported with oral tritherapy with 14% intolerance in the series of Borsch et al. [9] and up to 32% in the series of Borody and Carrick [5]. Thus, double drug therapy is now the modality of choice, but many questions remain unsolved and compliance is probably the main problem for long-term treatment. Various associations of amoxycillin, nitrofurans, tini-metronidazole, and bismuth salts are currently being used with a success rate varying between 60% and 80% for eradication. Boero et al. [4] reported that short course (1 week) therapy is equivalent to a 4-weeks treatment courses. Successive therapy has been shown to be less effective but the 5-day tritherapy described by Coelho et al. (1989, personal communication) is interesting and may lead to trying shorter treatments.

The second problem is the induction of resistance to antimicrobial agents. Posttreatment resistance to tinidazole was previously described by Marshall et al. [30] and confirmed by Burette et al. [10] who found almost 30% pretreatment resistance. The same phenomenon was observed, to at a lesser extent, by Mathai et al. [31] also for metronidazole. CBS has been claimed to prevent the occurrence of resistance to nitroimidazole [20]. Development of resistance to quinolones has been described by Glupczynski et al. [18].

Finally, another point is that the organism is not eradicated by drugs to which HP is and remains sensitive.

All forms of bismuth salts have been found to be equivalent for in vitro susceptibility [17]. As far as the potential toxicity of bismuth salts is concerned, Lambert et al. [25] reported that CBS and bismuth subsalicylate, in single dose or 4- and 8-week courses do not accumulate bismuth in antroduodenal mucosa or in gastric mucus. There is no prolonged action in the mucosa and bismuth is cleared from the stomach within 2–4 h after oral intake, but after 4- or 8-week courses, plasma concentration of bismuth is significantly increased. Cadranel [11] confirmed these data in children, demonstrating that bismuth plasmatic concentration decreases when the treatment is stopped, but the risk of tissular accumulation has not been clearly evaluated. The treatment of infected children to relieve symptoms and perhaps prevent future DU is a difficult question. The definition of NUD in children is, at least, as difficult as in adults and the natural evolution of infection during childhood is not precisely known. Eradication in all family members must be obtained to avoid quick reinfection. If mass screening is decided upon, the only reasonable screening test in terms of safety and cost would be serology. Amoxicillin is almost the sole antimicrobial that has been tested in children: the reported eradication rate is low (27% according to Oderda et al. [36]) but the association amoxycillin-tinidazole achieved eradication in 94% of 32 children [35].

Omeprazole, which may be the most-prescribed drug for peptic diseases in the near future, may be associated with antimicrobial agents. So far, there is no proof that Omeprazole alone is able to eradicate HP: Mainguet et al. [27] reported suppression, not eradication. False-negative culture after a course of Omeprazole may occur because of the overgrowth of contaminants. The potentialization of amoxycillin by Omeprazole has been confirmed by Lamoulliatte et al. [26] with a 53% eradication rate 4 weeks after the discontinuation of Omeprazole.

No data are currently available about topical or intermittent therapy, urease inhibitors, or other drugs.

Conclusions

Eradication of *H. pylori* may be an important step towards prevention of both gastric cancer and peptic ulcer of the stomach and duodenum. Duodenal ulcer remains a very frequent, expensive, and sometimes life-threatening disease. Despite performant endoscopic bipolar coagulation in hard bleeders and close medico-surgical cooperation, 13 of our patients died from acutely bleeding gastric duodenal ulcer in a 7-year period; they represent around 2% of all upper GI tract bleeders admitted in our hospital.

At present, we strongly feel ourselves right in trying to eradicate HP in duodenal ulcer disease because there is so far no 100% safe and definitive treatment of the disease. To confirm this feeling, one may follow the proposal of Dr. Pounder (1989, personnal communication) and set up a prospective randomized study with three arms: two with tritherapy amoxicillin plus metronidazole for 2 weeks and either BSS or CBS for 8 weeks and a third arm of ranitidine alone followed by maintenance treatment.

We would advise short-term bitherapy to avoid dropouts because of bad compliance or side effects and recommend systematic endoscopy at 6 and 12 months to rule out asymptomatic relapse under anti-H2, with perhaps an additionnal group being treated by Omeprazole which may soon be the most prescribed drug in the acute phase of DU. Whatever the antimicrobial combination used, the occurrence of resistance must be carefully checked and the follow-up of healed patients, the monitoring of reinfection and its consequences may provide a more precise approach to increasing our knowledge of the relationship of *H. pylori* infection and peptic ulcer.

References

1. Axon ATR (1986) Campylobacter pyloridis: what role in gastritis and peptic ulcer? Br Med J 293:772
2. Bamford KM, Collins JSA, Collins BJ, Hugues DF, Porter KG, Wilson TS (1989) Amoxycillin and colloidal bismuth subcitrate improve C. Pylori associated gastritis but not symptoms in patients with non ulcer dyspepsia. Poster session. 2nd European Meeting of ECPSG, Oct 12–14th, Ulm
3. Bayerdoerffer E, Pirlet T, Sommer A, Kasper G, Ottenjann R (1988) Ofloxacin in der Therapie "resistenter" Ulcera duodeni. Ein Pilotstudie. Z Gastroenterol 26:155–159
4. Boero M, Ponzetto A, Rosina F, Forzani I, Berti E, Morgando A, Forzani B, Bellis D, Verme G (1989) Rate of erdication of C. pylori with different drugs schedules. Poster session. 2nd European Meeting of ECPSG, Oct 12–14th, Ulm
5. Borody T, Carrick J (1988) Long term Campylobacter pylori recurrence following apparent eradication. 1st Meeting of ECPSG, Bordeaux
6. Borody T, Hennessy W, Daskapoulos G (1987) Double blind trial of De-Nol in non-ulcer dyspesia associated with Campylobacter pylori gastritis. Gastroenterology 92:1324
7. Borody T, Noonan S, Cole P, Hyland L, Morgan A, Lenne J, Moore-Jones D, Brandl S (1989) Duodenal ulcer recurrence in patients remaining C. pylori (CP)-negative long term post eradication. Gastroenterology 96:A 52
8. Borsch G, Opferkuch W (1989) Short- and medium-term results of oral triple therapy to erdicate C. pylori. Gastroenterology 96:A 53
9. Borsch G, Mai U, Opferkuch W (1988) Oral triple therapy may effectively eradicate Campylobacter pylori in man: a pilot study. Gastroenterology 94:A 44

10. Burette A, Glupczynski Y, Thibaumont F, Deprez C (1989) Evaluation of short term amoxicillin/tinidazole combination in the treatment of Campylobacter pylori infection. Poster session. 2nd European Meeting of ECPSG, Oct 12–14th, Ulm
11. Cadranel S (1989) Preliminary results with bismuthemia in children. Poster session. 2nd European Meeting of ECPSG, Oct 12–14th, Ulm
12. Chiverton SG, Hunt RH (1989) Initial therapy and relapse of duodenal ulcer: possible acid secretory mechanisms. Gastroenterology 96:632–639
13. Coghlan JG, Humphries H, Dooley C, Keane C, Gilligan D, McKenna D, Sweeney E, O'Morain C (1987) Campylobacter pylori and recurrence of duodenal ulcers: a 12 month follow-up study. Lancet II:1109–1111
14. Deltenre M, De Koster E, Nyst JF, de Reuck M, van Gossum M, Jonas C (1989) Traitement actuel de l'infection par Campylobacter pylori: pourquoi et comment? Med Mal Infect 19:89–94
15. Deltenre M, Nyst JF, Jonas C, Glupczynski Y, Deprez C, Burette A (1989) Données cliniques, endoscopiques et histologiques chez 1100 patients dont 574 colonisés par Campylobacter pylori. Gastroenterol Clin Biol 13:89 B–95 B
16. Eberhardt R, Kasper G, Dettmer A, Hochter W, Hagena D (1988) Effect of bismuth subsalicylate versus cimetidine on Campylobacter pylori and healing- and relapse-rate of gastric ulcer. Med Klin 83 (12):402–405
17. Glupczynski Y, Thibaumont F (1989) Comparative in vitro susceptibility of Campylobacter pylori to different bismuth salts. Poster session. 2nd European Meeting of ECPSG, Oct 12–14th, Ulm
18. Glupczynski Y, Labbe M, Burette A, Delmee M, Avesani V, Bruck C (1987) Treatment failure of ofloxacin in Campylobacter pylori infection. Lancet I:1096
19. Glupczynski Y, Burette A, Labbe M, Deprez C, de Reuck M, Deltenre M (1988) Campylobacter pylori-associated gastritis: a double-blind placebo controlled study. Am J Gastroenterol 83:365–372
20. Goodwin CS, Marshall BJ, Blincow ED, Wilson DH, Blackbourn S, Phillips S (1988) Prevention of nitroimidazole resistance in Campylobacter pylori by coadministration of colloidal bismuth subcitrate: clinical and in vitro studies. J Clin Pathol 41:207–210
21. Graham DY, Lew GM, Michaletz PA (1989) Randomised controlled trial of the effect of eradication of C. pylori on ulcer healing and relapse. Gastroenterology 96:A 181
22. Kang JY, Tay HH, Wee A, et al. (1988) The effect of colloidal bismuth subcitrate on symptoms and gastric histology in non-ulcer dyspesia – a double blind placebo controlled study. Gut 10:A 1440
23. Karnes WE, Samloff IM, Siurala M, Kekki M, Sipponen P, Walsh JH et al. (1989) Does atrophic gastritis represent the "end-stage" of C. pylori-associated gastritis? Gastroenterology 96:A 249
24. Lambert JR, Borromeo M, Korman MG, Hansky J (1989) Campylobacter pylori and non ulcer dyspepsia-randomized controlled trial. Gastroenterology 96:A 284
25. Lambert JR, Way DJ, King RG, Eaces ER, Lianeas K, Wan A (1989) Pharmacokinetics of De Nol and Pepto Bismol in the human gastric mucosa. Poster session. 2nd European Meeting of ECPSG, Oct 12–14th, Ulm
26. Lamouliatte H, Megraud F, de Mascarel A, Barberis C, Bernard PH, Cayla R, Quinton A (1989) Does omeprazole improve amoxicillin therapy towards Campylobacter pylori associated gastritis? Poster session. 2nd European Meeting of ECPSG, Oct 12–14th, Ulm
27. Mainguet P, Delmee M, Debongnie JC (1989) Omeprazole, Campylobacter pylori and duodenal ulcer (Letter to the editor). Lancet II:389–390
28. Malfertheiner P, Stanescu A, Baczako K, Bode G, Ditschuneit H (1988) Chronic erosive gastritis. A therapeutic approach with bismuth. Scand J Gastroenterol 23 (142):87–92
29. Marshall BJ, Warren JR (1984) Unidentified curved bacilli in the stomach of patients with gastritis and peptic ulceration. Lancet II:1311–1315
30. Marshall BJ, Goodwin CS, Warren JR et al. (1988) Prospective double blind trial of duodenal ulcer relapse after eradication of Campylobacter pylori. Lancet II:1437–1442
31. Mathai E, O'Riordan T, Tobin A, Beattie S, Cafferkey M (1989) Drug resistance in Campylobacter pylori. Poster session. 2nd European Meeting of ECPSG, Oct 12–14th, Ulm

32. McNulty CAM, Gearty JC, Crump B et al. (1986) Campylobacter pyloridis and associated gastritis: investigator blind, placebo controlled trial of bismuth salicylate and erythromycine ethylsuccinate. Br Med J 293:645–649
33. O'Connor HJ, Wyatt JI, Dixon MF, Axon ATR (1986) Campylobacter-like organisms and reflux gastritis. Gut 27:A 608
34. Oderda G, Dell'Olio D, Morra I, Ansaldi N (1989) Campylobacter pylori gastritis: long term results of treatment with amoxycillin. Arch Dis Child 64(3):326–329
35. Oderda G, Vaira D, Holton J, Ainley C, Dell'Olio D, Ansaldi N (1989) Combined amoxycillin and tinidazole for eradication of Campylobacter pylori (CP) associated gastritis in children. Effect on serum pepsinogen I (PG I), gastrin, and IgG antibody title. Gut 30:A 733
36. Rauws EAJ (1989) Campylobacter pylori. PhD Thesis, University of Amsterdam
37. Rauws EAJ, Langenberg W, Houthoff HJ, Zanen HC, Tytgat GNJ (1988) Campylobacter pyloridis-associated chronic antral gastritis. Gastroenterology 94:33–40
38. Rokkas T, Pursey C, Uzoechina E et al. (1988) Non ulcer dyspepsia and short term De-Nol therapy: a placebo controlled trial with particular reference to the role of Campylobacter pyloridis. Gut 29:1386–1391
39. Smith AC, Price AB, Borriello P, Levi AJ (1988) A comparison of ranitidine and tripotassium dicitrato-bismuth (TDB) in relapse rates of duodenal ulcer. The role of Campylobacter pylori. Gastroenterology 94:A 431

How to Eradicate *Helicobacter pylori* and Prevent Reinfection

C. O'MORAIN

The idea that peptic ulcer disease may have an infective cause is not new. In 1940 Freedberg and Barron [1] suggested that the success of bismuth compounds in peptic ulcer might be due to the suppression of gastric spirochaetes. The discovery of *H. pylori* and its strong association with duodenal ulcer disease has heighted interest in a possible infective cause for duodenal ulcer disease and in turn has led to investigations to discover whether eradication of this organism would succed where antacid therapy had failed. The prolonged remission from duodenal ulcer achieved if *H. pylori* is eradicated has intensified efforts to eradicate the organism and prevent reinfection.

Relationship Between *H. pylori* and Relapse

H. pylori is strongly associated with duodenal ulcer disease and histologically proven gastritis. In a study from our unit *H. pylori* has been found in 90% of patients with duodenal ulcer. Histological gastritis was present in 76% of duodenal ulcer patients and 95% of patients with histological gastritis were *H. pylori* positive [2]. These results are typical of those reported worldwide.

Linking *H. pylori* pathogenically to duodenal ulcer disease has proved very difficult. Endoscopic studies of asymptomatic people are not common and one cannot unequivocally prove that infection precedes ulceration. Studies of ulcer relapse therefore provide the logical method of resolving this problem since if a causative relationship does exist, eradication of *H. pylori* should alter the natural history of duodenal ulcer disease. Colloidal bismuth subcitrate (CBS) and bismuth subsalicylate (BSC) have been shown to inhibit *H. pylori* both in vitro and in vivo. In vitro studies of over 100 strains of *H. pylori* have been shown to have a minimum inhibitory concentration (MIC) 90 for CBS with a dosage range of 2–30 mg/ml. These levels are achieved in vivo in the gastric antrum for 4 h after administration. This explains our findings and those of others that administering CBS in a dosage spread over four times a day is far more effective in eradicating *H. pylori* than is a twice a day regime [3].

In a study from our unit of patients randomised to treatment with either CBS or cimetidine it was found that both were equally effective in healing duodenal ulcers after 4 weeks of treatment but that CBS reduced gastritis in 34% of patients and

Consultant Gastroenterologist, Department of Gastroenterology, Meath and Adelaide Hospitals, Trinity College Dublin, Dublin 8, Ireland

Helicobacter pylori, Gastritis and Peptic Ulcer
Edited by P. Malfertheiner and H. Ditschuneit

H. pylori in 39% of patients. Cimetidine had little or no effect on gastritis or *H. pylori* eradication [4]. Patients with healed ulcers were followed up for 1 year and endoscoped at the end of that year of earlier if they became symptomatic. Patients who were *H. pylori* positive at the end of the treatment had a 79% relapse rate compared with a 29% relapse rate in those patients who were rendered *H. pylori* negative. Of the patients who were *H. pylori* negative initially and relapsed, five had become *H. pylori* positive, all of these showed evidence of histological gastritis and three had developed duodenal ulcer [5]. There has been confirmation of this finding by four other independent studies showing that patients rendered *H. pylori* negative are less likely to relapse than are those who are *H. pylori* positive (Table 1) [6–9].

Table 1. Relapse of duodenal ulcer according to posttreatment *H. pylori* status

Reference	Studies (*n*)	Relapse rate after 1 year	
		H. pylori positive (%)	*H. pylori* negative (%)
Coghlan et al. [5]	39	79	27
Smith et al. [6]	36	88	25
Lambert et al. [7]	45	78	40
Borody et al. [8]	21	–	14
Marshall et al. [9]	68	78	20

It is not yet entirely clear whether *H. pylori* or the ulcer reappears first. Langenberg et al. have shown that despite apparent clearance of the *H. pylori*, recurrence up to 6 months later is probably due to recrudescence of the same organism as undentified by the DNA pattern [10]. In our unit we have found that while CBS temporarily suppresses *H. pylori* in up to 70% of treated patients, eradication adjudged by negative culture at 1 month posttreatment occurs in only 20%–30% of cases. Borody et al. [8] followed up patients in whom eradication was apparently achieved 1 month after treatment; for a further 12–25 months they found that reinfection was extremely uncommon with only 5% becoming *H. pylori* positive with in this time.

Bismuth Compounds

The association of bismuth compounds with the treatment of gastric disorders goes back over 200 years. The advent of H2 antagonists which are simple and convenient to take, highly effective, and whose mechanism of action was apparently logical rapidly displaced bismuth compounds in the treatment of ulcer disease in the 1970s. A report from France of nearly 200 cases of bismuth encephalopathy further hastened the relegation of bismuth therapy. By the early 1980s H2 antagonists had become the gold standard in the treatment of duodenal ulcer.

Seven trials compared CBS with H2 antagonists in the treatment of duodenal ulcers where no maintenance therapy was given and at least a 1-year endoscopic follow-up was available. The relapse rate was lower in the bismuth group in six of the studies, and in four statistical significance was reached. The above findings have been confirmed by Dobrilla et al. [11] who used pooled data from available trials to compare the relapse rates for duodenal ulcer patients whose had been healed with H2 antagonists with those whose ulcers had been healed by any other treatment. They showed that patients initially healed with CBS are less likely to relapse than their H2-treated counterparts by 36% at 6 months and 26% at 1 year. The data reviewed included patients treated with pirenzepine, sucralfate, carbenoxolone, and antacids. It is interesting to note that of all the mucosal protective agents only CBS has a significantly superior relapse profile to H2 antagonists.

The reason why CBS should be associated with a lower relapse rate remains unclear. If H2 antagonists induce relapse by a rebound effect when acid supression is lifted, then all the other agents should demonstrate the same apparent reduction in ulcer relapse. Furthermore, the excess should occur within weeks of discontinuing H2 antagonist therapy and relapse should be the same thereafter – this is not the case. It has been suggested that accumulated body stores of bismuth continue to exert a therapeutic effect long after dosage has ceased. As urinary excretion declines nonexpotentially to normal levels within 2 months of stopping treatment [12], it is difficult to conceive that immeasurable drug levels continue to exert benefit up to 1 year later. Improvement of associated histological gastritis and more complete ulcer healing with normalisation of duodenal mucosa seems a reasonable explanation. This would, however, apply equally to all types of cytoprotective agents and the available evidence does not support this concept.

Finally, one must consider the possibility that suppression of *H. pylori* which colonises the gastric antrum of over 90% of patients with duodenal ulcer and which is strongly associated with histological gastritis could in some way explains the lower relapse rate found with CBS.

CBS has cytoprotective effects on the gastric and duodenal mucosa which are independent of *H. pylori* [12, 13]. CBS binds to gastric mucin forming a glycoprotein bismuth complex which adheres to the ulcer crater and retards hydrogen ion diffusion. It also increases local PGE2 and while reducing the concentrations of local aggressive factors such as leukotreine C4 and pepsin. It is therefore important not to simply equate CBS-induced ulcer healing with *H. pylori* eradication. In the treatment of duodenal ulcer the other properties of bismuth are important and antibiotics alone as suggested possible treatment for *H. pylori* are unlikely to be effective in healing ulcers. Although surprisingly some studies suggest that they are equally effective in healing duodenal ulcer.

Antibiotic Therapy

Although *H. pylori* is sensitive in vitro to a wide range of antibiotics including penicillin, metronidazole, tinidazole, cephalosporins, some quinolones, gentamycin, tetracycline, erythromycin and rifampicin, the clinical efficacy of most of these agents however, is poor [18]. Erythromycin, for example, is not effective in the low pH environment of the stomach.

Most studies of eradicating *H. pylori* using antibiotics involve a wide variety of diseases including gastritis duodenitis and both duodenal and gastric ulcers. Studies looking at a single disease entity are probably more valid as it is possible that different strains are responsible for different conditions. Antibiotics alone have been used to eliminate *H. pylori*. Antibiotic monotherapy with nitrofurantoin, amoxycillin, and furazolodine over 4 weeks results in bacterial elimination rates similar to those obtained with bismuth compounds [19, 20]. However, prolonged administration of antibiotics results in unacceptable side effects. A combination of bismuth and antibiotics results in elimination rates between 33% and 94% [8, 21–23].

Over the last 2 years we have treated patients with duodenal ulcer with amoxycillin 500 mg t.d.s., metronidazole 200 mg t.d.s. or 400 mg t.d.s. (for a period of 1 week) or amoxycillin 500 mg t.d.s. and metronidazole 400 mg t.d.s. with CBS for 4 weeks. The most effective treatment among these regimens was CBS and metronidazole 400 mg t.d.s. and with this we obtained 80% eradication. A lower dose of metronidazole 200 mg t.d.s. for 1 week and amoxycillin on its own did not give any extra benefit over CBS alone. The addition of amoxycillin to metronidazole did not improve eradication rates. When we reexamined the patients to see why our dual therapy with metronidazole and CBS was unsuccessful in eradicating the organism we found that some of the organisms had developed resistance to metronidizole. Of 29 isolates examined two already had resistance prior to commencing treatment. Following treatment a further seven resistant pairs were discovered (Fig. 1). Five of these patients were treated with amoxycillin in addition to metronidazole. The isolates were still sensitive to

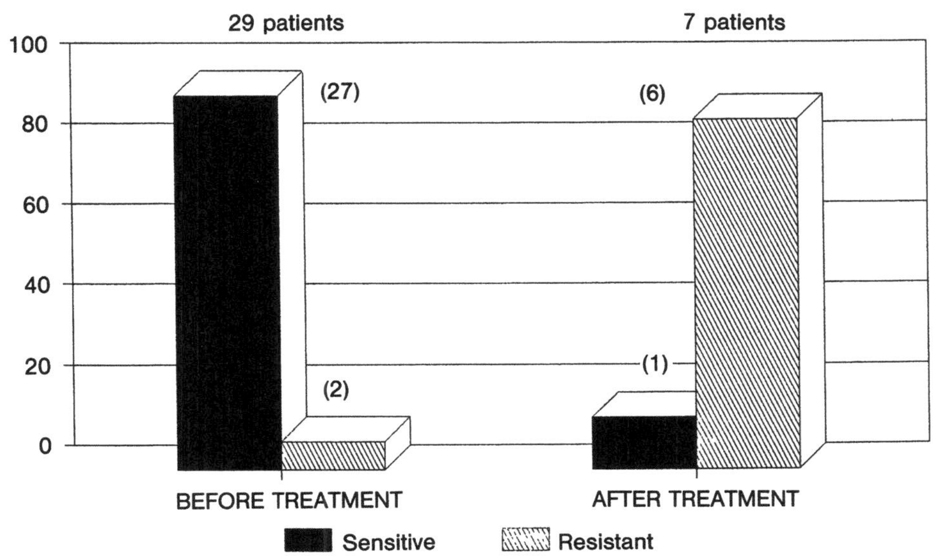

Fig. 1. Development of resistance to metronidazole with treatment

amoxycillin following treatment but had developed resistance to metronidizole. The addition of amoxycillin did not prevent development of the emergence of resistant strains.

We did not encounter any antibiotic toxicity. The 74% *H. pylori* eradication rates in our study are comparable with the results achieved by Borsch et al. [23, 24] who in a pilot study of peptic ulcer patients with *H. pylori*-positive gastritis used a 2-week combination of bismuth subsalicytate 600 mg t.i.d., metronidazole 500 mg t.i.d. and amoxycillin 500 mg t.i.d. and achieved 90% eradication 4 weeks after completion of treatment. H_2 antagonists were used by some of the patients in combination with the above regimen. No antibiotic toxicity was reported.

Resistance to tinidazole has already been reported, however, it was suggested that the combination of CBS with tinidazole would prevent the emergence of resistant strains – clearly this is not our experience. We feel that this is a most important aspect in the management of patients with peptic ulcer. We would suggest that *H. pylori* infection should be treated as we would treat other infections and that sensitivity should be evaluated before and after treatment. *H. pylori* is a slowly growing organism that presents difficulties in using conventional antibiotic disc diffusion tests to assess sensitivity to antibiotics and MIC assay (incorporating drugs in the agar plates), is the most accurate way of assessing sensitivity.

Triple Therapy

It may be that triple therapy may be indicated if another effective antibiotic drug is found in vivo. Metronidazole has some inherent advantages as its activity is not dependent on pH, it appears to be secreted into the gastric juices where it is able to get access to bacteria deep in the gastric pits [25]. Borody et al. [8] have found a combination of tetracycline and metronidazole and colloidal bismuth subcitrate to be the most effective method of eradicating *H. pylori.* This combination needs to be tested in larger groups of patients. However, they did use large doses of antibiotics and experienced some serious side effects. Two of their patients developed *Clostridium difficile* induced colitis. Increasing the complexity of treatment increases the risk of side effects.

Data on 38 patients whom we treated with a combination of antibiotics and CBS and rendered *H. pylori* negative following treatment are now available. At the end of 1 year 20 of these patients remain *H. pylori* negative and none of them has developed ulcers, 17 have reverted to *H. pylori* positive and of these 5 have developed duodenal ulcer and a further 7 gastritis. This suggests that the natural history may well be that recurrence of *H. pylori* first induces gastritis and this can go on to duodenal ulcer. The overall relapse rate of 5 out of 38 is far lower than what would be expected following treatment with H2 antagonists and compares favourably with patients maintained long term on H2 antagonists. There was a 40% reinfection rate which is higher than the experience of other units. Borody's experience [8] was as low as 10%, but again he was treating not only peptic ulcer disease, and Rauws who was treating gastritis experienced no relapse rate over an observation period of 1 year [21].

Reinfection or Recrudescence

It is not certain if the reappearance of *H. pylori* is due to reinfection or recrudescence. There is strong evidence that *H. pylori* is transmitted by person to person contact. A Toronto study [26] suggested that siblings and parents of children with *H. pylori*-positive gastritis have far greater levels of antibody than a control population, suggesting interfamilial spread. It is not surprising, therefore, that there is reinfection as patients return to their homes following successful treatment. Restriction endonuclease analysis of *H. pylori* shows that the same strain can infect a family. The reappearance of the same strain would not necessarily mean that it is a recrudescence of the same bacteria.

In conclusion, we feel that if patients have a *H. pylori*-related duodenal ulcer that microbiological tests should be routinely done to assess sensitivity to antibiotics. If reinfection occurs the organism should be cultured to ascertain if it has developed resistance to the antibiotics initially used.

References

1. Freedberg AS, Barron LE (1940) The presence of spirochetes in human gastric mucosa. Am J Dig Dis 7:433–445
2. Dooley CP, McKenna D, Humphreys H, Bourke S, Keane CT, Sweeney E, O'Morain C (1988) Histological gastritis in duodenal ulcer relationship to Campylobacter pylori and effect of therapy. Am J Gastroenterol 83:278–281
3. Coghlan JG, Gilligan D, Humphreys H, McKenna D, Sweeney E, Keane C, O'Morain C (1990) Efficacy of different dosage regimes in duodenal ulcer healing and eradication of Campylobacter pylori. Aliment Pharmacol Ther (in press)
4. Humphreys H, Bourke S, Dooley C, McKenna D, Power B, Keane CT, Sweeney EC, O'Morain C (1988) Effect of treatment on Campylobacter pylori in peptic disease, a randomised prospective trial. Gut 29:279–283
5. Coghlan JG, Gilligan D, Humphreys H, McKenna D, Dooley C, Sweeney E, Keane C, O'Morain C (1987) Campylobacter pylori and recurrence of duodenal ulcers – a 12 months follow-up study. Lancet II:1109–1111
6. Smith AC, Price AB, Borriello P, Leos AJ (1988) A comparison of ranitidine and tripotasium dicatrato bismuthate (TDB) in relapse rate of duodenal ulcer. The role of Campylobacter pylori (CP). Gastroenterology 94:A431
7. Lambert JR, Borromes M, Korman MC, Hanskey J, Eaves ER (1987) Effect of colloidal bismuth (De-Nol) on healing and relapse of duodenal ulcers: a role of Campylobacter pyloridis. Gastroenterology 92:A1489
8. Borody T, Cole P, Noonan S, Morgan A, Ossip G, Masey J, Bradle S (1988) Long-term Campylobacter pylori recurrence post-eradication. Gastroenterology 94:A43
9. Marshall BJ, Goodwin CS, Warren JR, Murray R, Blincow E, Blackburn SJ, Phillips M, Waters TE, Sanderson CR (1988) Prospective double blind trial of duodenal ulcer relapse after eradication of Campylobacter pylori. Lancet 1432:1440
10. Langenberg W, Rauws EAJ, Widjojokusumo A, Tytgat GNJ, Zanen HC (1986) Identification of Campylobacter pyloridis isolates by restriction endonuclease DNA analysis. J Clin Microbiol 24:414–417
11. Dobrilla G, Vallaperta P, Amplatz S (1988) Influence of ulcer healing agents on ulcer relapse after discontinuation of acute treatment; a pooled estimate of controlled clinical trials. Gut 29:181–187
12. Bader JP (1987) The safety profile of De-Nol. Digestion [Suppl 2] 32:53–59
13. Lee SP (1982) A potential mechanism of action of colloidal bismuth diffusion to hydrochlonic acid. Scand J Gastroenterol [Suppl 80] 17:17–21

14. Konturek SJ, Radeckie R, Plastucki, Drosowiez D (1987) Studies on the gastroprotective and ulcer healing effects of colloidal bismuth subcitrate. Digestion [Suppl 2] 37:8–15
15. Konturek SJ, Bilski J, Kwieclen N, Obtulowiez W, Kopp B, Olesky J (1987) DeNol stimulates gastric and duodenal alkaline secretion through a prostaglandin dependent mechanism. Gut 28:1557–1563
16. Ahmed A, Cairns SR, Vaira D, Hoult JRS, Salmon PR (1988) Colloidal bismuth subcitrate (DeNol) reduces leukotriene C_4 formation in duodenal ulcer. Gastroenterology 94:A3
17. Barron JH, Barr J, Batten J, Sidebotham R, Spencer J (1986) Acid pepsin and mucin secretion in patients with gastric and duodenal ulcers before and after colloidal bismuth subcitrate (DeNol). Gut 27:486–490
18. McNulty CAM, Dent JC, Wise R (1988) Susceptibility of Campylobacter pylori to twenty-one antimicrobial agents. Eur J Clin Microbiol Infect Dis 7:567–569
19. Morgan D, Kraft W, Bender M, Pearson A (1988) Nitrofurans in the treatment of gastritis associated with Campylobacter pylori. Gastroenterology 95:1178–1184
20. Rauws EAJ, Langenberg W, Houthoff HJ, Zamen HC, Tytgat GWJ (1988) Campylobacter pyloridis associated chronic antral gastritis. Gastroenterology 94:33–40
21. Rauws EAJ (1989) Therapeutic attempts of eradication of Campylobacter pylori. Eur J Gastroenterol Hepatol 1:34–41
22. Goodwin CS, Marshall BJ, Blincow ED, Wilson DH, Blackbourne S, Phillips M (1988) Prevention of nitronidazole resistance in Campylobacter pylori coadministration of colloidal bismuth subcitrate, clinical and in vitro studies. J Clin Pathol 41:207–210
23. Borsch G, Mai U, Muller K (1988) Monotherapy on polychemotherapy in the treatment of Campylobacter pylori related gastroduodenal disease. Scand J Gastroenterol [Suppl 142] 23:101–106
24. Borsch G, Mai U, Opferkuch W (1988) Oral triple therapy may effectively eradicate Campylobacter pylori in man, a pilot study. Gastroenterology 94:44A
25. Hollingsworth JA, Goldie J, Silette LY, Richardson H, Hunt RH (1987) Gastric secretion of antibiotics used for Campylobacter pyloridis. Gut 28:A1409
26. Drumm B, et al. (1990) Intrafamilial spread of C. pylori. N Engl J Med (in press)

Drug Resistance in *Helicobacter pylori*

E. Mathai, T. O'Riordan, A. Tobin, S. Beattie, M. Cafferky, and C. O'Morain

Introduction

Initial studies have suggested that eradication of *Helicobacter pylori* is associated with lower duodenal ulcer relapse rates [1, 2]. Hence there has been continued effort in identifying the optimum antibacterial regimen for total clearance of *H. pylori*. Even before the discovery of the association of *H. pylori* with gastroduodenal disease, bismuth had been used successfully in the treatment of duodenal ulcers [3]. It was thought that this effect was achieved merely by mucosal protection as the drug binds to gastric mucin producing a complex which adheres to the ulcer crater and retards hydrogen ion diffusion. H_2-receptor antagonists were introduced in the 1970s and were found to be equally as effective as bismuth in the immediate healing of duodenal ulcers. However, there was a significant difference in the ulcer relapse rates. Of duodenal ulcers 60%–90% relapsed within 1 year when treated with H_2-receptor antagonists but only 35%–76% of duodenal ulcers treated with bismuth relapsed during this time. Bismuth was also found to be better than H_2-receptor antagonists in healing the associated antral gastritis [4, 5]. Bismuth has since then been widely used in the treatment of duodenal ulcer.

However, bismuth alone does not eradicate *H. pylori* completely which is necessary for reducing relapses. Various antibiotics have been used alone or in combination with bismuth to achieve eradication of the organism. Monotherapy with antimicrobial agents results in bacterial elimination rates similar to those obtained with bismuth compounds alone, and the chances of development of drug resistance are higher with monotherapy [6]. The elimination rates are higher with combination therapy (up to 94%). The efficacy of such a regimen in healing the ulcer is the same as with bismuth alone, but the relapse rate is lower.

The choice of antibacterial agents to be used in combination with bismuth is still unclear. Although nitroimidazoles are only moderately active against *H. pylori* in vitro, [7] they have proved to be the most successful adjuvants to bismuth in eradication of the organism and in ulcer healing. A few strains are naturally resistant to metronidazole [8]. Further more, when drug therapy consists of nitroimidazole alone or in combination with H_2-receptor antagonists, strains of *H. pylori* are found to acquire drug resistance [9]. Combination of nitroimidazoles and bismuth are believed to prevent emergence of resistance [9]. Amoxycillin has

Department of Gastroenterology, Meath/Adelaide Hospitals, Trinity College, Dublin 8

Helicobacter pylori, Gastritis and Peptic Ulcer
Edited by P. Malfertheiner and H. Ditschuneit

also been found to be effective against *H. pylori* in vivo. Monotherapy with this agent has been associated with up to 91% clearance rates [7].

This study looks at the effects of various treatment regimens in eradicating *H. pylori* and healing duodenal ulcer and the associated gastritis. The susceptibility of *H. pylori* to metronidazole and amoxycillin before and after treatment was also assessed.

Methods

Patients

Ninety-eight patients with endoscopically diagnosed duodenal ulcers, histological antral gastritis and evidence of *H. pylori* infection were included in this study. Patients who had been on H_2-receptor antagonists, bismuth preparations, antibiotics, corticosteroids or nonsteroidal antiinflammatory drugs during the preceding month were excluded from the study.

The patients were allocated into four treatment groups. All received colloidal bismuth subcitrate (CBS) 120 mg four times a day for 4 weeks. In addition, patients in

- Group 1 had amoxycillin 500 mg thrice daily for the first week
- Group 2 had metronidazole 200 mg thrice daily for the first week
- Group 3 had metronidazole 400 mg thrice daily for the first week
- Group 4 had metronidazole 400 mg and amoxycillin 500 mg thrice daily for the first week

Patients were advised to take bismuth 30 min before meals and the antibacterial drug with food. Compliance was monitored using tablet counts.

Two antral biopsy specimens were taken from each patient during the initial endoscopy before treatment and 4 weeks after completion of treatment. One biopsy specimen was processed for histological examination while the other was sent for microbiological examination.

Histology

Specimens were fixed immediatley in 10% buffered formalin and transported to the laboratory where they were embedded in paraffin, sectioned and stained with haematoxylin and eosin. Gastritis was graded according to the classification proposed by Warren and Marshall: grade 0, normal epithelium with no inflammatory cell infiltration; grade 1, mild round cell infiltration; grade 2, massive round cell infiltration with or without polymorphs but without epithelial migration or destruction; and grade 3, polymorph infiltration of epithelial layers with or without glandular destruction. Grades 2 and 3 were regarded as positive for gastritis.

Microbiology

The biopsy specimen was transported to the laboratory in nutrient broth and cultured as described earlier. A smear stained by Gram stain was also examined for Helicobacter-like organisms.

Minimum inhibitory concentration (MIC)

Isolates from 29 patients in groups 3 and 4 were tested for their susceptibility to metronidazole and amoxycillin.

Susceptibility to metronidazole was assessed using a minimum inhibitory concentration (MIC) assay and disc diffusion tests. Disc diffusion was done with 5-mg discs on chocolate agar. A zone of inhibition 2 cm or more in diameter was considered indicative of sensitivity to the drug.

Metronidazole (Sigma) powder was diluted serially in distilled water. These dilutions were mixed at 45°–50 °C with sterile brain-heart infusion agar containing 10% horse blood to obtain concentrations of 32 mg/l, 16 mg/l, 8 mg/l, 4 mg/l, 2 mg/l, 1 mg/l and 0.5 mg/l. The agar was then poured onto the plates (about 20 ml/plate) and allowed to set. Plates were used within 1 week of preparation.

Isolates for sensitivity testing were suspended in nutrient broth to obtain a faint turbidity (approximately 10^6–10^7 colony forming units/ml). Twentyfive different strains were inoculated on one plate using a multipoint inoculator. Tests were done in triplicate. Brain-heart infusion agar with 10% horse blood but without any antibiotics was used as control. Plates were incubated and read after 48 and 72 h of incubation for the presence or absence of growth.

Amoxycillin sensitivity was tested using 10 mg amoxycillin discs. A zone of inhibition 5 mm or more in diameter was considered to indicate sensitivity to the drug.

Data Analysis

Results from the above study were analysed using the chi-squared and Fischer's exact tests.

In addition to the isolates from the patients in this study, a few isolates obtained after 1 January, 1989 were also tested. The sensitivity patterns of these isolates were compared with those of isolates obtained prior to this date.

Results

Sixty (65%) of the 98 patients in this study had *H. pylori* eradicated after treatment with CBS in combination with metronidazole, with or without additional amoxycillin (Table 1). A clearance rate of 57% was achieved with bismuth in combination with 200 mg metronidazole while bismuth with amoxycillin achieved 50% clearance. Bismuth in combination with 400 mg metronidazole eradicated *H. pylori* in 73% of patients. The addition of amoxycillin to this

Table 1. Eradication of *H. pylori* from gastric antrum with treatment

Group	Patients *n*	Eradicated *n* (%)	Not eradicated *n* (%)
CBS + Amox	18	9(50)	9(50)
CBS + M 200	23	13(57)	10(43)
CBS + M 400	26	19(73)	7(27)
CBS + M 400 + Amox	31	23(74)	8(26)

CBS, colloidal bismuth subcitrate; Amox, amoxycillin; M 200, metronidazole 200 mg; M 400, metronidazole 400 mg

regimen did not improve the eradication rate. There was no statistical difference between the groups.

Ulcer healing rates observed with the different combinations of drugs are listed in Table 2. About 85% of ulcers were healed 4 weeks after treatment with bismuth and 400 mg metronidazole with or without additional amoxycillin. There was no statistical difference between the different groups.

Table 2. Duodenal ulcer healing 4 weeks after completion of therapy

Group	Patients *n*	Healed *n* (%)	Not healed *n* (%)
CBS + Amox	18	14(78)	4(22)
CBS + M 200	23	14(61)	9(39)
CBS + M 400	26	22(85)	4(15)
CBS + M 400 + Amox	31	26(84)	5(16)

Abbreviations as in Table 1

Treatment with bismuth in combination with 400 mg metronidazole with or without amoxycillin achieved healing in 70% of antral gastritis (Table 3). Bismuth with 200 mg metronidazole healed only 52% of antral gastritis. However, this difference was not statistically significant.

Table 3. Effect of therapy on antral gastritis

Group	Patients *n*	Healed *n* (%)	Not healed *n* (%)
CBS + Amox	18	11(61)	7(39)
CBS + M 200	23	12(52)	11(48)
CBS + M 400	26	18(69)	8(31)
CBS + M 400 + Amox	31	22(71)	9(29)

Abbreviations as in Table 1

Table 4. Development of resistance to metronidazole with treatment

Treatment	Tested *n*	Sensitive *n* (%)	Resistant *n* (%)
Before	29	27(93)	2(7)
After	7	1(13)	6(87)

Of the isolates 93% were sensitive to metronidazole before treatment (Table 4). However six of the seven isolates obtained after treatment with this drug were resistant to the drug. Four of the six resistant isolates had acquired resistance during treatment.

All 30 isolates tested were sensitive to amoxycillin before treatment. Two isolates which persisted after treatment with amoxycillin remained sensitive to this drug. One of these isolates from a patient who had received metronidazole also had acquired resistance to metronidazole.

The sensitivity of the isolates from 1989 to metronidazole and that of the isolates before this year were compared. Of 30 isolates prior to 1989, 26 (87%) were sensitive to metronidazole compared with 24/41 (59%) in 1989.

Discussion

In this study the groups on metronidazole 400 mg thrice daily in combination with CBS achieved best rates for ulcer healing, healing of antral gastritis and eradication of *H. pylori.* While CBS 120 mg four times daily is as effective as 240 mg twice daily in healing of duodenal ulcers, *H. pylori* eradication is significantly better with the former. Hence all patients in this study received bismuth 120 mg four times daily. Bismuth was used for 4 weeks only as studies have shown no benefit in prolonging treatment beyond this period. Borsch et al. [10] used three drugs for 2 weeks while Borody et al. [11] used two drugs for 4 weeks and metronidazole for 10 days. There were slightly better clearance rates with these regimens but the toxic effects were greater when compared with the two-drug therapy.

To some extent the response was dose dependent in this study. Indices of healing were better in patients treated with metronidazole 400 mg thrice daily than in those treated with 200 mg thrice daily. The efficacy of the latter regimen is equivalent to that of bismuth and amoxycillin. There appears to be no additional benefit in using amoxycillin along with bismuth and metronidazole 400 mg thrice daily. However, four strains which persisted with triple therapy were seen only in smear and could not be grown on culture. Healing of duodenal ulcers and antral gastritis correlated with eradication of *H. pylori.*

The development of resistance of *H. pylori* has been reported before. Bayerdorffer et al. [6] observed resistance to oflaxacin in isolates from patients treated with bismuth and oflaxacin. Resistance to tinidazole has been observed by Goodwin et al. [9] in strains from patients on cimetidine and tinidazole. In the above study, development of resistance to metronidazole was reduced to a great extent by bismuth. In the present study 93% of the isolates were sensitive to metronidazole before treatment. Of the seven persistent isolates studied, two were resistant to the drug before treatment while four acquired resistance during therapy. Acquired resistance seemed to be more common than pretreatment resistance and could be an important factor in the outcome of treatment. Concomitant use of bismuth or amoxycillin did not prevent development of resistance. One persistent isolate from a patient on triple drug therapy had acquired metronidazole resistance but remained sensitive to amoxycillin. All strains tested before treatment and the persistent strains which were tested from a patient on bismuth and amoxycillin were also sensitive to amoxycillin. Therefore, metronidazole resistance seems to be the major reason for the persistence of *H. pylori* in the group treated with metronidazole. Persistence with amoxycillin therapy does not seem to be related to resistance. Hollingworth et al. [12] found that metronidazole was secreted in gastric mucus while amoxycillin was not. They postulated that those antibiotics not secreted in mucus are active only topically resulting in high but transient clearance of *H. pylori* with the organisms persisting in the pits. This could explain the persistence of positive strains even after treatment.

Anti-*H. pylori* treatment is justifiable in patients with duodenal ulcers since eradication of the organism correlated well with healing of antral gastritis and duodenal ulcers. Testing sensitivity of the organism before treatment and in treatment failure will facilitate proper management in that metronidazole, a drug that is not well tolerated, can be avoided in resistant cases. However, it is necessary to search for other agents capable of completely eradicating *H. pylori*.

Comparison of resistance patterns over the years shows an increase in resistance to metronidazole. This may reflect the increased use of this drug such as in the treatment of gastroduodenal abnormalities. The means of acquiring resistance are not clear. Goodwin et al. found that resistant strains of *H. pylori* are present in almost all populations; these are selected out by exposure to metronidazole, and then the search for another appropriate antibacterial agent is undertaken.

In conclusion, treatment with CBS and metronidazole 400 mg thrice daily results in 74% clearance of *H. pylori* at the end of 4 weeks after completion of treatment. The addition of amoxycillin to this regimen does not improve the results. *H. pylori* eradication correlates well with the healing of duodenal ulcer and gastritis. The results of therapy with bismuth with lower doses of metronidazole are comparable to those using a combination of bismuth and amoxycillin. Resistance to metronidazole develops even with combination therapy. Resistance to metronidazole is the main reason for treatment failure with this drug. Lack of response to amoxycillin is not due to the development of resistance. There is increasing resistance to metronidazole in *H. pylori* treatment probably due to the increasingly widespread use of this drug.

References

1. Coghlan JC, Gilligan D, Humphries H, McKenna D, Dooley C, Sweeney E, Keane C, O'Morain C (1987) Campylobacter pylori and recurrence of duodenal ulcers – a 12-month follow-up study. Lancet II:1109–1111

2. Marshall BJ, Warren JR, Blincow ED, Phillips M, Goodwin CS, Murray R, Blackbourm SJ, Waters TE, Sanderson TR (1988) Prospective double blind trial of duodenal ulcer relapse after eradication of C. pylori. Lancet II:1437–1441
3. Coghlan JG, Tobin A, O'Morain C (1989) Campylobacter pylori and ulcer treatment. In: Rathbone BJ, Heatley RV (eds) Campylobacter pylori and gastroduodenal disease. Blackwell, Oxford, pp 139–145
4. Marshall BJ, Armstrong JA, Francis G, Nokes NT, Wee SH (1987) The antibacterial action of bismuth in relation to C. pyloridis colonisation and gastritis. Digestion [Suppl 2] 37:16–30
5. Rauws EAJ, Tytgat GN (1989) Campylobacter pylori: treatment of gastritis. In: Rathbone BJ, Heatley RV (eds) Campylobacter pylori and gastroduodenal disease. Blackwell, Oxford, pp 225–231
6. Bayerdorffer E. Simon TH, Bastlein CH, Ottenjan R, Kasper G (1987) Bismuth/ofloxacin combination for duodenal ulcer. Lancet II:1467–1468
7. Hirschl AM, Pletschette M (1989) Antibiotic treatment of Campylobacter pylori infection. In: Rathbone BJ, Heatley RV (eds) Campylobacter pylori and gastroduodenal disease. Blackwell, Oxford, pp 217–224
8. McNulty CAM (1989) Bacteriological and pharmacological basis for the treatment of Campylobacter pylori infection. In: Rathbone BJ, Heatley RV (eds) Campylobacter pylori and gastroduodenal disease. Blackwell, Oxford, pp 209–216
9. Goodwin CS, Marshall BJ, Blincow ED, Wilson DH, Blackburn S, Phillips M (1988) Prevention of nitroimidazole resistance in C. pylori by coadministration of colloidal bismuth subcitrate: clinical and in vitro studies. J Clin Pathol 41:207–210
10. Borsch A, Mai U, Opferkuch W (1988) Oral triple therapy (OTT) may effectively eradicate C. pylori in man: a pilot study. Gastroenterology 94:A43
11. Borody T, Cole P, Noonan S, Morgan A, Ossip G, Masey J, Brandl S (1988) Long term C. pylori recurrence post-eradication. Gastroenterology 94:A43
12. Hollingworth J, Goldie J, Silletti L, Rishardson H, Hunt RH (1987) Gastric secretion of antibodies used for C. pyloridis. Gut 28:A1409

If, When, and How to Treat Children?

S. CADRANEL

Introduction

In 1986, the first cases of *Helicobacter pylori* (HP) infection in children were published [1, 2] and followed by other reports from several authors [3–5], stressing the similitude of findings in pediatric and adult patients. Secondary ulcers are not uncommon in children whereas primary ulcers are infrequent [6]; on the other hand, in this age group, secondary gastritis, due to a known precipitating cause, is often clinically diagnosed but rarely fully documented whereas primary gastritis is most often underdiagnosed. The availability of adapted miniaturized endoscopes has led pediatric gastroenterologists to give more attention to the possibility of gastric lesions in children complaining of chronic abdominal pain, a very common complaint in childhood. More and more pediatricians are becoming aware of the possible role of chronic gastritis as a cause of symptoms in recurrent abdominal pain and, as a matter of consequence, of the possible role of HP in childhood. Information is now being gathered about the incidence of HP in the pediatric population in various parts of the world. Because children are seldom exposed to the harmful effects of alcohol, tobacco, or other aggressive factors commonly encountered in adult life, the occurrence of HP infection in childhood, and its successful treatment, might help to better understand the role of this microorganism as a pathogen in the gastric mucosa.

Clinical Features

Symptoms

Similarly to findings in adults, a strong correlation between primary ulcer or gastritis and the presence of HP has also been found in the first reports of HP infection in children [1–4]. Four larger series have been reported in the literature, from Canada [7], Belgium [8], England [9], and Italy [10], in which the authors have conducted a prospective systematic detection of HP in all children undergoing an upper GI endoscopy or analysed the data retrospectively. The clinical features of these four series are summarized in Table 1. It is difficult to make a comparison between them because the indications may vary according to

Queen Fabiola Children's Hospital, Free University of Brussels, 15, rue J. J. Crocq 1020 Brussels, Belgium

Helicobacter pylori, Gastritis and Peptic Ulcer
Edited by P. Malfertheiner and H. Ditschuneit

Table 1. Symptoms in children with HP infection

	Drumm et al. 1987 [7]	Cadranel et al. 1988 [8]	Mahony et al. 1988 [9]	Oderda et al. 1989 [10]
No. of patients	71	114	111	42
Data available	67	114	38	42
Type of study	Prospective, unselected	Prospective, unselected	Retrospective, unselected	Prospective, selected
Indications for endoscopy				
Recurrent abdominal pain	39	26	19	15
Vomiting and GER	23	27	16	
Hematemesis, anemia or occult blood	18		2	
Peptic history	11		18	
Esophagitis		9		9
Dysphagia		7	1	
Crohn's disease	7	3		
Varia		31		
Symptoms in HP+	12	43	9	42
Epigastric pain	7	19	7	16
Vomiting		7	1	7
Nausea or dysphagia		5		2
Anorexia				1
Hematemesis (or anemia)		2	1	1
Peptic disease (or follow-up)	5	3		
Asymptomatic	?	7	?	?

local medical attitudes and there is not a clear-cut separation between indications for endoscopy, symptoms, and findings. However, epigastric pain seems to be one of the most frequent clinical symptom found in 38%–78% of the HP-positive patients according to the series.

Age

The different larger prospective series generally include infants aged less than 1 year to adolescents. In our series, the HP-positive patients range from 2 to 16 years, with nine additional older patients included because they were still being followed up for chronic diseases initiated in infancy (Table 2). The most numerous are to be found in the 9–16 year age group though as many as 15 children under the age of 5 were positive for HP.

Table 2. Age of HP-Positive Children

Years of age	0–5	6–10	10–16	>16
No. of patients	15	27	52	9

Ethnic Origins

The ethnic origin of the patients in our own series is given in Table 3. More than one-half of our HP-positive children are of Moroccan origin and the incidence of HP infection was as high as 57% in Moroccan children investigated. The same proportion is found in children of Turkish and Yugoslav origins. These are children from low social class background and their parents are first generation immigrants. The Bantu children, on the contrary, come from a different social environment, their parents being usually Zairese students, fellows of different Belgian universities. However, the incidence of HP infection seems to be also very high in this population. Therefore, there is a high incidence of 58% of HP infection in symptomatic children living in Belgium and whose parents come from three different regions of the world, contrasting with an incidence of only 24% in symptomatic children of Western European parents.

Table 3. Ethnic origins of HP-positive children

	HP+ (*n*)	(%)	Endoscopy (*n*)	CP+ (%)
Moroccans	57	55	100	57
Belgians	23			
Italians	2	27	191	14
Portuguese	2			
Turkish	8	10	17	58
Yugoslavs	2			
Latin Americans	1	1	1	–
Bantus	7	7	12	58
	102	100	321	

Endoscopic Findings

Ulcers

Primary ulcers are infrequent in childhood compared with in adulthood; however, the long-term prognosis of patients with peptic ulcers beginning in their teens is not favorable and progression towards chronic relapsing peptic disease in adulthood is noted in a large proportion of them [11]. The same association between primary ulcers and HP occurs in children as in adults [9, 10, 12, 13]. In our own series, during the last 4 years, seven teenagers, six boys and one girl aged 11–19 years, were diagnosed with a duodenal ulcer and all of them had a HP-positive culture of antral, fundic (and in five cases duodenal) biopsy. Furthermore, eight additional children, four boys and four girls, aged 4–18 years investigated previously for HP-associated gastritis developed small shallow ulcerations in the antrum or the duodenal bulb: in all cases HP was cultured from antral biopsies.

Micronodular Gastritis

In all the pediatric series already mentioned, ulcers are less frequent than gastritis, which is documented by histologic studies. Is there a particular endoscopic pattern predictive of HP infection? Our first report in 1986 [2] insisted on a micronodular aspect of the antrum. It is referred to as "antral nodularity", "antral micronodular pattern," "nodular antritis," "lymphoid nodular hyperplasia" or "nodular antral gastritis" in several reports [3, 8, 10, 14].

Endoscopically Normal Mucosa

In our previous series [8], a normal aspect of the antrum was found in only five HP-positive children, though only one of these mucosae was histologically normal. The same coincidence of normal endoscopy with abnormal pathology has been widely reported in adults but also in children [3, 14, 15] and stresses the need of systematic biopsies [9].

Epidemiology

Little is known about the source and spread of HP. Most authors reporting pediatric cases of HP infection whether associated with primary ulcers or primary gastritis usually report the high incidence of a family history of peptic diseases [8, 10]. How often do we find a HP infection in the family contacts of these children? How often are they symptomatic? What is the incidence of HP infection in the pediatric age group? As systematic endoscopy is not an acceptable method for epidemiologic studies, data rely mostly on serologic studies or on labelled ^{14}C- or ^{13}C-urea breath tests.

Mahony showed, in an endoscopically validated study [16], that there was a good correlation between IgG antibodies to HP and endoscopic findings. Similar results were obtained by Thomas et al. [17]. From the same group, using a serological method, the prevalence rate of HP infection was estimated, in 361 Gambian children, to be 14% of children under the age of 20 months, rising to 46% in those up to 60 months of age. The prevalence rate of HP infection was 53% in 77 children less than 3 years of age with chronic diarrhea and malnutrition, significantly more frequent than in age-matched healthy controls (26%). In a mass screening of 1000 sera from France, Algeria, Ivory Coast, and Vietnam, Megraud et al. [18] showed evidence for a worldwide distribution of HP infection and a high prevalence in developing countries where 50% of the population was HP-positive by the age of 10 years, rising to 75% from the third decade on. The same high prevalence was found in Peruvian children by Klein et al. [19] using a ^{13}C-urea breath test: 54% of children aged 1–10 years and 90 of parents from low income families were HP-positive contrasting with "only" 30% of children and 78% of parents from high income families.

The results of the screening of the families of our HP-positive gastritis children are given in Table 4: 90% of the fathers and 80% of the mothers, whether

Table 4. ^{14}C-urea breath test (UBT) in families of children with HP gastritis

	Fathers		Mothers		Siblings	
Total no.	35		38		102	
Symptomatic	19		15		19	
Asymptomatic		16		23		83
^{14}C-UBT positive	19	14	13	14	16	15 [a]
negative	0	2	0	6	0	5
Not done			2	3	3	63 [b]

[a] 10 > 15 years old
[b] 32 > 15 years old

symptomatic or not, had a positive ^{14}C-urea breath test, meaning that the incidence of HP-infection was, indeed, very high in this population. The same was true for young adult siblings in which the test was performed. Though our modified technique (1 μCi) delivers a very low radiation, it is not a suitable technique for epidemiologic investigation in asymptomatic children. Therefore, we are currently investigating the prevalence rate of HP infection in "healthy" schoolchildren using a ^{13}C-urea breath test.

Treatment

Attempts of eradication of HP in adults have often been hampered by a high rate of recolonization. In a very well-designed study from Amsterdam, Rauws et al. [20] showed that HP did not disappear spontaneously from the gastric mucosa. Treatment with amoxycillin, colloidal bismuth subcitrate, or the combination of both drugs was effective in eradicating HP from 68%, 45%, and 90% respectively of patients with nonulcer dyspepsia. Recolonization with the same bacterial subtype was frequent during the first month after the discontinuation of the therapy. Almost every pediatric group has been trying to eradicate HP either with amoxycillin alone [10, 21, 22], or with colloidal bismuth subcitrate alone [13], or a combination of both [13, 21] or a combination of amoxycillin with tinidazole [22], or, in our own series, with metronidazole. A tri-therapy has not yet been reported. As a matter of fact, it is very difficult, at the present moment, to compare all these studies, since the drugs, their combination, the doses, and the regimens used vary greatly. Besides, long-term follow-up is currently not available in these pediatric series.

Because of the alledged toxicity of bismuth compounds, some concern about its use in children has arisen in several countries. However, the symptoms occurred mainly in diarrheic children treated with bismuth subnitrate and were due to the transformation of nitrate into nitrite and not to the bismuth itself or following prolonged use of oral bismuth subgallate; encephalopathy was associated with elevated levels (> 100 μg/l) of bismuthemia.

We have been treating a series of 22 children with 50 mg/kg/day amoxycillin for 1 week and 7–8 mg/kg/day colloidal bismuth subcitrate for 1 month. Disappearance of HP occurred in only 50% of the cases but compliance was very poor. The levels of bismuthemia were under 20 μg/l in all of them except in three in whom they ranged between 25 and 28 μg/l and in one, who had been taking erroneously a double dose and in whom bismuthemia reached 50 μg/l. The results are reported in the poster session of the present symposium.

Currently we are treating for shorter periods of 2 weeks with a combination of amoxycillin (50 mg/kg/day) and metronidazole (20 mg/kg/day) and the preliminary results in 20 children show a very high clearing rate after 1 month (80%). The relapse rate is not yet known but could be reduced by repeating short periods of 1 week treatment every 3 months. Low socioeconomic background, rapid relief of painful symptoms together with the reluctance of undergoing frequent endoscopic reevaluations are, in our opinion, the main reasons for this noncompliance. Practical details such as palatability, side effects, and number of doses play also a role: for instance, many children complained about the bad taste of the colloidal bismuth subcitrate solution or about the difficulty of taking their antibiotic medication as a fluid solution while attending school or about nausea following ingestion of metronidazole. In order to avoid constant recontamination from family contacts an attempt aimed at eradication of HP in children, designed as a simultaneous treatment of all the members of the family, is currently under study in our unit.

Conclusion

There is now proof that HP is associated with gastric antral lesions in the child. Therefore, it seems reasonable to undertake a therapeutic trial only in those children with symptomatic complaints (or with positive endoscopic findings). What is the most appropriate therapy in symptomatic children? Collaborative multicentric therapeutic trials are now planned in different centers throughout Europe, Asia, and Africa in order to enrole an important number of pediatric cases over a short period with different drugs and combinations of drugs. In any case attempts aimed at eradication of HP should always be completed by a thorough investigation about the HP-infection status of other members of the family in order to evaluate the possibility of continuous reinfection. Bi-therapy seems sufficient to obtain clearing of the microorganism and repeated short periods (1 to 2 weeks) of treatment could be better accepted by the child and the family and probably yield as good results as longer tedious treatments for which the compliance is often so poor.

Is eradication of asymptomatic children a necessity? In other words should we try so hard to eradicate HP from these patients or does a symbiotic adaptation occur when infection is acquired early in life? Answering these questions is a stimulating challenge for epidemiologists and pediatric gastroenterologists. Epidemiology of HP infection in the pediatric population might probably be a most rewarding task for the understanding of the role of this common microorganism living in the stomach of (probably) more than one-half of the pediatric population in the developing world, without overt symptoms.

References

1. Hill R, Pearman J, Worthy P, Caruso V, Goodwin S, Blincow E (1986) Campylobacter pyloridis and gastritis in children. Lancet I:387
2. Cadranel S, Goossens H, De Boeck M, Malengreau, Rodesch P, Butzler JP (1986) Campylobacter pyloridis in children. Lancet I:735
3. Czinn S, Dahms BB, Jacobs GH, Kaplan B, Rothstein FC (1986) Campylobacter-like organisms in association with symptomatic gastritis in children. J Pediatr 109:80–83
4. Drumm B, Sherman P, Cutz E, Kharmali M (1987) Association of Campylobacter pylori on the gastric mucosa with antral gastritis in children. N Engl. J Med 316:1557–1561
5. Mahony MJ, Wyatt J, Littlewood JM (1987) Campylobacter associated gastritis in children. Gut 28:A1357
6. Hermier M, Descos B, Foasso MF (1983) Les ulceres gastro-duodenaux primaires de l'enfant. Etude multicentrique de 58 cas de diagnostic endoscopique ou chirurgical. Arch Fr Pediatr 40:681–687
7. Drumm B, O'Brien A, Cutz E, Sherman P (1987) Campylobacter pyloridis associated primary gastritis in children. Pediatrics 80:192–195
8. Cadranel S, Glupczinski Y, Labbe M, De Prez C (1988) Campylobacter pylori in children. In: Menge H, Gregor M, Tytgat GNJ, Marshall BJ (eds) Campylobacter pylori. Springer, Berlin Heidelberg NewYork, pp 110–117
9. Mahony MJ, Wyatt JI, Littlewood MJ (1988) Campylobacter pylori gastritis. Arch Dis Child 63:654–655
10. Oderda G, Dell'Olio D, Morra I, Ansaldi N (1989) Campylobacter pylori gastritis: long-term results of treatment with amoxycillin. Arch Dis Child 64:326–329
11. Murphy MS, Eastham EJ (1987) Peptic ulcer disease in childhood: long-term prognosis. J Pediatr Gastroenterol Nutr 6:721–724
12. Socha J, Celinska-Cedro D, Teisseyre M, Rasinski A (1988) Campylobacter pyloridis in children. Pediatr Res 24:A416
13. Drumm B, Sherman P, Chiasson D, Karmali M, Cutz E (1988) Treatment of Campylobacter pylori-associated antral gastritis in children with bismuth subsalicylate and ampicillin. J Pediatr 113:908–912
14. Eastham EJ, Elliott TSM (1987) Campylobacter pyloridis in children. Arch Dis Child 62:652
15. Oderda G, Vaira D, Holton J, Dowsett JF, Ansaldi N (1990) Serum pepsinogen I and IgG antibody to Campylobacter pylori in nonspecific abdominal pain in childhood. Gut 30:(in press)
16. Mahony MJ, Littlewood JM (1988) Diagnosis of Campylobacter pylori gastritis by IgG ELISA. Pediatr Res 24:416
17. Thomas JE, Eastham EJ, Elliott TSJ, Berkely D, Jones D (1988) Campylobacter pylori gastritis can be diagnosed serologically? Arch Dis Child 63:696
18. Megraud F, Brassens-Rabbe MP, Denis F, Touhami M, Hoa DQ (1989) Campylobacter pylori infection. Evidence for a worldwide distribution and a high prevalence in developing countries. Gastroenterology 96:A338
19. Klein PD, Graham DY, Opekun AR, Sekeley S, Evans DG, Evans DJ (1989) High prevalence of Campylobacter pylori infection in poor and rich Peruvian children determined by ^{13}C-urea breath test. Gastroenterlology 96:A260
20. Rauws EAJ, Langenberg W, Houthoff J, Zanen HC, Tytgat (1988) Campylobacter pyloridis-associated chronic active gastritis. Gastroenterology 94:33–40
21. De Giacomo C, Maggiore G, Licardi G, Scotta MS, Fiocca R (1988) Effects of antibacterial treatment of Campylobacter pylori-associated gastritis in children. Gastroenterology 95:1699
22. Oderda G, Holton J, Altare F, Vaira D, Ainley C, Ansaldi N (1990) Amoxycillin plus tinidazole for Campylobacter pylori gastritis in children: assessment by serum IgG antibody, peposinogen I and gastrin levels. Gut 30:(in press)

Bismuth – From the Element to the Tablet: The General Chemistry of Bismuth with Relevance to Pharmacy and Medicine *

H. H. Paradies

Introduction

There is now considerable evidence accumulating which links *Helicobacter pylori* (*H. pylori*) etiologically to histological gastritis [18, 21], and the strength of its close association with duodenal ulcer disease is beyond all doubt [9, 3]. Moreover, bismuth compounds, which are apparently biologically active in the treatment of *H. pylori*, are much more effective in preventing post-treatment relapse of duodenal ulcer disease than are other agents such as cimetidine which lack this bismuth-related bactericidal activity [19, 12].

There has been renewed interest in the chemical and physical structures of bismuth compounds in the light of their antimicrobial-related biological activities, especially in the treatment of *H. pylori*, and the relation of these to the gastroduodenal physiology of bismuth salts, which were first described in the 15th century (G. Agricola) and have been used ever since. It thus seems relevant to review some important new discoveries about bismuth salts and their chemistry in the solid state as well as in solution, and to inquire whether between there is any detailed structural relation between their pharmacodynamic relevance in the treatment of *H. pylori* infection and their gastroduodenal physiology. Other important points are intimately related to the solubility, aggregation, gelation, flocculation, and coagulation of bismuth salts and to the enhanced surface activities of bismuth gels with regard to the adhesion of microorganisms, the way in which they attract pharamcodynamically active low molecular weight compounds, and their interaction with biological entities, e.g., membranes, mucosa, and gastric parietal cells.

So far, bismuth molecules in the solid state and in solution as cationic, anionic, or neutral complexes exist entirely within the realm of the observable, whereas molecular models of these particular bismuth compounds represent the appearance of molecular behavior at some selected level of abstraction with regard to precipitation, the binding of protein molecules, and shielding of biological membranes or external surfaces of cells. The "metaphorical" molecular shape and appearance of bismuth salts in the context of their biochemical and medicinal

---

* This work was supported by a grant from the Deutsche Forschungsgemeinschaft (1981).

Märkische Fachhochschule, Biotechnologie und Physikalische Chemie, Frauenstuhlweg 31, D-5860 Iserlohn, FRG; Department of Chemistry, Schrenk Hall, University of Missouri, Rolla, 65401, USA

Helicobacter pylori, Gastritis and Peptic Ulcer
Edited by P. Malfertheiner and H. Ditschuneit

activities does work amazingly well considering the complexity of the chemical structures, e.g., their stereochemistry in the presence of organic enantiomeric ligands.

Chemistry

Bismuth naturally occurs as Bi_2S_3 associated with the sulfide ores of lead, copper, and tin(IV) oxide. It is known that bismuth(III) oxide (Bi_2O_3) is a very strong oxidizing agent ($Bi \rightleftharpoons Bi^{3+} + 3e^-$; $E_0 = -0.277$ V). In contrast to antimony(III) compounds, for bismuth it can be said that there is an extensive true cationic chemistry, which includes solid bismuth(III) compounds with inorganic ligands. However, some organic ligands, e.g., citrate, malate, tartrate, and their enantiomeric counter parts (*S*-or *R*-citrate, *S*- or *R*-lactate) do form stable anionic complexes in the solid state as well as in solution [24, 28]. Unlike the inorganic bismuth(III) salts, which are scarcely soluble in aqueous solutions between pH 1.0 and pH 7.0, the complex structures of bismuth(III) salts with organic ligands, e.g., the so-called colloidal bismuth subcitrate [38], are readily soluble in aqueous solutions of neutral pH. However, in acid medium, most of the organic bidentate or multidentate complexes of bismuth (III) precipitate (pH <5.0), leading to the formation of discernible insoluble bismuth complexes, the most important of which is BiOCl, bismuth(III) oxychloride.

Aqueous solutions of bismuth(III) contain well-defined hydrated cations, but there is no experimental evidence so far for the existence of a single aqua ion, designated $Bi(H_2O)_n^{3+}$ as is normally known for other cations, especially for the alkaline and alkaline earth cations, e.g., Na^+ K^+, Mg^+, Ca^{2+}, and the transition metal ions, e.g., Co^{2+}, Zn^{2+}, and Fe^{2+}, all of which are so important for many biochemical regulations including ion pumps, allosteric enzymes, and proton motive forces for synthesizing ATP (H^+-ATPase). In neutral perchlorate solutions, the main species is $[Bi_6O_6]^{6+}$ or its hydrated form, $[Bi_6(OH)_{12}]^{6+}$, although the amount of $[Bi_6(OH)_{12}]^{6+}$ is expected to be negligible in fairly acidic solutions of low Bi^{3+} concentrations [16, 1]. Moreover, the ultracentrifugation studies by Holmberg et al., [11], which explored the hydrolytic behavior of the Bi^{3+} ion in aqueous perchloric acid solution, discovered a monodisperse polymer. These polymeric complexes, which the authors designated $(Bi-O)_n^{n+}$ where n is probably six, are stable over a wide range of acidities and Bi^{3+} concentrations, including those pH ranges which are relevant for the stomach. In earlier investigations, Graner and Sillén [10] interpreted their data in terms of a successive formation of polymeric particles having the general formula $[Bi_nO_{n-1}]^{n+2}$, where the average degree of polymerisation n rapidly increases with bismuth concentration and pH. However, Olin [22] provided evidence using potentiometric techniques that $[Bi_6(OH)_{12}]^{6+}$, i.e., hydrated $[Bi_6O_6]^{6+}$, is the only polymeric species present. This is consistent with the results of the ultracentrifugation studies by Holmberg et al. [11]. Since these results indicated a fairly monodisperse polymeric species containing heavy bismuth atoms, presumably in a compact, regular arrangement, Levy et al. [16] used x-ray diffraction to investigate this bismuth(III) polymer in aqueous solution. They disclosed the octahedral geometry and presence of this polymeric species. This is also consistent with the vibrational spectrum of $[Bi_6(OH)_{12}[^{6+}$ reported by Maroni and Spiro [17]. The spectral features are satisfactorily interpreted as arising from the normal modes of cuboctahedral cage. In addition, using a normal coordinate analysis the authors disclosed Bi–Bi bonding to which they attribute the low-frequency Raman bands, i.e., the vibration of the cage.

In the solid state, bismuthoxy halides, nitrates, and hydrosulfates (BiOX; X = Cl, Br, NO_3, HSO_4) form large cationic polyelectrolytes with positive charges partly on the oxygen atoms and bismuth atoms. The crystal structures of these BiOX compounds (Fig. 1) in their dehydrated state are true two-layer structures. The molecular structure of the hydrated $BiONO_3$ form, which is a tetraoxotetrahydroxo bismuth(III) nitrate monohydrate, $[Bi_6O_4(HO)_4](NO_3)_6 \cdot H_2O$ as determined by Sundvall [35], crystallizes in the monoclinic space group $P2_{1/c}$ (no. 14) with cell dimensions of $a = 9.289(2)$, $b = 13.462(4)$, and $c = 19.527(5)$ Å, $\beta = 114.13(2)°$ with four molecules in the unit cell. Lazarini (1979), however, determined cell dimensions of the same compound as $a = 9.313(2)$, $b = 13.514(7)$, $c = 19.575(5)$ Å, $\beta = 114.12°$ with four molecules in the unit cell; very similar values were determined by Paradies et al. [25]. In the polycation of this particular structure, the six bismuth atoms are located at the corners of an octahedron with the O atoms and OH groups above the corner of the octahedral faces. Each oxygen atom of the hydroxyl groups has three bismuth neighbors which are arranged pyramidally at a longer distance, while the

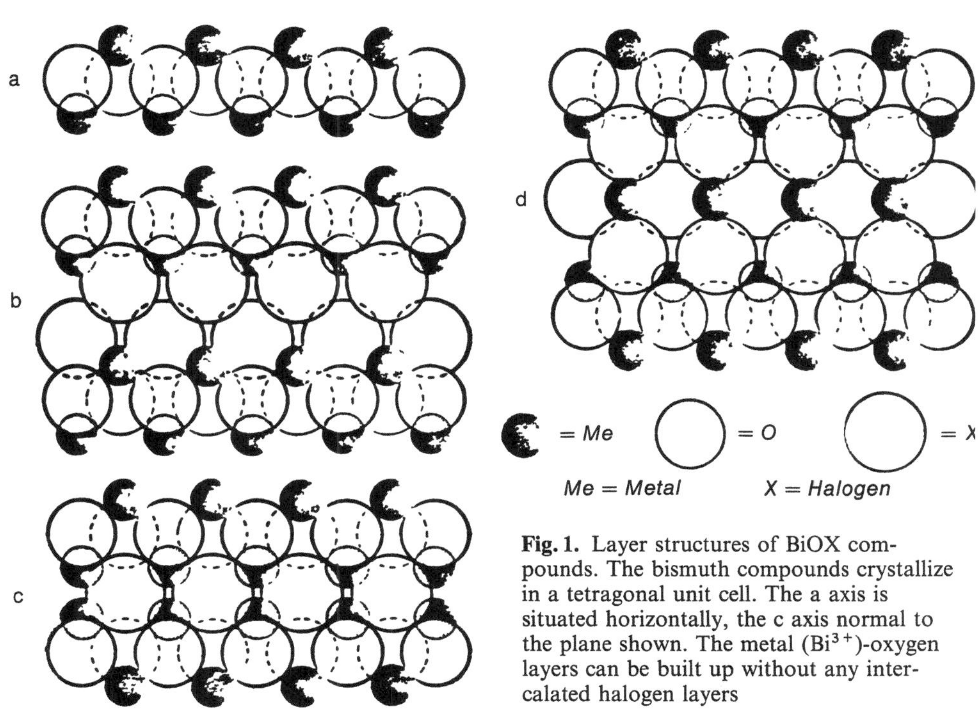

Fig. 1. Layer structures of BiOX compounds. The bismuth compounds crystallize in a tetragonal unit cell. The a axis is situated horizontally, the c axis normal to the plane shown. The metal (Bi^{3+})-oxygen layers can be built up without any intercalated halogen layers

remaining oxygen atoms are linked to three nearly coplanar bismuth atoms at a shorter distance. Furthermore, a NO_3^- group and the water molecule are bound to two neighboring bismuth atoms (Fig. 2 and Fig. 3). The first solid hydrolysis product of $Bi(NO)_3$ at pH 1.2, i.e., $[Bi_6O_4(OH)_4](NO_3)_6 \cdot 4H_2O$, contains polycations which seem to be the product of intramolecular condensation of the $[Bi_6(OH)_{12}]^{6+}$ ions present in aqueous solutions of basic bismuth salts. However, at pH values between 1.2 and 3.5, another hydrolysis product, $[Bi_6O_5(OH)_3](NO_3)_5$, can be observed. According to the crystal structure, it contains polycations in which pairs of $[Bi_6O_5(OH)_3]^{5+}$ groups are joined by two

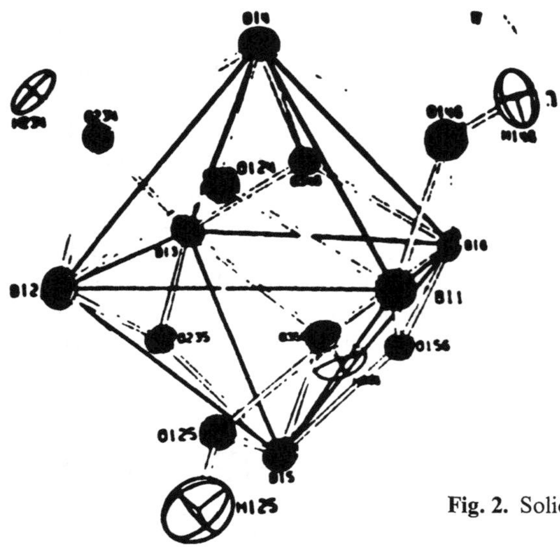

Fig. 2. Solid state structure of bismuth oxynitrate

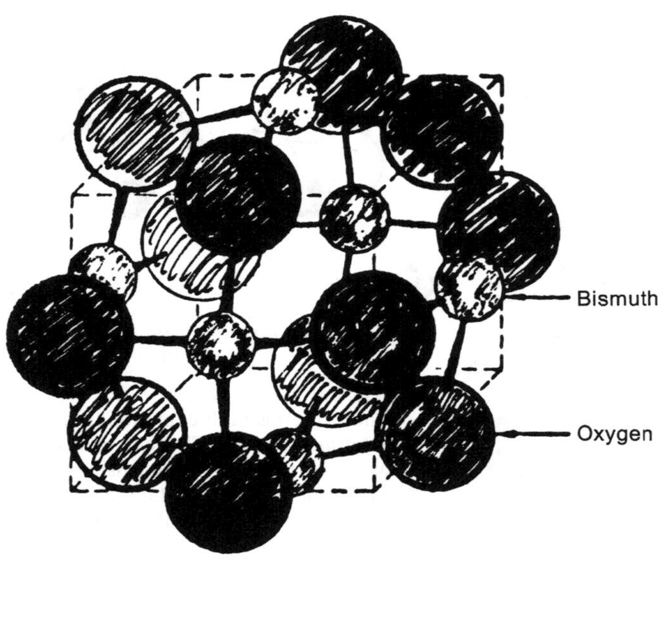

Fig. 3. Model of hydrated bismuth(III) oxy compounds in acidic solution, similar to the cubotahedral structure of $[Bi_6(OH)_{12}]^{6+}$ according to Levy et al. [16]

bridging oxygen atoms. The solid state structure and the unit cell volume of $[Bi_6(H_2O)(NO_3)O_4](NO_3)_3$, which is smaller than that of $[Bi_6O_4(OH)_4](NO_3)_6 \cdot 4H_2O$, is a consequence of much shorter distances between the polycations and NO_3^- anions or water molecules. The geometry of the $Bi_6O_4(OH)_4$ part of the polycation is very close to that found for the $[Bi_6O_4(OH)_4]^{6+}$ polycation or to the $[Bi_6O_5(OH)_3]^{5+}$ polycation.

Furthermore, the occurrence of the $[Bi_6O_4(HO)_4]^{6+}$ ion in two different basic bismuth(III) salts indicates that this complex is also present in hydrolyzed solution. This is consistent with the results we obtained very recently [29] and the results obtained by Sundvall [36] using a x-ray diffraction study of a hexanuclear complex of bismuth(III) in aqueous perchlorate solutions. These data are consistent with a $[Bi_6O_4(HO)_4]^{6+}$ unit in solution very similar to that discovered in the solid state of the hydrated nitrate or perchlorate. The refined value of the bismuth–bismuth distances in an assumed regular octahedral arrangement of bismuth is 3.69 Å. The oxygen atoms are situated above octahedral faces of the Bi_6 octahedron, and each type of oxygen is tetrahedrally arranged around the center of the complex. The $Bi-O^{2-}$ and $Bi-HO^{2-}$ bond lengths of the bismuth(III) in solution accounts for 2.21 Å and 2.37 Å, respectively. The x-ray scattering dimers of the aqueous solutions of bismuthoxy nitrate are shown in Fig. 4, as is the calculated peak shape for an idealized Bi_6 octahedron [29]. Figure 5 shows the experimental reduced intensities, $Si(S)_{obs}$ in dashed lines, and the calculated intensity curves, $Si(S)_{calc}$ derived from the crystal structure of hydrated bismuth(III) oxynitrate.

In addition, BiOX (Fig. 1) has coordination sites which can bind metallic cations, e.g., $(AlO_2)_n^{2-}$ or $(GaO_2)_n^{2-}$, and nitrates, SiO_3^-, $Si_2O_5^{2-}$ quite strongly. This was found also in solution, where it retained the regular structure of the hydrated bismuth(III) cation as shown in Figs. 2 and 3.

Two types of site can be distinguished: (i) a central one in which alkaline or alkaline earth cations are included, and (ii) external sites where metal cations as complexes can be fixed or buried.

Particle Size Distribution of Bismuth(III) Salts

By using small-angle x-ray scattering techniques, quite detailed information can be obtained about the size and shape of bismuth(III) salts in aqueous solutions at pH 1.2–3.5 in samples contaning identical independent particles [8]. Aqueous solutions of inorganic bismuth(III) salts containing the complex cation $[Bi_6(OH)_{12}]^{6+}$ or $[Bi_6O_4(OH)_4]^{6+}$ at pH 1.2–pH 3.5 were examined using the above-mentioned techniques and yielded the following values: radius of gyration of $\langle Rg \rangle$, 15.6 Å; molecular weight of $\langle M \rangle$, 1500; degree of smelling, 0.95; water content, 13.0% (w/w). However, when characteristic parameters are calculated from small-angle x-ray scattering data obtained from a polydispersed system, average values of $\langle Rg^2 \rangle$, $\langle M \rangle$ over the size distribution are obtained [30]. Highly polynuclear complexes were found for organic dentated complexes of bismuth was found to have a very large size distribution between $\langle D \rangle = 80 - 150$ Å. Similar results were obtained using inelastic light scattering techniques which yielded

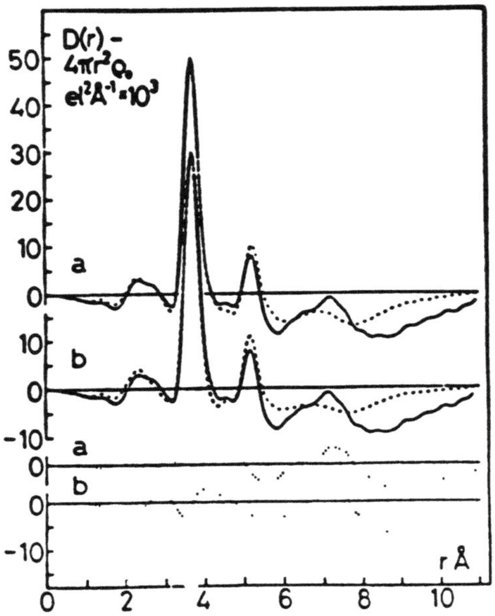

Fig. 4. Experimental distribution curve $D(R) - 4\pi R^2 \cdot \varrho_0$. *a*, The calculated distribution on the model derived by Sundvall [36]; *b*, the calculated distribution curve based on the crystal structure of bismuthoxy nitrate. The dashed lines correspond to data obtained from bismuthoxy-(*S*)-tartrate. The corresponding difference curves are given at the bottom of the figure (*dotted lines*)

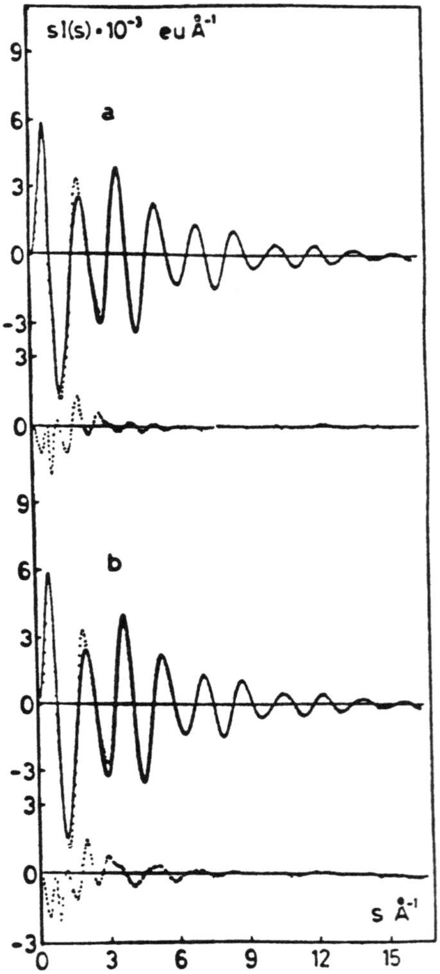

Fig. 5. Experimental, reduced intensities $Si(S)_{obs}$ (*dotted lines*) and the calculated intensity curves $Si(S)_{calc}$. *a*, Idealized B_6 octahedron, in accordance with Sundvall [36]; *b* based on the crystal structure of bismuthoxy-(*S*)-tartrate

values 20% higher than those obtained by x-ray scattering techniques (Paradies 1989, unpublished results). However, the narrow size distribution ⟨R⟩ of the bismuth(III) oxy salts at pH 1.2, their degree of hydration, molecular weight distribution, and size distribution, ⟨R⟩, imply that under the conditions used (25 °C; $\mu = 0.01$; pH 1.2–2.5) the $[Bi_6O_4(OH)_4]^{6+}$ is itself one of the main species and similar to the species found by Sundvall [36] in concentrated bismuth(III) perchlorate solutions. The same is true for the basic bismuth(III) nitrate, revealing a molecular complex of the form $Bi(NO_3)_2^+ (H_2O)_{10} \ldots Bi(NO_3)_4^-$, as determined from vapor pressure osmometry and the results of x-ray scattering techniques.

This again corroborates the fact that no hydrated bismuth(III) ion, as is normally found for other cations, e.g. Cu^{2+}, Mg^{2+}, etc., in the form of $[Bi(H_2O)_n]^{3+}$ exists.

When bismuth(III) solutions are kept at pH 1.2, between 25 °C–30 °C and in aqueous hydrochloric media, they develop large aggregates of molecular weights, approximately 9000–10 000, and even the low molecular weight particles have an average molecular weight of 1500. This study further indicates that the particles contain about 10 wt% water, and their shape is more an oblate ellipsoid of revolution having an axial ratio of 0.2–0.35. It can be assumed that these particles in solution follow a discontinous size distribution function, which seems to be mainly dependent on two parameters; one of these is associated with the large complex formation with time, and the other with the relative weights of the bismuth(III) units of which the complexes are composed. The notion that a very large particle exists in solution is consistent with observations of gel formation, ordered or unordered, and with the results light scattering measurements, which reveal an average particle diameter of 90–110 Å. The formation of such large particles of bismuth(III) in solution also corroborates the results of diffusion measurements taken at this particular pH and obtained in a force-free measuring cells. More experimental work is needed to clarify the development of the particular precipitates and their amorphous network. This is especially true in the presence of biological membranes or other biological structures which are relevant to the mode of action of bismuth(III) salts and the mechanism of formation of the larger bismuth(III) complexes at the ulcer site.

The cause of the extreme flocculation of organic bismuth(III) salts at pH 1.2–2.5 and at 37 °C, as measured by inelastic light scattering techniques, can be explained on the basis that the hydrophobic groups, e.g., salicylate (benzene ring), tend to escape from the aqueous phase, in which the two-nuclear bismuth(III) complexes are dissolved or dispersed. When colloidal bismuth(III) particles become hydrophobic in aqueous suspensions, they flocculate very fast because the hydrophobic surfaces of two different particles tend to stick together. For instance, it is easy to demonstrate that the addition of an anionic surfactant such as a glycerolipid to the suspension will tend to lead to the formation of a double layer of the glycerolipid molecules on the inorganic bismuth(III) cluster surface. The second glycerolipid layer is oriented together with the phosphoryl groups away from the surface thus rendering the surface hydrophobic again. Such bismuth(III) particles will then be negatively charged by the outer layer of phosphoryl charges and therefore will be flocculated.

In general it appears that before the negatively charged bismuth(III) particles can react with organic anions, the charge on the surface has to be reversed by adsorption of the polyvalent bismuth(III) ion. This reaction with biological anions is of course also important for bismuth(III) cations complexing with ATP, ADP, AMP, or cAMP.

Precipitation Phenomena of Bismuth(III) Salts

Very recent findings have discerned various phenomena related to pH-dependent bismuth(III) precipitation and particle growth, especially at pH 1.5–3.5 [26].

1. *Coalescence*, which is the degree of ordering of bismuth(III) particles in solution to which these particles are bonded or grown together, especially at the beginning of precipitation of the sheet-like structures of BiOCl (Fig. 6).
2. *Coacervation* of bismuth(III) particles, especially at pH 1.5–2.0. This means the formation of a liquid precipitate. Coacervation actually involves coagulation, however, with the further provision that the bridges between bismuth particles and bismuth (glycoprotein) or colloidal units are labile, so that they can exist in a sort of equilibrium, being formed and reestablished. In this way the coagulation can continue to contract its network to the greatest possible extent (Fig. 7).
3. Xerogel, which is a BiOX gel where the liquid medium has been removed. The structure is compressed and porosity reduced to at least some degree by the surface tension forces as the liquid is removed (Fig. 8).

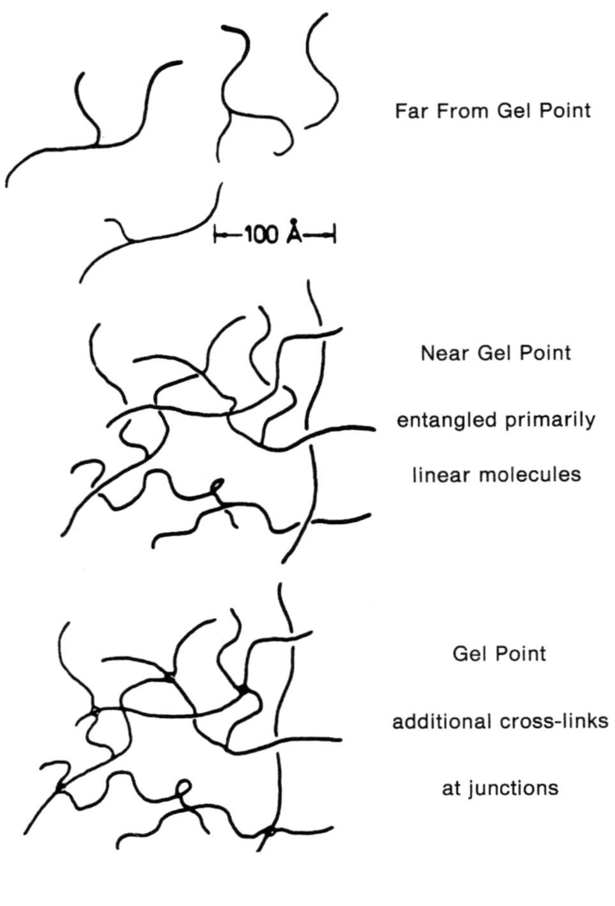

Fig. 6. A model of polymer growth and gel formation in an acid-catalyzed system

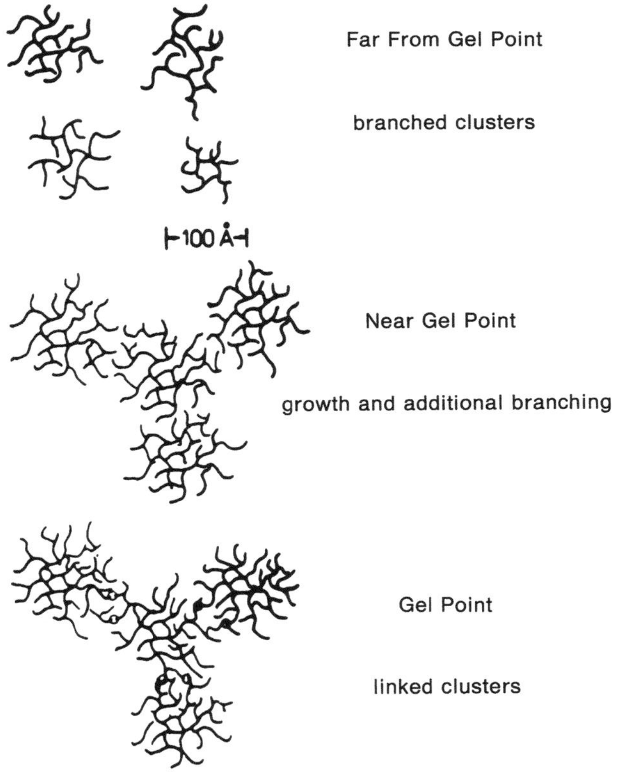

Fig. 7. A model of polymer growth and gel formation in a base catalyzed system for bismuthoxy compounds

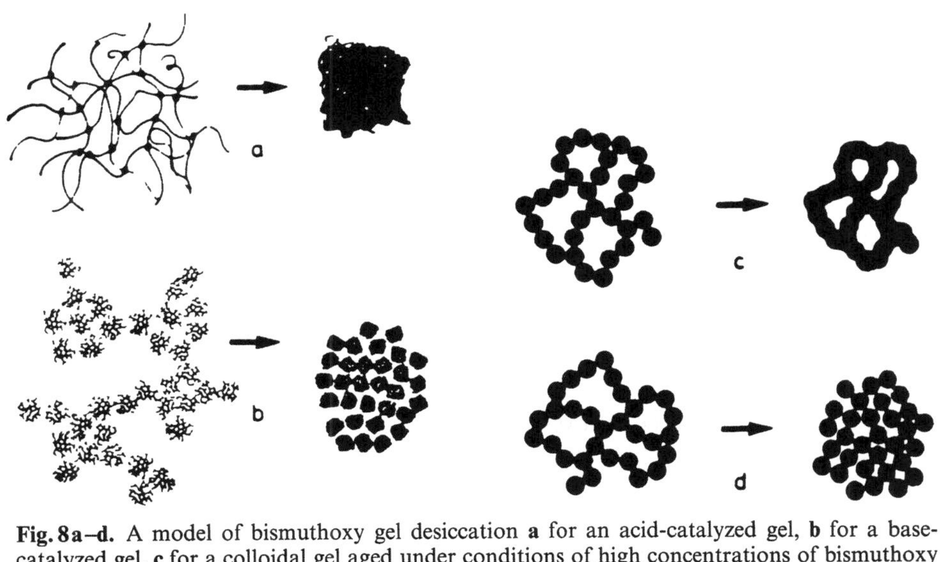

Fig. 8a–d. A model of bismuthoxy gel desiccation **a** for an acid-catalyzed gel, **b** for a base-catalyzed gel, **c** for a colloidal gel aged under conditions of high concentrations of bismuthoxy chloride in the presence of citrate, and **d** a colloidal bismuthoxy gel at basic pH. The gel is composed of weakly bonded particles

4. Aquagel, which is very similar to xerogel. The liquid has, however, been removed in such a way as to prevent any collapse or change in structure as the liquid is removed (Fig. 8). This is important for the polymeric structures formed when bismuth oxychloride (Fig. 1) adheres to cell membranes, the polymeric glycoprotein (MW $2.0 \pm 0.5 \times 10^6$), or to acinar gastric cells, or highly cooperative membrane structures, e.g., H^+-ATPase which are sensitive to Ca^{2+}, Mg^{2+} and the K^+, H^+-ATPase of the parietal cells, which is chloride sensitive.

In the context of these various forms of aggregationally related bismuth(III) precipitates, the aggregation naturally includes gelling. Here the particles are linked together in branched chains that fill the whole volume of a colloidal sol. This means that there is no increase in the concentration of bismuth(III) ions in any macroscopic region in the medium or nor is there any contact with membrane compartments. Instead, the overall medium becomes viscous and is then solidified by a coherent network of bismuth(III) particles. By a special mechanism of action, this network retains the liquid and preserves the structural water content of the mucosa, as observed by differential scanning calorimetry (Paradies, unpublished results). However, coagulation occurs accordingly where bismuth(III) particles come together as colloidal particles, a process which naturally occurs in solution at pH 1.5–5.0 as shown above. Relatively close packed clumps are observed in which the bismuth(III) complexes are more concentrated than in the original colloidal sol (diameters between 400 Å–800 Å), so the coagulation settles as a relatively dense precipitate. The formation of this precipitate at acidic pH is controlled mainly by typical flocculation kinetics if the bismuth(III) complexes are truly cationic, leaving low concentrations of bismuth ions for resorption. However, the kinetics of anionic bismuth(III) complexes, especially in the presence of organic ligands, i.e., citrates, lactates, malates, or tartrates, with respect to hydration is different, due to different Donnan potentials. As a result quite an amount of bismuth can be resorbed within a certain limited Bi^{3+} concentration. Accordingly, flocculation occurs primarily where the particles are linked together by bridges (Figs. 6, 7), or where the concentration of the flocculating agent, i.e., at pH 1.5–5.0, is sufficiently low in comparison with the cationic bismuth(III) complexes, so that the aggregated structure remains open and voluminous. So, large colloidal bismuth(III) particles, e.g., as sheets of BiOX [X = Cl, HSO_4^-, acetate, or bicarbonate (Fig. 1), though usually BiOCl at acidic pH is meant] are adsorbed on coated membrane surfaces and there form a visible film. This film is not, however, formed on the area previously exposed to small colloidal bismuth particles, which are invisible to the eye but detectable with x-ray photon spectroscopy [26]. In contrast, in a colloidal sol of bismuth(III) complexes with organic ligands, i.e., anionic bismuth complexes, there is a mixture of large and small particles of colloidal bismuth(III) and the smaller particles diffuse more rapidly to the membrane surface, thus excluding the slower larger colloidal bismuth(III) ones.

Some of the bismuth(III) structure complexes mentioned above are formed during interactions when coming into contact with solid or membrane surfaces. However, the development, kinetics, and stability of such structures are beyond the scope of this contribution [27]. Therefore, the different polymeric structures of bismuth(III) will be surveyed, in oder to characterize the most important features of the bismuth(III) polymers receiving attention with respect to gastritis, duodenal ulcers, and *H. pylori* infection.

The Structure of The Bismuth(III) Complexes as Gels

The structure of the cationic BiOX ($X = Cl, NO_3, HSO_4$) gels made up of anionic complexed bismuth(III) compounds, i.e., coordinated bismuth(III) citrate, tartrate, malate, lactate, and especially enantiomeric pure bismuth(III) compounds [26], depends on the pH of the solution during hydrolysis and the kinetics of the condensation reactions as shown below. Base-catalyzed bismuth(III) compounds produce gels that are apparently granular in texture and retain less organic or organic-like material, e.g., citrate, tartrate, etc. However, bismuth(III) salts catalyzed in acidic solution react to form a sort of gel which has a finer, denser structure and is not particulate. This occurs especially with inorganic bismuth(III) salts. The influence of reaction conditions on the growth of bismuth(III) films is strongly dependent on the nature of the solution and the environment from which the precipitation occurs or starts, respectively. Nucleation and epitaxial effects also play a crucial role under these circumstances, as ellipsometry measurements of bismuth(III) films indicate (Paradies, unpublished results). Another reaction condition which influences the growth of bismuth(III) films is the degree of surface roughness. The thickness of the film formed on a protein surface (glycoprotein) is different from the one developed on a surface consisting of bismuth(III) polymer with patches of BiOX deposits (hydrated or not) or glycoprotein (mucosa). This results in distinct areas of the bismuth film being covered. These areas have different surface tensions due to their structure, as indicated by the results of x-ray photon spectroscopy experiments [27], and some changes do occur among the hydrophobic regions within this biofilm. These physical changes are related to biological activities, i.e., adhesion of microorganisms, influences on the proton motive forces, environmental pH changes, metabolic changes, and hence alterations in chemotaxis and chemokinesis, as observed and revealed by microbial induced corrosion (MIC) processes [5, 6, 23, 28]. The morphology of the solid phase and the amount of retained organic material in the cases of citrate, tartrate, or lactate have a profound effect on the densification behavior of the bismuth(III) gels which are subsequently formed. In this way it is possible to interpret the sintering kinetics of bulk bismuth gel and thin films and especially of the pure cationic bismuth(III) salts in structural terms, using x-ray diffraction and x-ray scattering studies.

Bismuth(III) oxychloride can form a xerogel which contains large amounts of hydroxyl ions or, when derived from organic dentated bismuth composition, fair amounts of organic groups which are bound to the sheet-like network of BiOX (Fig. 1). Instead of hydroxyl ions, NO_3^- ions are easily bound to the network accordingly. This material, i.e., NO_3^- citrate, can be difficult to remove because very small pores can develop the gel along the infinite sheets notes within the network of bismuth(III) gels (Fig. 7). All these gels are hydrous and have a layered structure; they can accordingly intercalate molecules and ions. They then develop ion exchange properties similar to zeolites, often with high selectivity. They have a wide range of catalytic properties, where even transition metal ions can be intercalated. The structure is very much like that of "pillared clays" in which very large cations or anions are intercalated to form "pillars". This leads to expanded layers, changing diffusion, sorption, and electrochemical potentials (Donnan's equilibrium). These results were obtained mainly using x-ray scattering techniques [27] and can be accounted for as follows:

Acidic conditions promote the growth of comparatively linear, tightly cross-linked bismuth(III) polymers. These polymers become entangled at an early stage of growth, but do not form a gel until a sufficient density of cross-links has been formed (Fig. 6). The hydrolysis of the organic dentates of bismuth is completed long before condensation, but cationic inorganic bismuth(III) complexes form polymers of bismuth(III) lattices very fast when precipitation starts. By contrast under alkaline conditions, denser more heavily cross-linked clusters of bismuth(III) gels are formed and these do not interpenetrate (Fig. 7). Many free hydroxy groups persist unhydrolyzed after the gel has been point reached. Figure 8 shows a diagrammatic scheme of the microstructure of bismuth(III) dried gels. Dried gels can be produced in a variety of ways, e.g., by decreasing the amount of surrounding bulk water, by chelating of protons by H_2O through forming $(H_3^+O)_3$ which is subsequently released from the network of the bismuth(III) gel, or by cracking the hydrated gels, which produces smaller pores because of increased capillary pressure.

Although the acid-catalyzed gels are completely hydrolyzed in solution, the dried gels of bismuth(III) polymers contain a large number of chemically bound OX groups, which can be hydrated fairly well because of rehydrating during the drying (sintering) process of the well-hydrated mucosa surface. However, when the gels come into contact with biological structures or organelles, this causes a large weight loss, as illustrated in Fig. 9. These reactions produce new cross-links and seem to stiffen the structure. The shrinkage observed in acid-catalyzed and in alkaline-catalyzed bismuth(III) gels are different with respect to their kinetics and the densification of the network. When analyzing the isothermal sintering kinetics of the acid-catalyzed bismuth(III) gels, it was found that these kinetics are not linear due to changes in the viscosity of the gel. This change, which is approximately proportional to the viscosity, can be attributed only to a change in viscosity brought about by the surface energy of the gelatinous network. This does not undergo large variations, since neither crystallization nor growth in pore size was observed. Therefore, the increase in viscosity is attributed mainly two factors, namely loss of hydration and structural relaxation. Condensation increases the cross-link density, and structural relaxation increases the molecular density thereby creating a stiffer structure. Since the kinetics of dehydration and structural relaxation are not necessarily identical, these transition states of bismuth(III) gels can produce gels with similar after contents, but different degrees of relaxation.

The fact that the precipitation of soluble bismuth(III) salts or inorganic bismuth(III) salts at pH 1.5–3.5 occurs with any change in molecular confor-

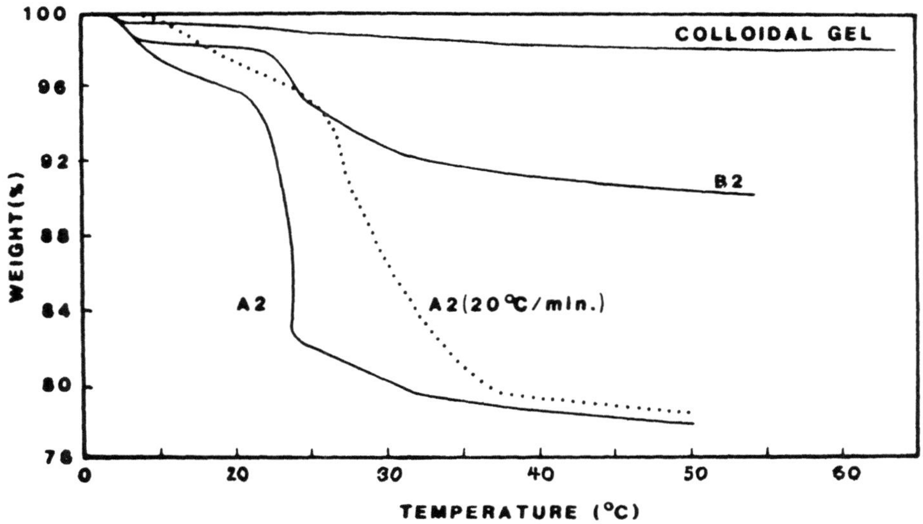

Fig. 9. Loss of original weight for a colloidal bismuthoxychloride-nitrate gel in relation to temperature. *A2*, Acidic solution, pH 1.5; *B2*, alkaline solution, pH 7.5

mation with respect to $[Bi_6O_4(HO)_4]^{6+}$ or the building unit BiO^+ $(Bi_6O_4(OH)_4(X^-{}_6)$ is not unexpected, because no conformational change of BiO^+ $(Bi_6O_4(OH)_4(X^-)_6)$ is expected, since no conformational change in BiO^+ was observed due to hydration. For example, for the soluble $[Bi_6O_4(HO_4)_4]^{6+}$ at pH 1.5 precipitation occurs (I = 0.10 M) within 1 min at 10 °C, but requires over 400 min at 32 °C. This large negative temperature coefficient is a characteristic feature of nucleation-controlled kinetics [37, 20]. The steady state rate of nucleation, r, can be expressed as

$$r \cong \exp[(-E + \Delta G^{++})/RT],$$

with E = the activation energy for transport across the interface, and ΔG^{++} = the free energy required to form a nucleus of critical size, and $\Delta G^{++} = \Delta H^{++} - T\Delta S^{++} = -RT\ln K$, with ΔH^{++} denoting the activation enthalpy, ΔS^{++} the corresponding entropy, and K the equilibrium constant. ΔG^{++} will vary inversely with the square of the free energy difference between a molecule in the homogeneous solution and in the precipitate, whereas ΔS^{++} and ΔH^{++} will depend on the type of bismuth(III) nucleus being formed. ΔS^{++} will be at constant T unity, since no change in molecular conformation is achieved.

On the basis of the x-ray results in the solid state and in solution [27] and assuming a cylindrically shaped nucleus comprised of rod-like molecules of length l and radius R, which is consistent with the molecular conformation of the sheet-like structures of BiOX (Fig. 1), we obtain, according to the nucleation theory [20]

$$\Delta G^{++} = \frac{8\pi R^4 l^2 \delta_n^2 \delta_e}{(\Delta G_n)^2}$$

with δ_n and δ_e denoting the interfacial free energies appropriate to the lateral and end surfaces, and ΔG_n denoting the free energy of dissolution. Because $\Delta G_n = 0$ at equilibrium dissolution temperature ($\sim$45 °C, T_S°C), precipitation cannot occur at this elevated temperature. By lowering the temperature, ΔG_n will increase proportionally to the undercooling (T_S°C$-T_p$°C), where T_p is the precipitation temperature at 37 °C. This factor is the cause of the negative temperature coefficient of the nucleation-controlled process. Our studies of dissolution temperatures for BiOX salts yielded estimated T_S°C in the range of 28 °C–40 °C. This precipitation occurs at measurable rates at undercooling of the order of 25 °C–30 °C. These values are typical of nucleation-controlled kinetics involving macromolecular systems as shown by Jackson and involving macromolecular systems as shown by Jackson and Mandelkern [13] and for complex proteins [30].

However, to correlate these findings with the mechanism of formation of the bismuth(III) layer at the ulcer site in vivo is quite speculative at the present time. In in vitro studies with organic dentates of bismuth(III) compounds, e.g., tripotassium dicitrate bismuthate, which has a pH between 9 and 10, visible precipitation occurred when the pH was lowered below pH 5.0, precipitation taking place predominantly at pH 2.5–3.0, and on lowering the pH further to pH 1.0, the precipitate redissolved according to the solubility constant of BiOX. It can be shown that the optimal pH range for bismuth precipitation is pH 2.5–3.0, similar to that for inorganic bismuth(III) salts, which can dissolve at lower pH according to the equation:

$$BiOX + 2H^+ \rightleftharpoons Bi^{3+} + H_2O$$

Thus the pH of fasting gastric juice in humans is below the optimal pH for precipitation, which is governed by the equilibrium constant, free amino acid residues and carboxy groups according to their pK_a of the glycoprotein hydroxy vesicles of the carbohydrate moiety of the glycoprotein and membrane fragments as biochemical factors mentioned above. Furthermore, it can thus be postulated that certain amino acids and hydroxy vesicles at the ulcer sites take up hydrogen ions by protonation. Here BiOX interferes, thereby locally elevating the pH towards that for the optimal precipitation of the BiOX layers.

Biochemical and Physiological Actions of Bismuth Salts

Suggestions for the mechanism by which *H. pylori* causes hypochlorhydria, including a reduction in parietal cell mass and alteration in gastric permeability to acid was put forward by Ramsey et al. [32]. Furthermore, in a volunteer study, Morris and Nicholson [21] showed that the very rapid recovery of acid production observed after the administration of doxycycline, given to try to eradicate the *H. pylori* also suggests that hypochlorhydria is directly associated with the presence of this organism. Cave and Vargas [2] measured the uptake of [^{14}C]aminopyrine by rabbit gastric epithial cells as an indirect assay for acid secretion from parietal cells. This assay is widely used as an indirect measure of the H^+ ion secretion by parietal cells, because the loss of polarity that occurs with isolated cells prevents direct measurements. Since [^{14}C]aminopyrine is a fairly weak base which diffuses into parietal cells, once inside these cells it is in a compartment with a pH below its pKa of pH 5.0 and is converted into mainly ionised form that only poorly diffuses across the membranes. Therefore, the accumulation of [^{14}C]aminopyrine reflects the acid production of the parietal cells. As a result, one can demonstrate alterations in stimulated acid production in the presence of *H. pylori*. Cave and Vargas found that *H. pylori* strains, isolated from the stomach of three patients with chronic gastritis, duodenal ulcer, and near normal mucosa, respectively, inhibited acid secretion at a concentration of 10^{-4} mol/l cimetidine [2]. They characterize an inhibitor protein of molecular mass of 12000–14000 Da. The concentration of the inhibitor was found to be very low, since fivefold dilution reduces the activity and tenfold dilution abolishes it. Surprisingly, the molecular mass of 12000–14000 Da is the same molecular mass correlated to the inhibitory protein subunit of the ε-subunit of the Ca- and Mg-sensitive H^+-ATPase of various bacterial strains [33, 34].

Gastric acid secretion results in a pH gradient of about 6.6 units across the secretory canaliculi of the parietal cells. That this depends directly on ATP and no other process can be shown in isolated, permeable gastric glands where in the presence of mitochondrial inhibition and anoxia, which collectively reduce acid secretion to zero, added ATP can to restore secretion [4]. Accordingly, this H^+ secretion is controlled by a H^+, K^+-ATPase, controlled by the presence of K^+, and Cl^-. Since Bi^{3+} can penetrate membranes, as observed in fully synthetic vesicles in vitro (Paradies unpublished results), and since the nonpermeable BiO_2^- $(BiO_3)^{3-}$ can influence the K^+- and Cl^--dependent $H^+ + K^+$-ATPase, which is cimetidine sensitive in a catalytic and regulatory way strongly dependent on corroborate the conclusions by Cave and Vargos [2]. They implied that the site of activity of the inhibitory protein of *H. pylori* could well be on the apical cell membrane involving the H^+, K^+-ATPase.

Furthermore, the competitive inhibition of the H^+-ATPase, which couples the ATP synthesis of ADP and inorganic phosphate through an electrochemical gradient, is correlated to the Bi^{3+} dentate shown in Fig. 10. By using NMR techniques to study synthetically produced Bi^{3+}-ATP complexes, a pseudoaxial and pseudoequatorial conformation of these Bi dentates can be discerned. Since ATP binds to the α-subunit of the bacterial H^+-ATPase [31], a large conformational change was observed after the Bi dentates bound to a specific site of the α-subunits. Here the K_d value was 1.5 μ*M*, whereas a value of $K_d = 0.1$ μ*M* was calculated for Mg-ATP [31]. This change in turn influences the whole complex enzyme system, including the molecular geometry of the H^+-ATPase [34]. Furthermore, upon binding of Bi-ATP, a considerable loss of structural flexibility of the α-subunit of H^+-ATPase has been observed using fluorescence methods. This results in a more compact (tight) structure of this subunit [29]. Moreover, it has been observed that the BiO^+ cation interferes with the transhydrogenase activity of the H^+-ATPase of *H. pylori* within the inner membrane close to the cell wall. Apparently, nucleotides, i.e., ADP, ATP and cAMP, seem to be the primary molecular target of this enzyme system, especially in order to change the state of hydration of vesicular systems, e.g., membranes, as well as to establish new hydrogen bonds between BiO^+ and certain amino acids residues. Furthermore,

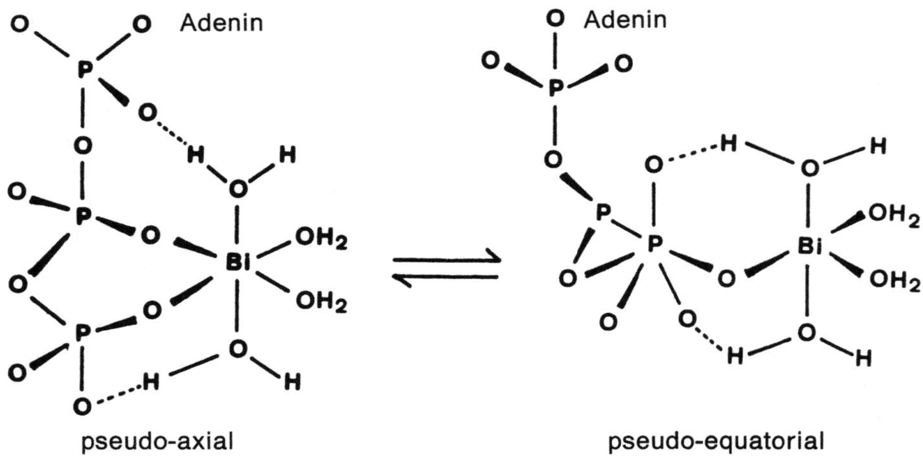

Fig. 10. Representation of two possible BiO-ATP diastereoisomers according to NMR results and chemical synthesis (Paradies, unpublished)

protonation of the bases of ATP, ADP, and cAMP and of bismuth(III) cations is an important role of acidic pH, which can be described as $BiO^+ \ldots {}^-OP$ and $NH^+ \ldots OP$ or $NH + \ldots, OH_2 \ldots Bi$, respectively.

One possible mechanism by which a bismuth coating helps to heal an ulcer can be the passive protection provided by the polymeric inorganic bismuth(III) layer against the digestive activity of acid and pepsin. Thus, normal healing processes are permitted to occur. Though the antipeptic activity of bismuth(III) compounds in the in vivo state seems to be insignificant [7], it remains possible that the bismuth(III) coating can act as a local antipeptic effect at the ulcer site. This is also supported by the investigations of Koo et al. [14].

Conclusion

Far more data are needed to establish conclusively the molecular mechanism of bismuth(III) salts with membranes and intrinsic enzyme systems within membranes, as well as the interaction of bismuth(III) salts with *H. pylori* and gastric parietal cells or the mucosa, respectively. Furthermore, with modern surface-sensitive techniques, e.g. x-ray reflectivity, ellipsometry, and x-ray photon spectroscopy, and with surface-enhanced Raman spectroscopy, it should be possible to delineate the bismuth(III) gel films within biological structures with respect to adhesion and in relation to the chemotaxis and chemokinesis of *H. pylori* and their possible attractants to the mucosa. Bismuth chemistry seems to encapsulate and convey a great deal of useful information, which is still not complete. However, it is of desirable, heuristic value in view of its new application, and it serves as a powerful fount of inspiration. Whether isolated molecules can properly be said to have any shape at all is irrelevant in this context.

Acknowledgements. The author would like to thank Mrs. Kirsten Schlösser for typing the manuscript, B.C.S. for critical and encouraging suggestions in the course of this contribution and patience, and Priv.-Doz. Dr. med. E. Gödde for correction and proof reading.

References

1. Baes AR, Mesmer RE (1979) The Hydrolysis of Metal Ions. John-Wiley, New York
2. Cave RD, Vargas M (1989) Effect of a Campylobacter pylori Protein On Acid Secretion by Parietal Cells. Lancet II:187–189
3. Coghlan JG, Humphries H, Dooly C et al. (1987) Campylobacter pylori and Recurrence of Duodenal ulcers: A 12 month follow up study. Lancet II:1109–111
4. Feller L, Jackson R, Malinowska D et al. (1982) Mechanistic Aspects of Gastric ($H^+ + K^+$)-ATPase. In: Ann New York, Sci, Vol 402, 146–163
5. Fischer W, Haenßel I, Paradies HH (1988) First results of microbial induced corrosion of copper pipes. In: Microbial Corrosion, Vol 1, eds. CAC Sequeira, AK Tiller. Elsevier 300–328
6. Fischer W, Haenßel I, Paradies HH (1989) The Role of exopolymers on metallic copper surfaces. Amer Chem Soc, Surface and Transport Phenomena in Bioparticle Adhension, Toronto, Abstract Nov. 12
7. Flawell-Matts SG, Swan CHJ, Koh J (1965) Double-blind trial of bismuth aluminate and magnesium trisilicate in peptic ulceration with simultaneous gastric analysis. Br Med J 1:753–756
8. Glatter O, Kratky O (1982) Small-angle x-ray scattering. Academic, New York
9. Goodwin CS, Armstrong JA, Marshall BJ (1986) Campylobacter pyloridis, gastritis, and peptic ulceration. J Clin Pathol 39:353–365
10. Graner F, Sillin LG (1947) On the hydrolysis of the Bi^{3+} ions. Repeated oxygen bridging: a new type of ionic equilibrium. Acta Chem Scand 1:631–640
11. Holmberg RW, Kraus KA, Johnson JS (1956) Hydrolytic behaviour of metal ions. VIII. Ultracentrifugation of bismuth(III) in perchlorate solution. J Am Chem Soc 78:5506–5512
12. Hamilton I, O'Connor H, Wood N et al. (1986) Healing and recurrence of duodenal ulcer after treatment with tripotassium dicitrato bismuthate tablets of cimetidine. Gut 27:106–110
13. Jackson JF, Mandelkern L (1968) Some macromolecular properties of poly(γ)glutamic acid random coils. Macromolecules 1:546–554
14. Koo J, Ho J, Lam SK, Wong J, Ong GB (1982) Selective coating of gastric ulcer by tripotassium dicitrato bismuthate in the rat. Gastroenterology 82:864–870
15. Lazarini FU (1979) Bismuth basic nitrate. Acta Crystallograph B35:448–450
16. Levy HA, Danford MD, Agran PA (1959) X-ray diffraction study of bismuth polymer in aqueous perchlorate solution. J Chem Phys 31:1458–1461
17. Maroni VA, Spiro TG (1966) The vibrational spectrum of the hydrolytic hexamer of bismuth(III). J Amer Chem Soc 88:1410–1412
18. Marschall B, Armstrong J, McGreiche D et al. (1985) Attempt to fulfil Koy's postulates for pyloric campylobacter. Med J Austral 142:436–439
19. Martin D, May S, Tweedle D et al. (1981) Difference in relapse rates of duodenal ulcer after healing with cimetidine or tripotassium dicitrato bismuthate. Lancet I:7–10
20. Mandelkern L (1964) Crystallization of polymers. McGraw Hill, New York
21. Morris A, Nicholson G (1987) Ingestion of Campylobacter pyloridis causes gastritis and raised fasting gastric pH. Am J Gastroenterol 82:192–199
22. Olin A (1957) Studies on the hydrolysis of metal ions. 19. The hydrolysis of bismuth(III) in perchlorate medium. Acta Chem Scand 11:1445–1450
23. Paradies HH (1988) Non-invasive techniques to study metal-biofilm interactions. In: Sequeira CAC, Tiller AK (eds) Microbial Corrosion, vol 1. Elsevier, Amsterdam, pp 444–445

24. Paradies HH (1989a) An X-ray diffraction study of bismuthoxy nitrate at different temperatures, pH, and ionic strength. J Phys Chem (to be published)
25. Paradies HH (1989b) A small-angle X-ray scattering Study of bismuthoxy nitrate and chloride in aqueous solutions. J Phys Chem (to be published)
26. Paradies HH (1989c) Wismutsalze – Das neue pharmakologische Prinzip. Zuckschwerdt, München
27. Paradies HH (1990a) The gel structure of bismuthoxy salts determined by X-ray diffraction. J Phys Chem (to be published)
28. Paradies HH (1990b) The Crystal Structure of bismuth-tartrate (S). J Appl Crystallograph (to be published)
29. Paradies HH, Kagawa Y (1981) Stability and flexibility of the α-subunit of F_1-ATPase from the thermophilic bacterium PS 3. FEBS Letters 136:3–7
30. Paradies HH, Werz G (1989) Solution properties of a cell wall protein from Acetabularia cliftonii. Protoplasma 93:443–461
31. Paradies HH (1981) Size and shape of the α-subunit of the (Ca, Mg)-dependent Adenosintriphosphatase from Escherichia coli in Solution in the Presence and Absence of ATP. Eur J Biochem 118:187–194
32. Ramsey EJ, Carey KV, Peterson WL (1979) Epidemic gastritis with hypochlorhydria. Gastroenterology 76:1449–1457
33. Schmidt UD, Paradies HH (1977) The structure of the ε-subunit from the coupling factor studied by means of small-angle X-ray scattering and inelastic light scattering. Biochem Biophys Res Commun 78:383–392
34. Schmidt UD, Paradies HH (1979) Size and molecular parameters of adenosine triphosphatase from Escherichia Coli. J Biol Chem 254:5257–5263
35. Sundvall B (1979) Crystal and molecular structure of tetraoxotetrahydroxo bismuth(III) nitrate monohydrate. Acta Chem Scand A 33:219–224
36. Sundvall B (1980) An X-ray diffraction study of the hexanuclear complex of Bi(III) in aqueous perchlorate solution. Determination of the oxygen positions. Acta Chem Scand A 34:93–98
37. Turnbull D, Fischer JC (1949) Rate of nucleation in controlled systems. J Phys Chem 17:71–73
38. Wieriks J, Hespe W, Jaitly KD et al. (1982) Pharmacological properties of colloidal bismuth subcitrate (CBS). Scand J Gastroenterol 17 [Suppl 80]:11–16

Safety Profile of Bismuth-Containing Medications

J. R. LAMBERT

Introduction

Bismuth compounds have been taken as medicinals for more than two centuries with the first recorded medical application in 1785. Early uses included local demulcent and protective action in the skin and gastrointestinal tract, to aid healing of ulcers, as well as to treat and prevent acute diarrhea. Bismuth salts are also employed as a contrast material in gastrointestinal radiology. While the above describe local effects, the systemic effects of bismuth have been used as treatment for syphilis and hypertension.

Many oral preparations of bismuth have been used and these appear to differ in their clinical efficacy as well as their pharmacokinetics. A number of bismuth compounds have been employed to treat a variety of gastrointestinal complaints including abdominal pain, flatulence, diarrhea, and constipation. The inorganic salts used included the subnitrate, subcarbonate, subgallate, tartrate, and subsalicylate. These preparations were mainly available over the counter and often contained other compounds; several of these are still available in some countries.

Colloidal bismuth subcitrate (CBS) has been successfully administered in the treatment of peptic ulcer disease. This preparation has proved as effective as the histamine H_2-antagonists in the treatment of gastric and duodenal ulcers with lower relapse rates following cessation of therapy. The lower relapse rates have been attributed to a beneficial effect on *Helicobacter (Campylobacter) pylori*-associated gastritis. Long-term eradication of *Helicobacter pylori* improves when CBS is coadministered with antibiotics [1].

Pharmacokinetics

Absorption

Following oral intake of bismuth compounds a significant increase in blood concentrations of bismuth occurs [2,3]. Table 1 summarizes peak plasma concentrations of bismuth after ingestion of bismuth compounds for more than 2 weeks [2, 4–8].

Monash University, Dept. of Medicine, Prince Henry's Hospital, St Kilda Road, Melbourne Vic 3004, Australia

Helicobacter pylori, Gastritis and Peptic Ulcer
Edited by P. Malfertheiner and H. Ditschuneit

Table 1. Plasma concentration posttreatment after 2 or more weeks therapy with bismuth compounds

Reference	Bismuth compound	Formulation	Range peak plasma concentration (μg/l)
Conso et al. (1975) [4]	Subnitrate	Tablet	17–38
Steffen et al. (1986) [5]	Subsalicylate	Tablet	1–10
Lee (1981) [2]	Colloidal bismuth subcitrate	Solution	2–35
Dekker et al. (1986) [6]	Colloidal bismuth subcitrate	Chewable tablet	3–33
McLean et al. (1988) [7]	Colloidal bismuth subcitrate	Chewable tablet	0–58
Gavey et al. (1989) [8]	Colloidal bismuth subcitrate	Swallowable tablet	4–40

Gavey and colleagues [8] calculated that 0.2% of ingested bismuth was absorbed, based on urinary excretion following cessation of CBS administration. This determination is questionable as it does not account for temporary storage of bismuth in body compartments or a possible enterohepatic circulation.

A rapidly occurring peak in blood bismuth concentration is observed after single oral ingestion of CBS (DeNol). The peak occurs between 15 and 60 min after administration of DeNol chewable tablets with peak values less than 50 μg/l [7]. The intake of DeNol swallowable tablets gave peaks in plasma ranging from 25 to 300 μg/l after a mean of 30 min (range, 15–105 min) [9]. Low as well as high absorption peaks could be found in the same person with multiple doses of the swallowable preparation. This suggests that bismuth absorption with the swallowable tablets of CBS is subject to high intra- and interindividual variation. Considerable care should be taken in relating the chronic toxicity described for bismuth compounds with the peak concentrations in blood and plasma shortly after ingestion. Multiple dosing studies with both CBS preparations show an apparent steady state reached by 3–4 weeks [2, 7, 8]. The elimination half-life after steady state was reached was between 20 and 30 days for urine and plasma for both CBS preparations [7, 8].

The large interindividual variation in bismuth absorption may be due to a number of factors. These include differences in dissolution of tablets, gastric pH, and gastrointestinal motility [2, 4, 10]. It has also been suggested that the chemical form of bismuth at least partly determines the rate of absorption [11, 12].

The site and mechanism of bismuth absorption in humans is unclear. Electron microscopy has demonstrated CBS particles in enterocytes of the duodenum and proximal small bowel in humans after oral intake. No particles were noted in the gastric mucosal cells [12]. In the rat the uptake of bismuth by the intestinal mucosa after oral CBS was highest in the proximal jejunum, ileum, and duodenum with low levels in the stomach [13]. The rapid appearance of bismuth in blood after oral intake of CBS in humans suggests bismuth absorption in the stomach [10].

Distribution

The absorbed bismuth is distributed in blood and other tissues. The chemical form in which bismuth is present in blood is unknown. Thomas et al. [5] showed an association of bismuth with the high molecular weight fraction ($\geqq$200000 daltons) including α_2-macroglobulin, IgM, and α-lipoprotein after incubation of human blood with bismuth subgallate. Russ et al. [14] showed that 17% of the radioactive bismuth citrate incubated with blood was associated with erythrocytes and the remainder was not specifically bound to serum proteins.

In humans and animals receiving bismuth, the highest concentration/g net weight was always found in the kidney [14–17]. After 14 months of administration of CBS to rats the organic concentrations bismuth, from high to low, were in the kidney, lung, spleen, liver, brain, and muscle [18]. Values ranged from 13.9 μg/g net weight in the kidney to 0.13 μg/g in the muscle.

The concentrations of bismuth in the brain of humans and animals receiving bismuth was higher than in controls [18–20]. In patients dying from bismuth encephalopathy the concentration in the gray matter was about twice as high as that in white matter, with the highest concentrations in the thalamus and cerebellar cortex [21].

The factors influencing tissue distribution are not known. The physicochemical form of the bismuth compound in the vascular space has some influence on distribution [14, 22].

Elimination

Elimination of bismuth from the body is via both the urine and feces. The majority of orally administered bismuth is not absorbed and excreted in feces. The exact contribution of the urine and fecal routes to elimination of absorbed bismuth is unclear and may in part be dependent on the salt and dose, as suggested using radiolabelled intravenous ^{206}bismuth salts in rats [23, 24]. Bismuth present in feces after cessation of oral therapy is excreted partly via the bile and partly via intestinal secretion. Preliminary evidence has suggested an enterohepatic circulation of the bismuth pool [2, 25].

Studies have suggested a two- or three-compartment model to describe the elimination kinetics of bismuth from the blood in humans [26] and in animal models [14, 27]. The kidney is an important compartment with the other compartments including the liver, muscle, and adipose tissue probably being less important.

Bismuth Toxicity in Humans

A large number of toxic effects have been attributed to bismuth compounds as shown in Table 2 [28–37]. The adverse affects on various organ systems may be associated with different bismuth compounds. As shown in Table 2 the same compounds may affect a number of organs.

Table 2. Toxic effects of bismuth compounds in humans

Organ system	Toxic effect	Bismuth salts	Reference
Kidney	Acute reversible renal failure	Triglycollamate, thioglycollanate, diallylacetate, CBS (100 tablets)	[28, 29]
Liver	Acute hepatitis	Thioglycollate	[30]
Bone	Osteoporosis Osteomalacia Fracture	Subnitrate Subcarbonate Subgallate	[31, 32]
Skin	Erythroderma	Tartrate, salicylate	[33]
Neuropsychiatric	Chronic "encephalopathy"	Subnitrate Subgallate, subsalicylate	[34–36] [37]

Bismuth encephalopathy has been reported mainly as case reports particularly from France with often incomplete descriptions of the clinical features and lack of blood or plasma concentrations of bismuth. Diagnosis is generally confirmed by the presence of elevated bismuth concentrations in the blood, plasma, serum, or CSF. In subjects with encephalopathy blood bismuth levels are usually more than 100 μg/l with most having levels greater than 500 μg/l at time of presentation [34, 38]. In some case reports a lack of relationship between clinical illness and blood concentrations was noted [37, 39]. This can be accounted for by one or more of the following explanations: blood bismuth levels have been inadequately assessed and samples have often been taken late after presentation; there is large interindividual variation in susceptibility to bismuth neurotoxicity; and bismuth intake has often been discontinued by subjects prior to hospitalization.

The cause of bismuth encephalopathy is unclear. When subjects with encephalopathy were compared with a healthy control group of bismuth consumers, no difference in age, sex, quantity of bismuth consumed, duration of drug use, or use of other drugs was observed. The only difference between the two groups was the problem of constipation in those who developed encephalopathy [40]. A hypothesis that intestinal absorption is facilitated in subjects with bismuth encephalopathy has thus been suggested [41].

The two commonly used bismuth compounds are CBS and bismuth subsalicylate (BSS). As anticipated from the pharmacokinetics of CBS there have been no reports of toxic effects in subjects taking the recommended dosage and duration of therapy. The steady state plasma concentration of bismuth with standard doses of CBS was less than 50 μg/l after any of the preparations [2, 7, 8]. The range of plasma concentrations is 1.5%–5% of those concentrations reported from patients with encephalopathy. The only case of encephalopathy linked to BSS occurred after unknown amounts of a noncommercial preparation were consumed by a patient with numerous gastrointestinal complaints [37]. Plasma concentration of bismuth after BSS therapy is below 50 μg/l [5, 42].

Conclusion

Bismuth salts have been in medical use for over 200 years with most therapy previously available over the counter for a variety of complaints. The current medical indications for specific bismuthcontaining preparations are now well established. CBS is effective in the treatment of peptic ulcer disease and, in combination with antibiotics, is beneficial against *Helicobacter pylori*-associated gastritis [1]. BSS is of benefit in treatment and prevention of infective diarrhea [5]. Bismuth compounds, when administered to subjects with normal renal and hepatic function for short duration, rarely result in toxicity.

The adverse publicity relating to toxicity from bismuth compounds has resulted mainly from unsupervised, indiscriminate use of these drugs for often inappropriate indications. A rational scientific evaluation of the clinical efficacy and pharmacokinetics of several bismuth compounds, particularly CBS and BSS, has established their usefulness in clinical medicine. Ongoing evaluation of new formulations of bismuth are essential to optimize their effect on the healing of peptic ulcer disease as well as eradication of *Helicobacter pylori*, while maintaining the low toxicity. Special attention must be paid to the gastric mucosal pharmacokinetics of bismuth compounds to ensure adequate delivery of bismuth [43].

References

1. Rauws EAJ, Langenberg W, Houthoff HJ, Zamen HC, Tytgat GNJ (1988) Campylobacter pyloridis-associated chronic active antral gastritis: a prospective study of its prevalence and the effects of antibacterial and antiulcer treatment. Gastroenterology 94:33–40
2. Lee SP (1981) Studies on the absorption and excretion of tripotassium-dicitrato-bismuthate in man. Res Commun Chem Pathol Pharmacol 43:359–364
3. Lechat P, Kisch R (1986) Les encephalopathies bismuthiques: reevaluation du risque. Gastroenterol Clin Biol 10:562–569
4. Conso F, Bourdon R, Gaultier M (1975) Bismuth sanguin et urinaire apres traitement bref par differents sels "insolubles" de bismuth. Eur J Toxicol 8:137–141
5. Steffen R, DuPont HL, Heusser R, Helminger A, Witassek F, Manhart MD, Schar M (1980) Prevention of traveller's diarrhea by the tablet form of bismuth subsalicylate. Antimicrob Agents Chemother 29:625–627
6. Dekker W, Dal Monte PR, Bianchi Porro G, van Bentem N, Boekhorst JC et al. (1986) An international multi-clinic study comparing the therapeutic efficacy of colloidal bismuth subcitrate coated tablets with chewing tablets in the treatment of duodenal ulceration. Scand J Gastroenterol [Suppl 122] 21:46–50
7. Froomes PRA, Wan AT, Keech AC, McNeil JJ, McLean AJ (1989) Absorption and elimination of bismuth from oral doses of tripotassium-dicitrato-bismuthate. Eur J Clin Pharmacol 37:533–536
8. Gavey CJ, Szeto M-L, Nwokolo CU, Sercombe J, Pounder RE (1989) Bismuth accumulates in the body during treatment with tripotassium dicitrato bismuthate. Alimen Pharm Therap 3:21–28
9. Nwokolo CU, Gavey CJ, Smith JTL, Pounder RE (1989) The absorption of bismuth from oral doses of tripotassium dicitrato bismuthate. Alimen Pharm Ther 3:29–39
10. Hespe W, Staal HJM, Hall DWR (1988) Bismuth absorption from the colloidal subcitrate. Lancet II:1258
11. Billon JP, Gernez G, Gourdin JC, Martin Ch, Palliere M (1976) Quelques caracteristiques physicochimiques de sels de bismuth utilises en pharmacie. Ann Pharm Franc 34:161–171

12. Coghill SB, Hopwood D, McPherson S, Hislop S (1983) The ultrastructural localisation of De-Nol®. (colloidal tripotassium dicitrato-bismuthate-TDB) in the upper gastrointestinal tract of man and rodents following oral and instrumental administration. J Pathol 139:105–114
13. Stiel D, Murray DJ, Peters TJ (1985) Uptake and subcellular localisation of bismuth in the gastrointestinal mucosa of rats after short term administration of colloidal bismuth subcitrate. Gut 26:364–368
14. Russ GA, Bigler RE, Tilbury RS, Woodard HQ, Laughlin JS (1975) Metabolic studies with radiobismuth. I: retention and distribution of ^{206}Bi in the normal rat. Radiat Res 63:443–454
15. Hall RJ, Farber T (1972) Drug residues in animal tissues: determination of bismuth in body tissues and fluids after administration of controlled doses. J Assoc Off Anal Chem 55:639–642
16. Hursh JB, Brown C (1969) Tissue distribution of ^{212}Bi in rats. Proc Soc Exp Biol Med 131:116–120
17. Sollmann T, Cole HN, Henderson K (1938) Clinical excretion of bismuth. VII: The autopsy distribution of bismuth in patients after clinical bismuth treatment. Am J Syph Gonorrhea Vener Dis 22:555–583
18. Lee SP, Lim TH, Pybus J, Clarke AC (1980) Tissue distribution of orally administered bismuth in the rat. Clin Exp Pharmacol Physiol 7:319–324
19. Hamilton El, Minski MJ, Cleary JJ (1972/73) The concentration and distribution of some stable elements in healthy human tissues from the United Kingdom. Sci Total Environ 1:341–374
20. Warren HV, Horsky SJ, Gould CE (1983) Quantitative analysis of zinc, copper, lead, molybdenum, bismuth, mercury and arsenic in brain and other tissues from multiple sclerosis and non-multiple sclerosis cases. Sci Total Environ 29:163–169
21. Escourolle R, Bourdon R, Galli A, Galle P, Jaudon MC et al. (1977) Etude neuropathologique et toxicologique de douze cas d'encephalopathie bismuthique. Rev Neurol (Paris) 133:153–163
22. Erdani S, Balzarini M, Taglioretti D, Romussi M, Valentini R (1964) The distribution of radiobismuth in the rat. Br J Radiol 37:311–314
23. Zidenberg-Cherr S, Parks NJ, Keen CL (1987) Tissue and subcellular distribution of bismuth radiotracer in the rat: considerations of cytotoxicity and microdosimetry for bismuth radiopharmaceuticals. Radiat Res 111:119–129
24. Chaleil D, Allain P, Emile J (1979) Etude de l'elimination urinaire et fecale du bismuth administre par voie parenterale. Pathol Biol (Paris) 27:417–420
25. McLean AJ, Froones PRA, Wan AT (1989) Biliary handling of bismuth subcitrate in Man. Clin Pharmacol Ther 44:150
26. De Bray J-M, Emile J, Moreau R, Daver A, Cheguillaume J et al. (1977) Pathogenie des encephalopathies bismuthiques: etude preliminaire a l'aide du bismuth$_{206}$ chez l'homme. Nouv Presse Med 6:1394
27. Vienet R, Bouvet P, Istin M (1983) Cinetique et distribution du ^{206}Bi chez le rat et le lapin: un modele. Int J Appl Radiat Isot 34:747–753
28. Randall RE, Osherhoff RJ, Bakerman S, Setter JG (1972) Bismuth nephrotoxicity. Ann Intern Med 77:481–482
29. Hudson M, Mowat NAG (1989) Colloidal bismuth subcitrate (De-Nol) self-poisoning – reversible toxicity. Br Med J (in press)
30. Karelitz S, Freedman AD (1951) Hepatitis and nephrosis due to soluble bismuth. Pediatrics 8:772–777
31. Gaucher A, Netter P, Faure G, Pourel J, Hutin MF et al. (1980) Les osteoarthropathies "bismuthiques": interet du dosage du bismuth osseux. Rev Rhum Mal Osteoartic 47:31–35
32. Murray JR (1978) Bismuth neurotoxicity, Med J Aust 1:279–280
33. Goyer RA (1986) Toxic effects of metals. Bismuth. In: Klaassen et al. (eds) Casarett and Doull's toxicology, 3rd edn. Macmillan, New York, p 620
34. Buge A, Supino-Viterbo V, Rancurel G, Metzger J, Dechy H et al. (1979) Correlations evolutives: cliniques electroencephalographiques, tomodensitometriques et toxicologiques, dans cinq cas d'encephalopathies bismuthiques. Sem Hop 55:1466–1472

35. Burns R, Thomas DW, Barron VJ (1974) Reversible encephalopathy possibly associated with bismuth subgallate ingestion. Br Med J 1:220–223
36. Moindrot J, Evreux JC, Savet JF, Aimard G (1977) Encephalopathies bismuthiques: bilan et commentaires. Lyon Med 238:537–541
37. Hasking GJ, Duggan JM (1982) Encephalopathy from bismuth subsalicylate. Med J Aust 2:167
38. Bes A, Caussanel JP, Geraud G, Jauzac Ph, Geraud J (1976) Encephalopathie toxique par les sels de bismuth. Rev Med Toulouse 12:801–813
39. Guard O, Soichot P, Dumas R, Besancenot J-B, Gaudet M (1977) Reflexions a propos de dix observations d'encephalopathie bismuthique. J Med Lyon 58:403–416
40. Martin-Bouyer G, Foulon G, Guerbois H, Barin C (1981) Epidemiological study of encephalopathies following bismuth administration per os: characteristics of intoxicated subjects – comparison with a control group. Clin Toxicol 18:1277–1283
41. Slikkerveet A, de Wolff FA (1989) Pharmacokinetics and toxicity of bismuth compounds. Med Toxicol 4:303–323
42. Ericsson ED, DuPont HL, Pickering HL (1980) Bismuth preparations for diarrhea. J Am Med Assoc 244:1435–1436
43. Lambert JR, Way DJ, King RG, Eaves ER, Liancas K, Wan A (1989) DeNol vs pepto-bismol bismuth pharmacokinetics in human gastric mucosa. Gastroenterology 96:A284

Is Nonulcer Dyspepsia Improved by Treating *Helicobacter pylori* Infection?

A. T. R. Axon

Clinical research in nonulcer dyspepsia (NUD) is not straightforward. Reported studies show variation in the definition of NUD, the population groups studied, the endpoints assessed and symptoms evaluated. In attempting to give an overview of the benefit of treating *Helicobacter pylori* (HP) in this condition it is relevant first to consider these aspects.

A recent international working party has defined NUD as upper abdominal or retrosternal pain, discomfort, heartburn, nausea, vomiting (or other symptoms referable to the proximal alimentary tract), lasting for longer than 4 weeks, unrelated to exercise, and for which no focal lesion or systemic cause was found responsible [1].

One difficulty with this definition is that it includes individuals whose primary complaint is heartburn. Although this is technically NUD, it is more likely to result from abnormalities of the gastro-oesophageal junction than from HP infection and probably should be excluded from studies assessing treatment of HP. Secondly, this definition does not adequately exclude those patients with irritable bowel syndrome with upper abdominal symptoms. Thirdly, some patients with peptic ulcer in remission may be included and, finally, care must be exercised in deciding what is meant by the presence or absence of a "focal" lesion, should this include severe gastritis or antral erosion?

When considering the selection of individuals for NUD trials, the varying prevalence of HP, in different age groups, countries, racial subsets, and social stata must be considered especially when comparing one study with another.

To date no convincing well-controlled study has shown HP prevalence to be higher in dyspeptic populations than normals, although the weight of subjective evidence favours this belief. Figure 1 shows the prevalence of HP in blood donors with an age-matched population attending a dyspepsia clinic in the same city. HP appears to be approximately 20% higher in the dyspepsia group compared with age-matched blood donors, but it is unlikely that these groups represent the same social background or racial mix. If, however, these figures are regarded as generally representative, HP is likely to be responsible for only about 20% of the dyspeptics with HP and therefore only about 10% of an unselected dyspeptic population (which will include HP-positive and HP-negative patients). It follows that any study attempting to show improvement by medication will need large numbers to show a difference.

Gastroenterology Unit, The General Infirmary, Leeds, LS1 3EX, U.K.

Helicobacter pylori, Gastritis and Peptic Ulcer
Edited by P. Malfertheiner and H. Ditschuneit

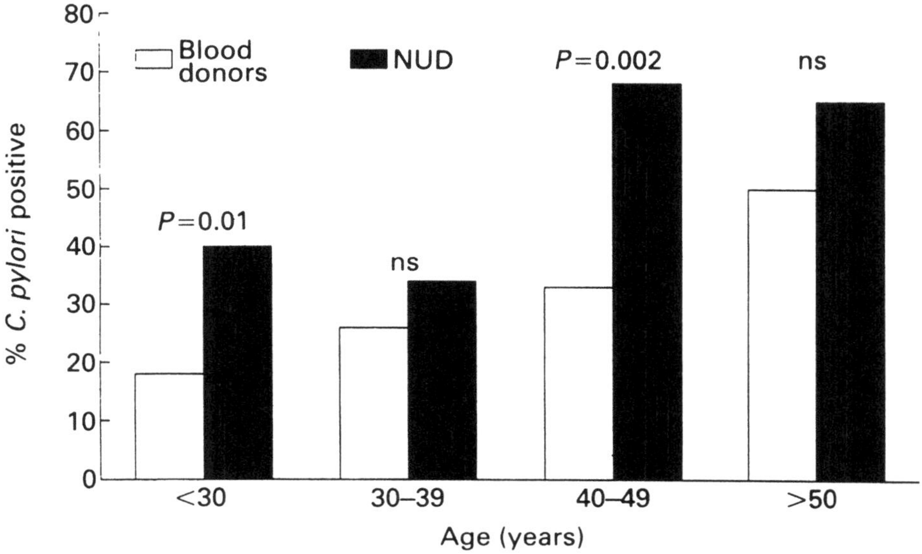

Fig. 1. Frequency of *H. (C.) pylori* in nonulcer dyspepsia (NUD) patients and blood donors [8]

In setting up a trial it would be helpful if a symptom complex could be identified which is associated with HP. A number of studies have addressed this question, but no consistent syndrome has been identified, and where a symptom has been identified, it has usually not been subjected to an age-specific analysis. This is essential when studying HP where there is a very strong association with age and thus a risk that a symptom apparently associated with HP is actually associated with age.

Table 1 lists the placebo-controlled studies analysed in this overview and includes the drug and the numbers studied. At first sight the numbers appear reasonable, but when analysed first for those that were HP positive and then for those treated with the active preparation, it is apparent that no single study contains sufficient numbers to provide a confident result especially, as previously argued, only around 20% would be expected to improve on anti-HP medication.

Table 1. Placebo controlled trials in nonulcer dyspepsia (NUD) of *H. pylori* (HP) positive patients treated with bismuth

Author	Treatment	No. studied	HP+	HP+ on active treatment
McNulty et al. [2]	BSS	50	50	18
Lambert et al. [3]	CBS	54	33	15
Borody et al. [4]	CBS	43	43	22
Rokkas et al. [5]	CBS	52	20	12
Kang et al. [6]	CBS	51	23	11
Loffeld et al. [7]	CBS	50	50	26

Table 2. Results of bismuth treatment in HP+ patients

Author	Outcome
McNulty et al. [2]	Trend in favour, N.S.
Lambert et al. [3]	Pain improved, $P<0.01$
Borody et al. [4]	Pain improved, $P<0.001$
Rokkas et al. [5]	Symptoms improved, $P<0.001$
Kang et al. [6]	Asymptomatic, $P<0.05$
Loffeld et al. [7]	No significant differences

Table 2 shows, somewhat surprisingly, that statistical differences were found between the groups. The possibility of a type 1 statistical error should be considered i.e. the greater number of symptoms assessed, the greater is the likelihood that a statistically significant factor difference will arise by chance.

It is possible to amalgamate the results in those studies where an endpoint of total symptomatic remission was used. If HP is responsible for NUD elimination or suppression of the organism should lead to total relief in the same way as ulcer healing is usually associated with complete resolution of symptoms.

Table 3. HP+ patients becoming asymptomatic

Author	CBS	Placebo
Rokkas et al. [5]	5/12	0/13
Kang et al. [6]	8/11	3/12
Loffeld et al. [7]	4/26	0/24
Total	17/49 (35%)	3/49 (6%)

Table 3 shows those publications which can be analysed in this way and if the totals are added up (a practice of dubious scientific credibility) it can be seen that 49 HP-positive NUD patients were treated with colloidal bismuth subcitrate (CBS), and 49 with placebo. It is relevant at this point to emphasize that these studies are not double-blind because bismuth causes darkening of the stool. Bearing in mind these strictures it appears that 35% in the bismuth group became asymptomatic compared with only 6% in the placebo group. It is unreasonable to apply statistics in this unscientifically designed assessment, but this result might imply that HP may be responsible for NUD in up to 30% of individuals, and suggests that a properly designed study using large numbers of patients should be undertaken. An alternative explanation for the above finding is that CBS may have a nonspecific benefit in patients with NUD unrelated to its effect on HP. Unfortunately few studies undertaken have investigated HP-negative nonulcer dyspeptics. Those that have done so are included in Table 4. There was no obvious trend in favour of CBS, but again the data are unsatisfactory.

Table 4. Does CBS improve HP− nonulcer dyspepsia?

Author	HP−	Result
Kang et al. [6]	28	N.S. (trend to placebo)
Rokkas et al. [5]	32	N.S. (4/13 vs 4/19 improved)
Lambert et al. [7]	30	N.S. (12/16 vs 9/14 improved)

References

1. Working Party Report (1988) Management of dyspepsia. Lancet I:576–579
2. McNulty CAM, Gearty JC, Crump B, et al. (1986) Campylobacter pyloridis and associated gastritis: investigator blind, placebo controlled trial of bismuth salicylate and erythromycin ethylsuccinate. Br Med J 293:645–649
3. Lambert JR, Borromeo M, Korman MG, Hansky J (1987) Role of Campylobacter pyloridis in non-ulcer dyspepsia – a randomized controlled study. Gastroenterology 92 (5/2):1488
4. Borody T, Hennessy W, Daskalopoulos G, Carrick J, Hazell S (1987) Double blind trial of DeNol in nonulcer dyspepsia associated with Campylobacter pyloridis gastritis. Gastroenterology 92 (5/2):1324
5. Rokkas T, Pursey C, Uzoechina E, et al. (1988) Non-ulcer dyspepsia short term De-Nol therapy: a placebo controlled trial with particular references to the role of Campylobacter pylori. Gut 29:1386–1391
6. Kang JY, Tay HH, Wee A, Guan R, Math MV, Yap I The effect of colloidal bismuth subcitrate on symptoms and gastric histology in non-ulcer dyspepsia – a double-blind placebo-controlled study.
7. Loffeld RJLF, Potters HVJP, Stobberingh E, Flendrig JA, van Spreeuwel JP, Arends JW (1989) Campylobacter associated gastritis in patients with non-ulcer dyspepsia: a double blind placebo controlled trial with colloidal bismuth subcitrate. Gut 30:1206–1212
8. Shallcross TM, Rathbone BJ, Heatley RV (1989) Campylobacter pylori and non-ulcer dyspepsia. In: Rathbone BJ, Heatley RV (eds) Campylobacter pylori and gastroduodenal disease. Blackwell, Oxford, pp 155–166

Five-Day Triple Therapy and 15-Day Double Therapy on *Helicobacter pylori* Eradication

L. G. V. Coelho, M. C. F. Passos, D. M. M. Queiroz, A. J. A. Barbosa, E. N. Mendes, G. A. Rocha, C. A. Oliveira, G. F. Lima Jr, Y. Chausson, and L. P. Castro

Introduction

Many studies have shown that *Helicobacter* (*Campylobacter*) *pylori* (HP) is sensitive to a number of different antibiotics in vitro [1–5] but their use in vivo has been disappointing [6]. The lack of concordance between in vitro sensitivity and in vivo efficacy in eradicating HP is remarkable.

At this point we must define the sometimes misleading words clearance and eradication. Clearance is the absence of the organism in antral biopsy specimens on histological or microbiological testing at the end of treatment. This is not difficult to achieve, but recrudescence of the original infection usually takes place. The organism probably lies deeply within gastric pits, protected from luminal chemotherapy by mucus and from blood-borne antibiotics by being extracellular. Eradication is defined as absence of the organism on testing 1 month after the end of the treatment period. If the patients are clear of the organism at this stage over 85% will remain negative at 12 months [7].

Although there have been many studies showing that HP is sensitive to a wide range of antibiotics, their use in vivo as monotherapy has been rather dissapointing [6]. Dual therapy using bismuth and amoxycillin has proved more effective in eradication giving an overall success rate of around 42%. Rauws et al. [8] used amoxycillin 375 mg t.i.d. with colloidal bismuth 120 mg q.i.d. for 4 weeks and achieved eradication in 7 out of 20 patients. The combination of bismuth subsalicylate (BSS) and amoxycillin has been studied by Wagner et al. [9]. They used BSS 600 mg t.i.d. with amoxycillin 750 mg t.i.d. for 15 days and eradicated HP in 4 out of 8 patients. O'Riordan et al. [10] compared a number of adjuvant antibiotic regimens designed to achieve optimal HP eradication in patients with duodenal ulcers and an associated HP-positive antral gastritis. The association of bismuth 120 mg q.i.d. plus amoxycillin 500 mg t.i.d. ($n = 18$); bismuth 120 mg t.i.d. plus metronidazole 200 mg t.i.d. ($n = 23$), and bismuth 120 mg t.i.d. plus metronidazole t.i.d. for 7 days eradicated HP in 50%, 57%, and 85% of the patients, respectively.

Preliminary use of triple therapy appears to be more successful. Borody and Carrick [11] treated 100 consecutive patients for 4 weeks with a combination of colloidal bismuth subcitrate (CBS; chewable De-Nol) q.i.d. tetracycline 500 mg q.i.d. for 2 weeks. An eradication rate of 94% was achieved at 8 weeks and 47 of 50

Departamento de Clínica Médica, Faculdade de Medicina da UFMG, Avenida Alfredo Balena, 190, 30.130 – Belo Horizonte, MG, Brazil

Helicobacter pylori, Gastritis and Peptic Ulcer
Edited by P. Malfertheiner and H. Ditschuneit

patients were still clear 12–25 months later. Börsch et al. [12] studied 105 patients, 71 of whom were followed-up for a mean period of 6–9 months. Patients received a combined oral medication of amoxycillin 500 mg t.i.d. and metronidazole 500 mg t.i.d. in addition to BSS 600 mg t.i.d. for various time periods, ranging from 0 to 2 weeks. The 7-day treatment regimen achieved an eradication rate of 80% (12/15) and was considered the shortest regimen for HP eradication. In a previous study [13] we have shown that furazolidone, similarly to bismuth salts, effects HP clearance, peptic ulcer healing, and reduction of ulcer relapse.

We describe a preliminary study of two different furazolidone-antibiotic associations on HP eradication using triple short-term and double medium-term regimens.

Patients and Methods

After giving written consent, 10 HP-positive subjects (6 asymptomatic volunteers and 4 duodenal ulcer patients) took part in the study. Six participants used furazolidone 100 mg, amoxycillin 500 mg, and metronidazole 250 mg orally, t.i.d. for 5 days. Four subjects used furazolidone 100 mg and metronidazole 250 mg every 6 h for 15 days. Gastroscopy and the ^{14}C-urea breath test were performed before and after (range 72–184 days) the end of the treatment in all subjects. Antral biopsy specimens were collected for histology, Gram staining, culture, and urease test, also for monitoring HP clearance. HP eradication was defined as HP-negative at culture. Antral mucosa histology was evaluated according to Whitehead. The presence of neutrophils in the lamina propria was graded for the assessment of the activity (PMN score, 0 to 3) of the gastritis.

Results

All participants completed the trial. In the 5-day triple therapy group, HP eradication was demonstrated by the negative tests performed 73–96 days ($x = 88$) after therapy in five (83.3%) out of six patients. Also, normal histology, with only rare PMNs in the lamina propria was seen in these five patients. The only one who persisted to be HP positive showed a striking improvement of the active gastritis (PMN score, +++ to +). Two subjects developed minor GI symptoms. In the 15-day double therapy group, HP was eradicated in two (50%) of the four subjects as demonstrated by the negative tests performed 72–184 days ($x = 128$) posttreatment. Although chronic gastritis persisted in the two HP-negative subjects, signals of activity were not seen. No side effects were seen. The duodenal ulcer patients that cleared HP remained asymptomatic and healed.

Conclusion

Five-day triple therapy using furazolidone, amoxycillin, and metronidazole showed itself to be a short, inexpensive, and effective regimen for HP eradication. A larger trial using this therapeutic regimen is currently under way.

References

1. Lambert T, Mégraud F, Gerbaud G, Courvalin P (1986) Susceptibility of Campylobacter pyloridis to 20 antimicrobial agents. Antimicrob Agents Chemother 30:510–511
2. Andreasen JJ, Andersen LP (1987) In vitro susceptibility of Campylobacter pyloridis to cimetidine, sucralfate, bismuth and sixteen antibiotics. Acta Pathol Microbiol Immunol Scand [B] 95:147–149
3. McNulty CAM, Dent J, Wise R (1985) Susceptibility of clinical isolates to Campylobacter pyloridis to 11 antimicrobial agents. Antimicrob Agents Chemother 28:837–838
4. Czinn S, Carr H, Aronoff S (1986) Susceptibility of Campylobacter pyloridis to three macrolide antibiotics (erythromycin, roxithromycin [RV 28905], and CP 6, 993) and rifampin. Antimicrob Agents Chemother 30:328–329
5. Shungu DL, Walin DR, Gilman RH, Gadebushe HH, Cerami AT, Gill C, Weissberger B (1987) Comparative susceptibilities of Campylobacter pylori to norfloxacin and other agents. Antimicrobiol Agents Chemother 31:949–950
6. McNulty CAM (1987) Treatment of Campylobacter infections in man. J Antimicrob Chemother 19:281–284
7. Axon ATR (1988) Treatment of Campylobacter pylori. Symposium on Campylobacter pylori; defining a cause of gastritis and peptic ulcer disease? 13th International Congress of Gastroenterology, Rome
8. Rauws EA, Langengerg W, Houthoff HJ, Zanen HC, Tytgat GNJ (1988) Campylobacter pyloridis-associated chronic active antral gastritis. A prospective study of its prevalence and the effects of antibacterial and antiulcer treatment. Gastroenterology 94:33–40
9. Wagner S, Freise J, Wittenberg H, Schmidt FW, Bär W (1988) Epidemiology and treatment of gastric Campylobacter pylori infection. In: Mégraud F, Lamouliatte H (eds) Gastroduodenal pathology and Campylobacter pylori. Elsevier, Amsterdam
10. O'Riordan T, Tobin A, Beattie S, O'Morain CO, Sweeney E, Keane C (1989) Adjuvant antibiotic treatment improves eradication of Campylobacter pylori in duodenal ulcer. Gastroenterology 95 (5): A 378
11. Borody T, Carrick J (1988) Long term Campylobacter pylori recurrence following apparent eradication. In: Mégraud F, Lamouliatte (eds) Gastroduodenal pathology and Campylobacter pylori. Elsevier, Amsterdam
12. Börsch G, Mai U, Opferkuch W (1989) Short- and medium-term results of oral triple therapy to eradicate C. pylori. Gastroenterology 96 (5): A 53
13. Coelho LGV, Queiroz DMM, Barbosa AJA, Mendes EN, Rocha GA, Oliveira CA, Lima Jr GF, Passos MCF, Castro LP (1989) Furazolidone x cimetidine in duodenal ulcer C. pylori positive patients. Gastroenterology 96 (5): A 91

Antibacterial Therapy: Effect on Healing and Relapse of Peptic Ulcer Disease

G. Börsch

Introduction

Helicobacter pylori (formerly *Campylobacter pylori*, *H. pylori*, HP) is rapidly evolving as the single most important factor in the pathogenesis of ulcer disease, at least gastritis-associated duodenal ulcer disease [13]. Unlike the underlying genetic predispositions of this condition, *H. pylori* has the tremendous practical advantage of being a treatable condition.

Ulcer treatment has previously – and successfully – focused on manipulating physiologic events such as secretion of acid or pepsin. *H. pylori* has now opened a fundamentally new era by making it both logical as well as tempting to use antimicrobials as a therapeutic regimen.

A survey of such treatment modalities will be the topic of this review. It will address any studies on duodenal and gastric ulcer disease, that will allow an analysis of the correlations between ulcer healing and relapse with treatment effects on *H. pylori*. Since conventional ulcer therapy based on H_2-receptor antagonists, antacids, sucralfate, or pirenzepine will in all likelihood fail to influence a patient's *H. pylori* state, only trials using bismuth salts and/or antimicrobial drug modalities will be taken into consideration. Also excluded from further discussion will be a small number of trials that have investigated the effect of antimicrobial therapy on peptic ulcer disease without monitoring the patients' pre- and posttherapeutic *H. pylori* state. These studies and also their shortcomings have been reviewed in detail elsewhere [1, 5, 11].

There is then, however, a dearth of well-conducted, sufficiently large, and adequately reported *H. pylori* ulcer trials. Important studies have recently been initiated which are still going on [15] or will possibly be finished in the near future. These may in due time corroborate and refine our current thinking, but they will in all likelihood not refute present basic assumptions on the causal role of *H. pylori* in ulcer disease.

Influence of *H. pylori* Therapy on the Kinetics of Ulcer Healing

The pertinent studies and their location in the literature, as identified by their principal author and the journal of publication of full reports or abstracts, are

Department of Medicine, Elisabeth-Hospital, Academic Teaching Hospital of the University of Essen Medical School, Moltkestr. 61, D-4300 Essen, FRG

Helicobacter pylori, Gastritis and Peptic Ulcer
Edited by P. Malfertheiner and H. Ditschuneit

Table 1. Studies addressing the effect of the state of *H. pylori* on healing and relapse of peptic ulcer disease

Coghlan et al. [10][a]
Marshall et al. [19][a]
Bayerdörffer et al. [2][a]
Lambert et al. [17]
Borody et al. [6]
Smith et al. [20]
Graham et al. [15]

[a] Published in full detail

summarized in Table 1. Detailed formal papers still represent the exception rather than the rule and are therefore marked by an asterisk. All studies indicated in Table 1 have been systematically analysed in regard to both major areas of this review: ulcer healing and relapse.

Before addressing the first topic of the review – ulcer healing – reference should be made to the fact that acceleration of healing is not a crucial test for the role of *H. pylori* in ulcer disease, neither in theory nor in practice. There exists some potential analogy to the inefficiency of cytoprotective prostaglandins, that is, low dose prostaglandins, to promote ulcer healing [7]. This is best explained by the assumption that factors leading to ulcer formation (for example, local epithelial breakdown in *H. pylori* induced gastroduodenitis) might be quite distinct from those that eventually govern ulcer healing (for example, speed and capacity for local epithelial repair).

Therefore, any lack of influence of *H. pylori* oriented therapeutic regimens on ulcer healing kinetics would certainly not disprove an important pathogenetic role of *H. pylori* in peptic ulcer disease. On the other hand, acceleration of ulcer healing by antibacterial therapy would nevertheless provide further support for an important causal effect of gastroduodenal *H. pylori* colonization on peptic ulcers.

Studies published by Graham et al. [15], Bayerdörffer et al. [2], Marshall et al. [18, 19], and Lambert et al. [17] have investigated the influence of *H. pylori* on duodenal ulcer healing. By adding 2 weeks of oral triple therapy, that is, bismuth subsalicylate (BSS), tetracycline 2 g, and metronidazole 750 mg per day, to ranitidine 300 mg alone significantly accelerated healing in the first 39 duodenal ulcer patients of a randomized controlled *H. pylori* oriented duodenal ulcer trial [15]. The 4-week healing rates were 81% after additional triple therapy versus a rather low 44% rate after ranitidine alone (Table 2).

Likewise, Bayerdörffer et al. [2] have described significantly faster healing of duodenal ulcer after adding ofloxacin 2 × 200 mg to ranitidine 300 mg nocte (Table 2). The latter data are somewhat difficult to explain, since ofloxacin almost never eradicates *H. pylori*. It may, however, reduce the bacterial density and could thus exert some influence on ulcer healing. The same authors have described a beneficial effect of combined ofloxacin/H_2-receptor antagonist therapy in refractory duodenal ulcers [3]. Since quinolones have the propensity to induce resistance of *H. pylori* in a high proportion of patients, these drugs are presently not considered to be useful agents in the therapy of *H. pylori* colonization.

Table 2. Effect of adding antimicrobial drugs to standard therapy on the healing kinetics of duodenal ulcers [2, 15, 18]

	Cumulative percentage healed – no. of weeks					
	2	4	6	8	10	12
Ranitidine [a]	10	44	n.d.	60	n.d.	83
R + triple	37	81*	n.d.	87	n.d.	100
Ranitidine [b]	44	68	88	96	96	100
R + ofloxacin	80	92	100	n.d.	n.d.	n.d.
Bismuth [c]	n.d.	n.d.	n.d.	n.d.	68	n.d.
B + tinidazole	n.d.	n.d.	n.d.	n.d.	74	n.d.
Cimetidine [c]	n.d.	n.d.	n.d.	n.d.	59	n.d.
C + tinidazole	n.d.	n.d.	n.d.	n.d.	76	n.d.

* $P < 0.05$; n.d., not done
[a] Graham et al. [15]: initial 39 duodenal ulcer (DU) patients
[b] Bayerdörffer et al. [2]: 50 DU patients
[c] Marshall [18]: 100 DU patients

The data of the largest randomized controlled *H. pylori* oriented trial in duodenal ulcer disease [18, 19] show a tendency for tinidazole to improve 10-week ulcer healing rates above those observed for either cimetidine (76% vs 59% healing) or colloidal bismuth subcitrate (CBS) alone (74% vs 68% healing; Table 2). A different look at these figures also reveals that after bismuth therapy with or without concomitant tinidazole administration, ulcers healed in 10 of 22 patients (45%) remaining positive for *H. pylori* posttherapeutically. Ulcer healing occurred much more frequently in 25 of those 27 that had become *H. pylori* negative (93%).

The sole evidence arguing for a lack of influence of the *H. pylori* state on ulcer healing is a brief remark by Lambert et al. [17] to the effect that the *H. pylori* state after 8 weeks of CBS therapy was unrelated to healing in their uncontrolled duodenal ulcer trial. However, numbers that would allow reader scrutiny of this statement are not given in the published abstract. All other papers forming the data base of this review fail to adress the issue of ulcer healing (Table 1).

Therefore, available data on the effect of *H. pylori* oriented therapy on ulcer healing kinetics suggest that, first, suitable antimicrobial therapy may accelerate healing of duodenal ulcers [2, 15]. Second, after bismuth therapy, clearance of *H. pylori* is probably associated with faster ulcer healing, as compared with persistence of *H. pylori* [18, 19]. Third, the effect of antimicrobials on the healing kinetics of gastric ulcers is presently unknown, and this is due to a current lack of published data.

Influence of *H. pylori* Therapy on Ulcer Relapses: Profound Modification of the Natural History of the Disease

The posttherapeutic *H. pylori* state has a most profound, almost dramatic impact on duodenal ulcer relapse rates. However, a reservation has to be made insofar as published observations have covered periods up to 12 or 18 months only. Also, the studies shaping our knowledge in this field are far from flawless in design, have small subject numbers, and with only two exceptions [10, 18] are yet to be published in full detail.

The pertinent data are summarized in Table 3. They clearly suggest that patients remaining *H. pylori* positive after ulcer therapy have high 12–18 month duodenal ulcer relapse rates of approximately 80%, while ulcer relapses in patients maintaining a state of therapeutically induced *H. pylori* eradication are extremely rare, probably not more than 8% or 10% per year, and sometimes even around zero. These findings certainly come close to suggesting a "no *H. pylori* – no duodenal ulcer" relationship [14].

The lower relapse rate of 10% in the Coghlan et al. [10] (Table 3) study applies to patients remaining persistently *H. pylori* negative, while the relapse rate of 27% involves all patients who were *H. pylori* negative after an index therapy of cimetidine or CBS and thus includes patients with bacterial recrudescence and reinfections. Likewise, the lower relapse rate of 8% in the Marshall et al. [18, 19] study applies to *H. pylori*-negative patients with anatomically proven duodenal ulcer relapse, while the higher relapse rate of 21% includes recurrence of any ulcer-like symptoms, not necessarily being associated with genuine ulcer craters.

Three other trials in this field merit further comment. A study by Eberhardt et al. [12] in 39 duodenal and 3 gastric ulcer patients showed higher healing and lower relapse rates after BSS 3×600 mg versus cimetidine 800 mg nocte, but this work is seriously flawed by technical difficulties leading to a *H. pylori* detection rate of only 61% in this small group of ulcer patients. This precludes a meaningful analysis of the influence of *H. pylori* on healing and relapse.

A study by Bianchi Porro and Lazzaroni [4] comparing ulcer recurrences after healing by H_2-receptor antagonists or by CBS and published as a letter to the editor reports the posttherapeutic, but not the initial *H. pylori* state. There was a high bacterial clearance, but a low real eradication rate after CBS. Therefore, too few patients were actually eradicated to form a truly *H. pylori*-negative group for follow-up of relapses. However, 94% of patients were *H. pylori*-positive on relapse, which clearly argues in favor and not against a causal role of *H. pylori* for ulcer recurrences.

A study by Humphreys et al. [16] includes cases with duodenal and gastric ulcers in addition to erosions or esophageal lesions, which also precludes a precise analysis. The data are nevertheless interesting insofar as they support a role of *H. pylori* in determining the healing rate after therapy with liquid CBS 4×5 ml. Six weeks of this drug induced healing in 82% of *H. pylori*-positive, but only 42% of *H. pylori*-negative lesions, while the results of cimetidine treatment were similar in both groups, 70% vs 67%.

Comment on the Influence of *H. pylori* Therapy on Relapses in Gastric Ulcer Disease

Concerning relapses in gastric ulcer disease, it is the author's personal view that many or even most gastric ulcers will behave exactly like duodenal ulcers, which is

Table 3. Duodenal ulcer relapses and state of *H. pylori* (HP)

Author	*n*	Therapy	HP-negative after therapy	Follow-up (months)	Ulcer relapses and posttherapeutic HP state	Ulcer relapse rate after 12–18 months HP+	HP–
Lambert et al. [17]	45	CBS [a]	12/45 (27%)	12	All relapses HP+	N.S. [d]	0%
Borody et al. [6]	21	Triple [a]	(94%)	18	3/21 (14%), all relapses HP+	N.S.	0%
Coghlan et al. [10]	66	CIM vs CBS	4/23 (17%) 12/23 (52%)	12	HP+ 19/24 (79%) HP– 1/10 (10%)	79%	27% (10%) [b]
Marshall et al. [19]	100	CIM/TIN CBS CBS/TIN	1/51 (2%) 7/22 (32%) 20/27 (74%)	12	HP+ 37/44 (84%) HP– 5/24 (21%)	84%	21% (8%) [c]
Smith et al. [20]	44	RAN vs CBS	0/23 (0%) 16/21 (76%)	18	HP+ 14/18 (78%) HP– 4/16 (25%) All relapses HP+	78%	25% (all relapses HP+)

[a] Uncontrolled study.
[b] Patients persistently remaining *H. pylori* negative, while the higher rate includes reinfections.
[c] Patients with genuine duodenal ulcer craters, while the higher rate includes symptomatic relapses.
[d] N.S. = Not stated.

based on my own observations of 30 patients with *H. pylori*-positive gastric ulcers treated by oral triple therapy [9]. However, any formal or controlled data on gastric ulcers to prove this assumption are currently unavailable in the literature.

Clearly, the relation between gastric ulcers and *H. pylori* will by necessity be less clear-cut than the relation between duodenal ulcers and *H. pylori* [8]. Undoubtedly, there are causative factors for gastric ulcers other than *H. pylori*, and therefore a considerable proportion of gastric ulcers develops independent of *H. pylori*. Such ulcers stay under life-long risk of acquiring a coincidental, but causally unrelated *H. pylori* colonization, potentially making these ulcers *H. pylori* associated, but not *H. pylori* induced. Presumably, treatment of *H. pylori* will in all likelihood not affect these patients' ulcer diathesis. In addition, such independent ulcer causes may also affect patients with previously established, but so far asymptomatic *H. pylori* colonization. Again, such ulcers would then be *H. pylori* associated, but not *H. pylori* induced, with treatment of *H. pylori* probably not affecting the natural history of these patients' ulcer disease. Therefore, recurrence of gastric ulcers despite eradication of *H. pylori* may be expected in a considerable proportion of gastric ulcer patients and would not disprove a decisive causal role of *H. pylori* for a subgroup of gastric ulcer disease.

Summary and Conclusions

The rather small data base on *H. pylori*-oriented therapy in peptic ulcer disease may best be summarized by stating that, first, suitable antimicrobial therapy may accelerate healing of duodenal ulcers; second, clearance of *H. pylori* may also be associated with acceleration of duodenal ulcer healing; and third, and above all, eradication of *H. pylori* will virtually eradicate duodenal ulcer relapses, at least for the 12- to 18-month periods that have so far been prospectively investigated, unless bacterial reinfection from outside sources occurs.

The same will in all likelihood also apply to many or most gastric ulcers [8, 9], even though controlled data to lend formal proof to this assumption are currently unavailable.

In conclusion, then, the data presented here clearly support a key role of *H. pylori* for peptic ulcers and highlight its potential to cure – and not merely to control – the most common, gastritis-associated variant of ulcer disease, termed GAUD [3, 13]. The available scientific information is now sufficient for *H. pylori* "believers" to accept this proposition of a potentially lasting cure of (previously) idiopathic, gastritis-associated ulcer disease as valid. Current *H. pylori* "nonbelievers" are still entitled to ask for larger patient numbers and better study designs, before finally giving in to the *H. pylori* ammendment of the ulcer equation. However, still being "in search of the flawless trial" should not become a permanent excuse for evading a timely personal decision on possibly flawed and imperfect, but nevertheless persuasive experimental findings as part of the rapidly growing data base which clearly points to a key role of *H. pylori* in ulcer pathogenesis.

However, for *H. pylori* eradication to become a standard procedure in ulcer therapy, better treatment regimens – in terms of simplicity, safety, and effectiveness – are urgently needed.

References

1. Bayerdörffer E, Ottenjann R (1988) The role of antibiotics in Campylobacter pylori associated peptic ulcer disease. Scand J Gastroenterol [Suppl 142] 23:93–100
2. Bayerdörffer E, Kasper G, Pirlet T, Sommer A, Ottenjann R (1987) Ofloxacin in der Therapie Campylobacter-pylori-positiver ulcera duodeni. Dtsch Med Wochenschr 112:1407–1411
3. Bayerdörffer E, Pirlet T, Sommer A, Kasper G, Ottenjann R (1988) Ofloxacin in der Therapie "resistenter" Ulcera duodeni. Z Gastroenterol 26:155–159
4. Bianci Porro G, Lazzaroni M (1988) Campylobacter pylori and ulcer recurrence. Lancet 1:593
5. Bianci Porro G, Lazzaroni M (1989) Campylobacter pylori and peptic ulcer therapy. Gastroenterol Clin Biol 13:107B–111B
6. Borody T, Cole P, Noonan S, Morgan A, Ossip G, Maysay J, Brandl S (1988) Long-term Campylobacter recurrence post eradication. Gastroenterology 94:A43
7. Börsch G (1987) Campylobacter pylori: new and renewed insights into gastritis-associated ulcer disease (GAUD). Hepatogastroenterology 34:191–193
8. Börsch G (1989) Ist Campylobacter pylori der Erreger der Ulkuskrankheit? Pro. Z Gastroenterol 27:121–126
9. Börsch G, Wegener M, Reitemeyer E, Mai U, Opferkuch W (1989) Clinical results and medium term follow-up studies after eradication of Campylobacter pylori. In: Mégraud F, Lamouliatte H (eds) Gastroduodenal pathology and Campylobacter pylori. Elsevier, Amsterdam, p 599
10. Coghlan JG, Gilligan D, Humphreys H, McKenna D, Dooley C, Sweeney E, Keane C, O'Morain C (1987) Campylobacter pylori and recurrence of duodenal ulcers – a 12-month followup study. Lancet II:1109–1111
11. Coghlan JG, Tobin A, O'Morain C (1989) Campylobacter pylori and ulcer treatment. In: Rathbone BJ, Heatley RV (eds) Campylobacter pylori and gastroduodenal disease. Blackwell, Oxford, p 232
12. Eberhardt R, Kasper G, Dettmer A, Höchter W, Hagena D (1988) Wirkung von Wismut-subsalicylat versus Cimetidin auf Campylobacter pylori, Ulkusheilung und -rezidivrate. Med Klin 83:402–405
13. Graham DY (1989) Campylobacter pylori as a pathogenic factor in duodenal ulcer: the case for. Scand J Gastroenterol [Suppl 160] 24:46–52
14. Graham DY (1989) Campylobacter pylori and peptic ulcer disease. Gastroenterology 96:615–625
15. Graham DY, Lew GM, Michaletz PA (1989) Randomized controlled trial of the effect of eradication of C. pylori on ulcer healing and relapse. Gastroenterology 96:A181
16. Humphreys H, Bourke S, Dooley C, McKenna D, Power B, Keane CT, Sweeney EC, O'Morain C (1988) Effect of treatment on Campylobacter pylori in peptic disease: a randomised prospective trial. Gut 29:279–283
17. Lambert JR, Borromeo M, Korman MG, Hansky J, Eaves ER (1987) Effect of colloidal bismuth (De-Nol) on healing and relapse of duodenal ulcers – role of Campylobacter pyloridis. Gastroenterology 92:1489
18. Marshall BJ (1989) Antimicrobial therapy of duodenal ulcer? Hold off for now. Reply. Gastroenterology 97:510
19. Marshall BJ, Goodwin CS, Warren JR, Murray R, Blincow ED, Blackbourn SJ, Phillips M, Waters TE, Sanderson CR (1988) Prospective double-blind trial of duodenal ulcer relapse after eradication of Campylobacter pylori. Lancet II:1437–1442
20. Smith AC, Price AB, Borriello P, Levi AJ (1988) A comparison of ranitidine and tri-potassium dicitrato-bismuth (T.D.B.) in relapse rates of duodenal ulcer. The role of Campylobacter pylori (C.P.) Gastroenterology 94:A431

Natural History and Long-Term Monitoring of Therapeutic Attempts to Eradicate *Helicobacter pylori*

E. A. J. Rauws and G. N. J. Tytgat

Introduction

Information about the natural history of *H. pylori*-associated gastritis is mainly obtained indirectly because without symptoms individuals are rarely endoscoped, and so infections may remain undiagnosed. To truly determine the natural history many persons, colonized or not colonized with *H. pylori* and with or without symptoms, should be followed up, including endoscopy and biopsies, without treatment. An alternative approach to study the significance of H. pylori is long-term follow-up of a substantial group of patients after eradication of these microorganisms.

H. pylori and Natural History

We prospectively followed a large group of *H. pylori* culture positive ($n = 50$) and negative ($n = 27$) patients up to more than 3 years after diagnosis, without treatment. Antral mucosal biopsy specimens were taken for culture and microbiological examination about every 3–6 months. If *H. pylori* was isolated, the strain was stored at $-70\,°C$. All formalin-fixed hematoxylin and eosin stained antral biopsy specimens were scored for severity of active chronic gastritis using a score scale for density of the inflammatory infiltrate in the lamina propria (0–2), density of polymorphonuclear leukocytes in the lamina propria (0–3), presence of intraepithelial polymorphonuclear leukocytes (0–3), and presence or absence of superficial erosions (0–2). For each parameter, score 0 was none, 1 was mild, 2 was moderate, and 3 was severe. Without exception, all culture-positive patients remained positive and all culture-negative patients remained negative for a follow-up period or more than 3 years. All the culture-positive persons initially had active chronic or chronic gastritis which remained roughly unchanged during the follow-up period as measured by the gastritis score. No changes in degree of atrophy or intestinal metaplasia were noted during this relatively short follow-up period. We did not see spontaneous disappearance of the microorganisms. Initially the culture-negative persons had normal gastric histology which remained constant during follow-up.

Dept. of Gastroenterology, Academic Medical Center, Meibergdreef 9, 1105 AZ Amsterdam, Netherlands

Helicobacter pylori, Gastritis and Peptic Ulcer
Edited by P. Malfertheiner and H. Ditschuneit

Restriction endonuclease analysis of *H. pylori* DNA showed that all the isolates from different patients showed different DNA restriction patterns, suggesting that they all differ in their subtypes [1]. The *H. pylori* isolates harvested during the follow-up study were also characterized by restriction endonuclease analysis of the bacterial DNA [2]. This made it possible to distinguish between persistence (same pattern) of the infection and reinfection (appearance of a different pattern). In all culture-positive individuals, over a median follow-up of 22 months (range 12–28 months), the DNA digestion patterns of the isolates remained identical. This proves, that if *H. pylori*-associated gastritis is present, the same *H. pylori* strain is likely to continue, for at least 2 years.

But, when, how, and why an individual becomes colonized remains to be elucidated. We only met a few patients who were *H. pylori* negative and became *H. pylori* positive, but in these cases endoscopic patient-to-patient transmission could be proven by DNA analysis of the isolated strains. This phenomenon will happen all over the world and is of course very important and may be a consequence of endoscopic follow-up.

Concluding Remarks

In the literature there is only scanty information on the natural history of *H. pylori* colonization and concomitant gastritis. It is not known if, and how often, spontaneous clearance of *H. pylori* infection and the associated gastritis occurs in those persons who acquired the disease naturally. We did not see spontaneous disappearance of the microorganisms in nonulcer dyspepsia (NUD) patients, in a maximal follow-up of 51 months. Spontaneous clearance especially of a small inoculum is, however, possible, as was shown by the spontaneous resolution in one volunteer self-inoculation experiment [3].

Follow-Up After Therapeutic Attempts to Eradicate *H. pylori*

Introduction

The link between *H. pylori* and gastritis led to treatment aimed at eradicating the microorganisms. Eradication of *H. pylori* is defined as the absence of microorganisms for at least 1 month after stopping treatment. Some authors use the word "eradication" incorrectly to describe only transient clearance of *H. pylori*.

We have treated many *H. pylori*-positive patients with various agents, and scored all antral biopsy specimens for the severity of gastritis [4].

Treatment Regimens

Patients with a gastric or duodenal ulcer were treated with cimetidine (200 mg q.i.d.) or colloidal bismuth subcitrate (CBS; 120 mg q.i.d.) for 4 weeks. Nonulcer dyspepsia patients with a positive culture for *H. pylori* and histological evidence of chronic active gastritis received either cimetidine (200 mg q.i.d.), sucralfate (1 gr q.i.d.), spiramycine (500 mg q.i.d.), tetracycline (250 mg q.i.d.), CBS (120 mg q.i.d.), or amoxicillin (375 mg t.i.d.).

Gastroscopy and Biopsies

The upper gastrointestinal endoscopy was performed after an overnight fast. From antral mucosa biopsy specimens were taken using a sterilized biopsy forceps. The biopsy specimens were taken from either the lesser or greater curvature about 2 cm from the pylorus. Two specimens were placed in 2 ml of phosphate-buffered saline at 4 °C for bacteriologic examination. Two specimens were fixed in 10% formalin for histopathologic examination.

Results of Treatment

Neither H_2-receptor antagonists, sucralfate, spiramycine, nor tetracycline had any effect on the presence of *H. pylori*. The effects of therapy on *H. pylori* cultures of antral mucosal biopsy specimens are summarized in Table 1.

In patients where *H. pylori* was successfully eradicated (*H. pylori* not detectable at least 1 month after stopping treatment), we observed hardly any recolonization for now more than 3 years after initial treatment. This very low recolonization rate during follow-up suggests either a low rate of exposure to *H. pylori* or that conditions in the stomach are unfavorable for colonization by *H. pylori*.

The effect of therapy on the gastritis score (severity of gastritis) parallels the effect of therapy on the presence of *H. pylori*. The gastritis scores remained unchanged after treatment with H_2-receptor blockers, sucralfate, spiramycine, and tetracycline. But after CBS, amoxicillin, or the combination of CBS and

Table 1. Follow-up of culture results of 378 *H. pylori* culture-positive patients with chronic active or chronic gastritis after various therapeutic regimens

Treatment	No. of patients	Negative immediately after treatment	Negative after (Months) 1	3	6	>12
H_2-receptor blocker	97	0	–	–	–	–
Sucralfate	33	0	–	–	–	–
Spiramycine	6	0	–	–	–	–
Tetracycline	7	0	–	–	–	–
Metronidazole	10	2	1	1	1	1
CBS	115	36	11[b]	10	10	10
Amoxicillin	45	29	14	13[a]	13	12[b]
CBS + amoxicillin	65	44	27	26[a]	26	23[a b]

[a] Recurrence, recolonization with the same strain
[b] Cases of accidental inoculations (that is Colonization by other than initial strain (identical to previously endoscoped *H. pylori*-positive patient)

amoxicillin, a significant decrease in the severity of gastritis is demonstrated. After successful eradication of *H. pylori*, the acute polynuclear component of the inflammatory reaction disappears rapidly, whereas the mononuclear component takes longer to diminish and disappear, as has been shown in our long-term follow-up study [4]. After real eradication of *H. pylori* there is a complete disappearance of the inflammatory changes during follow-up, but the gastritis score worsens again if recolonization of *H. pylori* occurs. Recolonization usually occurs within 1 month after apparently successful eradication (Table 1). Late recolonization is rare and sometimes caused by iatrogenic person-to-person transmission of the microorganism.

These data suggest that long-standing *H. pylori*-induced active chronic gastritis may ultimately evoke chronic atrophic gastritis and gastric atrophy after 20–40 years of follow-up [5, 6]. To what extent this may be true is still controversial. Many long-term studies of patient cohorts will be necessary to answer this question. If *H. pylori* colonization is etiologically linked to the sequence: active chronic gastritis → atrophic gastritis → gastric atrophy → dysplasia → carcinoma, this sequence of events can be averted via long-term eradication of *H. pylori*.

References

1. Langenberg W, Rauws EAJ, Widjojokusomo A, Tytgat GNJ, Zanen HC (1986) Identification of Campylobacter pyloridis isolates by restriction endonuclease DNA analysis. J Clin Microbiol 24:414–417
2. Langenberg W, Rauws EAJ, Houthoff HJ et al. (1988) Follow-up study of individuals with untreated Campylobacter pylori-associated gastritis and of noninfected persons with nonulcer dyspepsia. J Infect Dis 157:1245–1249
3. Marshall BJ, Armstrong JA, McGechie DB, Glancy RJ (1985) Attempt to fulfil Koch's postulates for pyloric campylobacter. Med J Aust 142:436–439
4. Rauws EAJ, Langenberg W, Houthoff HJ, Zanen HC, Tytgat GNJ (1988) Campylobacter pyloridis-associated chronic antral gastritis. Gastroenterology 94:33–40
5. Siurala M, Vuorinen Y (1963) Follow-up studies of patients with superficial gastritis and patients with a normal gastric mucosa. Acta Med Scand 173:45–52
6. Ihamäki T, Saukkonen M, Siurala M (1978) Longterm observation of subjects with normal mucosa and with superficial gastritis. Results of 23–27 years follow-up examination. Scand J Gastroenterol 13:771–776

Optimal Conditions for the Design of a Clinical Trial in *Helicobacter pylori* Related Duodenal Ulceration

R. E. Pounder and C. U. Nwokolo

Introduction

The general principles involved in the design of clinical trials relating to peptic ulceration are well known and have recently been reviewed in detail by Lauritsen and Rask-Madsen [1]. None of the studies that are quoted as evidence for the beneficial effect of eradication of *H. pylori* in duodenal ulceration comes close to the standards recommended in that article [2–6].

Existing Studies Related to Eradication of *H. pylori* and Duodenal Ulcer Management

Review of the data of the five major studies [7–11] usually quoted to prove the beneficial role of eradication of *H. pylori* in the management of duodenal ulceration demonstrates the remarkably small number of patients that have been involved in these studies, confusion between eradication and clearance of *H. pylori*, and lack of comparison with the best existing strategy for the management of duodenal ulceration (full-dose H_2-blockade for acute ulcer healing, followed by low-dose maintenance treatment). Duodenal ulceration is a long-term illness, and any modern trial should involve follow-up for at least 1 year.

It is probably impossible for studies in this area to be performed under double-blind conditions. The different types of bismuth formulation which cause the darkening of stools during bismuth therapy, the need for multiple antibiotics in eradication regimens, and the necessary comparison with an H_2-blocker would all combine to produce such a complicated trial, if the "double dummy" technique were used, that compliance would be compromised inevitably. For these reasons, we believe that it is acceptable to reject the need for double-blind conditions – if all the end-points involved in a trial are verifiable by independent observers during later analysis.

Recent studies at the Royal Free Hospital have demonstrated that the bioavailability of bismuth varies between the different formulations [12]. Two De-Noltabs contain 214 mg of bismuth, Pepto-Bismol 90 ml contains 897 mg of bismuth, and two Roter tablets contain 456 mg of bismuth – but only the De-

Academic Department of Medicine, Royal Free Hospital School of Medicine, London NW3 2QG, United Kingdom

Helicobacter pylori, Gastritis and Peptic Ulcer
Edited by P. Malfertheiner and H. Ditschuneit

Noltabs are associated with significant systemic absorption of bismuth. As the bioavailability of bismuth from the three compounds differs, it is important to define exactly which bismuth salt is used in any clinical study. The situation is even more complicated with the introduction of a range of bismuth-containing drugs of uncertain bioavailability and unreported clinical benefit in the Federal Republic of Germany. Does bismuth act as a luminal agent, or is systemic absorption necessary for either short-term or long-term benefit [13, 14]?

Proposal for a Trial in Duodenal Ulceration

To aid the discussion of the problems associated with the design of a trial to test the clinical benefit of eradication of *H. pylori*, we would like to propose the outline of a clinical trial for the management of duodenal ulceration using two different strategies for the eradication of *H. pylori*, in comparison with acute and maintenance treatment using H_2-blockade.

Patients

The trial should include all patients with duodenal ulceration presenting for endoscopy – except those treated with any of the trial drugs during the preceding 6 months, those with plasma urea greater than 10 mmol/l, those who need maintenance treatment because of aggressive ulceration, and those unfit for a clinical trial.

Clinical trials in peptic ulceration have often used a highly refined population of patients, to such an extent that final results bear little relevance to everyday clinical practice. Even this proposal suggests the exclusion of those patients who have had treatment with any of the trial drugs during the preceding 6 months – obviously such an exclusion is necessary for people who have been exposed to certain bismuth-containing compounds, because of persistence of bismuth in the body [13]. However, such an exclusion may eliminate many patients, because of the widespread use of H_2-blockade by general practitioners. It may be reasonable to modify this exclusion criterion, eliminating patients who have been treated with bismuth in the preceding 6 months, and those who have received H_2-blockade during the preceding month.

First Endoscopy (−7 Days to Day 0 of Trial) with Antral Biopsies for Histological Analysis

The introduction of video endoscopy provides the opportunity for a remarkable improvement in this aspect of clinical trial management. For the first time the index endoscopy and follow-up endoscopies can be reviewed by independent observers, to ensure that only adequate duodenal ulceration is admitted to a study and to provide a later objective assessment of ulcer healing. This type of quality control could provide the independent end-points that would allow the study to be performed without double-blind conditions. Antral biopsy specimens should be taken for histological analysis, allowing later independent assessment of *H. pylori* status and microbiological culture for in vitro sensitivity of *H. pylori* [14].

^{13}C-Urea Breath Test (−7 Days to Day 0)

The ^{13}C-urea breath test does appear to be the best-available test of active infection with *H. pylori* [14, 15]. It is suitable for use in all subjects without risk of radiation hazard, but it does have one problem – that of cost. Developments in the near future may reduce the cost by identifying the minimum oral dose of ^{13}C-urea and the least number of breath samples that are necessary. Certainly, this technique is suitable for analysis by a central laboratory; containers of exhaled breath can be sent by post.

Eight Weeks of Treatment with One of Three Regimens

1. De-Noltab i q.d.s. plus antibiotics
2. Pepto-Bismol tab i q.d.s. plus antibiotics
3. Ranitidine 300 mg nocte

It appears that bismuth alone achieves eradication in only a minority of patients infected with *H. pylori* and that the addition of an antibiotic regimen is necessary. Obviously, the final choice of the antibiotic will depend on the investigator, but at the present time the optimum regimen appears to be amoxycillin 250 mg t.d.s. with metronidazole 400 mg t.d.s. for 2 weeks [6]. It is planned that all patients will receive an acute course of treatment for 8 weeks, even if ulcer healing is demonstrated at 4 weeks. This is a reasonable strategy: first, in clinical practice at 4 weeks the gastroenterologist has no idea which patients has healed, and it is only by 8 weeks that 95% of ulcers can be expected to have healed; second, if some patients have 4 weeks of acute treatment and others have 8 weeks of acute treatment, it produces a doubling of a number of groups that enter the maintenance phase – thereby increasing the total number of subjects required for a confident discrimination between the groups at the various end-points. Compliance can be a problem in any medical treatment but the use of alarm wrist watches, which can be preprogrammed to sound at fixed intervals during the day, should encourage regular medication.

Second Endoscopy (Day 24–32) and Possible Third Endoscopy (Day 52–60)

The purpose of this endoscopy is to document the state of the duodenum in terms of ulcer healing and also to provide antral biopsy specimens for histological assessment of clearance of *H. pylori*. Patients unhealed after 28 days of treatment will be subjected to further endoscopy at 56 days – continued lack of ulcer healing at that stage would result in the patient being withdrawn from the trial.

All Healed Patients Enter Follow-Up Phase (Week 0–52)

This phase of the study would be managed entirely by the post, with the patient never having any direct contact with the investigator. Once a fortnight each patient would receive a letter and diary card: he or she would maintain a daily record of ulcer symptoms, and be asked each fortnight to confirm one of two statements:

Either: "My duodenal ulcer is causing me little or no trouble, and I am happy to continue with the study."
Or: "My duodenal ulcer is causing me so much trouble that I wish to withdraw from the study."

The patients in the ranitidine group would receive further ranitidine 150 mg tablets each fortnight and be advised to take one tablet every night; the remaining patients would receive no maintenance treatment.

Boyd et al. [16] have shown that in the long-term management of duodenal ulceration, the intermittent endoscopy of asymptomatic patients identifies what is probably irrelevant asymptomatic ulceration – and it is much more valuable to measure success in long-term ulcer treatment by an assessment of symptoms.

The proposed design allows the patient to withdraw from a study at any time during 1 year of follow-up, which will allow the analysis of success or failure in terms of the symptomatic management of peptic ulceration.

^{13}C-Urea Breath Test (Week 3–5)

All patients should be recalled for a ^{13}C-urea breath test approximately 1 month after cessation of the original 2-month course of treatment, to determine eradication of *H. pylori* infection.

Final Endoscopy (Treatment Failure or Week 52)

All patients should undergo a final endoscopy either at the time of symptomatic relapse or at the end of 1 year of follow-up. This endoscopy will allow examination of the duodenum for assessment of either symptomatic or asymptomatic ulceration, and will also allow a final antral biopsy to assess eradication of *H. pylori.*

^{13}C-Urea Breath Test (Treatment Failure or Week 52)

The final ^{13}C-urea breath test will determine whether there is persisting infection with *H. pylori* or reinfection after initial eradication.

Trial End-Points

This study provides six independent end-points for the comparison of the three trial regimens:

1. Relief of acute symptoms (days 0–56)
2. Cumulative acute ulcer healing at day 28 and day 56
3. Cumulative *H. pylori* clearance at day 28 and day 56
4. *H. pylori* eradication at week 4 and end of trial (biopsy and ^{13}C-urea criteria)
5. Long-term treatment success: symptom relief, symptomatic ulceration, and asymptomatic ulceration (life-table analysis)
6. Adverse events and complications

Calculation of Statistical Significance

A large number of patients will be required to achieve a statistically significant analysis of the final result, because of the very high success rates that can be anticipated in the group receiving ranitidine: cumulative duodenal ulcer healing with ranitidine 300 mg nocte is 84% at 4 weeks and 95% at 8 weeks [17]; after 1 year of maintenance treatment with ranitidine 150 mg nocte, approximately 90% of patients can be expected to be symptom-free and 80% ulcer-free (the 10% difference is due to the detection of asymptomatic ulceration at the week 52 endoscopy) [16]. Thus, the end-points for the ranitidine strategy range between 80% and 95%. If the average end-point is 85%, to achieve a four-in-five chance of detecting a significant minus 10% difference with another treatment (75%) needs 587 patients in each group, and to have a four-in-five chance of detecting a significant plus 10% benefit (95%) requires 500 patients in each group. Hence, the proposed three-arm study should aim to enrol 1761 duodenal ulcer patients, and even that number does take account of losses due to protocol violation. Similarly large studies have been performed in peptic ulceration – but these calculations demonstrate the problem of proving "no significant difference" between an established effective treatment and a new management strategy.

References

1. Lauritsen K, Rask-Madsen J (1987) Clinical trials in peptic ulcer disease – problems of methodology and interpretation. Aliment Pharmacol Ther 1:91–123
2. Bartlett JE (1988) Campylobacter pylori: fact or fancy? Gastroenterology 94:229–238
3. Dooley CP, Cohen H (1988) The clinical significance of Campylobacter pylori. Ann Intern Med 108:70–79
4. Graham DY (1989) Campylobacter pylori and peptic ulcer disease. Gastroenterology 96:615–625
5. Borsch GMA (1989) Clinical significance of Campylobacter pylori. Eur J Gastroenterol Hepatol 1:27–33
6. Rauws EAJ (1989) Therapeutic attempts at eradication of Campylobacter pylori. Eur J Gastroenterol Hepatol 1:34–41
7. Martin DF, Hollanders D, May SJ, et al. (1981) Difference in relapse rates of duodenal ulcer after healing with cimetidine or tripotassium dicitrato bismuthate. Lancet I:7–10

8. Coghlan JG, Gilligan D, Humphries H, et al. (1987) Campylobacter pylori and recurrence of duodenal ulcers – a 12 month follow-up study. Lancet II:1109–1111
9. Bayerdorffer E, Simon TH, Bastlein CH, Ottenjann R, Kasper G (1987) Bismuth/ofloxacin combination for duodenal ulcer. Lancet II:1467–1468
10. Rauws EAJ, Langenberg W, Houthoff HJ, et al. (1988) Antibacterial action of bismuth in relation to Campylobacter pyloridis-associated chronic active antral gastritis: a prospective study of its prevalence and the effects of antibacterial and antiulcer treatment. Gastroenterology 94:33–40
11. Marshall BJ, Goodwin CS, Warren JR, Murray R, Glincow ED, Blackbourn SJ, Phillips M, Waters TE, Sanderson CR (1988) Prospective double-blind trial of duodenal relapse after eradication of Campylobacter pylori. Lancet II:1437–1442
12. Nwokolo CU, Mistry P, Pounder RE (1990) Reabsorption of bismuth and salicylate (oral communication)
13. Gavey CJ, Szeto M-L, Nwokolo CU, Sercombe J, Pounder RE (1989) Bismuth accumulates in the body during treatment with tripotassium dicitrato bismuthate. Aliment Pharmacol Ther 3:21–28
14. Nwokolo CU, Gavey CJ, Smith JTL, Pounder RE (1989) The absorption of bismuth from oral doses of tripotassium dicitrato bismuthate. Aliment Pharmacol Ther 3:29–39
15. Graham DY, Klein PD, Evans DJ, Evans DG, Alpert LC, Opekun AR, Boutton TW (1987) Campylobacter pyloridis detected noninvasively by the ^{13}C-urea breath test. Lancet I:1174–1177
16. Boyd EJS, Penston JG, Johnston KA, Wormsley KG (1988) Does maintenance therapy keep duodenal ulcer healed? Lancet II:1324–1327
17. Jones DB, Howden CW, Burget DW, Kerr GD, Hunt RH (1987) Acid suppression in duodenal ulcer: a meta-analysis to define optimal dosing with antisecretory drugs. Gut 28:1120–1127

Treatment of *Helicobacter pylori* Infection: a Comment

M. DELTENRE

Thanks to the results of therapeutic trials *Helicobacter pylori* is known to be a prominent factor in the development of chronic active gastritis and the recurrence of duodenal ulcers.

Acute achlorhydric gastritis may be the first manifestation of *H. pylori*-related diseases, but no controlled trial is currently available about the impact of any treatment during the invasive phase which may be overlooked because of the absence of severe symptoms.

Obviously, recurrent duodenal ulcer (DU) is a multifactorial disease, but *H. pylori* is now recognized as a key factor; the suppression of *H. pylori* might accelerate the healing of the ulcerous lesion, but the most important fact is the striking reduction of the relapse rate of DU after *H. pylori* eradication, as proven by several independent studies. Relapse rates of DU after *H. pylori* eradication vary between 6% and 25%, and the mean relapse rate (around 15%) approaches the long term relapse rate after selective vagotomy. In this respect, a prospective, randomized study comparing H2 blockers maintenance treatment and *H. pylori* eradication would be useful.

As fas as non ulcer dyspepsia (NUD) (or perhaps should we say idiopathic dyspepsia) is concerned, A.T.R. Axon pointed out that among NUD patients the prevalence of HP is approximately 20% higher than in normal individuals (age-matched blood donors), and varies with age, race, and social condition. There is no specific symptomatic profile of *H. pylori*-positive NUD, and the majority of *H. pylori* carriers with gastritis are asymptomatic. The symptomatic benefit of *H. pylori* suppression or eradication in NUD remains controversial. Because of randomization, very few *H. pylori*-positive patients have been treated with active drugs. Moreover, the respective role of *H. pylori* suppression versus other potential actions of the treatment on the symptoms is not clearly precised. A sharp definition of NUD, a convenient scoring technique for symptomatology, and long-term follow-up are needed to answer those questions.

Nevertheless, the problem of eradication has not been solved; despite a large in vitro sensitivity of *H. pylori* to many antibiotics, monotherapies have been shown to be ineffective. Tritherapies were shown to *H. pylori*, but brought an unacceptable rate of intolerance and side effects. Various bitherapies with amoxycillin, nitrofurans, Tinidazole, metronidazole, or bismuth salts are successful (60%–80% eradication rate). A 1-week course seems to be equivalent to 4 weeks

Department of Gastroenterology, Hoopital Universitaire Brugmann, Place van Schuchten 4, 1020 Brussels, Belgium

Helicobacter pylori, Gastritis and Peptic Ulcer
Edited by P. Malfertheiner and H. Ditschuneit

treatment and successive therapy reveals no advantage. An interesting trial from Brazil suggests that a short course of tritherapy could be useful. The possible roles of topical or intermittent therapy and urease inhibitors have not been discussed. Potentialization of amoxycillin by omeprazole remains an open question and many problems have not been solved, e.g., compliance or reinfection rate (6% within 6 months according to G. Borsch, see p. 441).

One of the most important questions is the induction of *H. pylori* resistance to antimicrobial drugs. Posttreatment resistance to tinidazole or metronidazole has been confirmed as well as induced resistance to quinolones. Pretreatment resistance to tinidazole has been also clearly described, showing the necessity for testing sensitivity before therapy.

Why is there no eradication of *H. pylori* with drugs to which *H. pylori* is and remains sensitive? All forms of bismuth salts are found to be equivalent for in vitro *H. pylori* susceptibility, but local concentration of the therapeutic agent (for instance amoxycillin) ont the gastric mucosa is important. Bismuth sub-salicylate (BSS) is absorbed less than colloidal bismuth subcitrate (CBS) and may apparently suppress better, but might be inferior to CBS q.i.d. in terms of eradication. Finally, the (controversial) description of coccoid forms, perhaps induced by antimicrobials but possible also by antacids or H2 blockers, might be an elegant explanation for early reactivation after transient suppression.

As far as toxicity of bismuth salts is concerned, there is no accumulation of CBS or BSS (single dose, 4- and 8-week courses) in the antroduodenal mucosa or gastric mucus. Bismuth is cleared rapidly from the stomach, but after 4 or 8 weeks of treatment, a reversible increase in plasma bismuth concentration has been observed, also in children. The risk of tissular accumulation is not clearly evaluated. Nobody can yet answer the question of treatment for infected children. One may speculate about the prevention of future DU disease or even gastric cancer, but so far there is no clear demonstration of any positive effect of eradicating *H. pylori* in those fields. For NUD, the question is at least as difficult here as in adults and the natural evolution of infection during childhood remains unknown.

In conclusion, the eradication of *H. pylori* seems to be a very reasonable way to reach a definitive cure of DU disease. Short term bitherapy could be a solution, but primary or acquired resistance must be carefully checked. For NUD, long term, randomized, placebo-controlled studies are needed, since there is so far no proof that the eradication of *H. pylori* and subsequent healing of chronic gastritis would be beneficial in terms of symptoms.

Perspectives

Helicobacter pylori: Future Direction in Research *

D. Y. GRAHAM

Introduction

My task is to synthesize the material of current knowledge, integrate the new information into what is known, and to identify areas that appear most productive for additional research. I envision research as occurring in two broad categories. The first category is forefront research, defined as efforts riding the crest of the current wave of technology and ideas; the second I term "puddle" research. This is research done after the wave is receded; even then we find small tidal pools full of interesting questions, by analogy starfish and urchins. It is often from such puddles that the impetus for the next wave of research comes. One form of research is not better than the other and most do both.

The field of *Helicobacter pylori* research is rapidly maturing. Much of the research that we hear and see reported at meetings such as this, confirms what we already think we know. Usually close insection of what we "know" reveals that there are many major gaps in our knowledge. There are now sufficient data to allow us to make many projections that should come very close to the mark. At the same time, there are a number of areas of research that remain largely unaddressed but are becoming sufficiently defined so that they can be tackled in an organized fashion.

I will divide my article into two parts with some overlap. I will discuss those things that I know (or I think that I know) and those items that I do not know but would like to. The latter exercise is designed to uncover areas for fruitful investigation. I recognize that not all will agree with the tenets and hypotheses that I will enumerate as they often represent synthesis and projections of data from a variety of sources. Thus, while some of the ideas and concepts outwardly will appear clean and crisp, they may not be based upon firm underpinnings. Nevertheless that is what I have been asked to attempt and I will start with the areas of epidemiology and transmission.

---

* This work was supported by the Department of Veterans Affairs, by grant DK 39919 from the National Institute of Diabetes and Digestive and Kidney Diseases, and by the generous support of Hilda Schwartz.

Departments of Medicine and Molecular Virology, Veterans Affairs Medical Center and Baylor College of Medicine, Houston, Texas

Helicobacter pylori, Gastritis and Peptic Ulcer
Edited by P. Malfertheiner and H. Ditschuneit

Epidemiology and Transmission

The epidemiology of *H. pylori* infection appears to parallel quite closely that of other enteric infections known to be transmitted by the fecal-oral route. The closest current parallel is with the epidemiology of hepatitis A (HAV) infection [7]. We believe the age-specific frequency of anti-HAV antibody in a population provides a reasonable gauge of fecal-oral exposure and predicts the age-specific rate of onset of *H. pylori* infection in that population, i.e., rapid, intermediate, or slow. To date, this hypothesis has been proven correct.

Study of the factors previously found to be important in the epidemiology of gastritis, of hepatitis A, and of poliomyelitis (before the introduction of polio vaccine), all of which are essentially identical, should form the framework for studies of the epidemiology of *H. pylori* [7]. Details concerning the factors that should be important such as socioeconomic status, sanitation, housing density, ethnic group, etc., need to be addressed and verified. Recent knowledge and concepts suggest that testing sera obtained from uncharacterized populations for anti-*H. pylori* antibody will provide little new data, insights, or understanding. Studies need hypotheses rather than "I wonder what the frequency of *H. pylori* antibody is in a population." Previous *H. pylori* correlations need to be reassessed as the marked differences in frequency of *H. pylori* infection between socioeconomic classes and ethnic groups were not taken into account. For example, the reported possible increase in prevalence of *Hy. pylori* infection in abattoir workers may have been an artifact based upon unrecognized differences between groups that were compared [20].

One particularly striking finding is that the motile spiral form of *H. pylori* can survive for at least 1 week in river water. This finding increases the likelihood that transmission of *H. pylori* is fecal-oral. The relevance of the observation that coccoid forms can survive in river water for 1 year or more is unknown as, in my opinion, they are unlikely to be infectious for man. The rapid turnover of the gastric epithelium probably precludes their prolonged residence in the stomach, although they may be responsible for some instances of early relapse.

Diagnostic Tests

I believe there have been a few genuine advances in the area of diagnostic tests. The currently available major tests for the diagnosis of *H. pylori* infection were developed soon after the problem of cultivating the organism was solved. These tests have proven to be robust, i.e., they can withstand considerable change and still provide reasonably accurate information. A number of minor modifications of culture methods, of the biopsy urease test, and of the breath test were presented. Each modification was attempted with the aim of "simplifying" or "optimizing" a diagnostic test. In reality such modifications or adaptions are those particularly suited for an individual laboratory or country but usually they do not represent general advances. Data predicting the reported "advance" was often present within the graphs and tables of previous publications. For example, in the urea breath test, the rate of hydrolysis of urea is linear over a considerable time [8]. The choice of timing and the number of samples to be taken may seem quite arbitrary as more information is present in the test than is collected. One simply must decide how much information one needs to balance cost and patient and technician time against overall accuracy. For example, the potential gains by eliminating one or two samples must be compared with how much one might lose. The duration of our current ^{13}C-urea breath test is 1 h [15]. Although it is clear from our published data that one breath sample following substrate ingestion would be the minimum required, we know that to do so would reduce the accuracy of the test slightly by increasing the rate of the false-positive results. This knowledge is based on the performance of hundreds of tests in all age groups. Improvements do not come easily.

Another good example is the biopsy urease test. Changes in the recipe for biopsy urease tests continue to be introduced as "improvements" while the most important parameter is probably the size of the biopsy specimen and not the actual ingredients. Most do not even attempt to take the biopsy size into account. In addition, biopsy urease tests require fiber-optic endoscopy and, as the availability of accurate serologic tests and breath tests increases, endoscopy may soon be relegated to the status of a second-line diagnostic test.

Considerable work continues in attempting to grow large quantities of *H. pylori* in liquid media. Some progress has been made; more is required.

The discovery that a commercial monoclonal antibody directed against *Campylobacter jejuni* was useful in identifying *H. pylori* in paraffin-fixed tissue was, in my opinion, a significant advance [2]. The number of bacteria that must be present before special stains, such as the Warthin-Starry stain, are definitely positive is quite high, limiting the usefulness of histology for determining whether the infection has been eradicated. Theoretically, immunohistochemical stains using the monoclonal antibody would allow identification of a much lower bacterial load and increase the usefulness of histology for determining whether therapy has been truly successful. The commercial availability of the monoclonal antibody should lead to rapid and widespread testing of its actual usefulness. The potential problem of cross-reactivity with *C. jejuni* subsp. doylei will also need to be addressed.

Pathophysiology of Duodenal Ulcer Disease

Most now believe that *H. pylori* plays a critical role in the development of duodenal ulcer disease [6]. The extent of gastric metaplasia in the duodenal bulb is becoming widely recognized. Many previous studies were done without acknowledging that the duodenal bulb in a patient with duodenal ulcer disease is both structurally and functionally different from one in a normal individual. For example, in duodenal ulcer disease the surface area of villus mucosa of the bulb is decreased as a variable proportion of the mucosa is gastric and not intestinal. We have also rediscovered and confirmed that areas of gastric metaplasia in the bulb are capable of producing acid [1]. It is unknown how many of the previously described abnormalities in duodenal function are due to *H. pylori*-induced differences in bulb anatomy and physiology. In addition, we do not know the degree to which these changes are reversible, but preliminary evidence suggests that they are reversible, at least partially [6]. Previously reported abnormalities in bulb function, such as abnormal bicarbonate secretion in duodenal ulcer disease [16], will need to be reinvestigated following eradication of the *H. pylori* infection.

The newest information relates *H. pylori* infection and antral gastrin release. It is now clear that meal-stimulated gastrin release is exaggerated in patients with *H. pylori* infection. This abnormality is seen in both asymptomatic individuals and in patients with duodenal ulcer. Several laboratories, including our own, have shown ablation of this exaggerated gastrin release following treatment designed to eradicate the infection. Gastrin is a trophic hormone for parietal cells. Exaggerated meal-stimulated gastrin release has been previously postulated to be responsible for the increased number of parietal cells characteristic of patients with duodenal ulcer disease. This hypothesis can now be tested by measuring the number of parietal cells before and after eradication of *H. pylori* infection in patients with duodenal ulcer. Such studies are under way in a number of laboratories.

Areas Needing Additional Work

Virulence Factors

There is a considerable amount of work being done on putative virulence factors [10]. Proof that a suspected virulence factor has such a function in vivo is difficult without a good animal model as the traditional method of proof is to compare the effect of challenge with a strain with the factor with one lacking it. The importance of motility has been strongly suggested by studies in the gnotobiotic pig [3]. The fact that *H. felis* infection of the mouse leads to gastritis complete with a polymorphonuclear response suggests that it will be a good model for evaluation of various putative virulence factors, although it is not clear that the interactions of *H. felis* and the mouse will be directly transferable to man. A better model may be *H. pylori* in the germ-free puppy but only the actual experiments will decide.

Catalase, which possibly provides protection from leukocyte-derived oxygen metabolites, and urease-negative strains of *H. pylori* have been described although it is not clear that lack of these enzymes is a stable property of the strain. For example, we infected an *H. pylori*-negative chimpanzee with a urease-negative strain of *H. pylori* and later recovered urease-positive organisms from the animal. Several groups have now cloned the *H. pylori* urease gene and will be able to produce definite urease-negative mutants by deletion or site-specific mutagenesis of the urease gene followed by homologous reconstruction. It should soon be possible to directly evaluate the importance of urease as a virulence factor without the confounding variables associated with mutants that are otherwise uncharacterized.

The possible damaging role of ammonium ion produced by urease has been questioned on a number of grounds. For example, the urease-producing *H. felis* colonized the cat stomach but without histologic abnormalities and ammonium is used to load tissue culture cells with hydrogen ion without damaging effect [18].

The oral administration of ammonium chloride is not known to be associated with any particular gastric damage and ammonium chloride has a long history as a drug. Stuart Hazell suggested that formation of chloramines by the interaction of ammonium ion and neutrophils may be the mechanism by which urease becomes a virulence factor promoting mucosal damage [12]. Normally, ammonium ion produced by urease hydrolysis of urea would react with gastric acid from the relatively nontoxic product, NH_4Cl. In the chloramine model, interaction with neutrophil-produced hypochlorous acid with ammonium ion would produce the mono-*N*-chloramine, NH_2Cl which is toxic to the mucosa and should also be toxic to the bacteria [11]. *H. pylori* catalase might reduce the amount of neutrophil-produced H_2O_2 in the vicinity of the bacterium and thus protect it from the damaging effects of NH_2Cl. In contrast, the chloramine would be produced in the region of the polymorphonuclear leukocytes and might damage them preventing them from successfully evicting the bacteria. Studies evaluating the sensitivity of *H. pylori* would be of interest. The intriguing suggestion concerning the possible role of mono-*N*-chloramine in urease-related toxicity deserves further evaluation.

A toxin that produces nonlethal vaculation of cells in tissue culture has been described and several laboratories are working toward understanding whether it is a laboratory phenomenon or has some relation to disease. One criterion for in vivo production of the proposed virulence factor is the presence of antibody to it in infected individuals. This has been described for the vaculating toxin. An association with disease can be shown by demonstration of the toxin, or toxin-producing strains, in patients with the disease (e.g., duodenal ulcer) and not in infected but asymptomatic individuals. Studies in gnotobiotic piglets did not confirm a pathogenic role for the toxin [3]. Studies are under way in several laboratories comparing the frequency of antibody against the toxin in *H. pylori* infected patients with and without ulcer disease.

Serologic Tests

There is considerable need for additional work in the area of serologic testing. It has been demonstrated that serologic tests using crude antigens, although validated in developed countries, do not provide reliable data in developing countries (and also probably not in some populations in the developed country) [17]. It has been suggested that such tests must be standardized for each population and the failure to do so may lead to a very high frequency of false-positive results [17]. We have had the opportunity to compare the results of our second generation ELISA [4] with the results of a first generation test that used crude *H. pylori* lysates as antigens. We were able to confirm that the first generation test yielded a poor positive predictive value when used in populations from developing countries (unpublished observations).

Serologic tests based on defined antigens with no known cross-reactivity are needed. One candidate is to base tests on the high molecular weight cell-associated antigens of *H. pylori* [4]. Other antigens are needed as one would feel more comfortable relying on the results of two tests that used entirely different antigens to confirm the diagnosis of *H. pylori* infection. Another possible candidate antigen is the fibrillar hemagglutinin (adhesin) of *H. pylori* but this will not be possible until the full range of antigenic types of the adhesin is known [5].

It has been suggested that the combination of positive anti-*H. pylori* IgG and IgA yields increased specificity [19]. Studies using immunoblots have not confirmed that hypothesis and it seems unlikely that such an approach will lead to improved positive or negative predictive value [21]. Current tests for an IgA, IgM, or IgE response have not used techniques such as antibody capture to overcome the problems encountered because of the presence of a large excess of serum IgG directed towards the antigens in question [19]. Such modifications should increase the specificity and sensitivity of detection of such antibodies.

One hopes that the European Helicobacter Study Group will take on the task of developing standardized reagents for the serologic testing of *H. pylori*.

Therapy to Eradicate *H. pylori* Infection

We clearly need improved therapy to eradicate *H. pylori* infection. Although considerable work has been done in the area of therapy during the past 2 years, no breakthroughs have yet been either identified or confirmed. Current therapies have been associated with side effect rates of 7%–10%. The relatively high frequency of side effects must be reduced if we are to recommend widespread use of anti-*H. pylori* therapy in diseases for which we already have safe and effective therapies. The use of omeprazole to block acid secretion may simplify therapy by eliminating low pH as a barrier to effective antibiotic therapy.

What Is Happening in the Stomach and Why?

Here I will pose questions for which I would like to know the answers. What is happening in the stomach and duodenum? What virulence factor(s) causes the inflammatory response? Is it preventable? How does the infection usually begin – abruptly, slowly, or both? We have only limited data from two human challenge studies and from a few examples of iatrogenic infection. We do not have a clear idea of how the disease starts, how it progresses, or what governs the pattern of spread. We also do not know the pattern of reinfection or relapse. The latter is easier to study because many laboratories are following groups of patients after attempted eradication of *H. pylori*. Our impression is that relapse appears slowly as if the bacteria have difficulty in reestablishing the high level of infection present before treatment. Similarly, in the months following onset, the disease seems to wane as both the distribution and number of bacteria decrease as does the inflammatory response. What are the patterns of disease onset? What mechanism is responsible for the waning of the infection? How much of the control of the infection is humoral (e.g., antibodies neutralizing secreted toxins) and how much cellular? Which cells?

How Is Progress Made?

In general, progress is made using little data and considerable insight. This synthesis of available data leads to a projection about "how things really are." Such a projection, formally called a theory, is often presented as a fact. For example, several recent theories are that the age-of-acquisition of an *H. pylori* infection determines the clinical spectrum of the disease or diseases that result [9], and that the epidemiology of *H. pylori* closely parallels that of hepatitis A infection [7]. A good theory leads to testable hypotheses which ultimately serve to confirm, refute, or modify the theory. A good theory, even if ultimately proven wrong, relates many previously disparate facts so clearly that one wonders why they had not made the same synthesis earlier. Many ideas and theories will be found to be rediscoveries. For example, my concept of the urea breath test came in a flash of understanding while I was acting as a reviewer of the manuscript of Hazell et al. [13]. We subsequently found that the urea breath test was a rediscovery of concepts that were well described in the past, only to have been lost [15]. As Thomas Carlyle said, "All that mankind has done, thought, gained or been: it is lying as in magic perservation in the pages of books." How much more is simply waiting for us in the library! We must also remember to acknowledge our intellectual debts for concepts and theories gained from the work of others.

References

1. Carrick J, Lee A, Hazell S, Ralston M, Daskalopoulos G (1989) Campylobacter pylori, duodenal ulcer, and gastric metaplasia: possible role of functional heterotopic tissue in ulcerogenesis. Gut 30:790–797
2. Cartun RW, Al-Annouf N, Morin SG, Pedersen CA, Ludwig ME (1989) Immunocytochemical identification of Campylobacter organisms in formalin-fixed, paraffin-embedded tissue utilizing a commercially available monoclonal antibody (MAb) (Abstr). Gastroenterology 96:A75
3. Eaton KE, Morgan DR, Karkowka S (1989) Campylobacter pylori virulence factors in gnotobiotic piglets. Infect Immun 57:1119–1125
4. Evans DJ Jr, Evans DG, Graham DY, Klein PD (1989) A sensitive and specific serologic test for detection of Campylobacter pylori infection. Gastroenterology 96:1004–1008
5. Evans DJ Jr, Evans DG, Smith KE, Graham DY (1989) Serum antibody responses to the N-acetylneuraminyllactose-binding hemagglutinin of Campylobacter pylori. Infect Immun 57:644–647
6. Graham DY (1989) Campylobacter pylori and peptic ulcer disease. Gastroenterology 96:615
7. Graham DY (1990) Campylobacter pylori in human populations: the present and predictions of the future based on the epidemiology of polio. In: Menge H, Gregor M, McNulty CAM, Marshall BJ, Tytgat G (eds) 2nd International Symposium on Campylobacter pylori. Springer, Berlin Heidelberg New York (in press)
8. Graham DY, Klein PD, Evans DJ, Evans DG, Alpert LC, Opekun AR, Boutton TW (1987) Campylobacter pyloridis detected noninvasively by the ^{13}C-urea breath test. Lancet I:1174–1177
9. Graham DY, Adam E, Klein PD, Evans DJ Jr, Evans DG, Hazell SL, Alpert LC, Michaletz PA, Yoshimura HH (1989) Epidemiology of Campylobacter pylori. Gastroenterol Clin Biol 13:84B–88B
10. Graham DY, Evans DG, Evans DJ Jr (1989) Campylobacter pylori: the organism and its clinical revelance. J Clin Gastroenterol [Suppl 1] 11:543–548

11. Grisham MB, Jefferson MM, Melton DF, Thomas EL (1984) Chlorination of endogenous amines by isolated neutrophils: ammonia-dependent bactericidal, cytotoxic, and cytolytic activities of the chloramines. J Biol Chem 259:10404–10413
12. Hazell SL (1990) Urease and catalase as virulence factors of Campylobacter pylori. In: Menge H, Gregor M, McNulty CAM, Marshall BJ, Tytgat G (eds) 2nd International Symposium on Campylobacter pylori. Springer, Berlin Heidelberg New York (in press)
13. Hazell SL, Lee A, Brady L, Hennessy W (1986) Campylobacter pyloridis and gastritis: association with intercellular spaces and adaptation to an environment of mucus as important factors in colonization of the gastric epithelium. J Infect Dis 153:658–663
14. Heinz FX, Roggendorf M, Hofmann H, Knuz C, Deinhardt F (1981) Comparison of two different enzyme immunoassays for detection of immunoglobulin M antibodies against tick-borne encephalitis virus in serum and cerebrospinal fluid. J Clin Microbiol 14:141–146
15. Klein PD, Graham DY (1989) Campylobacter pylori detection by the ^{13}C-urea breath test. In: Rathbone B, Heatley V (eds) Campylobacter pylori and gastroduodenal disease. Blackwell, Oxford, pp 94–106
16. Isenberg JK, Selling JA, Hogan DL, Koss MA (1987) Impaired proximal duodenal mucosal bicarbonate secretion in patients with duodenal ulcer. N Engl J Med 361:374
17. Mitchell HM, Lee A, Berkowicz J, Barody T (1988) The use of serology to diagnose active Campylobacter pylori infection. Med J Aust 149:604–609
18. Olender EJ, Woods D, Kozol R, Fromm D (1986) Salicylate effects on proton gradient dissipation by isolated gastric mucosal surface cells. Proc Soc Exp Biol Med 183:177–185
19. Perez-Perez GI, Dworkin BM, Chodos JE, Blaser MJ (1988) Campylobacter pylori antibodies in humans. Ann Intern Med 109:11–17
20. Vira D, d'Anastasio CD, Holton J, Dowsett JF, Londei M, Bertoni F, Beltrandi E, Grauenfels P, Salmon PR, Gandolfi L (1988) Campylobacter pylori in abattoir workers: is it a zoonosis? Lancet II:725–726
21. von Wulffen M, Grote HJ, Gatermann S, Loning T, Berger B, Buhl C (1988) Immunoblot analysis of immune repsonse to Campylobacter pylori and its clinical associations. J Clin Pathol 41:653–659

Subject Index

MIX
Papier aus verantwortungsvollen Quellen
Paper from responsible sources
FSC® C105338

If you have any concerns about our products,
you can contact us on
ProductSafety@springernature.com

In case Publisher is established outside the EU,
the EU authorized representative is:
Springer Nature Customer Service Center GmbH
Europaplatz 3, 69115 Heidelberg, Germany

Printed by Libri Plureos GmbH
in Hamburg, Germany